The Economics of Non-Market Goods and Resources

Volume 14

Series editor

I.J. Bateman, School of Environmental Sciences, University of East Anglia, Norwich, UK

More information about this series at http://www.springer.com/series/5919

Robert J. Johnston · John Rolfe
Randall S. Rosenberger · Roy Brouwer
Editors

Benefit Transfer of Environmental and Resource Values

A Guide for Researchers and Practitioners

Springer

Editors
Robert J. Johnston
Department of Economics and George Perkins Marsh Institute
Clark University
Worcester, MA
USA

Randall S. Rosenberger
Department of Forest Ecosystems and Society
Oregon State University
Corvallis, OR
USA

John Rolfe
School of Business and Law
CQUniversity
Rockhampton, QLD
Australia

Roy Brouwer
Department of Environmental Economics, Institute for Environmental Studies
Vrije Universiteit
Amsterdam
The Netherlands

ISSN 1571-487X
The Economics of Non-Market Goods and Resources
ISBN 978-94-017-9929-4 ISBN 978-94-017-9930-0 (eBook)
DOI 10.1007/978-94-017-9930-0

Library of Congress Control Number: 2015936159

Springer Dordrecht Heidelberg New York London

Printed on acid-free paper

Springer Science+Business Media B.V. Dordrecht is part of Springer Science+Business Media (www.springer.com)

Preface

As its title suggests, this book is about benefit transfer, particularly as applied to environmental and natural resource decisions. Benefit transfer uses economic study results from one situation and extrapolates them to another similar situation. It is the only means to provide empirical economic information, such as estimates of benefits or costs, when primary studies are not available, considered infeasible or simply too expensive. Without benefit transfer, the value of many environmental goods and services would remain unacknowledged (or at least unquantified) in decision-making processes, leading to decisions based on incomplete information. Among its many uses, benefit transfer is a virtually indispensable—and some have argued nearly universal—component of large-scale cost–benefit analysis.

Primary studies are generally viewed by academics (and the journals in which they publish) as being a more accurate way to inform decisions. Benefit transfer is a second-best solution used when constraints on time, funding, analytical methods, or data prevent the use of primary studies to provide needed information. Yet benefit transfer also prolongs and magnifies the impact of primary economic research, extending the relevance of primary studies beyond the original setting and time frame in which they were conducted. Many valuation studies that were largely irrelevant to (or not intended to influence) real-world decisions when first published can be given new life and relevance when used as a basis for subsequent benefit transfer. Hence, benefit transfer not only provides policy makers with the pragmatic tools to conduct detailed analysis of policies and decisions; it also improves the returns on investment in primary valuation studies. In this way, it helps us understand the value of information used to inform decisions, along with tradeoffs between the accuracy and cost of this information.

For these and other reasons, the relevance of benefit transfer is indisputable. It is the valuation technique most commonly applied to inform policy and other decisions. Yet benefit transfer is also among the most misunderstood approaches in policy analysis. There is a frequent divergence between the increasingly sophisticated transfer methods recommended by the scientific literature and the often simple approaches used to inform decisions, support advocacy, and calculate values within off-the-shelf decision support tools. In some cases simpler methods are justified. More complex approaches do not always generate more accurate results, and in some cases there may be insufficient time, resources, or expertise to support more sophisticated transfer methods. Yet the widespread use of the least-rigorous (and generally least accurate) transfer methods may also promulgate similarly widespread misunderstanding of environmental and resource values. Challenges in the application of more sophisticated transfer approaches are further increased by the size, complexity, and relative disorganization of the benefit transfer literature—much of which is not designed to directly support policy application. There are also disagreements within the academic literature—for example with regard to the relative importance of theoretical versus empirical factors when evaluating transfer methods.

Recognizing the critical importance of benefit transfer and the breadth of associated scholarly and policy work, this book has been designed to provide a comprehensive review of transfer methods, issues, and challenges, covering topics relevant to both researchers and practitioners across different continents. Among the primary themes of the book is the availability of a range of rigorous transfer methods suitable for broad application, choices between these methods, and implications of these choices for transfer validity and reliability (or accuracy). We target a wide audience, including undergraduate and graduate students, practitioners in economics and other disciplines looking for a one-stop handbook covering benefit transfer topics, and others who wish to apply or evaluate benefit transfer methods. Early chapters provide accessible introductory and "how to" materials suitable for those with little economic training. These chapters also detail how benefit transfer is used within the policy process. Later chapters cover intermediate and advanced topics better suited to valuation researchers, graduate students, and those with similar knowledge of economic and statistical theory, concepts, and methods. While no single volume can provide all relevant information on a topic, our intent is for this book to provide the most complete coverage of benefit transfer methods available in a single location.

The motivation for this book arose through many hours of discussion with academics, policy-makers and others regarding the need for broadly accessible guidance for those seeking to use or evaluate benefit transfers. It is our sincere hope that this book will both advance the practice of benefit transfer worldwide and spur future research to improve it.

Robert J. Johnston
John Rolfe
Randall S. Rosenberger
Roy Brouwer

Acknowledgments

This volume reflects the work of four editors and 38 authors, two technical editors at Clark University (Mimi Berberian and Lu Ann Pacenka), and the administrative support of Pamela Dunkle, also at Clark. We acknowledge the assistance and guidance of many individuals at Springer, and particularly Fritz Schmuhl, who was instrumental in motivating the project. We also thank Ian Bateman (University of East Anglia) for his support. Each of us—the authors and editors alike—owe a debt of gratitude to all of the colleagues with whom we work and rely upon for feedback, suggestions, and constructive criticism, as well as to previous researchers who laid the foundations for the valuation and benefit transfer methods being used today. Last but certainly not least, we acknowledge the support and patience of our partners and families, without whom none of this would be possible.

Contents

Part IV Spatial and Geographical Considerations

Part V Bayesian Methods

Part VI Status and Prospects

Contributors

Jorge E. Araña Instituto de Turismo y Desarrollo Económico Sostenible (TiDES), Universidad de Las Palmas de Gran Canaria, Las Palmas, Spain

Ian J. Bateman Centre for Social and Economic Research on the Global Environment, School of Environmental Sciences, University of East Anglia, Norwich, UK

Jeffrey Bennett Crawford School of Public Policy, Australian National University, Canberra, ACT, Australia

Kevin J. Boyle Virginia Tech, Blacksburg, VA, USA

Luke M. Brander Department of Environmental Economics, Institute for Environmental Studies, Vrije Universiteit, Amsterdam, The Netherlands; Kadoorie Institute, University of Hong Kong, Pokfulam, Hong Kong

Roy Brouwer Department of Environmental Economics, Institute for Environmental Studies, Vrije Universiteit, Amsterdam, The Netherlands

Richard T. Carson University of California, San Diego, CA, USA

Florian V. Eppink Landcare Research, Auckland, New Zealand

Silvia Ferrini Centre for Social and Economic Research on the Global Environment, School of Environmental Sciences, University of East Anglia, Norwich, UK; University of Siena, Siena, Italy

William L. Huth Department of Marketing and Economics, University of West Florida, Pensacola, FL, USA

Robert J. Johnston Department of Economics and George Perkins Marsh Institute, Clark University, Worcester, MA, USA

Sapna Kaul University of Utah, Salt Lake City, UT, USA

Geoffrey Kerr Faculty of Environment, Society and Design, Lincoln University, Christchurch, New Zealand

Carmelo J. León Instituto de Turismo y Desarrollo Económico Sostenible (TiDES), Universidad de Las Palmas de Gran Canaria, Las Palmas, Spain

Henrik Lindhjem Norwegian Institute for Nature Research (NINA), Vista Analysis, Oslo, Norway

John B. Loomis Department of Agricultural and Resource Economics, Colorado State University, Fort Collins, CO, USA

Jordan J. Louviere University of South Australia Business School, Sydney, Australia

Klaus Moeltner Department of Agricultural and Applied Economics, Virginia Tech, Blacksburg, VA, USA

O. Ashton Morgan Department of Economics, Appalachian State University, Boone, NC, USA

Ståle Navrud School of Economics and Business, Norwegian University of Life Sciences (NMBU), Ås, Akershus, Norway

Jon P. Nelson Department of Economics, Pennsylvania State University, University Park, PA, USA

Christopher F. Parmeter University of Miami, Coral Gables, FL, USA

George R. Parsons School of Marine Science and Policy, Department of Economics and Department of Applied Economics and Statistics, University of Delaware, Newark, DE, USA

Daniel J. Phaneuf Department of Agricultural and Applied Economics, University of Wisconsin, Madison, WI, USA

Mahesh Ramachandran Cape Cod Commission, Barnstable, MA, USA

John Rolfe School of Business and Law, CQUniversity, Rockhampton, QLD, Australia

John M. Rose Institute for Choice, University of South Australia, Sydney, Australia

Randall S. Rosenberger Department of Forest Ecosystems and Society, College of Forestry, Oregon State University, Corvallis, OR, USA

Marije Schaafsma Geography and Environment & Centre for Biological Sciences, University of Southampton, Southampton, UK

Philipp Schägner Department of Landscape Architecture and Environmental Planning, Technical University Berlin, Berlin, Germany

Joffre Swait Institute for Choice, University of South Australia, Sydney, Australia

Pieter J.H. van Beukering Department of Environmental Economics, Institute for Environmental Studies, Vrije Universiteit, Amsterdam, The Netherlands

George Van Houtven RTI International, Durham, NC, USA

Alfred Wagtendonk Department of Environmental Economics, Institute for Environmental Studies, Vrije Universiteit, Amsterdam, The Netherlands

Lisa A. Wainger Center for Environmental Science, University of Maryland, Solomons, MD, USA

William J. Wheeler U.S. Environmental Protection Agency, National Center for Environmental Economics, Washington, DC, USA

John C. Whitehead Department of Economics, Appalachian State University, Boone, NC, USA

Jill Windle School of Business and Law, CQUniversity, Rockhampton, QLD, Australia

Acronyms and Abbreviations

ACF	Australian Conservation Foundation
AERE	Association of Environmental and Resource Economists
BT	Benefit transfer
CAFE	Clean Air for Europe
CAFO	Concentrated animal feeding operations
CBA	Cost-benefit analysis
CCW	Countryside Council for Wales
CE	Choice experiments
CM	Choice modeling
COAG	Council of Australian Government
COST	Cooperation in Science and Technology
CPI	Consumer price index
CS	Consumer surplus
CV	Contingent valuation
CWA	Clean Water Act
dBA	A-weighted decibels
DCE	Discrete choice experiments
DEFRA	Department for Environment, Food and Rural Affairs
DG	Directorate-General
DNREC	Delaware Department of Natural Resources and Environmental Control
EIS	Environmental Impact Assessment
ELS	Effluent limitation guideline
EO	Executive Order
EPBC	Environmental Protection and Biodiversity Conservation Act
ES	Environmental services
ESRC	Economic and Social Research Council
EuroFOREX	European Forest Externalities
EVRI	Environmental Valuation Reference Inventory
ExternE	External Costs of Energy

GAB	Great Artesian Basin
GCP	Gross cell product
GDP	Gross domestic product
GHG	Greenhouse gases
GIS	Geographic information system
GLWG	Great Lakes Water Quality Guidance
GPL	Green and pleasant land
GR	Golden Rule
HCM	Human capital method
HW	Hedonic wage
LNE	Department of Environment, Nature and Energy
MA	Meta-analysis
MB	Marginal benefit
MEA	Millennium Ecosystem Assessment
ML	Maximum likelihood
MLM	Multilevel modeling
MNL	Multinomial Logit
MP&M	Metal products and machinery
MRM	Meta-regression model
NERC	National Environment Research Council
NGO	Non-governmental organization
NIEA	Northern Ireland Environment Agency
NPDES	National Pollutant Discharge Elimination System
NRCS	National Resources Conservation Service
NRDA	National Resource Damage Assessment
NWPCAM	National Water Pollution Control Assessment Model
OCPSF	Organic chemicals, plastics and synthesis fibers
OLS	Ordinary least squares
OW	Office of Water
PPD	Posterior predictive distribution
PPP	Purchasing power parity
RE	Random effects
RIA	Regulatory impact analysis
RIS	Regulatory impact statements
RP	Revealed preference
RPA	Resource Planning Act
RUM	Random utility model
SAC	Special areas of concentration
SOC	Soil organic carbon
SP	Stated preference
SPF	Site prediction function
SSUT	Strong structural utility theoretic
TC	Travel cost
TEEB	The Economics of Ecosystems and Biodiversity
TGF	Trip generation function

TMDL	Total maximum daily load
UDV	Unit day values
UKNEA	UK National Ecosystem Assessment
USDA	U.S. Department of Agriculture
US EPA	U.S. Environmental Protection Agency
VSL	Value of a statistical life
WAG	Welsh Assembly Government
WFD	Water Framework Directive
WHO	World Health Organization
WTA	Willingness to accept
WTP	Willingness to pay

List of Tables

Part I
Introduction and Policy Perspectives

Part I
[illegible]

Chapter 1
Introduction: Benefit Transfer of Environmental and Resource Values

John Rolfe, Robert J. Johnston, Randall S. Rosenberger and Roy Brouwer

Abstract The goal of this handbook is to provide comprehensive coverage of contemporary methods, issues and challenges in benefit transfer, addressing topics relevant to both researchers and practitioners. This initial chapter provides an overview of benefit transfer and establishes the background for the handbook. It begins with a brief summary of benefit transfer methods and applications, including some of the key challenges faced by researchers and analysts. This sets the context for the remainder of the book. The chapter then provides background on the motivations for using benefit transfer and the historical development of transfer methods. These motivations, and the many of the controversies and challenges surrounding the development of benefit transfer, are summarized into three themes: pragmatic need, accuracy and validity. The chapter concludes with a summary of the handbook structure and a brief discussion about the potential for future development of benefit transfer.

Keywords Benefit transfer · Value transfer · Valuation · Reliability · Validity

J. Rolfe (✉)
School of Business and Law, CQUniversity, Rockhampton, QLD, Australia
e-mail: j.rolfe@cqu.edu.au

R.J. Johnston
Department of Economics and George Perkins Marsh Institute, Clark University, Worcester, MA, USA
e-mail: rjohnston@clarku.edu

R.S. Rosenberger
Department of Forest Ecosystems and Society, Oregon State University, Corvallis, OR, USA
e-mail: R.Rosenberger@oregonstate.edu

R. Brouwer
Department of Environmental Economics, Institute for Environmental Studies, Vrije Universiteit, Amsterdam, The Netherlands
e-mail: roy.brouwer@vu.nl

R.J. Johnston et al. (eds.), *Benefit Transfer of Environmental and Resource Values*, The Economics of Non-Market Goods and Resources 14,
DOI 10.1007/978-94-017-9930-0_1

1.1 Introduction

Benefit transfer may be defined as the use of research results from pre-existing primary studies at one or more sites or policy contexts (often called study sites) to predict welfare estimates or related information for other, typically unstudied sites or policy contexts (often called policy sites). It is the process of taking study results from one situation and extrapolating them to another similar situation. The goal is to provide empirical estimates for the particular issue of interest when time, funding or other constraints prevent the use of primary research to generate these estimates. Given the ubiquity of such constraints in the policy process, benefit transfer is often the only option available for providing needed information. As a result, benefit transfer is an indispensable component of virtually all large-scale benefit cost analyses in the United States, European Union and elsewhere.

Most benefit transfer research and use occurs within applied economics, where the practice is applied to economic measures such as willingness to pay (WTP) estimates, elasticity measures or demand relationships. Within these applications, the use of benefit transfer has been most prevalent in the sub-discipline of environmental economics. This is also the sub-discipline in which the most intense debates over benefit transfer—including potential biases and appropriate protocols—have occurred. However, benefit transfer is also important and widely used in other fields of applied economics such as recreation, transport and health.

Benefit transfer is superficially attractive because it often requires less time and money than primary research. In many cases it is also simpler to use. For analysts or policy makers who do not have the skills or resources to conduct primary studies, the ability to take values or benefit functions "off the shelf" can be appealing. At the same time, the validity and accuracy of benefit transfer rely on a number of conditions. Moreover, in many situations benefit transfer may not be a suitable substitute for primary research. Questions regarding when and how to conduct benefit transfer have preoccupied the literature since the beginning of formal research into benefit transfer methods.

Boyle and Bergstrom (1992) were among the first to recommend ideal criteria for benefit transfer. Under these ideal criteria, benefit transfer is valid only when source and target sites, populations and welfare measures are identical. These requirements have been tempered over time as researchers have searched for pragmatic guidelines to incorporate variations in study settings. For example, Bennett (2006) suggests a more widely applicable set of criteria, including five key requirements for process validity:

- the biophysical conditions in the source case must be similar to those in the target case;
- the scale of environmental change considered in the source must approximate the target;
- the socioeconomic characteristics of the population impacted by the change investigated in the source must approach those of the target population;

- the frame or setting in which the valuation was made at the source must be close to that of the target;
- the source study has to have been conducted in a technically satisfactory fashion.

There are two major difficulties with these and similar guidelines for benefit transfer. The first is that most cases of benefit transfer involve variations between at least sites and populations, if not frame and scale. As a result, few applied transfers meet these or similar criteria. Combined with the possibility of unrepresentative or unreliable source studies, the potential for invalid and unreliable[1] benefit transfers appears large. As many benefit transfers are conducted solely for policy purposes and are not subject to formal peer review, it is difficult to judge the extent to which transfer practices diverge from recommendations. However, there appears to be a substantial divergence between best practices identified in the research literature and those applied within policy analysis (Johnston and Rosenberger 2010). Moreover, despite the extent of use, the practice remains characterized by frequent misuse and misunderstanding.

Another challenge is a lack of clear protocols regarding how benefit transfer should proceed when a condition (ideal or otherwise) is not fully met. Despite attempts at methodological guidance, consensus about protocols remains elusive, leading to recurring questions regarding transfer reliability and validity. At a practical level, this has meant that practitioners often make informal and sometimes uninformed judgments about the applicability and importance of different types of similarity and recommended transfer practices. For example, while the benefit transfer literature commonly finds that transfers of benefit functions outperform transfers of fixed (unit) values when sites are not very similar, fixed value transfers are nonetheless applied frequently over dissimilar sites. This may occur with an implicit or explicit claim that value transfer is "good enough" for the required application or due to a lack of familiarity with more sophisticated approaches. Such practices clearly violate ideal transfer criteria. However, because virtually all transfers violate these ideal criteria to some degree, and because the need for accuracy varies across policy contexts, it has been difficult to identify and discourage transfers that fail to meet even minimal criteria of good practice. The size, complexity, conflicting viewpoints, and relative disorganization of the benefit

---

[1]Transferred errors from the original primary studies are denoted *measurement errors*, while *generalization errors* are generated by the transfer process itself. As described in Chap. 2, benefit transfers are generally evaluated in terms of predictive accuracy (sometimes termed reliability) and transfer validity. A transfer accurately predicts a value estimate, or is reliable, when the generalization error is small. Benefit transfer validity, in contrast, can be viewed from two perspectives. Statistical validity of benefit transfer requires that value estimates or other transferred quantities are statistically identical across study and policy contexts (i.e., there is no statistically significant transfer error). This is the most common use of the term in the benefit transfer literature. However, transfer validity may also be interpreted as a requirement that measurement errors are minimized, or that there are no systematic errors in the underlying primary study estimates.

transfer literature also provide an imposing obstacle to the use of updated methods by practitioners (Johnston and Rosenberger 2010).

This handbook is designed to provide a comprehensive reference for those seeking information on contemporary benefit transfer methods, debates, applications, challenges and frontiers. Among other contributions, it provides an introduction to benefit transfer theory and methods, discussion of international policy perspectives, coverage of contemporary methods and recent advances, illustrations of associated case studies, descriptions of methodological debates and controversies, and perspectives on the state of the art and future prospects. A primary goal of the book is to describe the range of alternative transfer methodologies suitable for broad application, issues that should be considered when choosing among available transfer methods, and implications of these choices for benefit transfer validity, reliability and policy guidance. Subsequent chapters detail both the information that can be provided through high-quality transfers as well as the potential challenges and pitfalls. Empirical illustrations are drawn from a range of worldwide applications, to demonstrate how transfer methods may be applied and tested in different situations.

This initial chapter provides a broad overview of benefit transfer and establishes the background for the handbook. It begins with a general overview of benefit transfer, including some of the key challenges faced by researchers and analysts. This sets the context for the remainder of the book. The chapter then provides background on the motivations for using benefit transfer and the historical development of transfer methods. These motivations, and the many controversies and challenges surrounding the development of benefit transfer are categorized according to three themes: pragmatic need, accuracy and validity. The chapter concludes with a summary of the handbook structure and a brief discussion about the potential for future development.[2]

1.2 The Need for Benefit Transfer and Its Development

Economic evaluations, such as benefit-cost analysis, typically require the quantification benefits and costs for different types of policy impacts. The theoretical foundation of welfare economics enables the measurement of these benefits and costs in a logically consistent and directly comparable form. This foundation also provides the underpinning framework within which benefit transfer is performed. Within economic evaluations, policy makers often face challenges related to the expertise, time and money required to evaluate anticipated benefits and costs. Primary studies are often prohibitively complex, time-consuming and expensive.

[2]An important point to note is that this chapter is intended to provide a very broad overview to help readers without specialized knowledge. There is no detailed description of benefit transfer or review of the literature included; readers will find this detail in subsequent chapters.

This is particularly true for non-market values, defined as values arising from goods and services that are not exchanged directly in organized markets. Due to this lack of market exchange and observable prices, specialized revealed and stated preference techniques are required to estimate values (Champ et al. 2003; Freeman et al. 2014). Benefit transfer provides a means to estimate non-market values and other needed economic information when constraints in time, funding, or informational requirements prevent the use of primary studies. As transferred values are ideally quantitative, consistent with theoretically well-defined welfare measures, and directly comparable to other estimates of costs and benefits, it is appropriate for these values to be incorporated into benefit-cost analysis and other economic evaluations.

The relevance and application of benefit transfer in applied economics grew during the 1980s alongside the development of benefit-cost analysis and non-market valuation. Although benefit-cost analysis (also termed cost-benefit analysis) had been widely used to evaluate policy options since the 1930s, concerns over a narrow focus on financial effects led to development of "extended" benefit-cost analysis in the 1970s and 1980s. This enabled benefit-cost analyses to incorporate impacts on environmental, social and other factors that had previously been excluded. This extended focus was accompanied by the development of non-market valuation techniques that quantify values for outcomes such as changes in environmental condition, recreational access and activities, human health, transport quality and risk prevention, among others.

There are a number of different non-market valuation techniques. These are generally classified into revealed preference and stated preference techniques. The former group, which includes the travel cost (or recreation demand) and hedonic pricing methods, analyzes available data from indirectly-related market transactions or behavior. The latter group, which includes continent valuation and choice modeling, asks respondents to state their preferences, via a survey instrument, for tradeoffs in a hypothetical market or situation. These and other non-market valuation methods have been subject to extensive research and development over the past four decades. Numerous publications summarize these methods and the associated literature (e.g., Champ et al. 2003; Freeman et al. 2014; Hanley and Barbier 2009). Although non-market valuation techniques vary in specialization and purpose, they are similar in that they are expensive and time-consuming to apply, and require technical expertise to generate accurate results.

The need for benefit transfer in applied economics was generated from the intersection of the demand and supply for economic information (e.g., on benefits and costs). Demand for economic information has increased with the growth of applied economics in fields such as environment, recreation, transport and health. Of particular note has been growth in the need for inputs into benefit-cost analysis, within which values for impacts on environmental assets, recreation opportunities or other non-market outcomes were required. For example, Executive Order 12291 (1981) in the United States mandated that all new major regulations be subject to benefit-cost analysis; the resulting assessments often require analysts to quantify non-market benefits and costs. Later initiatives such as the multinational Water

Framework Directive in Europe (first enacted in 2000) further increased the demand for benefit and cost estimates to be used in policy analysis.

On the supply side, there was rapid growth in the use of non-market valuation techniques, such as the travel cost method during the 1960s and 1970s and the contingent valuation method during the 1970s and 1980s. These studies were often conducted for methodological purposes rather than to support benefit-cost analysis, and hence did not *directly* meet the growing demand for economic information. However, this research did increase the pool of source studies for potential use in benefit transfer. At the same time, the opportunity cost of performing primary non-market valuation for direct policy application has remained high, not only because of the time and resources required to perform these studies, but also because of the growing awareness of potential biases and the need to generate results in technically sophisticated ways. For example, there has been increasing awareness that the validity and accuracy of applied valuation depends on the exacting development of empirical methods; this development often requires a significant commitment of time and resources.

This increase in available information on non-market values, coupled with the pragmatic need for value information, set the foundation for developing low-cost and pragmatic benefit transfer methodology. Early efforts at methodological guidance for benefit transfer included that of Freeman (1984). However, the first major academic focus on benefit transfer occurred with a special edition of *Water Resources Research* (28[3]) in 1992, in which a number of high-profile contributors debated the use, protocols and limitations to benefit transfer. Since that seminal publication, benefit transfer has been an important topic in the applied economics literature.

The motivations for developing and improving benefit transfer methods can be grouped into three categories, including two related to the demand for welfare estimates, and one related to the supply of them. Motivations in the first category relate to the demand for timely, accurate and cost-effective welfare estimates by those conducting policy analyses. These are the *pragmatic* justifications for benefit transfer. Substantial effort has been invested in identification of the appropriateness of benefit transfer for different purposes, establishment of criteria for ideal benefit transfers, and development of protocols for use. As noted below, the pragmatic focus dominated much of the earliest benefit transfer research.

Along with pragmatic needs for transferred information are increasingly sophisticated demands for this information to be *accurate*. These demands reflect growing awareness of the potential for biases and errors in transferred information, at least among many users. Reflecting this awareness, an expanding literature now addresses the types and magnitudes of errors that are likely with different types of benefit transfers. Included in this work is the development and testing of transfer methods better able to adjust for differences among study and policy sites, and hence increase transfer accuracy. Additional efforts have been made to develop transfer methods better linked to utility theory, and that can more effectively capitalize on information from large numbers of primary studies on similar phenomena. This includes the development of meta-analytic and structural benefit transfers that synthesize data from multiple prior studies.

On the supply side, there has been considerable effort to improve the underlying primary study methods that provide transferrable information. This includes research to improve methodologies for non-market valuation, to enhance the transferability of resulting functions and estimates, and to enhance the acceptance of these methods for policy analysis. There has been similar effort to develop benefit transfer techniques that make the most effective use of available primary study information, and to evaluate the range of situations over which results from any single primary study can be applied. These efforts can be broadly characterized as improving and *validating* both the primary valuation techniques and the ways that study results are adapted for benefit transfer. Some of this effort has also focused on demonstrating the appropriateness and accuracy of underlying non-market value estimates, and whether/when the resulting benefit functions can be used reliably to predict value estimates in other situations (thereby overlapping the accuracy concerns noted above). Other research has evaluated the capacity of primary study results to support the scaling and adjustment often conducted as part of benefit transfer.

1.3 Changes in Focus Over Time

Given these broad motivations for benefit transfer research, it is worth reviewing how the practice has been defined and studied over the past two decades. The initial focus on the pragmatic need for benefit transfer is reflected in the terminology used. Desvousges et al. (1992, p. 675) note that in 1982 the United States Environmental Protection Agency suggested that "off-the-shelf methodologies and studies can serve as the basis for benefit cost analysis," and go on to define benefit transfer in terms of "the use of existing studies." Brookshire and Neill (1992) review the various descriptions of benefit transfer in the special edition of *Water Resources Research*, noting that different authors in the special edition describe benefit transfer as "the application of secondary data to a new policy issue" (Boyle and Bergstrom 1992), the process of "extrapolate[ing] the results of benefit assessments done elsewhere" (Atkinson et al. 1992), and the "use (of) … existing studies … to suggest some likely limits on willingness to pay" (Luken et al. 1992). The terms that have perhaps endured best come from Desvousges et al. (1992), who define benefit transfer as taking study results from a "study" site and applying them to "policy" site. Over the following two decades that description has remained consistent, albeit with minor adaptations by different researchers (e.g., Johnston and Rosenberger 2010).

Much of the initial focus of research into benefit transfer was aimed at the mechanics of performing the transfer. This pragmatic focus is reflected in Brookshire and Neill (1992), in which they discuss the use of expert opinion as a key option for transferring values, tempered by consideration of "reasonableness." Key reasons advanced for performing benefit transfers included the lack of controversy around some issues, the pragmatic environment that surrounded many policy decisions in which accuracy was not very important, and the cost advantages. Additional evidence is provided by these researchers' identification of the lack of source studies and data sets as a key impediment to benefit transfer, and highlighting the importance of researcher judgment in performing valuation exercises and benefit transfers.

The issues of accuracy and validity were also emphasized, however, even in early research. Brookshire and Neill (1992) note that many contributors to the special issue were concerned about accuracy, although these concerns focused largely on the quality and validity of source studies rather than errors in the transfer process itself. Validity concerns focused largely on the potential use of contingent valuation for transfer. Brookshire and Neill (1992, p. 652) argue presciently that the "next logical step might well be the introduction of benefit transfer studies based on the contingent valuation method." In the two decades that followed, contingent valuation and its more recent counterpart, choice modeling, became common in benefit transfer applications.

In more recent years, focus has shifted away from the pragmatic aspects of benefit transfer (which are well-established) to issues including: (1) a more systematic understanding of transfer accuracy and its determinants, (2) the relative importance of formal utility-theoretic structure within the transfer itself, (3) the development of more nuanced and sophisticated empirical methods to recognize and capitalize on patterns in the underlying primary data, thereby supporting more accurate transfers, (4) improved methods for data synthesis and meta-analysis, (5) more sophisticated understanding of commodity and welfare consistency, (6) identification of issues such as excessive scale changes that are associated with invalid or inaccurate transfers, and (7) evaluation of general conditions for which different types of transfer methods are best suited. These and related issues are the primary focus of subsequent chapters in this book.

1.4 Challenges and Controversies

Over the past two decades the literature on the use and challenges of benefit transfer has grown rapidly, prompted in large part by concerns over validity and accuracy (Johnston and Rosenberger 2010). While the goals of benefit transfer have remained largely unchanged, the applications, understanding and limitations have developed substantially. In terms of applications, benefit transfer is now used to predict a variety of economic measures beyond willingness to pay and other welfare estimates. There is now much more detailed understanding of benefit transfer methods and the factors that influence transfer accuracy. Many of these are addressed in later chapters. The limitations of benefit transfer are also better understood.

As noted above, the *pragmatic* focus that characterized earlier applications of benefit transfer has largely been subsumed by research to improve accuracy and demonstrate validity. With better understanding and knowledge, researchers have realized that many earlier attempts at benefit transfer—often using the transfer of unadjusted unit values, administratively approved values, or values adjusted using expert opinion—were very inaccurate. Yet despite this recognition the use of such approaches by policy makers and others remains common. In many ways the divergence between transfer practices that are recommended in the literature and those that are commonly applied in policy situations has widened. Challenges to improving the pragmatic use of benefit transfer include the complexity of the current benefit transfer literature and the conflicting viewpoints within it, as well as the lack of universally accepted protocols. The technical knowledge needed to perform contemporary benefit transfer has also increased, raising questions about whether it is possible for benefit transfer to be conducted by non-specialists.

For example, the simple unit value transfers that were often thought to be acceptable in the 1980s and early 1990s—largely based on pragmatic considerations —are now considered inadequate for most applications. These increasing concerns for *accuracy* have led to recent recommendations for the use of more sophisticated methods such as benefit function transfers, structural transfers, and meta-analytic transfers that are simultaneously more accurate (in many applications[3]) and difficult to apply. These recommendations have not prevented the use of unit value and other simplistic benefit transfers in policy and advocacy situations, but have widened the gap between recommended and applied methods. While we now have the technical ability to conduct benefit transfers that are far more accurate than those implemented in past decades, the use of less accurate methods remains common in practice. To address this divergence, there is a need for transfer approaches that are both accurate and useful/accessible to non-experts. Johnston and Rosenberger (2010) note that the literature is almost universal in its call for more sophisticated techniques to address different shortcomings, but that the myriad of advances that have been made are sometimes inconsistent and scattered across the literature. Their review provides a good guide to the state of knowledge in this area.

Issues related to the *validity* of benefit transfer can be grouped into three main areas. Attention to each of these areas has varied over recent decades. The first category includes concerns about the validity of the underlying nonmarket valuation techniques, and stated preference methods in particular. For example, controversy over the contingent valuation method became heated in the United States during natural resource damage assessments for the 1989 Exxon Valdez oil spill in Alaska. A similar public controversy occurred in Australia when the technique was used to assess protection values for the area that is now a part of Kadadu National Park. The seemingly high values that were estimated by these research efforts generated intense debate over the validity of the approaches used. While the NOAA Blue Ribbon Panel (Arrow et al. 1993) and subsequent research (see Carson 2012 and Kling et al. 2012 for summaries) has largely validated stated preference methods, some mistrust of these methods and their results has continued. Similar concerns over theoretical and empirical validity—while less pointed than those over

[3]Bateman et al. (2011) provide a useful discussion of cases in which simpler forms of benefit transfer may outperform more sophisticated or complex methods.

stated preferences—have been voiced for other types of nonmarket valuation. Naturally, these validity concerns carry over to benefit transfers relying on the results of such valuation.

Second, the common lack of a structural theoretical foundation for benefit transfer (other than that underlying the original primary studies) continues to be a concern. The difficulty of establishing a theoretical foundation is complicated by the divergence of purposes and uses of benefit transfer. For example, transfer methods may be used not only to transfer value estimates but also elasticity estimates and other measures. Beginning with Smith et al. (2002), some authors have proposed transfer methods grounded in a strong structural utility-theoretic foundation. However, widespread adoption of these methods has been limited, in part due to the technical difficulty of these approaches, the potential sensitivity of transfer results to the assumptions required to establish the structural utility foundation, and the lack of empirical evidence that such approaches improve transfer accuracy.

Whereas the first two concerns described above relate to the theoretical validity of benefit transfer, the third concern in this area relates to statistical validity. This is closely related to the accuracy concerns described above. Convergent validity studies[4] continue to show that empirical validity tests of value transfers are often not satisfied, even when the source and target study situations are very similar. Understanding the reasons for such findings, and how the empirical validity of benefit transfer can be improved, continues to be an important focus.

In summary, benefit transfer practitioners face a number of challenges. While the key arguments for the use of benefit transfer are pragmatic, simplistic applications often lead to invalid and/or inaccurate results. Despite increasing research in this area, and the fact that benefit transfers are an almost universal component of large-scale benefit cost analyses, there are no generally accepted protocols for use. Advances are being made to improve the accuracy of benefit transfer applications, but these advances are often disjointed and sometimes inconsistent across the literature. Moreover, they are increasingly technical and often beyond the expertise of policy analysts. At the same time, despite these efforts and advances, questions of reliability and validity continue to arise. These observations lead to four stylized facts of benefit transfer, which serve as the launching point for this volume: (1) benefit transfer is, and will continue to be, a common component of policy analysis worldwide, (2) consensus best practice guidelines for benefit transfer are elusive but needed, (3) despite numerous recent advances in benefit transfer methods, additional work is needed to ensure validity and accuracy, and (4) efforts are required to make accurate benefit transfer methods more broadly accessible to policy analysts.

[4]As described in Chap. 2, out-of-sample predictive performance, or convergent validity testing, is most often used to test transfer accuracy and validity. That is, benefit transfer is tested in circumstances where a policy site study has been conducted. Benefit transfer estimates from source sites are then compared to the estimate provided by original research at the policy site. Empirical benefit transfer validity requires that value estimates or other transferred quantities are statistically identical across study and policy contexts (i.e., there is no statistically significant transfer error).

1.5 Aims and Organization of the Book

The key aims of this book are to consolidate available knowledge about benefit transfer and to provide a framework for its application to encourage more systematic and reliable use. The book is designed to meet the needs of (1) policy analysts who wish to apply or evaluate benefit transfer methods, (2) students and others who wish to learn about these methods, and (3) experts in benefit transfer who wish to remain abreast of the rapidly expanding literature. Different chapters are targeted to individuals with different interests and levels of prior expertise. Table 1.1 summarizes the topical emphasis and difficulty level of each chapter.

The volume begins with a detailed, practical overview of benefit transfer methods in Chap. 2 (written by the editors). This chapter is designed as a primer to benefit transfer methods; a simple "how to" guide for those less familiar with transfer theory and methods. It describes the different types of benefit transfer and their application, the foundations of benefit transfer in welfare economics and valuation, and the issues surrounding (and advanced techniques to address) common problems and challenges. Together with this introductory chapter, Chap. 2 provides a practical starting point for the rest of the book.

Following this methodological introduction, Chaps. 3, 4 and 5 describe the development and use of benefit transfer in three sectors of the developed world: North America in Chap. 3 (by Loomis), Europe in Chap. 4 (Brouwer and Navrud), and Australia and New Zealand in Chap. 5 (Rolfe, Bennett and Kerr). This is followed by Chap. 6 (Wheeler), which provides a detailed perspective on the use of benefit transfer within regulatory rulemaking in the United States. These chapters provide a historical context to the application of benefit transfer methods in each region and introduce the reader to the issues, challenges and applications that have taken place. These chapters also establish the policy contexts within which benefit transfers occur across the world, and highlight relationships between policy applications and methodological development. Finally, these chapters characterize differences in the ways that benefit transfer has developed and been used in different regions of the world.

Chapters in the second section of the book address the application of benefit transfer techniques to different types of underlying data and primary study methods, along with the reliability and validity of resulting transfers. Chapter 7 (Whitehead, Morgan and Huth) introduces the reader to benefit transfer using stated preference techniques, covering methodologies used to generate source data and the ways that these data are used for transfers. The focus on methodology is developed further with Chap. 8 (Rolfe, Windle and Johnston), which provides guidance on the use of unit value transfers with limited information, focusing again on transfers of stated preference results. Chapter 9 (Johnston, Ramachandran and Parsons) illustrates empirical methods for benefit transfer that combine revealed and stated preference data. Chapters 10 (Rolfe, Windle and Bennett) and 11 (Carson, Louviere, Rose and Swait) provide more advanced discussions of the use of choice models (or experiments)

Table 1.1 Chapter emphasis and technical difficulty

Chapter	Topical emphasis	Difficulty
	Introduction	Introductory
	Methods	Intermediate (suitable for advanced undergraduate students)
	Applications	
	Use in policy	Advanced (suitable for graduate students and advanced practitioners)
	Summary	
Part I Introduction and Policy Perspectives		
1. Introduction: Benefit Transfer of Environmental and Resource Values	Introduction	Introductory
2. Introduction to Benefit Transfer Methods	Introduction, methods	Introductory
3. The Use of Benefit Transfer in the United States	Use in policy	Introductory
4. The Use and Development of Benefit Transfer in Europe	Use in policy	Introductory
5. Applied Benefit Transfer: An Australian and New Zealand Policy Perspective	Use in policy	Introductory
6. Benefit Transfer for Water Quality Regulatory Rulemaking in the United States	Use in policy	Introductory
Part II Methods and Applications		
7. Benefit Transfers with the Contingent Valuation Method	Methods, applications	Intermediate
8. Applying Benefit Transfer with Limited Data: Unit value Transfers in Practice	Methods, applications	Intermediate
9. Benefit Transfer Combining Revealed and Stated Preference Data: Nourishment and Retreat Options for Delaware Bay Beaches	Methods, applications, use in policy	Intermediate
10. Benefit Transfers: Insights from Choice Experiments	Methods	Intermediate
11. Frontiers in Modeling Discrete Choice Experiments: A Benefit Transfer Perspective	Methods	Advanced
12. Benefit Transfer for Ecosystem Service Valuation: An Introduction to Theory and Methods	Introduction, methods	Introductory
13. Ecosystem Services Assessment and Benefit Transfer	Methods, applications	Intermediate
14. Benefit Transfer Validity and Reliability	Methods, applications	Intermediate

(continued)

Table 1.1 (continued)

Chapter	Topical emphasis	Difficulty
Part III Meta-analysis		
15. Meta-analysis: Statistical Methods	Methods	Intermediate
16. Meta-analysis: Rationale, Issues and Applications	Methods, applications	Advanced
17. Meta-analysis: Econometric Advances and New Perspectives Toward Data Synthesis and Robustness	Methods	Advanced
Part IV Spatial and Geographical Considerations		
18. Spatial and Geographical Aspects of Benefit Transfer	Methods	Introductory
19. Reliability of Meta-analytic Benefit Transfers of International Value of Statistical Life Estimates: Tests and Illustrations	Methods, applications	Advanced
20. GIS-based Mapping of Ecosystem Services: The Case of Coral Reefs	Methods, applications	Advanced
Part V Bayesian Methods		
21. A Bayesian Model Averaging Approach to the Transfer of Subjective well-being Values of Air Quality	Methods, applications	Advanced
22. Optimal Scope and Bayesian Model Search in Benefit Transfer	Methods, applications	Advanced
23. Structural Benefits Transfer Using Bayesian Econometrics	Methods, applications	Advanced
Part VI Status and Prospects		
24. Benefit Transfer: The Present State and Future Prospects	Summary	Introductory

for benefit transfer. Chapters 12 (Johnston and Wainger) and 13 (Ferrini, Schaafsma and Bateman) address methods and applications of benefit transfer for ecosystem service valuation—an area in which benefit transfer is increasingly used and misused. Finally, Chap. 14 (Rosenberger), discusses benefit transfer validity and reliability. This chapter addresses the types of transfer errors that have occurred in past evaluations, as well as patterns in transfer reliability and validity that have emerged from this literature. This chapter also addresses the relevance of selection biases for transfer accuracy and validity.

The final three sections of the book cover advanced material related to three large topics: meta-analysis, spatial and geographical considerations, and Bayesian approaches. A substantial proportion of recent research in benefit transfer methods has occurred in these areas. Three chapters address topics in meta-analysis and benefit transfer. These include a discussion of statistical methods in Chap. 15 (Nelson), empirical applications in Chap. 16 (Rolfe, Brouwer and Johnston), and robustness testing in Chap. 17 (Boyle, Kaul and Parmeter). Spatial and geographical considerations are also covered by three chapters, including an overview of spatial and geographical aspects of benefit transfer in Chap. 18 (Schaafsma), international benefit transfers in Chap. 19 (Lindhjem and Navrud), and the use of GIS value mapping in Chap. 20 (Brander and co-authors). The three chapters devoted to advanced Bayesian methods include a discussion of Bayesian model averaging in Chap. 21 (Araña and León), optimal scope and Bayesian model search in Chap. 22 (Moeltner) and Bayesian structural transfers in Chap. 23 (Phaneuf and Van Houtven).

The final chapter in the handbook is again written by the editors, and returns to the themes identified by the first section of the book. It draws together the contributions of prior chapters and discusses needs for future development. This chapter emphasizes that while benefit transfer is not straightforward and that significant expertise and attention to methodology is required, advances in benefit transfer techniques and improved understanding of these techniques offer the promise of widespread improvements in transfer accuracy and validity.

These chapters can be mixed and matched in different ways, for different purposes. For example, Parts I, II and VI (perhaps minus the advanced material in Chap. 11) could form the core of an advanced undergraduate course (or course section) on benefit transfer. A graduate course could supplement these chapters with the more advanced materials in Parts III, IV and V. Noneconomists seeking a primer in benefit transfer methods and applications could also focus attention on a subset of chapters. For example, noneconomists interested in an introduction to benefit transfer for ecosystem service valuation could devote primary attention to Chaps. 1, 2, 12 and 13. Finally, many chapters are explicitly designed to target topics relevant to more advanced benefit transfer researchers and practitioners. Examples include meta-analysis (Part III, Chaps. 19, 20, 22), spatial/geographical considerations (Part VI), and Bayesian methods (Part V). Taken together, this handbook is meant to consolidate information on contemporary benefit transfer issues, methods and advances in a single, accessible volume. While no book can cover all possible topics of interest, we have sought to compile a more comprehensive and current treatment of benefit transfer topics than is available in any other volume. The goal is to enable broader dissemination of this information to both researchers and practitioners.

This handbook does not resolve all the issues for benefit transfer, and many challenges remain. Many of these challenges reflect recurring themes across the chapters. Among these is the identification of factors that influence benefit transfer performance, and how adoption of different methodologies in different situations can improve the accuracy, reliability and stability of transferred values. A related theme is the need for consensus protocols and best practice guidelines, including guidance over the suitability of different types of transfer methods (particularly more advanced methods such as meta-analysis, Bayesian approaches, and structural benefit transfers) for different types of applications. The many unsolved challenges of benefit transfer, together with the likelihood that such methods will continue to be an indispensable part of policy analysis worldwide, underscore the need for future research and consensus in this area. Our hope is that this book will provide a foundation for future work.

References

Arrow, K., Solow, R., Portney, P. R., Leamer, E. E., Radner, R., & Schuman, H. (1993). Report of the NOAA panel on contingent valuation. *Federal Register, 58*, 4601–4614.

Atkinson, S., Crocker, T., & Shogren, J. (1992). Bayesian exchangeability, benefit transfer and research efficiency. *Water Resources Research, 28*, 715–727.

Bateman, I. J., Brouwer, R., Ferrini, S., Schaafsma, M., Barton, D. N., Dubgaard, A., et al. (2011). Making benefit transfers work: Deriving and testing principles for value transfers for similar and dissimilar sites using a case study of the non-market benefits of water quality improvements across Europe. *Environmental and Resource Economics, 50*, 365–387.

Bennett, J. (2006). Introduction. In J. Rolfe & J. Bennett (Eds.), *Choice modelling and the transfer of environmental values* (pp. 1–9). Cheltenham, UK: Edward Elgar.

Boyle, K., & Bergstrom, J. (1992). Benefit transfer studies: Myths, pragmatism and idealism. *Water Resources Research, 28*, 657–663.

Brookshire, D., & Neill, H. (1992). Benefit transfers: Conceptual and ethical issues. *Water Resources Research, 28*, 651–655.

Carson, R. (2012). Contingent valuation: A practical alternative when prices aren't available. *Journal of Economic Perspectives, 26*, 27–42.

Champ, P. A., Boyle, K. J., & Brown, T. C. (Eds.). (2003). *A primer on nonmarket valuation*. Dordrecht: Springer.

Desvousges, W., Naughton, M., & Parsons, G. (1992). Benefit transfer: Conceptual problems in estimating water quality benefits using existing studies. *Water Resources Research, 28*, 675–683.

Freeman, A. M. III. (1984). On the tactics of benefit estimation under executive order 12291. In V. K. Smith (Ed.), *Environmental policy under Reagan's executive order: The role of benefit cost analysis* (pp. 167–186). Chapel Hill, NC: The University of North Carolina Press.

Freeman, A. M. III., Herriges, J. A., & Kling, C. L. (2014). *The measurement of environmental and resource values: Theory and methods*. Washington, DC: Resources for the Future.

Hanley, N., & Barbier, E. B. (2009). *Pricing nature: Cost-benefit analysis and environmental policy*. Cheltenham, UK: Edward Elgar.

Johnston, J., & Rosenberger, R. (2010). Methods, trends and controversies in contemporary benefit transfer. *Journal of Economic Surveys, 24*, 479–510.

Kling, C. L., Phaneuf, D. J., & Zhao, J. (2012). From Exxon to BP: Has some number become better than no number? *Journal of Economic Perspectives, 26*, 3–26.

Luken, R., Johnson, F., & Kibler, V. (1992). Benefits and costs of pulp and paper effluent controls under the Clean-Water-Act. *Water Resources Research, 28*, 665–674.

Smith, V. K., van Houtven, G., & Pattanayak, S. K. (2002). Benefit transfer via preference calibration: "Prudential algebra" for policy. *Land Economics, 78*, 132–152.

Chapter 2
Introduction to Benefit Transfer Methods

Robert J. Johnston, John Rolfe, Randall S. Rosenberger and Roy Brouwer

Abstract This chapter provides an introductory overview of benefit transfer methods. It begins with a discussion of the different types of benefit transfer (such as unit value transfer and benefit function transfer), including a review of these different approaches and the relative advantages and disadvantages of each. This is followed by a summary of foundations in welfare economics and valuation. Included in this methodological introduction are a discussion of stated and revealed preference valuation and how the results from each may be used for benefit transfer. Following this introductory material are discussions of the theoretical and informational requirements for benefit transfer, the steps required to implement a benefit transfer, the challenges of scaling, and sources of data. The chapter concludes with brief discussions of transfer validity and reliability, advanced techniques for benefit transfer, and common problems and challenges.

Keywords Unit value transfer · Benefit function transfer · Meta-analysis · Preference calibration · Reliability · Validity · Non-market valuation · Methodology · Value transfer

R.J. Johnston (✉)
Department of Economics and George Perkins Marsh Institute, Clark University, Worcester, MA, USA
e-mail: rjohnston@clarku.edu

J. Rolfe
School of Business and Law, CQUniversity, Rockhampton, QLD, Australia
e-mail: j.rolfe@cqu.edu.au

R.S. Rosenberger
Department of Forest Ecosystems and Society, College of Forestry, Oregon State University, Corvallis, OR, USA
e-mail: R.Rosenberger@oregonstate.edu

R. Brouwer
Department of Environmental Economics, Institute for Environmental Studies, Vrije Universiteit, Amsterdam, The Netherlands
e-mail: roy.brouwer@vu.nl

R.J. Johnston et al. (eds.), *Benefit Transfer of Environmental and Resource Values*, The Economics of Non-Market Goods and Resources 14,
DOI 10.1007/978-94-017-9930-0_2

2.1 Introduction

As described in the first chapter, benefit transfer is defined as the use of research results from pre-existing primary studies at one or more sites or policy contexts (often called study sites) to predict welfare estimates such as willingness to pay (WTP) or related information for other, typically unstudied sites or policy contexts (often called policy sites). It has also been described as the "application of values and other information from a 'study' site where data are collected to a 'policy' site with little or no data" (Rosenberger and Loomis 2000, p. 1097), or the "practice of … adapting value estimates from past research … to assess the value of a similar, but separate, change in a different resource" (Smith et al. 2002, p. 134). That is, research results generated at one site or context are extrapolated or transferred to another site or context; or conversely, information needed at a policy site is inferred from an existing body of research.

Although benefit transfer is often discussed in the context of welfare estimates, other model outputs may be transferred. These can include predicted quantities such as site visits, commodity demand, elasticities, or the size of affected populations. Similarly, while transfers often occur across different geographical locations, this is not a requirement. For example, transfers may occur across different affected populations or types/scales of policy outcomes at the same site (Morrison and Bergland 2006; Chap. 9 in this book).

Benefit transfers are most often used when time, funding, data availability or other constraints preclude original research, so that preexisting estimates must be used instead. Although the use of high-quality primary research to estimate values is preferred in most cases (Allen and Loomis 2008), the realities of the policy process, particularly time and budget constraints, often dictate that benefit transfer is the only feasible option (Griffiths and Wheeler 2005; Iovanna and Griffiths 2006; Johnston and Rosenberger 2010). Given these realities, benefit transfer has become a central component of virtually all large-scale benefit-cost analyses (Smith et al. 2002). Hence, while benefit transfers are subject to a variety of potential errors, the literature increasingly recognizes the need for the resulting information (Bergstrom and De Civita 1999; Desvousges et al. 1998; Johnston and Rosenberger 2010; Rosenberger and Johnston 2009; Rosenberger and Loomis 2003; Smith et al. 2002).

The use of benefit transfer was common as early as the 1980s. Although authors as early as Freeman (1984) began work on methods, it was not until the early 1990s that benefit transfer was broadly recognized as a distinct area of research, with formal attention to methods, procedures and protocols (Johnston and Rosenberger 2010; Rosenberger and Loomis 2003). The 1992 Association of Environmental and Resource Economics (AERE) and U.S. EPA workshop, and subsequent special section of *Water Resources Research*, 28(3), are often credited as the first broad discussions of benefit transfer methodology in the literature. Early criteria for ideal benefit transfers were provided by Boyle and Bergstrom (1992), and the first generation of more comprehensive methodological guides included that of

Desvousges et al. (1998). Later compilations of benefit transfer research included Florax et al. (2002) and Navrud and Ready (2007c). The published benefit transfer literature now includes hundreds of published articles on numerous topics.

Despite the large and increasing scholarly literature in this area and the ubiquity of benefit transfer in policy analysis, the method remains subject to misuse and misunderstanding. Consensus protocols remain elusive, leading to recurring questions regarding reliability and validity. There is a divergence between transfer practices recommended by the scholarly literature and those commonly applied within policy analysis (Johnston and Rosenberger 2010; Chaps. 3, 4, 5 and 6). In addition, some areas of scholarly inquiry routinely apply substandard or unreliable benefit transfer methods.[1] As noted in Chap. 1, these common shortcomings in benefit transfer applications are among the primary motivations for this book.

This chapter provides an introductory overview of benefit transfer methods. It begins with a discussion of the different types of benefit transfer (such as unit value transfer and benefit function transfer), including a review of these different approaches and the relative advantages and disadvantages of each. This is followed by a brief summary of foundations in welfare economics and valuation. Included in this methodological introduction is a discussion of stated and revealed preference valuation and how the results from each may be used for benefit transfer. Following this introductory material are in-depth discussions of the theoretical and informational requirements for benefit transfer, the steps required to implement a benefit transfer, the challenges of benefit scaling, and sources of data for benefit transfers. The chapter concludes with brief discussions of transfer validity and reliability, advanced techniques for benefit transfer, and common problems and challenges. These final discussions introduce material covered in greater depth in other chapters of this book. We emphasize that this chapter is not meant to review the benefit transfer literature; those interested in a more comprehensive literature review are directed to Johnston and Rosenberger (2010) and later chapters in this book.

2.2 Types of Benefit Transfer

Although there have been varying classifications of benefit transfer methods, most recent works distinguish two primary approaches: unit value transfers and benefit function transfers (Johnston and Rosenberger 2010; Rosenberger and Loomis 2003). Unit value transfers involve the transfer of a single number or set of numbers from preexisting primary studies. Unit values can be transferred "as is" or adjusted using a variety of different approaches (e.g., for differences in income or purchasing power, or according to expert opinion). Function transfers, in contrast, derive information using an estimated, typically parametric function derived from original

[1] For example, a significant proportion of the ecosystem services valuation literature is subject to this critique.

research, a meta-analysis that synthesizes results from multiple prior studies, or preference calibration that constructs a structural utility model using results from two or more prior studies. Function transfers typically outperform unit value transfers in terms of accuracy (Kaul et al. 2013; Rosenberger and Stanley 2006), although this is not always the case (Brouwer 2000; Brouwer and Bateman 2005; Ready et al. 2004). For example, unit value transfers can perform satisfactorily if the study and policy contexts are very similar (Bateman et al. 2011). The following sections summarize methods used for each of the primary types of benefit transfer.

2.2.1 Unit Value Transfer

As a foundation for subsequent discussion, we begin with a simple conceptual model for unit value transfer. Although we focus on the transfer of welfare estimates, parallel approaches apply for other estimated outcomes such as elasticities. Welfare measures for environmental resources, such as WTP estimates, are estimated primarily using information derived from individuals expressing their level of welfare based on tradeoffs observed through choices they either make (revealed preferences) or would make under hypothetical situations (stated preferences) (Freeman et al. 2014). Empirical studies generally report an aggregate or central tendency measure (e.g., mean or median) of welfare for a representative individual in the study sample, for a particular change in a good or service (henceforth, "good"). For example, studies in the valuation literature frequently report mean WTP for a given (marginal) change in the good (e.g., per unit WTP) for a particular sample of individuals. We denote this marginal welfare measure $\bar{y}_{js}$ with the subscript j identifying the site at which the study was conducted and s denoting the population sampled by the primary study, or to which the welfare estimate applies.

For illustration, we consider a common context for benefit transfer in which the analyst requires information on a parallel, but unknown, welfare estimate for a similar change and good at a different but similar site $i \neq j$ and population $r \neq s$, which we denote $\hat{y}_{ir}$. While we illustrate transfers across both sites and populations, parallel approaches apply for transfers only across sites (predicting an unknown value $\hat{y}_{is}$) or only across populations (predicting $\hat{y}_{jr}$). We assume that no primary study has been conducted for site $i \neq j$ and population $r \neq s$, so that benefit transfer must be used to generate needed welfare information. From this underlying model, transferred unit quantities can include: (1) a single unadjusted value, (2) a value somehow adjusted according to attributes of the policy context or using expert opinion, (3) a measure of central tendency such as a mean or median value from a set of prior studies, or (4) a range of estimates from a set of prior studies.

The simplest, and often least accurate form of transfer uses a single unadjusted value. In this case, one simply assumes that per person (or household) WTP at the study site is equal to that at the policy site, $\hat{y}_{ir}^{BT} = \bar{y}_{js}$, where WTP is relative to the same marginal quantity at both sites (e.g., per unit), and the superscript BT identifies $\hat{y}_{ir}^{BT}$ as a benefit transfer estimate. Assuming a population of size W that realizes these benefits, an aggregate transferred welfare estimate is given by $W * \hat{y}_{ir}^{BT}$.[2] Note that this represents a transfer of WTP for the *same or similar quantity of the good* at both sites. Any significant "scaling up" or "scaling down" of benefits to account for quantity differences between the study and policy site requires strong assumptions, including that per unit WTP is invariant to the total quantity of the good consumed (i.e., utility is linear with respect to quantity). Issues related to scaling and aggregation are discussed in greater detail below.

The second form of unit values transfer adjusts the transfer estimates according to attributes of the policy context or using expert opinion. For example, one might wish to adjust a unit value to account for differences in currency value, income or other factors. Adjusted unit value transfer is distinguished from benefit function transfer in that the adjustments in question do not rely on functions provided by the original primary studies, but are conducted ex post using an adjustment function determined by the benefit transfer analyst. In this case, $\hat{y}_{ir}^{BT} = f(\bar{y}_{js})$, where the function $f(\bar{y}_{js})$ is the ex post adjustment function. This function may be determined using objective (e.g., differences in currency value) or subjective (e.g., expert opinion) factors. For example, one might use an appropriate price index, P, to account for differences in real currency value between the time period during which the primary study was conducted and the period for which benefit estimates are required. In this case, $\hat{y}_{ir}^{BT} = f(\bar{y}_{js}) = P * \bar{y}_{js}$. Scaling for the relevant population of beneficiaries, W, occurs as described above.

As above, these types of scaling adjustments often involve strong assumptions, the consequences of which analysts should be aware. For example, the simple (e.g., linear) scaling of WTP estimates according to aggregate measures of income or purchasing power parity implies strong assumptions about the structure of preferences. As a result, this type of ex post scaling or adjustment will not always increase transfer accuracy. In some cases, it may be the source of additional transfer error.

A variant of adjusted unit value transfer is the use of administratively approved values (Rosenberger and Loomis 2003). In this case, transferred estimates are not provided through a formal, quantitative adjustment of a prior benefit estimate, but are rather values derived using a subjective and sometimes arbitrary process within a government agency, typically based on some combination of "empirical evidence from the literature, expert judgment, and political screening" (Rosenberger and Loomis 2003, p. 456). Although this is among the least formal, systematic or theoretically defensible approaches to benefit transfer, it may be required when conducting work to meet certain agency needs.

---

[2]Determining the relevant extent of the market, or size of the affected population is not always straightforward. Moreover, the size or location of the affected population can also be correlated with the size of $\bar{y}_{js}$. For example, WTP for a given change in a non-market good often declines with distance from the affected area (Bateman et al. 2006). Hence, projecting unit values to a larger population or spatial area than that in the original primary study can lead to substantial errors.

Unit value transfer methods (3) and (4) are straightforward extensions of the approaches described above. The primary difference is that the analyst uses information from multiple prior studies rather than a single study. To illustrate this case, assume that the analyst has access to $k = 1 \ldots K$ primary studies that each provide comparable estimates of $\bar{y}_{js}$. We denote the resulting WTP estimates $\bar{y}_{js}^k$. From these estimates one can either conduct an adjusted or unadjusted unit value transfer using a measure of central tendency, for example mean $\bar{y}_{js}^k$ over available primary studies. Rosenberger and Loomis (2003) demonstrate a simple example of this type of transfer. One can also conduct sensitivity analyses using the range of values from $\min(\bar{y}_{js}^k)$ to $\max(\bar{y}_{js}^k)$.

The primary advantages of unit value transfers are ease of implementation and minimal data requirements. Moreover, if the study and policy sites (and relevant changes in the good) are very similar, unit value transfers can perform acceptably (Bateman et al. 2011). However, in general, the assumptions implied by unit value transfers lead to larger errors than are observed with otherwise similar benefit function transfers (Kaul et al. 2013; Rosenberger and Stanley 2006). For additional discussion of transfer accuracy as related to unit value versus benefit function transfer, see Chap. 14.

2.2.2 Benefit Function Transfer

Benefit function transfers use a benefit function derived from a primary study or set of prior studies to calculate a welfare estimate calibrated to selected characteristics of a policy site (Loomis 1992; Rosenberger and Loomis 2003). There are two primary requirements for a benefit function transfer. The first requirement is a parameterized function that enables one to calculate the empirical outcome of interest, as a function of variables that include conditions observable at the policy site. Second, information on at least a subset of these variables is required for the policy site, in order to adjust the transferred function from the study site context to the policy site context. In principle, the ability to adjust benefit estimates according to observable differences between the study and policy contexts can lead to more accurate transfer estimates (i.e., lower transfer errors), and can perhaps relax the requirement for close similarity between the study site and a policy site across all relevant dimensions (Loomis 1992; Rosenberger and Loomis 2003; Rosenberger and Stanley 2006). However, there is a fair degree of consensus that site similarity remains an important determinant of transfer accuracy, even for benefit function transfer (Johnston and Rosenberger 2010).

The primary difference between alternative forms of benefit function transfer is the source of the benefit function. The simplest form of benefit function transfer uses an estimated function from a single primary study to calculate a calibrated welfare estimate for the policy site. This is often denoted single-site benefit function transfer. Functions used for benefit function transfer can be drawn from many different types of studies; common sources include recreation demand models, contingent valuation studies and choice experiments (Johnston and Rosenberger 2010; Rolfe and Bennett 2006). Functions are frequently drawn from within a single country, but may also be estimated from data that span multiple countries (Brander et al. 2007; Johnston and Thomassin 2010; Lindhjem and Navrud 2008; Ready and Navrud 2006).

Continuing the notation above, we can illustrate a benefit function in general form as

$$\hat{y}_{js} = g\left(\mathbf{x}_{js}, \hat{\boldsymbol{\beta}}_{js}\right), \tag{2.1}$$

where $\hat{y}_{js}$ is a predicted welfare estimate, $\mathbf{x}_{js}$ is a vector of variables representing the determinants of welfare estimate $\hat{y}_{js}$ at site j for population s, and $\hat{\boldsymbol{\beta}}_{js}$ is a conforming vector of estimated parameters. For example, a very simple linear benefit function would be

$$\hat{y}_{js} = \hat{\beta}_{js0} + \sum_{k=1}^{K} \hat{\beta}_{jsk} x_{jsk} + \hat{\varepsilon}_{js}, \tag{2.2}$$

where K is the number of non-intercept variables in the model and $\hat{\varepsilon}_{js}$ is a residual or error assumed to have a normal distribution with zero mean. If this function is parameterized so that one estimated parameter vector applies to all sites and populations (i.e., no systematically varying slopes or intercepts across sites or population groups), then (2.2) simplifies to

$$\hat{y}_{js} = \hat{\beta}_{0} + \sum_{k=1}^{K} \hat{\beta}_{k} x_{jsk} + \hat{\varepsilon}_{js}, \tag{2.2a}$$

More sophisticated benefit functions may allow for non-linear effects of independent variables on the welfare estimate of interest.

For single-study benefit function transfer, all information in (2.1) would be gathered from a single primary study. Elements in $\mathbf{x}_{js}$ might include observable characteristics of the site, individual (or population), and good, including the quantity and/or quality of the good for which welfare effects are estimated. In general, the analyst will have policy site information for only some elements of $\mathbf{x}_{js}$. To accommodate this, we partition vector $\mathbf{x}_{js}$ into $\mathbf{x}_{js} = [\mathbf{x}_{js}^{1}\ \mathbf{x}_{js}^{2}]$, where $\mathbf{x}_{js}^{1}$ are variables for which the analyst has policy site data, and $\mathbf{x}_{js}^{2}$ are variables for which no policy site data are available. Parallel values for $\mathbf{x}_{js}^{1}$ at the policy site are given by $\mathbf{x}_{ir}^{1}$. As before, we assume that the analyst uses benefit transfer to predict a parallel but unknown welfare estimate for a similar change and good at a different but similar site $i \neq j$ and population $r \neq s$, which we denote $\hat{y}_{ir}$, with the associated benefit transfer estimate given by $\hat{y}_{ir}^{BT}$.

Given the simple model above, a benefit function transfer estimate of $\hat{y}_{ir}^{BT}$ may be calculated as

$$\hat{y}_{ir}^{BT} = g\left(\left[\mathbf{x}_{ir}^{1}\mathbf{x}_{js}^{2}\right], \hat{\boldsymbol{\beta}}_{js}\right). \tag{2.3}$$

That is, the analyst uses the parameterized function $g(\cdot)$ to calculate a benefit transfer estimate of value, substituting updated values of those variables for which policy site information is available ($\mathbf{x}_{ir}^{1}$). For variables with no updated policy site information, $\mathbf{x}_{js}^{2}$, original values from the study site are typically used. The result is a benefit transfer estimate, $\hat{y}_{ir}^{BT}$, that is adjusted for observable differences between the two valuation contexts (the study and policy site). The appendix to this chapter provides a simple textbook illustration of the difference between an applied unit value and benefit function transfer, for a hypothetical transfer of recreational value. Rosenberger and Loomis (2003) illustrate a benefit function transfer for a simple, real-world example.

In principal, benefit function transfers can be used to adjust or calibrate benefit transfer estimates for differences in such factors as the quantity or quality of the good being valued, the characteristics of individuals or populations (e.g., income, education), or other site characteristics such as the price, quality or availability of substitutes. However, Bateman et al. (2011, p. 384) argue that these functions should be "constructed from general economic theoretic principles to contain only those variables about which we have clear, prior expectations." An additional limitation is that function-based adjustments, for example adjusting for differences in socioeconomic characteristics of affected populations, will not always improve transfer accuracy (Brouwer 2000; Johnston and Duke 2010; Spash and Vatn 2006).

Single-site function transfers also require the strong assumption that the underlying parameterized valuation function, $g(\cdot)$, is identical at the study and policy sites. To account for potential differences in benefit functions across sites, one may also conduct *multiple-site benefit function transfer*, in which functions from different studies and/or sites are each used independently to derive distinct benefit function transfer estimates, with the results combined to provide a range of feasible values for the policy site. This approach differs from meta-analysis (described below), in which data from different studies and/or sites are combined statistically to generate a single "umbrella" benefit function that is used subsequently to generate a transfer estimate of value (Rosenberger and Phipps 2007). In contrast, multiple-site benefit function transfer generally involves the use of multiple, independent single-site transfers, the results of which are then somehow condensed into a single estimate (e.g., a mean value) or range of estimates.

2.2.2.1 Meta-analysis

Meta-analysis may be defined as the quantitative synthesis of evidence on a particular empirical outcome, with evidence gathered from prior primary studies. Meta-analysis in environmental economics is most often accomplished using statistical analysis, called meta-regression models (MRMs). Within these models, the dependent variable in a classical or Bayesian MRM is a comparable empirical outcome drawn from existing primary studies, with independent moderator variables representing observable factors that are hypothesized to explain variation in the outcome across observations. Observations used within meta-analysis (or the metadata) may be drawn from both the published and unpublished literature. Nelson and Kennedy (2009) provide a summary of meta-analysis in environmental and resource economics. Broad discussions of its use for benefit transfer are provided by numerous sources including Bergstrom and Taylor (2006), Florax et al. (2002), Johnston and Besedin (2009), Johnston and Rosenberger (2010), Rosenberger and Johnston (2009), Rosenberger and Loomis (2003) and Smith and Pattanayak (2002), among others.

When used for benefit transfer, MRMs are most often applied to identify and test systematic influences of study, economic, and resource attributes on WTP, characterize results of the literature addressing certain types of nonmarket values, and generate reduced-form benefit functions for direct transfer applications. All of these are grounded in the ability of meta-analysis to characterize a parameterized value surface reflecting multi-dimensional patterns in estimated WTP variation across multiple empirical studies (Johnston and Rosenberger 2010; Rosenberger and Phipps 2007). As described by Rosenberger and Johnston (2009, p. 411), if "empirical studies contribute to a body of WTP estimates (i.e., metadata), and if empirical value estimates are systematically related to variations in resource, study and site characteristics, then meta-regression analysis may provide a viable tool for estimating a more universal transfer function with distinct advantages over unit value or other function-based transfer methods." Many authors have noted the potential of MRMs to provide more robust, accurate benefit transfers compared to alternative methods (Johnston and Rosenberger 2010). Empirical evidence suggests that this may be true in many instances (Kaul et al. 2013; Rosenberger and Stanley 2006). Other works, however, have advised caution in the use of MRMs for benefit transfer (e.g., Bergstrom and Taylor 2006; Poe et al. 2001; Smith and Pattanayak 2002).

An empirical meta-analysis benefit function may be illustrated using similar notation to that introduced above. We now assume a case in which the analyst has access to a large number ($n = 1 \ldots N$) of studies, allowing her to estimate aggregate or central tendency measures (e.g., mean or median) of welfare for a particular good, from prior analyses conducted at different sites j and over different populations s.[3] Following the notation introduced above, we denote these welfare or WTP estimates as $\hat{y}_{js}$. These N welfare estimates then serve as the measured effect size or dependent variable in a statistical MRM represented in general form by,

$$\hat{y}_{js} = h(\mathbf{x}_{js}, \hat{\boldsymbol{\mu}}_{js}), \tag{2.4a}$$

[3]It is also possible for a single primary study to report multiple estimates for a single site and population, for example when multiple model specifications are estimated.

where $j = 1 \ldots J$ and $s = 1 \ldots S$. Here, $\mathbf{x}_{js}$ is a $1 \times K$ vector of variables representing resource, study and site characteristics (or moderator variables) hypothesized to explain the variation in welfare estimates $\hat{y}_{js}$ across sites j and populations s. $\hat{\boldsymbol{\mu}}_{js}$ is a conforming vector of parameters reflecting the estimated effect of each moderator variable on $\hat{y}_{js}$. In the simplest case a single parameter estimate will apply to each moderator variable in the dataset (across all j and s), so that the vector $\hat{\boldsymbol{\mu}}_{js} = \hat{\boldsymbol{\mu}}$.

The general form of (2.4a) allows for estimation using a variety of common linear-in-the-parameters functional forms, including linear, semi-log, log-linear and translog functional forms; all are common in the meta-analysis literature. For example, replacing the subscripts j and k with a single subscript n that identifies individual observations in the metadata, a simple linear econometric form for Eq. (2.4a) would be

$$\hat{y}_n = \mu_0 + \sum_{k=1}^{K} \hat{\mu}_k x_{nk} + \hat{\varepsilon}_n, \tag{2.4b}$$

where $\hat{\varepsilon}_n$ is the equation error or residual.

When estimating models such as (2.4b), analysts must account for a variety of potential statistical complications including sample selection effects, primary data heterogeneity, heteroskedasticity and nonindependence of multiple observations from individual studies (Nelson and Kennedy 2009; Rosenberger and Johnston 2009). Development of metadata also involves empirical challenges, including the reconciliation of variables across different primary studies (Bergstrom and Taylor 2006; Johnston and Rosenberger 2010; Nelson and Kennedy 2009). As noted by Nelson and Kennedy (2009) and others, many MRMs in the literature violate good practice guidelines for econometric estimation, although some of these guidelines are subject to debate (Johnston and Rosenberger 2010).[4] For conciseness these issues are not discussed further here, but are discussed and illustrated in later chapters.

Mirroring the methods for benefit function transfer presented above, prediction of an aggregate welfare measure for the policy site, $\hat{y}_{ir}^{MRM}$, using meta-regression model (2.4a) replaces moderator effects, $\mathbf{x}_{js}$, with analogous measures at the policy site $\mathbf{x}_{ir}$, such that

$$\hat{y}_{ir}^{MRM} = h(\mathbf{x}_{ir}, \hat{\boldsymbol{\mu}}_{js}). \tag{2.5}$$

The result is a predicted welfare estimate, $\hat{y}_{ir}^{MRM}$, calibrated to specific conditions at the policy site. When a variable in $\mathbf{x}_{ir}$ is unobservable at the policy site, it is often replaced by an associated mean value of that variable from the metadata. A step-by-step illustration of this process for a case study addressing water quality improvements is

[4] An example is the appropriateness of pooling otherwise commensurable Marshallian and Hicksian welfare measures within a single MRM (Johnston and Moeltner 2014; Londoño and Johnston 2012).

shown by Johnston and Besedin (2009). Other simple examples are provided by Rosenberger and Loomis (2003) and Chap. 12.

Variables that identify methodological aspects of primary studies included in the metadata (i.e., the methods used by each study to estimate values) are generally not observable for the policy site, because no research has been conducted there. Methodological factors shown to influence WTP in past MRMs include study type, survey implementation method, response rate, question format, treatment of outliers/protests, econometric methods, and other factors (Johnston et al. 2006a; Rolfe and Brouwer 2013; Stapler and Johnston 2009). In such cases, analysts either select values for these methodological variables based on levels they consider to be appropriate,[5] or use mean values for these variables from the metadata (Moeltner et al. 2007; Stapler and Johnston 2009).

The validity of any meta-analysis and the resulting benefit transfers depends on the quality, extent and unbiasedness of the underlying primary data (Nelson and Kennedy 2009; Rosenberger and Johnston 2009). Hence, it is critical that analysts use appropriate approaches to collect, evaluate and screen information gathered from the literature, and that methods used for this process are transparent. Section 2.5 discusses broader issues related to data sources and selectivity. Stanley et al. (2013) provide a concise review of recommended steps in meta-analysis data collection and reporting. We note that while these steps often complicate the selection of source studies for a meta-analysis, they have key advantages in that input studies of lower quality or relevance can be identified prior to use.

2.2.2.2 Preference Calibration or Structural Benefit Transfer

Benefit transfers, in general, lack a micro-level utility-theoretic foundation (Smith et al. 2002). Although all benefit transfers should draw on prior primary studies with a strong grounding in welfare theory, transfers themselves are almost always a purely empirical exercise; no additional constraints are placed on the transfer to ensure compliance with theory. As such, most benefit transfers and meta-analyses are considered to have a "weak" structural basis in utility theory (Bergstrom and Taylor 2006). Such critiques apply to traditional unit and benefit function transfers, as well as to MRM functions. In contrast to these other approaches, structural benefit transfer (or preference calibration) is distinguished by a strong and formal basis in an explicit, structural utility function. This assumed utility structure is used to combine and transfer information drawn from multiple prior studies or information sources (e.g., Pattanayak et al. 2007; Smith and Pattanayak 2002; Smith et al. 2002, 2006).

Structural benefit transfer requires the analyst to specify a specific, structural preference or utility function able to describe an individual's choices over a set of

[5] Johnston et al. (2006a) illustrate the potential risks of this approach related to the sensitivity of resulting transfer estimates.

market and nonmarket goods, presuming standard budget-constrained utility maximization. One then derives analytical expressions that determine a relationship between each available benefit measure from existing primary studies and the assumed utility function, inasmuch as possible guided by economic theory. Expressions also should "assure the variables assumed to enter the preference function are consistently measured across each study and linked to preference parameters" (Smith et al. 2002, p. 136). Finally, empirical methods are used to calibrate parameters to the specified utility-theoretic structure. That is, parameters of a benefit function (or system of functions) are solved so that the resulting preferences (and subsequent benefit transfers) are consistent with the empirical results of the available prior studies, given the assumed utility structure. In some cases preference or utility parameters may be solved algebraically based on the specified utility structure; in other cases some form of iterative optimization is required.

Unlike some other forms of benefit transfer, structural benefit transfer generally cannot be accomplished without significant expertise in welfare theory and mathematical economics. Because of the great variability in ways that structural benefit transfer may be accomplished and the complexity of the approach, it is not possible to provide a concise, general illustration of the method in this chapter. Readers are directed to Pattanayak et al. (2007), Smith and Pattanayak (2002), Smith et al. (2002, 2006) and Chap. 23 in this book for additional information and applications.

2.2.3 Choosing Among Different Types of Benefit Transfer

The choice among different types of benefit transfer is dictated by a number of different factors, including the type of information and number of studies that are available, the type of value that is required, the general similarity (or correspondence) between the study and policy contexts, the level of analyst expertise, the time and resources available to develop transfer methods, and the precision necessary for different types of policy decisions (Bergstrom and De Civita 1999; Navrud and Pruckner 1997). In general, benefit function transfers are preferred unless the study and policy contexts are very similar (Bateman et al. 2011; Kaul et al. 2013; Rosenberger and Stanley 2006), although the multiple dimensions over which sites may be similar or dissimilar can complicate such assessments (Colombo and Hanley 2008; Johnston 2007). Unadjusted unit value transfers, however, are generally treated with skepticism and considered one of the least appropriate forms of transfer (Johnston and Rosenberger 2010). An exception is the literature on the value of a statistical life (VSL), which emphasizes unit value transfers, noting that the appropriateness of function-based adjustments of VSLs is unclear (Brouwer and Bateman 2005; Mrozek and Taylor 2002; Viscusi and Aldy 2003).

The choice of single-site benefit function transfer versus meta-analysis depends on factors that include the availability of sufficient studies for MRM estimation and the availability of a single, closely matching study-site function (Stapler and Johnston 2009). The probability of finding a good fit between a single study site and a policy site is usually low (Boyle and Bergstrom 1992; Spash and Vatn 2006). In such cases the ability of MRMs to estimate a multidimensional value surface that combines information from many prior studies can lead to improved transfer accuracy (Rosenberger and Phipps 2007). However, the development of metadata and estimation of suitable MRMs requires greater time and expertise than is typically required for other forms of benefit transfer (structural benefit transfer is an exception). In general, MRMs are most appropriate when: (1) there is a large valuation literature addressing the nonmarket good in question, (2) there is no empirical study for a single, closely matching policy context, and/or (3) the analyst desires flexibility to estimate benefits for different policy contexts or outcomes (e.g., scales of improvement in the nonmarket good).

Structural benefit transfer methods have not yet been widely adopted for applied work. Advantages of structural transfer include the imposition of strong theoretical consistency on the use of prior information (Bergstrom and Taylor 2006) and greater transparency in assumptions. The method also has limitations, including potential sensitivity of model results to the assumed utility structure. The preference calibration method is also more complex than most alternative transfer methods, and the literature has yet to demonstrate clearly that this increased complexity leads to improvements in transfer accuracy (Johnston and Rosenberger 2010). Hence, the choice of strong structural versus weak structural transfer methods (Bergstrom and Taylor 2006) often depends on the analyst's level of expertise and predisposition regarding the relative importance of a strong structural utility foundation for benefit transfer.

Finally, while benefit transfer is often the only feasible option for estimating values required for policy analysis, analysts may sometimes have a choice between primary research and benefit transfer. This choice can be particularly relevant for smaller projects or policies, for which the cost of a high quality primary valuation study can be large compared to potential policy benefits. Allen and Loomis (2008, p. 9) model the choice of primary studies versus benefit transfer for a case study of recreational benefit estimation, and conclude that "only in the case of very small projects … would original research not yield positive returns in terms of better decisions." Hence, where primary research is feasible (at least of a certain minimum quality), it is almost always preferred.

2.3 Underlying Principles of Economic Valuation

Regardless of the transfer method used, a benefit transfer can only be as good as the underlying primary studies. Theoretically valid benefit transfers require a basis in theoretically valid primary valuations. There is a large and mature literature on valuation theory and methods (e.g., Bockstael and McConnell 2010; Champ et al. 2003; Freeman et al. 2014; Haab and McConnell 2002; Hanley and Barbier 2009;

Just et al. 2004). Here, we focus on the basic theoretical foundation for benefit or value estimation, although similar theoretical guidelines apply to the estimation of most empirical quantities used within benefit transfer.

Economic benefits and costs may be realized by individuals or firms (e.g., businesses). They are always quantified in comparative terms, relative to a well-defined baseline, and reflect the welfare (or well-being) of individuals or groups. For individuals, benefits are generally measured as the maximum amount of other goods that the individual is willing to forego in order to obtain another good that is desired. This reflects the individual's WTP. Although WTP is often denominated in money units, it can be expressed in any unit of exchange. Value may also be quantified in terms of willingness to accept (WTA), defined as the minimum amount that a person or group would be willing to accept in order to give up a specified quantity of a good that is already possessed. Economic values or benefits, therefore, are a simple reflection of tradeoffs: what individuals or groups are willing to give up in order to obtain something else, either in or out of organized markets. The resulting values are denoted *market* and *nonmarket values*, respectively. Economists' ability to monetize market or nonmarket benefits in this way relies on the concept of substitutability—that the welfare gained through increases in one commodity can be offset by decreases in other commodities.

Economic values are meaningful only for a particular quantity of a market or nonmarket commodity, relative to a specific baseline. If these changes are large (i.e., non-marginal), value estimation must account for the fact that per unit values for any commodity generally diminish as one obtains more of that commodity (this is called diminishing marginal utility). For example, a recreational angler is generally willing to pay more per fish to increase her catch from 0 to 1 fish than from 9 to 10 fish; the value of a marginal fish depends on how many fish have already been caught (Johnston et al. 2006b).

Different valid measures can be used to quantify economic values. Among the most common is *consumer surplus*, which may be interpreted as the difference between what an individual or group would be willing to pay for a commodity (measured off the estimated demand curve) and what is actually paid, summed over all units. A parallel measure for firms is *producer surplus*, which is similar to economic profits.[6] Theoretically exact welfare measures of surplus for individuals are called *Hicksian welfare measures*; these include compensating and equivalent surplus (Bockstael and McConnell 2010; Freeman et al. 2014; Just et al. 2004). For example, an individual's WTP for a fixed change in a public good such as air quality is a compensating surplus measure. Hicksian welfare measures, however, are difficult or sometimes impossible to measure using data on observed behavior. For this reason, economists will frequently use alternative estimates that can provide close approximations to exact Hicksian welfare measures. Among the most common of

[6]The difference between producer surplus and economic profits lies in the treatment of fixed costs of production.

these is Marshallian (consumer) surplus, which measures consumer surplus as the area below the estimated demand function but above the market price for a good (or zero if an amenity is not provided through markets or is otherwise unpriced).

Economists have developed different methods for quantifying market and non-market values such as these (Champ et al. 2003; Freeman et al. 2014; Hanley and Barbier 2009; Holland et al. 2010). Although the methods for measuring these values differ, these valuation techniques are all based on an internally consistent model of human welfare that allows benefit and cost measures to be aggregated and compared. The model assumes that, after considering the pros and cons of all options, people will make choices that they expect to provide the greatest long-term satisfaction or utility. The theoretical basis of this model allows one to link estimated monetary values (e.g., benefits, costs, and WTP) with the well-being of individuals, households, or groups.

The choice of valuation method(s) is determined by the type of values that are likely to be present. *Revealed preference methods* are based on the analyses of observed human behavior. Examples include recreation demand models and hedonic property value models (Bockstael and McConnell 2010; Champ et al. 2003; Freeman et al. 2014). Such methods can only measure *use values*—or values related to the consumptive or nonconsumptive use of a commodity. *Stated preference methods* are based on the analysis of responses to carefully designed survey questions; examples include *contingent valuation* and *choice experiment* (*or choice modeling*) *methods* (Bateman et al. 2002; Rolfe and Bennett 2006). Stated preference methods, while sometimes more controversial because of their reliance on survey responses, are able to measure both use and *nonuse values*. Nonuse values (also called passive use values) may be defined as values that do not require use of a commodity or related behavior. An example would be the value that individuals often hold for the continued existence of threatened wildlife species, apart from any direct or indirect use of that species.

All valuation methods have advantages and disadvantages (see discussions in, for example, Champ et al. 2003; Freeman et al. 2014). Both revealed and stated preference methods have been used extensively over the past three decades, have been extensively tested and validated by researchers, and are grounded in extensive published literatures. Both are widely accepted by government agencies as reliable for estimating economic values.

Just as there are a large number of valid techniques for estimating economic values, there are also a large number of techniques that—while producing results in monetary units—do not quantify economic benefits in ways that are consistent with well-defined surplus (or welfare) measures for consumers or producers (Holland et al. 2010). These methods generally have little or no grounding in economic theory or structural relationship to human welfare. Benefit transfers of such results will lead to similarly invalid and misleading estimates. Common examples of these techniques include *replacement cost methods*, which quantify the "value" of a nonmarket good or service based on the cost of "replacing" that good or service using technological or other means; *damage cost methods*, which seek to quantify the protective value of natural resources (such as wetlands that protect homes from flooding) based on the monetary damages they prevent; and *embodied energy methods*, which seek to estimate values based on the total energy required to produce goods and services (Holland et al. 2010). Except in rare circumstances, neither replacement nor damage cost approaches are suitable for quantifying economic value. Embodied energy approaches never generate valid economic value estimates.

When one conducts a benefit transfer, any errors in the original value estimates are also transferred. Hence, the transfer of an invalid measure of economic value will lead to an invalid transfer estimate. Regardless of the type of benefit transfer applied, it is crucial that the original primary study estimates represent valid measures of economic value.

2.4 Scaling Benefit Estimates

One of the most misunderstood and misused aspects of benefit transfer involves the scaling of benefits over populations, affected areas or quantities of change (Rolfe et al. 2011). A good general discussion of the role of scope and scale in valuation is provided by Rolfe and Wang (2011). Scaling, or multiplication of per unit values by a different quantity, population or area than was evaluated by the original source study (or studies), requires strong and often unrealistic assumptions. These include the invariance of per unit values to scale, an assumption that holds only in rare circumstances or for small changes in scale. For example, due to geographical proximity effects such as distance decay (that values tend to decline as one moves further from an affected area; Bateman et al. 2006) and diminishing marginal utility, per unit values tend to be higher in small local case studies than regional or national ones (Rolfe et al. 2011; Rolfe and Windle 2008). Hence, unit values should not be scaled to significantly larger or smaller geographic areas (or scales) without adjustments (Johnston and Duke 2009).

Common violations of accepted practice in benefit transfer involve the scaling of benefit measures in attempts to quantify the total benefits of an environmental asset at a planetary, nation/statewide, or ecosystem scale (Bockstael et al. 2000; Toman 1998). These attempts ignore diminishing marginal utility and the fact that economic values are meaningful only for clearly specified changes in a good or service, rather than an entire environmental asset (Bockstael et al. 2000). That is, they ignore the errors that can occur when benefit transfers scale benefit estimates. Perhaps the most commonly cited of these analyses is Costanza et al. (1997), which attempts to use benefit transfer to quantify the value of planetary ecosystem services (Bockstael et al. 2000). Many subsequent analyses have used similarly flawed benefit transfers in an attempt to value large ecological assets.

This section uses very simple graphical tools to demonstrate the importance of scale in benefit transfer and the errors that result when scale differences are ignored.

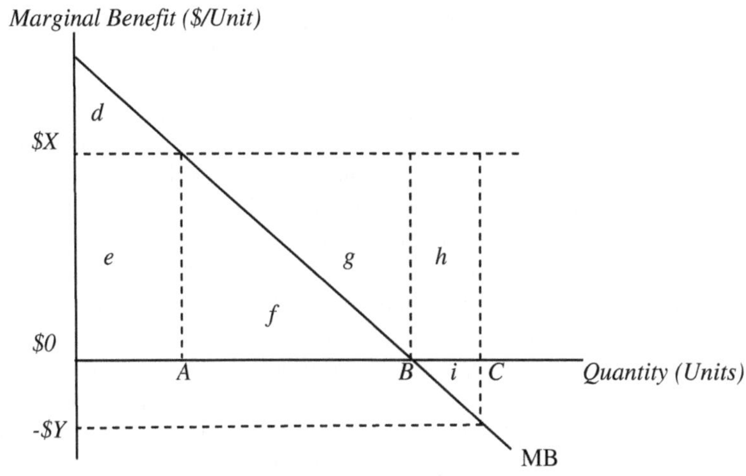

Fig. 2.1 Marginal benefits and scale over quantities

For illustration, we focus on three types of scaling that are often abused in benefit transfers—scaling over quantities, populations and geographic areas. Although the illustrations are stylized and simple, they are grounded in established theoretical expectations with strong support in empirical research.

We begin with Fig. 2.1, which illustrates a standard downward sloping marginal benefit curve (MB) for a market or nonmarket good. For a market good, this marginal benefit curve would be equivalent to a market demand curve. For any given quantity of the good, the MB curve shows a representative individual's marginal benefit (or WTP) for the last, or marginal unit consumed. For example, if an individual consumes *A* units of this good, the marginal benefit of the last unit is *$X*. The total benefit of consumption for all units consumed (not only the marginal unit) is the area underneath the MB curve from zero to the total quantity consumed. So, for example, the total benefit of consuming *A* units (assuming they were obtained at zero cost) would be area $d + e$. In contrast, the total benefit of consuming *A* units, assuming that the individual pays a price of *$X* per unit, would be area *d*. Assuming that Fig. 2.1 is a market (or Marshallian) demand curve, these are interpreted as measures of consumer surplus.

Drawing from this standard model, assume that a primary valuation study estimates a value of *$X* per unit, based on a consumption quantity of *A* units. Assuming perfectly matching study and policy sites, this unit value would reflect an accurate benefit transfer (zero error) of marginal value, as long as the quantity is the same at both sites (that is, no scaling). Now assume, however, that a benefit transfer attempts to scale this unit value to a larger quantity of the good, for example *B* units. At *B* units, the true marginal value of the good is zero, but the scaled benefit transfer would continue to predict *$X* per unit. The true total benefit of consuming *B* units (assuming zero cost) is $d + e + f$; the area under the MB curve up to *B* units.

The scaled unit value transfer, however, would predict a value of *$X* multiplied by *B*, or a total area of $e + f + g$. Hence, both the transferred marginal and total values are biased.

The bias becomes more severe for larger scaling, for example to *C* units. At this quantity, marginal net benefit drops below zero to *negative $Y*.[7] However, the scaled unit value transfer continues to predict a constant value of *$X*. In terms of consumer surplus for all units consumed, the true value at *C* units of consumption (assuming the good is obtained at zero cost) is area $d + e + f - i$. The scaled, transferred value, however, is $e + f + g + h$. As the scaling of unit values increases in magnitude, the error (the difference between the true and transferred value) also increases. From this diagram, it is easy to see the fallacy in the claim that scaled unit values provide a "reasonable" approximation of actual values. In fact, scaled unit values may not even have the same algebraic sign as true values. Scaling by a small amount, however, can sometimes generate reasonable approximations of value, depending on the slope of the marginal benefit curve.

Similar errors to those shown in Fig. 2.1 occur when one seeks to scale up values calculated per unit area (e.g., ecosystem service values per acre) to much larger areas than the original primary study. In this case, the total quantity of the good is correlated with the landscape area. For example, as shown by Johnston and Duke (2009), the marginal per acre WTP for farmland preservation is much greater when one considers smaller total acreages of preservation.[8] Among the few exceptions to this rule—at least in some cases—are market goods such as agricultural products sold on world markets. In such cases, production on additional acres in any local area (assuming equal productivity) is unlikely to influence world price to a large degree, and hence per acre agricultural production values can be scaled to a certain extent. One can make similar arguments for other goods or services that are valued primarily based on their global consequences. An example is the per ton value of carbon sequestration, which is likely approximately constant (albeit difficult to estimate) for a wide range of potential quantities.

Related difficulties with geographical scaling are shown by Fig. 2.2, which illustrates a similar per unit, marginal benefit curve. Here, however, we envision a case in which there is a fixed quantity increase in a nonmarket good at a particular location. The graph shows how the benefit of this change diminishes as the distance to the affected location increases. This is similar to the function derived empirically by Hanley et al. (2003). For example, it is well established that individuals are often willing to pay less for environmental improvements that are at a greater distance from their homes (Bateman et al. 2006; Jørgensen et al. 2013; Schaafsma et al. 2012).

---

[7]A good example of this pattern would be water levels in a river, which often have positive marginal values up to a point where flooding occurs, at which point marginal values for additional water become negative.

[8]In addition, non-linearities and thresholds in ecological systems can lead to nonconvexities when one considers ecosystem conservation at different geographical scales. This further complicates any scaling up or down of certain types of environmental values.

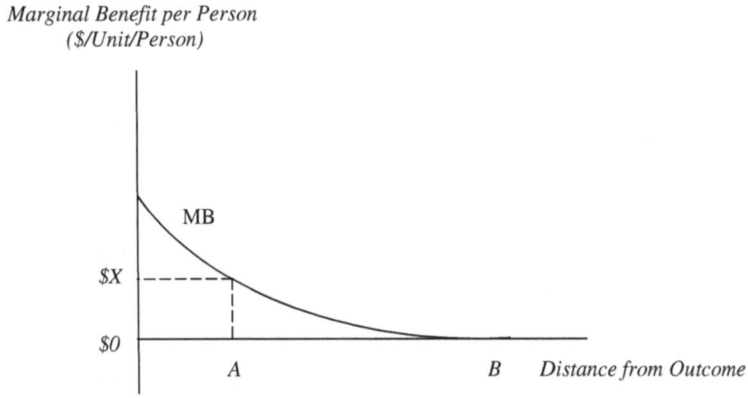

Fig. 2.2 Marginal benefits and scale over distance (or populations over greater areas)

Given this MB curve, we assume that a primary study has estimated a marginal unit value of *$X* per person, for a study over a population at distance *A* from the affected location. Based on such studies, benefit transfers will sometimes scale unit values over larger populations or areas, extending greater distances from the affected location—for example distance *B*. The graph shows the errors that result from such scaling. Continuing to scale the fixed unit value *$X* to a population at distance *B* overlooks the fact that marginal value drops to nearly zero at this distance. If one were to aggregate these transferred unit values over greater and greater distances, the error in aggregated values would increase with the distance of the aggregation (or the size of the total area).

These simple diagrams show the risks involved in scaling unit value estimates over different quantities, areas or populations than those considered by the original primary study. It is rarely the case that values—whether per person, per unit, or per unit area—are invariant to different types of scale. Simple linear scaling up of values—except in rare circumstances—will typically lead to significant errors. Among the advantages of benefit function transfer in such cases is the potential ability to model or predict the entire MB function, thereby providing a function-based mechanism to adapt value estimates to the resulting scale differences. Meta-analysis can also provide a possible means to adjust across scales, if the underlying studies in the metadata reflect studies conducted at different scales.[9] However, even in such cases, it is important that scaling not occur beyond the range of the data in the original primary studies; doing so risks the same types of scaling errors that occur with unit value transfers.

[9]For an example, see Johnston et al. (2005).

2.5 Site, Context and Commodity Similarity

Among the primary requirements for accurate benefit transfer is correspondence, or similarity between the site, valuation context and populations at the study site and those at the policy site (Loomis and Rosenberger 2006). This includes similarity in factors such as the availability of substitutes and complements to the good in question. The degree and dimensions of similarity that are required, however, can vary across different types of transfers (Colombo and Hanley 2008; Johnston 2007). Challenges of site similarity are even greater for international benefit transfers, given potential differences in such factors as currency conversion, user attributes, wealth/income measures, cultural differences and extent of the market (Johnston and Thomassin 2010; Ready and Navrud 2006). Commodity consistency is another critical prerequisite for valid transfer (Boyle and Bergstrom 1992; Johnston et al. 2005; Loomis and Rosenberger 2006; Smith et al. 2002). That is, accurate transfers require an understanding of the welfare-influencing quantities or qualities of goods at affected sites, both in primary studies from which values are estimated and in policy sites for which estimates are needed.

Even studies of seemingly similar nonmarket goods may estimate values for differing underlying quantities or qualities. For example, improvements in water quality within reservoirs used as a source of drinking water are, for welfare estimation purposes, a different type of commodity than improvements in water quality within streams used solely for recreation. Even though the chemical change in the water itself might be similar, the mechanism through which these changes influence utility differs. As a result, there is no theoretical or empirical expectation that WTP for these two changes should be related. Given such possibilities, benefit transfer requires analysts to consider similarity not only in the biophysical dimensions of affected goods, but also the welfare dimensions.

It is sometimes possible to reconcile (or match ex post) commodity definitions across studies. For example, this is almost always required to develop metadata for valuation MRMs (Johnston et al. 2005; Smith and Pattanayak 2002). However, reconciliation that promotes sufficient uniformity is not always feasible (Smith et al. 2002; Van Houtven et al. 2007), and analysts are often delinquent in such areas (Nelson and Kennedy 2009). Appropriately specified MRMs may also be able to account for some systematic patterns that differentiate welfare measures for similar, but not identical, commodities by including appropriate variables on the right-hand side of regression models. For example, the MRM of Johnston et al. (2006b) includes variables that allow marginal WTP per fish among recreational anglers to vary depending on the type of fish species, the fishing mode, and the catch rate, among other factors. The extent to which such adjustments can be accomplished in a defensible manner, however, is limited (Bergstrom and Taylor 2006). The task is made more difficult "as [the] complexity of changes in environmental quality and natural resources increase[s]" (Navrud and Ready 2007a, p. 3).

Given the many dimensions over which valuation commodities and contexts can be similar and dissimilar, benefit transfers typically require the analyst to make (hopefully informed) judgments regarding whether commodities at the study and policy site are "close enough" to support valid and accurate benefit transfer. These judgments should also be influenced by the degree of accuracy required of different types of transfer (Bergstrom and De Civita 1999; Navrud and Pruckner 1997). Theory can provide only limited assistance for these choices. Conservative decisions in such cases can be important to reducing generalization errors.

2.6 Data Sources and Selectivity

The accuracy of benefit transfer depends on the type and quality of primary studies used to generate transfer estimates. When one selects primary studies for benefit transfer, there are implicit assumptions that the underlying body of valuation literature provides an unbiased sample of the population of empirical estimates and that these estimates provide an unbiased representation of true resource values. If these assumptions do not hold, the result will be systematic transfer biases. These are often called selection biases. Such concerns are most often noted for the case of meta-analysis but apply to all types of transfer (Rosenberger and Johnston 2009). Benefit transfers can only be as good or as unbiased as the sample of data from which they are derived, or to the extent that any biases can be corrected during the transfer process. Among the first steps in any high quality benefit transfer is a comprehensive review of the literature to find suitable high quality studies.

The methods used to select and screen studies for benefit transfer are important. Rosenberger and Johnston (2009) identify four potential sources of selection bias that can influence benefit transfers, including research priority selection, methodology selection, publication selection, and [metadata] sample selection. For each, there are a variety of steps that can be taken to minimize the potential for these biases to carry over into empirical benefit transfers. For example, there are a variety of methods that can be used to identify, measure and correct for publication selection bias, including approaches that weight empirical estimates by their standard errors to give greater importance to estimates that have been estimated with greater accuracy (Florax et al. 2002; Stanley 2005, 2008). These and other approaches to address potential selection biases are discussed in Chap. 14.

At the same time, avoidance of measurement error in benefit transfers requires that primary studies are of a certain minimum quality. As noted above, for example, studies that do not generate theoretically valid estimates of value cannot serve as the basis for a valid benefit transfer. Hence, studies should be screened to ensure the fundamental validity of the estimated welfare measures according to economic theory. The quality of empirical research methods is also important, but is more difficult to quantify. Although publication in peer-reviewed journals can be a potential signal of study quality, presence in a peer-reviewed journal does not guarantee suitability for benefit transfer. For example, some studies may be published on the strength or novelty of their methodological or theoretical contributions, despite a weak empirical application. Such studies may not be suitable for benefit transfer. Other studies may have inadequate reporting of data or methods (Loomis and Rosenberger 2006). In still other cases, there may be biases inherent in the types of paper accepted for publication, leading to systematic biases in the published literature (Stanley 2005, 2008; Chaps. 14 and 15). Signals of quality also vary across types of valuation. For example, the use of extensive focus groups and survey pretesting is an important signal of quality in stated preference research, but is largely irrelevant for many revealed preference methods. Overall, the benefit transfer literature has reached consensus that primary study quality is necessary to avoid measurement errors in benefit transfer, but has not reached consensus on clear protocols to evaluate quality. Those with expertise in economic valuation can help identify studies suitable for benefit transfer, but even among experts there may not be total agreement.

Increasingly, valuation databases such as the Environmental Valuation Reference Inventory (EVRI, http://www.evri.ca) can help practitioners identify research studies suitable for transfer (Johnston and Thomassin 2009; McComb et al. 2006; Morrison 2001). EVRI is a specialized internet database that contains summaries of empirical studies of the economic value of environmental costs and benefits and human health effects. The database is now the largest international nonmarket valuation database in existence, including data from thousands of nonmarket valuation studies. Although EVRI has many potential uses, the primary goal of the database is to assist policy analysts with benefit transfer. Similar but smaller databases include ENVALUE (http://www.environment.nsw.gov.au/envalue/), developed by the New South Wales Environment Protection Authority. Academics tend to view these and other databases as a useful starting point for research or policy analysis and as an important source of information regarding the valuation literature, but treat with skepticism transfers that rely solely on database information (Johnston and Thomassin 2009; Morrison 2001). Accurate benefit transfers require careful and systematic attention to relationships between primary study attributes and those of the intended policy target. As noted by Morrison (2001, p. 54), "analysts should not expect to be able to simply download value estimates for a cost-benefit analysis from these [valuation] databases, unless the cost-benefit analysis is particularly rudimentary and of little policy significance." Hence, valuation databases cannot substitute for practitioner expertise and detailed analysis of original primary studies.

Yet even with the increasing size of the valuation literature and existence of expanding valuation databases, many of the primary challenges to benefit transfer relate to a lack of accessible, unbiased information (Johnston and Rosenberger 2010). Despite a large research literature in environmental valuation, there is a lack of studies providing high-quality, policy-relevant, replicable, empirical estimates of nonmarket values for many environmental commodities. Instead, the published literature is dominated by studies illustrating methodological advances, often at the expense of useful empirical estimates. Similarly, the literature often fails to provide sufficient information on study attributes and data to promote defensible transfer. Recognizing these challenges, a number of authors have called for additional emphasis into the provision of high-quality, well-annotated empirical estimates of nonmarket values (Loomis and Rosenberger 2006).

Given the difficulty in finding suitable data, practitioners may be tempted to use one of the increasing array of prepackaged software tools or databases to conduct or support benefit transfers. This should be considered only with extreme caution. These ready-made tools are often grounded in large-scale, data-intensive models built around spatial (GIS) modeling of ecosystem functions.[10] Although these tools have many useful purposes, among their shortcomings for welfare analysis is that the embedded benefit transfer methods and assumptions are often obscured; the typical lay user sees only the final transferred estimate of value. Also, the ecological components of these models are typically better developed than the economic components. Many of these tools rely on simple unit value transfers of the type often treated with skepticism by benefit transfer experts. Only through a dedicated, expert exploration of the model code or documentation are these underlying transfer methods and assumptions visible. The ability of such tools to support accurate benefit transfers depends on the validity of the underlying ecological and economic models and the coherence with which these models are combined. Before using such tools for benefit transfer, users should verify that the embedded methods, assumptions and data conform to recommended practices. Users should universally avoid any benefit transfer tool for which the underlying methods and functions are proprietary or otherwise unavailable for such a critical evaluation. Additional discussion of these issues is provided in Chap. 12.

2.7 Measuring Transfer Accuracy

Benefit transfers are subject to a variety of potential errors, many of which are related to the issues discussed throughout this chapter. These errors are often grouped into two general categories. Transferred errors from the original primary studies are denoted *measurement errors*. These are differences between the true underlying value and a primary study estimate (Brouwer and Spaninks 1999; Rosenberger and Stanley 2006). These are distinct from *generalization errors*, which are the errors related to the transfer process itself. Generalization errors are related to such factors as commodity inconsistency, benefit scaling and a lack of site similarity, among others (Rosenberger and Phipps 2007). Most evaluations of transfer accuracy focus on generalization error, since it is assumed that original research provides unbiased estimates, or that biased studies have been eliminated by quality control during the selection of studies for transfer.

[10]One of the best known and well-developed of these tools is InVEST (Integrated Valuation of Ecosystem Services and Tradeoffs), although many others have been developed over recent years. Among the advantages of InVEST is documentation that clarifies the transfer methods that are used. Many other tools lack such clarity, and are effectively "black boxes" in terms of transfer methods and data.

Benefit transfers are generally evaluated in terms of predictive accuracy (sometimes called reliability) and transfer validity. A transfer accurately predicts a value estimate, or is reliable, when generalization error is small. Benefit transfer validity, in contrast, requires that value estimates or other transferred quantities are statistically identical across study and policy contexts (i.e., there is no statistically significant transfer error).[11] In actual transfer applications the true value of the topic of interest is obviously unknown (otherwise benefit transfer would not be required). For this reason, out-of-sample predictive performance, or convergent validity testing, is most often used to test accuracy and validity. That is, benefit transfer is tested in circumstances where a policy site study has been conducted. Benefit transfer estimates are then compared to the estimate provided by original research at the policy site. Because the need for accuracy and validity varies across applications, there is no universal test or maximum error that dictates the acceptability of benefit transfer.

Evaluations of benefit transfer errors across the literature include Brouwer and Spaninks (1999), Kaul et al. (2013), and Rosenberger and Stanley (2006). The findings of these analyses are broadly similar and mostly intuitive. Results of Kaul et al. (2013), for example, suggest that (1) benefit function transfers tend to outperform unit value transfers, (2) transfers of values for quantities are more accurate than those for qualities, (3) geographic site similarity influences transfer error, (4) contingent valuation estimates are associated with systematically lower transfer errors than other nonmarket valuation techniques, and (5) the combination of data from multiple studies can reduce transfer errors. Chapter 14 in this book provides extensive discussion of the measurement and interpretation of transfer accuracy and validity.

2.8 Steps in a Benefit Transfer

There are various ways in which one may categorize the steps in a benefit transfer, and the steps will depend somewhat on the transfer method and policy context. These caveats aside, the following section attempts to briefly summarize the main steps involved. Readers are also directed to Desvousges et al. (1998) and Rosenberger and Loomis (2003), who provide alternative discussions of benefit transfer steps.

2.8.1 Define the Benefit Transfer Context

The first step in any benefit transfer is to define the valuation and policy context in which benefit transfer will potentially occur, and to determine the type of economic

[11]Transfer validity may also be viewed in terms of the underlying validity of the primary study estimates (i.e., lack of measurement error), although this is a less common use of the term.

information required. Answers to a variety of questions are typically sought at this stage. What is the circumstance for which values are required? What general type of information (e.g., value estimates) is required? What types of policy changes, commodities and populations will be affected, and what types of information are likely to be available on these effects? What will be the primary use of economic information, and what does this use imply for required accuracy? What time and resources are available for analysis? This step typically involves discussions with decision makers and reviews of background documents to establish the general context under which benefit transfer might occur, and the broad parameters of the analysis.

2.8.2 Establish the Need for Benefit Transfer

As noted above, high-quality primary studies are generally preferred to benefit transfer if feasible. Hence, the next stage in a potential benefit transfer is to assess whether benefit transfer is indeed necessary, or whether a primary study of sufficient quality is feasible. Factors influencing this decision include but are not necessarily limited to: (1) time and resources available for analysis relative to that required for a primary study, (2) availability of information for a primary study, (3) approvals or policy process constraints which restrict the collection of primary data or use of primary analysis, (4) the accuracy and other needs of the policy context and users of the information, (5) the size of policy impacts relative to the cost of a primary study (Allen and Loomis 2008), and (6) the availability of a suitable body of evidence from which one can conduct a defensible benefit transfer (Rosenberger and Johnston 2009).

2.8.3 Define the Policy, Good and Population

Assuming that benefit transfer is required, the next step is to define relevant aspects of the policy in question, the good(s) to be valued, and the affected population whose values are desired. In some cases these are clear ex ante, but not always. For example, while potential policy actions are often known, the effects on valued goods may not always be clear. Hence, even before the considerations of economic values begins, it is often necessary to clarify the specific types of market or non-market goods that will be affected (i.e., what are the aspects of policy effects that will directly influence people's welfare, or for which they would likely be willing to pay). This may involve consultations with biophysical scientists, policy makers and stakeholders, an examination of available biophysical data to predict policy effects and a review of the economic literature to assess how similar policy outcomes were valued in other settings.

Once the policy outcomes and goods have been identified, the relevant population for benefit assessment must also be identified. Three issues are particularly relevant for defining the population for benefit transfer. The first question is whether there are any policy, institutional or legal constraints for the policy analysis that dictate the population to be considered (i.e., the political jurisdiction; whose benefits count?). For example, benefit-cost analysis for state government programs is often limited to state residents, regardless of whether residents in other states are affected. The second question is the extent of the relevant economic jurisdiction or market, or where values are nonzero regardless of the political jurisdiction. Unlike the political jurisdiction, determining the extent of the economic market in a benefit transfer study is often difficult, because this information may not be provided by primary valuation studies (Desvousges et al. 1998; Loomis and Rosenberger 2006). Loomis (2000) provides a detailed contrast between economic markets and political jurisdictions, and when each is likely to be relevant for analysis. Desvousges et al. (1998) also provide a useful discussion of the extent of the market for benefit transfer studies.

A third question is whether values are likely heterogeneous in any systematic way over the population that is relevant for policy analysis and benefit transfer. A common example would be spatial heterogeneity or distance decay in WTP (Bateman et al. 2006; Hanley et al. 2003; Johnston and Ramachandran 2014; Loomis and Rosenberger 2006; Chap. 18). Policy makers may also require welfare estimates for different subgroups within the overall population.

Note that these steps are similar to those within any benefit cost analysis or primary valuation exercise (Boardman et al. 2006; Champ et al. 2003). However, an additional consideration for benefit transfer is that the answers to these questions may help define the type of benefit transfer that is most appropriate. For example, spatial heterogeneity in WTP may require adjustments possible only through a benefit function transfer in which spatial variables are incorporated; unit value transfers are unlikely to provide accurate transfers in this case.

2.8.4 Define and Quantify Policy Options and Changes in Goods

Benefit transfer validity depends on a clear quantification of the marginal changes to be valued. Based on the characterization of the general policy and affected goods in the prior step, the next step in benefit transfer is to specify the specific policy options that will be evaluated and the exact quantities of goods (including baselines and marginal changes) for which values will be estimated. This may include changes in quantities, qualities or sometimes both. Often these quantities or qualities are not determined by the analyst but are provided by the policymaker ex ante. However, in other cases economic models from the literature may be used to help quantify the change in valued goods that might occur under various policy options.

Care must be taken to ensure that these quantities and qualities are associated with non-overlapping and final effects on utility, so that double counting is avoided (Boyd and Banzhaf 2007; Johnston and Russell 2011).

Some valuation contexts will require welfare estimates for only a single policy option and set of changes in relevant goods. Others will require evaluation of multiple policies and changes. A related aspect of benefit transfer is whether there is *uncertainty* in the policy outcomes which must be accounted for; for example using expected values or sensitivity analysis (Desvousges et al. 1998; Holland et al. 2010). In such cases, benefit transfer may require information both on the possible policy outcomes and the probability of these outcomes (i.e., a probability distribution). All of these factors must be determined before gathering data and implementing transfers.

2.8.5 Gather and Evaluate Valuation Data and Evidence

This is typically among the most time-consuming components of a benefit transfer; it includes a comprehensive review of available data and evidence on the outcome to be evaluated from the published and unpublished literature. Typically this includes a comprehensive literature review to first identify prior empirical studies that address the general type of policy effects and goods under study. The resulting set of studies is then screened carefully for quality, relevance and correspondence to the specific policies and changes to be predicted by the transfer. Correspondence is evaluated in terms of numerous factors, including the general policy context, exact goods or services being valued, scope and scale of the analysis, policy site attributes (e.g., physical site characteristics, location, population attributes, availability of substitutes and complements), and the time period of the analysis. Table 2.1 lists a set of general criteria that can be used to help evaluate the methodological quality of primary studies. Desvousges et al. (1998) provide additional discussion of this topic. Stanley et al. (2013) provide guidelines for the systematic reporting of such literature reviews; although designed for meta-analysis, similar guidelines can apply to any gathering of data for benefit transfer.

The analyst must also identify the type of values or other quantities that are estimated by each study. As noted above, total economic values (or WTP) for any type of outcome may be comprised of multiple components (e.g., market versus nonmarket values; use versus nonuse values; different types of nonmarket use values). Given these different types of benefits, no single valuation methodology can generally measure and distinguish all possible aspects of the value for most types of environmental changes.[12] Different methodologies may be used to evaluate values for similar outcomes, each designed to measure a different aspect of value (Champ et al. 2003; Freeman et al. 2014; Holland et al. 2010). Analysts must

[12]For a practical example, see Johnston et al. (2002).

Table 2.1 General criteria for evaluating the quality of primary studies for benefit transfer

1.	Detailed and transparent reporting of data and methods
2.	Detailed reporting of site and population characteristics
3.	Foundation in economic theory
4.	Quality of underlying biophysical data or modeling
5.	Restrictiveness and realism of assumptions
6.	Clear specification of goods and quantities/qualities
7.	Empirical methods and development (e.g., use of accepted valuation methods)
8.	Modeling detail (i.e., model includes all elements suggested by theory)
9.	Data collection methods
10.	Sample sizes and representativeness
11.	Statistical techniques and model specifications
12.	Evidence of selectivity bias
13.	Robustness of results
14.	Evidence of peer review or other recognized quality indicators

exercise caution when comparing or aggregating benefits generated by different valuation methods, because these values may not be theoretically equivalent, and may sometimes overlap. Those unfamiliar with the nuances of different types of values and valuation methods should consult an expert in economic valuation to assist in the collection and interpretation of valuation data and evidence. Additional discussion of benefit transfer data, including approaches to avoid selection biases, is provided in Sect. 2.6.

2.8.6 Determine Benefit Transfer Method(s)

Based on the information provided by the prior research stages, the analyst must then determine the benefit transfer methods that are most appropriate to policy needs and available data. The choice of transfer method is covered in Sect. 2.2.3.

2.8.7 Design and Implement Transfer(s)

Methods to design and implement the transfer will depend almost entirely on the transfer method(s) applied. General methods for unit value, benefit function, meta-analysis and structural benefit transfer are described in Sect. 2.2. Johnston and Rosenberger (2010) provide an extensive literature review of prior benefit transfer analyses that apply different types of approaches. Desvousges et al. (1998) and Rosenberger and Loomis (2003) give contrasting examples of different types of benefit transfer. Many examples are provided by later chapters in this book.

2.8.8 Aggregate Values over Populations, Areas, and Time

Once per unit values are estimated, they must often be aggregated over relevant populations, geographical areas and time periods. Although aggregation can be straightforward in some cases, it is also an area in which large and often overlooked errors can be introduced. In the simplest possible case in which marginal values are homogeneous (i.e., approximately identical across the population), or in which the benefit transfer provides an accurate estimate of mean value across the population, aggregation across populations can be as simple as multiplying a representative mean value per person by the size of the population. However, there are a large number of complications which can occur (e.g., the treatment of households versus adults versus children when aggregating benefits; whether the sample of the primary study is indeed representative), so that simple multiplication by population size is no longer appropriate. In general, these aggregation issues are similar to those encountered in any benefit-cost analysis, as described by Boardman et al. (2006). The discussion in Sect. 2.4, is also relevant to these aggregations.

Similar considerations and caveats apply to aggregation across geographical areas. However, here there is an important distinction between the aggregation of benefits over populations living in different areas versus scaling of benefits provided by environmental changes in different (or different-size) areas. For the former, similar rules for aggregation over any population apply, although the analyst should also correct for any systematic differences in values across populations living in different areas or distances from affected sites (Bateman et al. 2006; Johnston and Duke 2009; Johnston and Ramachandran 2014; Jørgensen et al. 2013; Martin-Ortega et al. 2012; Schaafsma et al. 2012). For the latter (scaling of benefits provided by changes in different areas), see the discussion in Sect. 2.4.

The aggregation or comparison of benefits over time requires *discounting*.[13] People will not typically pay one dollar today for the opportunity to obtain one dollar in the future; future benefits are worth less than an otherwise identical benefit received in the present. As a result, future benefits must be discounted in order to make them comparable to benefits today. Assuming that time is counted in discrete units and that discounting is calculated accordingly, a simple formula for the present value (PV) of a future payment of *\$X*—what that future payment is worth today—is given by:

$$PV = \frac{\$X}{(1+r)^t},$$

where r is the discount rate per time period in decimal notation (i.e., 5 % = 0.05) and t is the number of periods into the future when the payment will be received. Aggregating all discounted future benefits associated with a project over all time periods generates the *net present value* of benefits. An alternative method based on

[13]It can also require adjustments for systematic differences in values over time (cf. Brouwer 2006).

continuous (non-discrete) discounting calculates present value as $\$X(e^{-rt})$, where e is the exponential operator; results will be similar, but not identical, to the discrete discounting method above.

Although discounting is the standard means of aggregating benefits over time, it can lead to unintended consequences when assessing projects with very long time horizons. For example, if one uses common discount rates between 4 and 10 %, then benefits in the distant future (e.g., 50 to 100+ years) often have little impact on present value. For this reason, researchers have proposed a number of alternative discounting approaches for projects with long-duration effects. However, for most projects and policies where benefits are evaluated over a limited time horizon (e.g., 40 years or less), standard discounting procedures will likely generate the most accurate reflection of true social benefits and costs. Additional discussion of methods and complications associated with the aggregation of benefits over time is provided by Boardman et al. (2006) and Portney and Weyant (1999).

2.8.9 Conduct Sensitivity Analysis and Test Reliability (Where Possible)

The penultimate step in benefit transfer is sensitivity and reliability analysis. As is the case with any model, sensitivity analysis quantifies the sensitivity of results to changes in the modeling approach and uncertainty about key parameters or data, including different potentially influential assumptions and model specifications (Boardman et al. 2006; Desvousges et al. 1998; Holland et al. 2010). For example, one might aggregate benefits under a variety of different discount rates to evaluate the impact on present values. One might also estimate MRMs using different functional forms, using different subsets of the data, or using a different treatment of outliers (see Chap. 19 for a technical discussion of these steps applied to meta-analysis). Benefit transfer can also be conducted using a variety of fundamental approaches, for example, unit value versus single-site benefit function transfer, to evaluate effects on transfer estimates. Monte Carlo simulation analysis can provide a systematic way to evaluate the sensitivity of model results to uncertainty regarding key model parameters or data (Desvousges et al. 1998; Holland et al. 2010); an application to the impacts of methodological variables on benefit transfer is illustrated by Johnston et al. (2006a). Among other goals, sensitivity analysis can help the users of benefit transfer outputs to understand the confidence they can and should have in transfer results, based on the relative robustness of those results to methodological choices made by the analyst.

Where possible, it is also useful to provide information characterizing the potential reliability of benefit transfer results (or the accuracy). Because the true value is unknown, a variety of indirect methods must be used. As described in Sect. 2.7, convergent validity tests may be used to evaluate the performance of similar types of transfer in cases for which a primary study has been conducted, and hence transfer errors can be calculated. When an MRM is used, one can use leave-one-out, cross-validation convergent validity tests to characterize predictive performance (e.g., Brander et al. 2007; Stapler and Johnston 2009).[14] For additional discussion of this topic, see Sect. 2.7 and Chap. 14.

2.8.10 Report Results

The final step in a benefit transfer is the reporting of benefit transfer results. In general, this reporting follows the same guidelines applicable to any analysis of economic benefits. Given that the accuracy of benefit transfer depends critically on the procedures and data that are applied, transparent description of these factors is crucial to good reporting. Minimum features that should be reported in a benefit transfer include, but are not limited to: (1) a full description of the steps of the transfer; (2) the policy site, populations and goods; (3) reasons for assumed correspondence among the site, populations and goods within the study and policy contexts; (4) quantities or qualities for which values are estimated, including the specific units in which these are measured; (5) data sources used; (6) the specific type of value that is transferred, e.g., WTP, consumer surplus, etc.; (7) methods used to collect and screen data; (8) transfer methods; (9) statistical methods and assumptions; (10) any scaling that is conducted and implied assumptions; (11) final transferred unit and aggregated estimates of value or other outcomes; (12) results of any sensitivity analyses, robustness tests and accuracy evaluations. Additional reporting requirements may apply for particular types of analyses (for example meta-analysis, as described by Stanley et al. 2013).

2.9 Advanced Techniques

In addition to fundamental approaches, there are a number of advanced techniques that are used for benefit transfer. Most of these extend or supplement the approaches outlined above. Although this section does not provide a comprehensive list of advanced benefit transfer techniques that have been proposed, it highlights a few

[14]Assume that one has metadata with $n = 1 \ldots N$ unique observations. The first step is the omission of the nth observation from the metadata. The MRM is then estimated (using the original model specification) for the remaining $N - 1$ observations. This is iterated for each $n = 1 \ldots N$ observation, resulting in a vector of N unique parameter estimates, each corresponding to the omission of the nth observation. For each $n = 1 \ldots N$ model runs, the nth observation is an out-of-sample observation corresponding to the vector of parameter estimates resulting from that iteration. Parameter estimates for the nth model iteration are then combined with independent variable values for the nth observation to generate a WTP forecast for the omitted observation. The result is N out-of-sample WTP forecasts, each drawn from a unique MRM estimation. Transfer error is assessed through comparisons of the predicted and actual WTP value for each of the N observations.

principal areas in which substantial work has been conducted. Examples include Bayesian model search, updating and averaging that may be used for such purposes as addressing the possible sensitivity of transfer results to model specification, incorporating prior information and expectations, enabling estimation of MRMs with small samples, and evaluating the commensurability of different types of data (Johnston and Moeltner 2014; León et al. 2002; Leon-Gonzalez and Scarpa 2008; Moeltner and Rosenberger 2008, 2012; Moeltner et al. 2007). There has been significant work on the use of advanced choice modeling techniques for benefit transfer (Johnston and Rosenberger 2010; Morrison and Bergland 2006; Rolfe and Bennett 2006); this is discussed in detail in Chaps. 10 and 11. Researchers have proposed a variety of approaches to extend, improve and advance classical statistical methods for meta-analysis (Nelson and Kennedy 2009; see also see Chaps. 15, 16 and 17), including approaches to avoid selection biases (Rosenberger and Johnston 2009). Methods have also been proposed to improve validity and reliability testing in benefit transfer (Johnston and Duke 2008; Kristofersson and Navrud 2005; Muthke and Holm-Mueller 2004), and particularly for meta-analysis (Kaul et al. 2013); such methods are discussed in Chap. 14. Finally, there has been work to extend the spatial and geographical aspects of benefit transfer (Bateman et al. 2006; Johnston and Duke 2009; Martin-Ortega et al. 2012); these are discussed in Chap. 18.

Although these and other advanced methods can often improve benefit transfer accuracy and robustness, users are cautioned that greater complexity or flexibility does not always imply improved performance (cf. Bateman et al. 2011; Johnston and Duke 2010; Rosenberger and Stanley 2006). As noted by Navrud and Ready (2007b, p. 288), "[s]imple approaches should not be cast aside until we are confident that more complex approaches do perform better."

2.10 Conclusion

Benefit transfer is one of the most commonly used, but also easily misused components of benefit-cost analysis. Despite the common presumption that benefit transfer is a simple and easy approach to valuation, accurate benefit transfer requires significant attention to methods and data. Fortunately, many determining factors of transfer accuracy are within direct control of the analyst, and the benefit transfer literature now provides guidance on these factors. This chapter is an attempt to describe some of the most important of these, and provide at least basic guidance on methods recommended by the benefit transfer literature. Given the dependence of transfer accuracy on the many choices made by the analyst, perhaps the most important aspect of any benefit transfer is transparency, including the provision of detailed information on the data used, methods applied, and assumptions made. Clear reporting can help ensure that users are aware of both the strengths and limitations of the underlying methods and data, as well as interpretations of the resulting benefit estimates. Conversely, benefit transfers for which the methods and data are not clearly stated should be treated with caution.

Although benefit transfers generally require less time and resources than comparable primary studies, they do not necessarily require less expertise—methods such as meta-analysis and structural benefit transfer, for example, can require a level of expertise that parallels or even exceeds that required to conduct primary valuation research. Even simpler methods such as single-site benefit function transfer and unit value transfers require considerable expertise to evaluate such influential factors as the choice of transfer method, site and commodity correspondence, the suitability of functions or values for transfer, the quality and interpretation of primary studies, the aggregation of benefit estimates, and many others. Although the chapters in this book provide considerable information on theory, methods and data, new producers or consumers of benefit transfer are urged to seek the assistance of those with relevant expertise.[15] Doing so can help ensure that transfer errors are minimized and that the resulting estimates reflect the best possible use of existing information.

As noted by Loomis and Rosenberger (2006, p. 349), "the pace and widespread activity in non-market valuation makes the future of benefit transfer promising. There will continue to be more and better empirical studies to base our benefit transfers on in the future." The availability of an increasing body of high-quality primary studies, however, does not guarantee the accuracy of benefit transfer. At the same time that the body of valuation literature is increasing, the body of benefit transfers—of both high and low quality—is also increasing. Only through careful attention to (improved) benefit transfer methods will researchers be able to optimally leverage this body of work to provide the most accurate and useful policy guidance.

Appendix

Illustration of Unit Value and Benefit Function Transfer

To illustrate the mechanics of a very simple benefit transfer, consider the following stylized example. Assume that a published study reports the results of a simple, linear travel cost model predicting the number of visits to a local wildlife refuge

[15]There is also an increasing array of national and international agency publications in the U.S., EU and elsewhere that provides guidance for benefit transfer (e.g., Commonwealth of Australia 2002; Pearce et al. 2006; UK Environment Agency 2004; U.S. Environmental Protection Agency 2007, 2009).

Table 2.2 Stylized travel cost recreation demand model results

Ordinary least squares parameter estimates (dependent variable: TRIPS)		
Variable	Parameter estimate	Prob > T
INTERCEPT	5.5000	0.0256
TRAVCOST	−0.5000	0.0001
INCOME	0.0001	0.0996
VIEWINGS	0.5000	0.0021
SUBCOST	0.0500	0.0852
N (number of observations)	116	
R^2	0.67	

Variable definitions	
TRIPS	Number of trips per season, to the refuge, by each individual in the sample
TRAVCOST	Cost of travel to the site, including the opportunity cost of time, for each visitor
INCOME	Annual income of each individual
VIEWINGS	Expected viewings of rare bird species, per average visit
SUBCOST	Cost of traveling to the nearest substitute wildlife refuge

Mean values for model variables	
Variable	Average value
TRIPS	4.5 trips per season
TRAVCOST	$10 per visit
INCOME	$20,000 per person
VIEWINGS	3.0 per trip
SUBCOST	$10 per visit

(Site A), with statistical model results reported in Table 2.2 (assume a simple ordinary least squares model).[16]

Assume that there is a *nearby wildlife refuge* (Site B) that is similar to Site A. However, the average number of rare bird viewings at Site B is higher than those at Site A. Assume that average viewings at Site B are 6.0 per visit. Assume also that the analyst wishes to use benefit transfer to estimate consumer surplus at Site B (the policy site), based on the study published from data at Site A (the study site).

[16]Note that this is a very simple model used for basic illustration purposes only. Linear OLS models such as this are rarely suitable for applied recreation demand modeling. Most recreation demand research applies more sophisticated approaches such as count data or random utility models (Bockstael and McConnell 2010; Haab and McConnell 2002). For an applied example see Rosenberger and Loomis (2003).

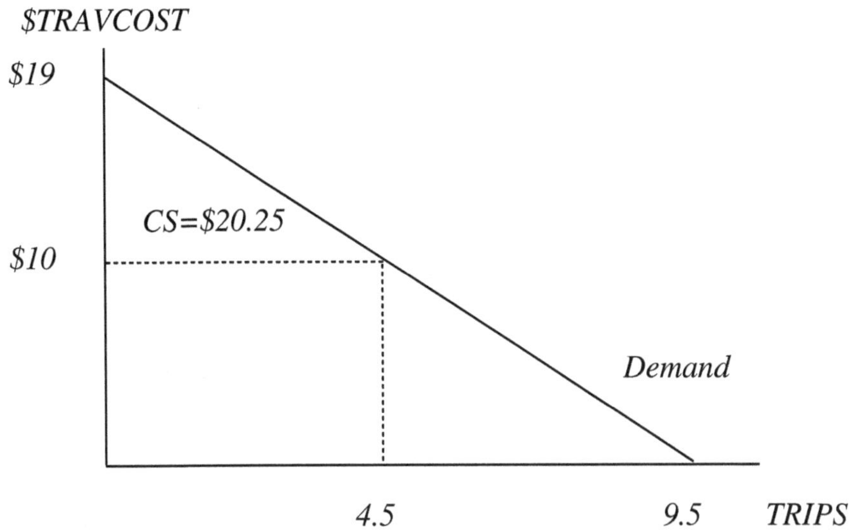

Fig. 2.3 Illustrative travel cost demand function and consumer surplus (CS)

To conduct our benefit transfer, we first use data at Site A to calculate the original study site demand curve and mean per visitor consumer surplus (CS).

$$
\begin{aligned}
TRIPS &= 5.5 - 0.5(TRAVCOST) + 0.0001(INCOME) \\
&\quad + 0.5(VIEWINGS) + 0.05(SUBCOST) \\
TRIPS &= 5.5 - 0.5(TRAVCOST) + 0.0001(20{,}000) \\
&\quad + 0.5(3) + 0.05(10) \\
TRIPS &= 9.5 - 0.5(TRAVCOST)
\end{aligned}
$$

The result is shown in Fig. 2.3, which illustrates the travel cost demand curve and associated consumer surplus. Here, the consumer surplus estimate of $20.25 reflects the access value of Site A, or the total value that each visitor receives from all visits to Site A, each year. Following standard practice, this is estimated as the area above the average travel cost ($10 per trip) and below the estimated travel cost demand curve.

To conduct a unit value transfer of this estimate to Site B, one would simply assume that the same consumer surplus estimate applies to both sites, so that the annual per visitor consumer surplus at Site B would be approximated as $20.25. This unit value estimate does not account for the difference in rare bird *VIEWINGS* between Site A and B.

To conduct a simple benefit function transfer of this estimate to Site B, one would estimate a new demand function using the updated information on *VIEWINGS* from Site B.

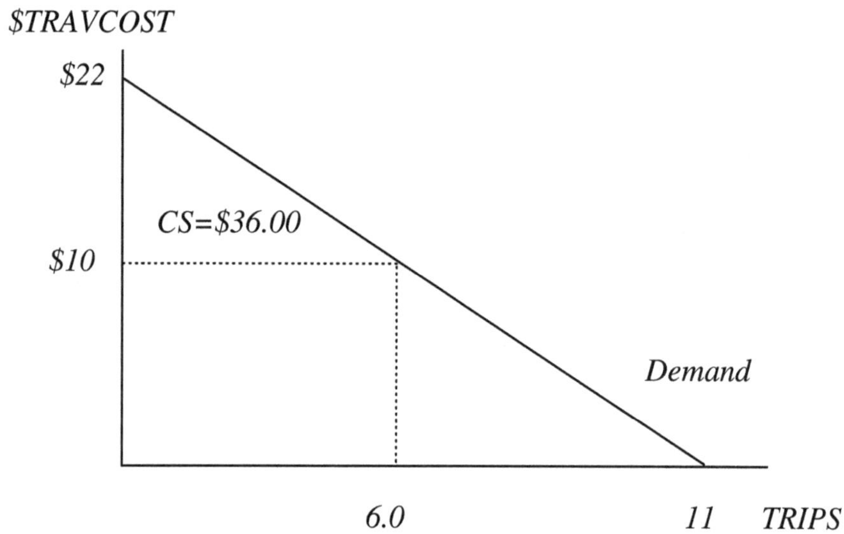

Fig. 2.4 Benefit function transfer of travel cost model results

$$\begin{aligned} TRIPS &= 5.5 - 0.5(TRAVCOST) + 0.0001(INCOME) \\ &\quad + 0.5(VIEWINGS) + 0.05(SUBCOST) \\ TRIPS &= 5.5 - 0.5(TRAVCOST) + 0.0001(20{,}000) \\ &\quad + 0.5(6) + 0.05(10) \\ TRIPS &= 11.0 - 0.5(TRAVCOST) \end{aligned}$$

Given this updated demand curve (Fig. 2.4), the benefit function transfer estimate of consumer surplus for Site B visitors is \$36.00 per year. The consumer surplus difference (\$36.00 vs. \$20.25) reflects ability of benefit function transfer to calibrate for the difference between *VIEWINGS* at the two sites, and hence predict a higher access value for Site B, all else equal. Although more sophisticated models (cf. Haab and McConnell 2002) may require more complex calculations to implement unit value or benefit function transfers, the general process is similar.

References

Allen, B. P., & Loomis, J. B. (2008). The decision to use benefit transfer or conduct original valuation research for benefit-cost and policy analysis. *Contemporary Economic Policy, 26*, 1–12.

Bateman, I. J., Carson, R. T., Day, B., Hanemann, M., Hanley, N., Hett, T., et al. (2002). *Economic valuation with stated preference techniques: A manual*. Cheltenham, UK: Edward Elgar.

Bateman, I. J., Day, B. H., Georgiou, S., & Lake, I. (2006). The aggregation of environmental benefit values: Welfare measures, distance decay and total WTP. *Ecological Economics, 60*, 450–460.

Bateman, I. J., Brouwer, R., Ferrini, S., Schaafsma, M., Barton, D. N., Dubgaard, A., et al. (2011). Making benefit transfers work: Deriving and testing principles for value transfers for similar and dissimilar sites using a case study of the non-market benefits of water quality improvements across Europe. *Environmental & Resource Economics, 50*, 365–387.

Bergstrom, J. C., & De Civita, P. (1999). Status of benefit transfer in the United States and Canada: A review. *Canadian Journal of Agricultural Economics, 47*, 79–87.

Bergstrom, J. C., & Taylor, L. O. (2006). Using meta-analysis for benefits transfer: Theory and practice. *Ecological Economics, 60*, 351–360.

Boardman, A. E., Greenberg, D. H., Vining, A. R., & Weimer, D. L. (2006). *Cost-benefit analysis: Concepts and practice* (3rd ed.). Upper Saddle River, NJ: Prentice Hall.

Bockstael, N. E., & McConnell, K. E. (2010). *Environmental and resource valuation with revealed preferences. A theoretical guide to empirical models*. Dordrecht: Springer.

Bockstael, N. E., Freeman, A. M., Kopp, R. J., Portney, P. R., & Smith, V. K. (2000). On measuring economic values for nature. *Environmental Science and Technology, 34*, 1384–1389.

Boyd, J., & Banzhaf, S. (2007). What are ecosystem services? The need for standardized environmental accounting units. *Ecological Economics, 63*, 616–626.

Boyle, K. J., & Bergstrom, J. C. (1992). Benefit transfer studies: Myths, pragmatism and idealism. *Water Resources Research, 28*, 657–663.

Brander, L. M., van Beukering, P., & Cesar, H. (2007). The recreational value of coral reefs: A meta-analysis. *Ecological Economics, 63*, 209–218.

Brouwer, R. (2000). Environmental value transfer: State of the art and future prospects. *Ecological Economics, 32*, 137–152.

Brouwer, R. (2006). Do stated preference methods stand the test of time? A test of the stability of contingent values and models for health risks when facing an extreme event. *Ecological Economics, 60*, 399–406.

Brouwer, R., & Bateman, I. J. (2005). Benefits transfer of willingness to pay estimates and functions for health-risk reductions: A cross-country study. *Journal of Health Economics, 24*, 591–611.

Brouwer, R., & Spaninks, F. A. (1999). The validity of environmental benefits transfer: Further empirical testing. *Environmental & Resource Economics, 14*, 95–117.

Champ, P. A., Boyle, K. J., & Brown, T. C. (Eds.). (2003). *A primer on nonmarket valuation*. Dordrecht: Kluwer.

Colombo, S., & Hanley, N. (2008). How can we reduce the errors from benefits transfer? An investigation using the choice experiment method. *Land Economics, 84*, 128–147.

Commonwealth of Australia. (2002). *Australians and National Resource Management 2002*. Braddon, Australia: National Land and Water Resources Audit.

Costanza, R., d'Arge, R., de Groot, R., Farber, S., Grasso, M., Hannon, B., et al. (1997). The value of the world's ecosystem services and natural capital. *Nature, 387*, 253–260.

Desvousges, W. H., Johnson, F. R., & Banzhaf, H. S. (1998). *Environmental policy analysis with limited information: Principles and applications of the transfer method*. Cheltenham, UK: Edward Elgar.

Florax, R. J. G. M., Nijkamp, P., & Willis, K. G. (Eds.). (2002). *Comparative environmental economics assessment*. Chelthenham, UK: Edward Elgar.

Freeman, A. M. III. (1984). On the tactics of benefit estimation under Executive Order 12291. In V. K. Smith (Ed.), *Environmental policy under Reagan's Executive Order: The role of benefit-cost analysis* (pp. 167–186). Chapel Hill, NC: The University of North Carolina Press.

Freeman, A. M. III., Herriges, J. A., & Kling, C. L. (2014). *The measurement of environmental and resource values: Theory and methods*. Washington, DC: Resources for the Future.

Griffiths, C., & Wheeler, W. (2005). Benefit-cost analysis of regulations affecting surface water quality in the United States. In R. Brouwer & D. Pearce (Eds.), *Cost-benefit analysis and water resources management* (pp. 223–250). Cheltenham, UK: Edward Elgar.

Haab, T. C., & McConnell, K. E. (2002). *Valuing environmental and natural resources: The econometrics of nonmarket valuation*. Cheltenham, UK: Edward Elgar.

Hanley, N., & Barbier, E. B. (2009). *Pricing nature: Cost-benefit analysis and environmental policy*. Cheltenham, UK: Edward Elgar.

Hanley, N., Schlapfer, F., & Spurgeon, J. (2003). Aggregating the benefits of environmental improvements: Distance-decay functions for use and nonuse values. *Journal of Environmental Management, 68*, 297–304.

Holland, D. S., Sanchirico, J. N., Johnston, R. J., & Joglekar, D. (2010). *Economic analysis for ecosystem based management: Applications to marine and coastal environments*. Washington, DC: Resources for the Future Press.

Iovanna, R., & Griffiths, C. (2006). Clean water, ecological benefits, and benefits transfer: A work in progress at the U.S. EPA. *Ecological Economics, 60*, 473–482.

Johnston, R. J. (2007). Choice experiments, site similarity and benefits transfer. *Environmental & Resource Economics, 38*, 331–351.

Johnston, R. J., & Besedin, E. Y. (2009). Estimating willingness to pay for aquatic resource improvements using benefits transfer. In H. W. Thurston, M. T. Heberling, & A. Schrecongost (Eds.), *Environmental economics for watershed restoration* (pp. 95–122). New York: CRC Press.

Johnston, R. J., & Duke, J. M. (2008). Benefit transfer equivalence tests with non-normal distributions. *Environmental & Resource Economics, 41*, 1–23.

Johnston, R. J., & Duke, J. M. (2009). Willingness to pay for land preservation across states and jurisdictional scale: Implications for benefit transfer. *Land Economics, 85*, 217–237.

Johnston, R. J., & Duke, J. M. (2010). Socioeconomic adjustments and choice experiment benefit function transfer: Evaluating the common wisdom. *Resource and Energy Economics, 32*, 421–438.

Johnston, R. J., & Moeltner, K. (2014). Meta-modeling and benefit transfer: The empirical relevance of source-consistency in welfare measures. *Environmental & Resource Economics, 59*, 337–361.

Johnston, R. J., & Ramachandran, M. (2014). Modeling spatial patchiness and hot spots in stated preference willingness to pay. *Environmental & Resource Economics, 59*, 363–387.

Johnston, R. J., & Rosenberger, R. S. (2010). Methods, trends and controversies in contemporary benefit transfer. *Journal of Economic Surveys, 24*, 479–510.

Johnston, R. J., & Russell, M. (2011). An operational structure for clarity in ecosystem service values. *Ecological Economics, 70*, 2243–2249.

Johnston, R. J., & Thomassin, P. J. (2009). Evaluating the environmental valuation reference inventory (EVRI): Results from a survey of actual and potential users. *AERE (Association of Environmental and Resource Economists) Newsletter, 29*, 33–38.

Johnston, R. J., & Thomassin, P. J. (2010). Willingness to pay for water quality improvements in the United States and Canada: Considering possibilities for international meta-analysis and benefit transfer. *Agricultural and Resource Economics Review, 39*, 114–131.

Johnston, R. J., Grigalunas, T. A., Opaluch, J. J., Diamantedes, J., & Mazzotta, M. (2002). Valuing estuarine resource services using economic and ecological models: The Peconic Estuary system study. *Coastal Management, 30*, 47–66.

Johnston, R. J., Besedin, E. Y., Iovanna, R., Miller, C., Wardwell, R., & Ranson, M. (2005). Systematic variation in willingness to pay for aquatic resource improvements and implications for benefit transfer: A meta-analysis. *Canadian Journal of Agricultural Economics, 53*, 221–248.

Johnston, R. J., Besedin, E. Y., & Ranson, M. H. (2006a). Characterizing the effects of valuation methodology in function-based benefits transfer. *Ecological Economics, 60*, 407–419.

Johnston, R. J., Ranson, M. H., Besedin, E. Y., & Helm, E. C. (2006b). What determines willingness to pay per fish? A meta-analysis of recreational fishing values. *Marine Resource Economics, 21*, 1–32.

Jørgensen, S. L., Olsen, S. B., Ladenburg, J., Martinsen, L., Svenningsen, S. R., & Hasler, B. (2013). Spatially induced disparities in users' and non-users' WTP for water quality improvements–Testing the effect of multiple substitutes and distance decay. *Ecological Economics, 92*, 58–66.

Just, R. E., Hueth, D. L., & Schmitz, A. (2004). *The welfare economics of public policy: A practical approach to project and policy evaluation*. Cheltenham, UK: Edward Elgar.

Kaul, S., Boyle, K. J., Kuminoff, N. V., Parmeter, C. F., & Pope, J. C. (2013). What can we learn from benefit transfer errors? Evidence from 20 years of research on convergent validity. *Journal of Environmental Economics and Management, 66*, 90–104.

Kristofersson, D., & Navrud, S. (2005). Validity tests of benefit transfer—Are we performing the wrong tests? *Environmental & Resource Economics, 30*, 279–286.

León, C. J., Vazquez-Polo, F. J., Guerra, N., & Riera, P. (2002). A Bayesian model for benefit transfer: Application to national parks in Spain. *Applied Economics, 34*, 749–757.

Leon-Gonzalez, R., & Scarpa, R. (2008). Improving multi-site benefit functions via Bayesian model averaging: A new approach to benefit transfer. *Journal of Environmental Economics and Management, 56*, 50–68.

Lindhjem, H., & Navrud, S. (2008). How reliable are meta-analyses for international benefit transfers? *Ecological Economics, 66*, 425–435.

Londoño, L. M., & Johnston, R. J. (2012). Enhancing the reliability of benefit transfer over heterogeneous sites: A meta-analysis of international coral reef values. *Ecological Economics, 78*, 80–89.

Loomis, J. B. (1992). The evolution of a more rigorous approach to benefit transfer: Benefit function transfer. *Water Resources Research, 28*, 701–705.

Loomis, J. B. (2000). Vertically summing public good demand curves: An empirical comparison of economic versus political jurisdictions. *Land Economics, 76*, 312–321.

Loomis, J. B., & Rosenberger, R. S. (2006). Reducing barriers in future benefit transfers: Needed improvements in primary study design and reporting. *Ecological Economics, 60*, 343–350.

Martin-Ortega, J., Brouwer, R., Ojea, E., & Berbel, J. (2012). Benefit transfer and spatial heterogeneity of preferences for water quality improvements. *Journal of Environmental Management, 106*, 22–29.

McComb, G., Lantz, V., Nash, K., & Rittmaster, R. (2006). International valuation databases: overview, methods and operational issues. *Ecological Economics, 60*, 461–472.

Moeltner, K., & Rosenberger, R. S. (2008). Predicting resource policy outcomes via meta-regression: Data space, model space, and the quest for 'optimal scope'. *The B. E. Journal of Economic Analysis & Policy*. doi 10.2202/1935-1682.2028.

Moeltner, K., & Rosenberger, R. S. (2012). *Cross-context benefit transfer: A Bayesian search for information pools*. Paper presented at the Annual Meetings of the European Association of Environmental and Resource Economists, Prague, June 27–30.

Moeltner, K., Boyle, K. J., & Paterson, R. W. (2007). Meta-analysis and benefit transfer for resource valuation–Addressing classical challenges with Bayesian modeling. *Journal of Environmental Economics and Management, 53*, 250–269.

Morrison, M. (2001). Non-market valuation databases: How useful are they? *Economic Analysis & Policy, 31*, 33–55.

Morrison, M., & Bergland, O. (2006). Prospects for the use of choice modelling for benefit transfer. *Ecological Economics, 60*, 420–428.

Mrozek, J. R., & Taylor, L. O. (2002). What determines the value of life? A meta-analysis. *Journal of Policy Analysis and Management, 21*, 253–270.

Muthke, T., & Holm-Mueller, K. (2004). National and international benefit transfer with a rigorous test procedure. *Environmental & Resource Economics, 29*, 323–336.

Navrud, S., & Pruckner, G. J. (1997). Environmental valuation–to use or not to use? A comparative study of the United States and Europe. *Environmental & Resource Economics, 10*, 1–26.

Navrud, S., & Ready, R. (2007a). Review of methods for value transfer. In S. Navrud & R. Ready (Eds.), *Environmental value transfer: Issues and methods* (pp. 1–10). Dordrecht: Springer.

Navrud, S., & Ready, R. (2007b). Lessons learned for environmental value transfer. In S. Navrud & R. Ready (Eds.), *Environmental value transfer: Issues and methods* (pp. 1–10). Dordrecht: Springer.

Navrud, S., & Ready, R. (Eds.) (2007c). *Environmental value transfer: Issues and methods*. Dordrecht: Springer.

Nelson, J. P., & Kennedy, P. E. (2009). The use (and abuse) of meta-analysis in environmental and resource economics: An assessment. *Environmental & Resource Economics, 42*, 345–377.

Pattanayak, S. K., Smith, V. K., & Van Houtven, G. (2007). Improving the practice of benefits transfer: A preference calibration approach. In S. Navrud & R. Ready (Eds.), *Environmental value transfer: Issues and methods* (pp. 241–260). Dordrecht: Springer.

Pearce, D., Atkinson, G., & Mourato, S. (2006). Cost benefit analysis and the environment: Recent developments. Paris, France: Organisation for Economic Co-operation and Development (OECD).

Poe, G. L., Boyle, K. J., & Bergstrom, J. C. (2001). A preliminary meta-analysis of contingent values for ground water revisited. In J. C. Bergstrom, K. J. Boyle, & G. L. Poe (Eds.), *The economic value of water quality* (pp. 137–162). Northampton, MA: Edward Elgar.

Portney, P., & Weyant, J. (Eds.). (1999). *Discounting and intergenerational equity*. Washington, DC: Resources for the Future Press.

Ready, R., & Navrud, S. (2006). International benefit transfer: Methods and validity test. *Ecological Economics, 60*, 429–434.

Ready, R., Navrud, S., Day, B., Dubourg, R., Machado, F., Mourato, S., et al. (2004). Benefit transfer in Europe: How reliable are transfers between countries? *Environmental & Resource Economics, 29*, 67–82.

Rolfe, J., & Bennett, J. (Eds.). (2006). *Choice modelling and the transfer of environmental values*. Cheltenham, UK: Edward Elgar.

Rolfe, J., & Brouwer, R. (2013). Design effects in a meta-analysis of river health choice experiments in Australia. *Journal of Choice Modelling, 5*, 81–97.

Rolfe, J., & Wang, X. (2011). Dealing with scale and scope issues in stated preference experiments. In J. Bennett (Ed.), *The international handbook on non-market environmental valuation* (pp. 254–272). Cheltenham, UK: Edward Elgar.

Rolfe, J., & Windle, J. (2008). Testing for differences in benefit transfer values between state and regional frameworks. *Australian Journal of Agricultural and Resource Economics, 52*, 149–168.

Rolfe, J., Bennett, J., Johnston, R. J., & Kerr, G. (2011). 'Yes we can…': Using benefit transfer to inform environmental policymaking. *EERH Policy Brief.* Canberra, Australia: Environmental Economics Research Hub, Australian National University.

Rosenberger, R. S., & Johnston, R. J. (2009). Selection effects in meta-analysis and benefit transfer: Avoiding unintended consequences. *Land Economics, 85*, 410–428.

Rosenberger, R. S., & Loomis, J. B. (2000). Using meta-analysis for benefit transfer: In-sample convergent validity tests of an outdoor recreation database. *Water Resources Research, 36*, 1097–1107.

Rosenberger, R. S., & Loomis, J. B. (2003). Benefit transfer. In P. A. Champ, K. J. Boyle, & T. C. Brown (Eds.), *A primer on non-market valuation* (pp. 445–482). Dordrecht: Kluwer.

Rosenberger, R. S., & Phipps, T. T. (2007). Correspondence and convergence in benefit transfer accuracy: Meta-analytic review of the literature. In S. Navrud & R. Ready (Eds.), *Environmental value transfer: Issues and methods* (pp. 23–43). Dordrecht: Springer.

Rosenberger, R. S., & Stanley, T. D. (2006). Measurement, generalization and publication: Sources of error in benefit transfers and their management. *Ecological Economics, 60*, 372–378.

Schaafsma, M., Brouwer, R., & Rose, J. (2012). Directional heterogeneity in WTP models for environmental valuation. *Ecological Economics, 79*, 21–31.

Smith, V. K., & Pattanayak, S. K. (2002). Is meta-analysis a Noah's Ark for non-market valuation? *Environmental & Resource Economics, 22*, 271–296.

Smith, V. K., Van Houtven, G., & Pattanayak, S. K. (2002). Benefit transfer via preference calibration: "Prudential algebra" for policy. *Land Economics, 78*, 132–152.

Smith, V. K., Pattanayak, S. K., & Van Houtven, G. (2006). Structural benefit transfer: An example using VSL estimates. *Ecological Economics, 60*, 361–371.

Spash, C. L., & Vatn, A. (2006). Transferring environmental value estimates: Issues and alternatives. *Ecological Economics, 60*, 379–388.

Stanley, T. D. (2005). Beyond publication bias. *Journal of Economic Surveys, 19*, 309–345.

Stanley, T. D. (2008). Meta-regression methods for detecting and estimating empirical effect in the presence of publication selection. *Oxford Bulletin of Economics and Statistics, 70*, 103–127.

Stanley, T. D., Doucouliagos, H., Giles, M., Heckemeyer, J. H., Johnston, R. J., Laroche, P., et al. (2013). Meta-analysis of economics reporting guidelines. *Journal of Economic Surveys, 27*, 390–394.

Stapler, R. W., & Johnston, R. J. (2009). Meta-analysis, benefit transfer, and methodological covariates: Implications for transfer error. *Environmental & Resource Economics, 42*, 227–246.

Toman, M. (1998). Why not to calculate the value of the world's ecosystem services and natural capital. *Ecological Economics, 25*, 57–60.

UK Environment Agency. (2004). *Assessment of benefits for water quality and water resources schemes in the PR04 Environment Programme*. London: UK Environment Agency.

U.S. Environmental Protection Agency. (2007). Advisory on EPA's issues in valuing mortality risk reduction. EPA-SAB-08-001. Washington, DC: U.S. Environmental Protection Agency, Science Advisory Board, Office of the Administrator.

U.S. Environmental Protection Agency. (2009). Valuing the protection of ecological systems and services: A report of the EPA Science Advisory Board. EPA-SAB-09-012. Washington, DC: U.S. Environmental Protection Agency, Science Advisory Board, Office of the Administrator.

Van Houtven, G., Powers, J. & Pattanayak, S. K. (2007). Valuing water quality improvements in the United States using meta-analysis: Is the glass half-full or half-empty for national policy analysis? *Resource and Energy Economics, 29*, 206–228.

Viscusi, W. K., & Aldy, J. E. (2003). The value of a statistical life: A critical review of market estimate throughout the world. *Journal of Risk and Uncertainty, 27*, 5–76.

Chapter 3
The Use of Benefit Transfer in the United States

John B. Loomis

Abstract This chapter provides an overview of applications and trends in benefit transfer for the United States. Benefit transfer has been widely applied in the United States for several decades. Initially it was used to value recreation at public water resource projects and on public lands. Now it is used extensively for benefit-cost analysis of new environmental regulations, and for monetizing natural resource damages and small oil spills. Groups that use benefit transfer, once primarily federal agencies, have grown to now include state agencies, consulting firms, and non-governmental organizations. As the underlying body of information continues to expand through primary research, new benefit transfer methods are developed to take advantage of this growth in information. Benefit transfer has become an important valuation tool in the United States where applications are expected to continue in the future.

Keywords Agencies • Non-market valuation • Recreation valuation • Ecosystem services • Meta-analysis

3.1 Introduction

U.S. federal and state agencies have used benefit transfers, in one form or another, for decades. Initially, agencies used point estimate transfers for recreation valuation in major water-related development projects. As the economics profession has broadened the type of economic benefits measured to include passive use (or nonuse) values and human health, benefit transfer has expanded to these types of benefits as well. The use of benefit transfer now spans a wide variety of economic analyses,

J.B. Loomis (✉)
Department of Agricultural and Resource Economics, Colorado State University, Fort Collins, CO 80523-1172, USA
e-mail: John.Loomis@colostate.edu

R.J. Johnston et al. (eds.), *Benefit Transfer of Environmental and Resource Values*, The Economics of Non-Market Goods and Resources 14,
DOI 10.1007/978-94-017-9930-0_3

including but not limited to (a) recreation; (b) natural resource damage assessment; (c) regulatory analysis of environmental standards; and (d) ecosystem services.

This chapter surveys these various uses and types of benefit transfers conducted in the United States. However, this chapter is not an exhaustive discussion of all past applications. In part this is due to space constraints given the plethora of past applications, but also because many benefit transfers lack sufficient documentation.

3.2 Benefit Transfer of Recreation Use Values

As early as the 1960s the U.S. Army Corps of Engineers and the U.S. Bureau of Reclamation relied upon "Unit Day Values" (UDV) of recreation developed and authorized by the U.S. Water Resources Council. These values were used to estimate the benefits of recreation resulting from new reservoirs that were being built during the 1960s. The UDVs were periodically updated as information from primary recreation studies (e.g., travel cost method and contingent valuation method) increased. The last significant update (other than for inflation) was performed in 1979 (U.S. Water Resources Council 1979), and reissued in 1983 (U.S. Water Resources Council 1983). The U.S. Department of Agriculture's Natural Resources Conservation Service (USDA/NRCS) makes available to other federal and state agencies the latest (inflation-updated) UDV estimates at its website (USDA/NRCS n.d.). This site also contains databases of existing travel cost method and contingent valuation method studies that can be used for benefit transfer. Their intent with this portal is to make publicly available information on existing studies so that a "most similar" study value can be found or average value transfer can be performed.

The U.S. Forest Service also has a long tradition of using the existing valuation literature to develop "unit day values," which the agency calls Resource Planning Act (RPA) values for recreation, including hunting, fishing and wildlife viewing. These recreation values and their underlying databases were updated and expanded on over the past few decades (Sorg and Loomis 1984; Walsh et al. 1992; Rosenberger and Loomis 2001; Loomis 2005; Rosenberger 2011). RPA values have been used by the agency in National Forest Plans and Environmental Impact Statements. The most recent documentation and value estimates have been made publicly available (Recreation Use Values Database 2011).

3.3 Benefit Transfer for Benefit Cost Analyses of Environmental Regulations

A U.S. federal agency that has used benefit transfer for decades is the U.S. Environmental Protection Agency (EPA). EPA performs benefit-cost analyses of its proposed air and water quality regulations, and has relied heavily upon either an average benefit transfer or most similar study benefit transfer for the last three decades (Iovanna and Griffiths 2006). The types of benefits evaluated include human health, recreation, ecosystem services, and passive use values (Iovanna and Griffiths 2006). Passive use values include the existence value to current generations and bequest values to future generations for improved environmental quality.

Benefit transfer is specifically discussed as a viable valuation method in EPA's Guidelines for Economic Analysis (U.S. EPA 2010a). Benefit transfer is suggested for all of the types of benefits in EPA's economic analyses, including the estimation of the value of a statistical life (VSL). Many of EPA's benefit transfers have focused on the benefits of improved water quality through regulation of emissions from pulp and paper mills, toxins, storm water runoff, confined animal feeding operations, and food processing facilities. In economic analyses of the food processing and cooling water intake regulations, ratios of use values to passive use values were utilized (Iovanna and Griffiths 2006; Johnston and Rosenberger 2010; Griffiths et al. 2012). See Griffiths et al. (2012) for a summary of benefit transfer applications by EPA to water quality regulations.

EPA is beginning to rely upon meta-analysis (van Houtven et al. 2007), given that the number of primary valuation studies in the non-market valuation literature has grown and benefit transfer methods have expanded to include meta-regression analysis (Johnston and Rosenberger 2010). One of the more high profile uses of benefit transfer has been to evaluate the benefits of improving water quality from reducing nutrients, and hence eutrophication, algal blooms and related limitations on human uses. EPA used a meta-analysis benefit transfer in its analysis of improved water quality in Florida (U.S. EPA 2010b).

State agencies also have used benefit transfers for a variety of purposes. One that is particularly noteworthy is the assessment of benefits derived through reductions of nutrients in surface waters. States are required by EPA under the Clean Water Act to set specific numerical criteria for the maximum amount of nutrient loadings (e.g., phosphorous and nitrogen) in surface and ground water. The benefits of reducing nutrient levels include improved water-based recreation due to clear water and less algae. Applications of benefit transfers for these assessment purposes are on the rise with state agencies. State agencies often conduct sensitivity analyses on the levels of benefits and costs due to different levels of stringency for alternative standards. While some states (e.g., Utah) have chosen to conduct primary studies on recreation and total economic values, several other states (e.g., Colorado) have chosen benefit transfer.

The Colorado Department of Public Health and Environment (2011) specifically mentions and references benefit transfer in its requests for proposals to conduct valuation work. For example, in one contract the consulting firm identified primary studies to be used in the benefit transfer of improving surface water quality in Colorado (Harvey Economics 2011). Montana also used a simple benefit transfer approach by scaling down national estimates of nutrient-water quality linkages based on Montana's population and land area (Mathieus et al. 2010).

3.4 Benefit Transfer for Natural Resource Damage Assessment

One area of benefit transfer that was sanctioned for use by federal agencies in the mid-1980s was in connection with Natural Resource Damage Assessment (NRDA) of toxins, heavy metals, etc. from hazardous waste sites and mines. Called Type A assessments, these were software programs developed for assessments of the Great Lakes, and for damages such as oil spills to coastal and marine environments (see the Code of Federal Regulations, Title 43, Part 11, Subpart D Natural Resource Damage Assessments 2012). This model used benefit transfer to recommend a value of $11 per beach recreation trip (Chapman et al. 1998). Various state agencies also rely on benefit transfer and NRDA Type A models (see Ando and Khanna (2004) for a summary).

A high profile court case that estimated damages from a moderately sized oil spill in the Southern California area used benefit transfer to estimate lost recreation values due to beach closures. In particular, a most-similar study point estimate of beach recreation in Florida was transferred to Southern California after adjusting for inflation (Chapman et al. 1998). The losses for sport fishing and boating also were estimated by benefit transfer. Of course, there was debate about the magnitude of these economic values between the oil company's economist and the government's economists. Values per beach trip and surfing trip in the range of $11–$19 were the primary estimates brought forward. In the jury verdict the consumer surplus of $13 per day was used for assessing recreation damages, which resulted in a jury award of $12.75 million in damages (Chapman et al. 1998).

3.5 Use of Benefit Transfer by U.S. Non-Governmental Organizations

Over the last two decades, conservation and environmental organizations have gone from being hostile toward the application of economics to public lands issues to embracing it. For example, since the mid-2000s the Wilderness Society has had an economics program staffed by several Ph.D. economists who rely on benefit transfers for most of their economic analyses. One example is their study comparing the economic value of timber versus non-timber uses of national forests in California. This study relied on most-similar study benefit transfer for recreation, ecosystem services, and passive use values of California's national forests (The Wilderness Society 2002).

Another non-governmental organization that has been proactive in using non-market valuation when commenting on environmental impact statements is Defenders of Wildlife. These supporters led the development of the Benefit

Transfer Toolkit (Loomis et al. 2008). This publicly available toolkit provides average values, databases and meta-analyses for fishing, hunting, wildlife-viewing, endangered species and open space. Defenders of Wildlife also commissioned benefit transfer studies as part of their comments on the environmental impact statement regarding expansion of sea otter habitat (Loomis 2006).

3.6 Design of Ex Ante Studies to Facilitate Future Benefit Transfers

As state and federal agencies have applied benefit transfer, they have become aware of the difficulty of finding studies that meet the ideal technical criteria of Boyle and Bergstrom (1992), which include identical commodities being valued, identical characteristics of affected populations, and the same assignment of property rights that lead to theoretically appropriate welfare measures. Often no primary studies are found that value the same resource, or if such studies do exist, then they are not in similar geographic locations. For example, in the previously discussed California beach oil spill damage assessment, the most-similar beach recreation valuation study was in Florida. As such, agencies have begun to commission studies with the specific intent of using them for benefit transfer in the future. Two examples are discussed below.

As part of an oil spill damage settlement, the State of California funded a prospective study of what the damages from an oil spill would be on three types of beaches commonly found along the California coast. Carson et al. (2004) conducted a contingent valuation survey of California households' willingness to pay (WTP) to avoid oil spills on sandy beaches, rocky shorelines, and saltwater marshes. In addition to the three beach types, typical coastal wildlife also were described to increase the generalizability of the survey's results. The study was designed in large part to provide defensible benefits that could be used (i.e., transferred) to estimate the damages from future oil spills along the California coast.

The U.S. National Park Service commissioned a series of studies of representative types of park units that might be exposed to oil spills. Padre Island National Seashore in Texas was chosen as a recreational beach that could potentially be closed due to an oil spill (Parsons et al. 2009). As the authors note, this beach was chosen so that "the results should be useful in damage assessment and benefit cost analyses applied to the Texas Gulf Coast, and through transfer to other coastal areas" (p. 214). The Padre Island study was conducted using a linked site choice and trip frequency model of day trips. Another example is the primary valuation study using a contingent valuation survey of visitors to Fort Sumter in South Carolina (Leggett et al. 2003) as representative of southeastern U.S. historical sites.

3.7 Use of Benefit Transfer for Valuation of Ecosystem Services

One of the most visible uses of benefit transfer has been for valuation of ecosystem services. The term "ecosystem services" is a shorthand expression for the beneficial goods and services that nature provides to humans. These beneficial goods and services include clean water, pollination, wildlife viewing opportunities, etc. (see Brown et al. 2007; Costanza et al. 1997 for more discussion). Likely the most extraordinary, and controversial, application of benefit transfer was Costanza et al.'s (1997) effort to value the world's ecosystem services. This ambitious effort was undertaken with reliance on transferring existing values per unit from other (often site-specific) valuation studies. Depending on one's perspective, this study pushed (or surpassed) the theoretical and practical boundaries of benefit transfers.

However, most benefit transfers to value ecosystem services have been more site-or area-specific. An example of one such benefit transfer was on Mount Hood National Forest outside of Portland, Oregon (Ervin et al. 2012). The quantity and value of three ecosystem goods–timber, water, and hydropower–flowing from this national forest were estimated. The main ecosystem service valued, however, was the recreation provided. Location-specific market values were used for measuring timber sold and hydropower produced on the forest, thus there was no need for benefit transfer. However, water quantity and recreation were valued using the U.S. Forest Service's administratively approved Resource Planning Act values (i.e., a point estimate benefit transfer). The analysis showed that individually each of the non-timber ecosystem benefits was two-to-three times the size of timber benefits, and collectively they were nearly an order of magnitude larger than timber benefits alone. The impact of this simple benefit transfer was significant in that it changed the perceived role of the Mount Hood National Forest in the local economy. These results led to more collaboration among stakeholder groups, in turn leading the U.S. Forest Service to develop a new strategic plan for managing the forest. As Ervin et al. (2012) note, although their benefit transfer was simplistic and did not value all ecosystem services coming from the Mount Hood National Forest, enough important ones were measured to effect change in how the forest was perceived and managed.

Broad access to the valuation literature was an initial motivation to the development of valuation databases (e.g., Environmental Valuation Reference Inventory (n.d.) and Recreation Use Values Database (2011)). Their value rests largely in gathering information into a single place, thus reducing the costs of searching, reviewing, and screening a broader body of literature. These databases range from bibliographies to providing key details of primary studies needed for simple (i.e., point estimate) or complex (i.e., meta-regression analysis) transfer applications. For listings of many different valuation databases, see McComb et al. (2006), and ARIES (n.d.).

The latest approach to conducting benefit transfers of ecosystem services has been to develop software packages that provide point estimates (e.g., values per visitor day, per acre, per household) for different ecosystem services or habitat types. One of the best known software packages is InVEST: Integrated Valuation of Environmental Services and Tradeoffs (Natural Capital Project n.d.). It contains values per acre derived from benefit transfer linked to land cover types, and then allocates these values across a landscape using GIS that also accounts for their proximity to human population centers. InVEST serves as an example of the type of valuation system that may be popular with land managers, especially those that seek "turn-key" valuation of ecosystem services. For example, the U.S. Department of Defense has commissioned a study to apply InVEST as a means to measure the economic value of ecosystem services on military lands. Much like the testing that has taken place to measure benefit transfer errors (see Chap. 14 of this book), there is a need to test the relative error by using secondary data approaches like InVEST against primary research applications. This specific line of research has yet to begin, but it does offer an important opportunity to learn more about the tradeoffs between ecosystem valuation transfer and primary valuation.

Some software packages not only provide point estimates, but also have programmed meta-analyses that are made amenable to benefit transfer. For example, the Defenders of Wildlife's Benefit Transfer Toolkit has developed spreadsheets for a variety of meta-analysis equations. Some of these equations were derived from meta-analyses previously published (e.g. Woodward and Wui 2001 for wetlands) or available in the grey literature (e.g. Ph.D. dissertations or agency reports), whereas others were estimated specifically for the Benefit Transfer Toolkit (Loomis et al. 2008).

3.8 Conclusions

Benefit transfer applications have grown over the last three decades as the demand for non-market values in public land management and regulatory impact analyses have increased. Users of benefit transfers have broadened from federal agencies to include state agencies, consulting firms, and non-governmental organizations. Correspondingly, there has been a growth in the number of primary studies for which benefit transfer estimates can be drawn. Further, the methods of benefit transfer have evolved from simple point estimate transfers to include meta-regression analysis transfer functions that draw on a set of primary valuation studies. Benefit transfer has become accepted as a viable option when there is insufficient time and money to perform an original study. In fact, this may ultimately be a drawback of benefit transfer in the long run. As policy makers and managers (and their cost conscious consultants) become aware of the option of benefit transfer, they may over-emphasize its use, leading to the loss of additional original valuation studies–the very foundation that makes benefit transfers possible. Further, increases in emphasis on benefit transfers may result in inappropriate applications that may not be credible, but rather may result in some "incredible" values. As researchers and practitioners of benefit transfer, it is important to convey not only its advantages, but also its limitations, to policy makers.

References

ARIES. (n.d.). *Artificial Intelligence For Ecosystem Services*. Accessed October 31, 2012, from http://www.ariesonline.org/.

Ando, A., & Khanna, M. (2004). Natural resource damage assessment methods: Lessons in simplicity from state trustees. *Contemporary Economic Policy, 22*, 504–519.

Boyle, K., & Bergstrom, J. (1992). Benefit transfer studies: Myths, pragmatism, and idealism. *Water Resources Research, 28*, 657–663.

Brown, T., Bergstrom, J., & Loomis, J. (2007). Defining, valuing, and providing ecosystem goods and services. *Natural Resources Journal, 47*, 329–376.

Carson, R., Conaway, M., Hanemann, M., Drosnick, J., Mitchell, R., & Pressler, S. (2004). *Valuing oil spill prevention: A case study of California's central coast*. Boston, MA: Kluwer Academic Publishers.

Chapman, D., Hanemann, M. & Ruud, P. (1998). The American trader oil spill: A view from the beaches. *Association of Environmental and Resource Economist (AERE) Newsletter, 18*, 12–25.

Colorado Department of Public Health and Environment. (2011). *Request for proposals: Cost/benefit study of the impacts of potential nutrient controls for Colorado point source dischargers*. Denver, CO: Colorado Department of Public Health and Environment. Retrieved October 31, 2012, from http://www.cwrpda.com/Funding%20Programs/CO%20Nutrient%20Cost-Benefit%20Study%20RFP%20Final%203-3-11.pdf.

Costanza, R., d'Arge, R., de Groot, R., Farber, S., Grasso, M., Hannon, B., et al. (1997). The value of the world's ecosystem services and natural capital. *Nature, 387*, 253–260.

Environmental Values Reference Inventory (n.d.). Accessed October 31, 2012, from https://www.evri.ca/Global/Splash.aspx.

Ervin, D., Larsen, G. & Shinn, C. (2012). Simple ecosystem service valuation can impact national forest management. *Association of Environmental and Resource Economist (AERE) Newsletter, 32*, 17–22.

Griffiths, C., Klemick, H., Massey, M., Moore, C., Newbold, S., Simpson, D., et al. (2012). U.S. Environmental Protection Agency valuation of surface water quality improvements. *Review of Environmental Economics and Policy, 6*, 130–146.

Harvey Economics. (2011). Draft memorandum: Task 3.2.1 data compilation and literature acquisition to support the analysis of public health and environmental benefits. Denver, CO: Harvey Economics. 16p. Retrieved October 31, 2012, from http://projects.ch2m.com/CWQFwebsite/Workgroups/Content/nutrient_criteria/Meetings/CostBenefitStudy/Appendix%20D/D2%20-%20HE%20Lit%20Rev%20Memo.pdf.

Iovanna, R., & Griffiths, C. (2006). Clean water, ecological benefits, and benefit transfer: A work in progress at U.S. EPA. *Ecological Economics, 60*, 473–482.

Johnston, R., & Rosenberger, R. (2010). Methods, trends and controversies in contemporary benefit transfer. *Journal of Economic Surveys, 24*, 479–510.

Leggett, C., Kleckner, N., Boyle, K., Duffield, J., & Mitchell, R. (2003). Social desirability bias in contingent valuation surveys administered through in-person interviews. *Land Economics, 79*, 561–575.

Loomis, J. (2005). *Updated outdoor recreation use values on national forests and other public lands*. General Technical Report PNW-GTR-658. Portland, OR: U.S. Department of

Agriculture, Forest Service, Pacific Northwest Research Station. 26p. Retrieved October 31, 2012, from http://www.fs.fed.us/pnw/pubs/pnw_gtr658.pdf.

Loomis, J. (2006). Estimating recreation and existence values of sea otter expansion in California using benefit transfer. *Coastal Management, 34*, 387–404.

Loomis, J., Kroeger, T., Richardson, L., & Casey, F. (2008). A benefit transfer toolkit for fish, wildlife, wetlands and open space. *Western Economics Forum, 7*, 33–43.

Mathieus, G., Suplee, M., & Blend, J. (2010). *Final report to the environmental quality council on progress toward numeric nutrient standards for Montana's surface water.* Helena, MT: Montana Department of Environmental Quality.

McComb, G., Lantz, V., Nash, K., & Rittmaster, R. (2006). International valuation databases: Overview, methods and operational issues. *Ecological Economics, 60*, 461–472.

Natural Capital Project (n.d.). *InVEST: Integrated Valuation of Environmental Services and Tradeoffs*. Stanford, CA: Stanford University. Accessed October 31, 2012, from http://www.naturalcapitalproject.org/InVEST.html#How.

Natural Resource Damage Assessments (2012). 43 C.F.R. pt. 11. Accessed October 31, 2012, from http://ecfr.gpoaccess.gov/cgi/t/text/text-idx?c=ecfr&tpl=/ecfrbrowse/Title43/43cfr11_main_02.tpl).

Parsons, G., Kang, A., Leggett, C., & Boyle, K. (2009). Valuing beach closures on the Padre Island National Seashore. *Marine Resource Economics, 24*, 213–235.

Recreation Use Values Database (2011). Accessed October 31, 2012, from http://recvaluation.forestry.oregonstate.edu.

Rosenberger, R. (2011). *Overview of recreation use values database*. Corvallis, OR: Oregon State University. Retrieved October 31, 2012, from http://recvaluation.forestry.oregonstate.edu.

Rosenberger, R. & Loomis, J. (2001). *Benefit transfer of outdoor recreation use values: A technical document supporting the Forest Service strategic plan (2000 revision).* General Technical Report RMRS-GTR-72. Fort Collins, CO: U.S. Department of Agriculture, U.S. Forest Service, Rocky Mountain Research Station. 59p. Retrieved October 31, 2012, from http://www.fs.fed.us/rm/pubs/rmrs_gtr072.pdf.

Sorg, C. & Loomis, J. (1984). *Empirical estimates of amenity forest values: A comparative review*. General Technical Report RM-107. Fort Collins, CO: U.S. Department of Agriculture, U.S. Forest Service, Rocky Mountain Forest and Range Experiment Station. 23p.

The Wilderess Society (2002). *California's national forests and the California economy*. 6p. Retrieved October 31, 2012, from http://www.sierraforestlegacy.org/Resources/Conservation/FireForestEcology/ForestEconomics/Economics-WildernessSociety02.pdf.

USDA/NRCS (n.d.) *Tools*. Washington, DC: U.S. Department of Agriculture, Natural Resources Conservation Service. Rerieved October 31, 2012 from http://www.nrcs.usda.gov/wps/portal/nrcs/main/national/technical/econ/tools#Recreation.

U.S. EPA. (2010a). *Economic analysis of final water quality standards for nutrients for lakes and flowing waters in Florida.* Washington, DC: U.S. Environmental Protection Agency, Office of Water, Office of Science and Technology. 194 p. Retrieved October 31, 2012 from http://water.epa.gov/lawsregs/rulesregs/upload/florida_econ.pdf.

U.S. EPA. (2010b). *Guidelines for preparing economic analyses*. Washington, DC: U.S. Environmental Protection Agency, Office of Policy, National Center for Environmental Economics. 252p. Retrieved October 31, 2012, from http://yosemite.epa.gov/ee/epa/eerm.nsf/vwAN/EE-0568-50.pdf/%24file/EE-0568-50.pdf.

U.S. Water Resources Council. (1979). Procedures for evaluation of national economic development (NED): Benefits and costs of water resources planning (level c) final rule. *Federal Register, 44*, 72892–72977.

U.S. Water Resources Council. (1983). *Economic and environmental principles and guidelines for water and related land resources implementation studies.* Washington, DC: U.S. Government Printing Office. Retrieved October 31, 2012, from ftp://ftp-fc.sc.egov.usda.gov/Economics/priceindexes/Data/PrinciplesandGuidelinesLocalSite.pdf.

van Houtven, G., Powers, J., & Pattanayak, S. (2007). Valuing water quality improvements in the United States using meta-analysis: Is the glass half-full or half-empty for national policy analysis? *Resource and Energy Economics, 29*, 206–228.

Walsh, R., Johnson, D., & McKean, J. (1992). Benefit transfer of outdoor recreation demand studies: 1968–1988. *Water Resources Research, 28*, 707–713.

Woodward, R., & Wui, Y. (2001). The economic value of wetland services: A meta-analysis. *Ecological Economics, 37*, 257–270.

Chapter 4
The Use and Development of Benefit Transfer in Europe

Roy Brouwer and Ståle Navrud

Abstract This chapter provides an overview of benefit transfer in Europe. It does so by linking demand for transfer values to important pieces of European regulation and legislation. We also discuss national and European projects that have further developed benefit transfer guidelines and applied the resulting benefit estimates to specific environmental issues. The goal is to provide a general perspective on the use and continual improvement of benefit transfer within European policy making, focusing on applications within the last two decades.

Keywords Benefit transfer · European Union · Value function transfer · Air pollution · Noise · Forest non-market benefits · Water resources · Ecosystem services and biodiversity

4.1 Introduction

Benefit transfer, or more generally "value transfer" as both a benefit and cost estimate can be transferred in space and time (Brouwer 2000; Navrud and Ready 2007), has been used and researched extensively in Europe. One of the longest-running and best-known examples is the External Costs of Energy (ExternE) project funded by the European Commission from the early 1990s until 2005, with contributions from numerous scientists from European universities and research

R. Brouwer (✉)
Department of Environmental Economics, Institute for Environmental Studies, Vrije Universiteit, Amsterdam, The Netherlands
e-mail: r.brouwer@vu.nl

S. Navrud
School of Economics and Business, Norwegian University of Life Sciences (NMBU), Ås, Norway
e-mail: stale.navrud@nmbu.no

R.J. Johnston et al. (eds.), *Benefit Transfer of Environmental and Resource Values*, The Economics of Non-Market Goods and Resources 14,
DOI 10.1007/978-94-017-9930-0_4

institutes (www.externe.info). The methodology for calculating the external environmental costs of different types of non-renewable (oil, coal, gas, nuclear) and renewable (wind, hydro, biomass) energy sources and fuel cycles was developed in a series of projects based on the damage function approach. This methodology was later incorporated in integrated assessment models, including web-based applications like the EcoSense model. Monetary valuation was an integral part of the methodology and focused on health endpoints (morbidity and mortality) due to air pollution, amenity losses from noise, corrosion of materials in buildings (including cultural heritage), visibility, transmission lines and crop losses. The available valuation results were summarized and, based on expert judgment, synthesized into what were considered reliable estimates, appropriate for policy use in cost-benefit analysis (Bickel and Friedrich 2005).

Despite the inaccuracies involved, benefit transfer remains one of the most attractive valuation methods in Europe, particularly for the estimation of market and non-market values for use within cost-benefit analysis (CBA) of environmental policy. Although it is generally acknowledged that the transfer of a constant unit value may lead to large errors in some cases, and more sophisticated adjustment procedures than these "unadjusted value transfers" have been proposed, unit value transfer remains the most widely applied approach in Europe. In fact, some countries have developed lists of "indicator values" for different ecosystem services, i.e. constant unit values (e.g. per hectare). These are updated to account for price-level changes but without consideration of important spatial characteristics of the ecosystem services or the population of beneficiaries. This is in contrast to academic efforts to develop spatially sensitive values for ecosystem services using geographical information systems (GIS) (e.g., Brander et al. 2012). The rapid emergence of numerous projects in the context of The Economics of Ecosystems and Biodiversity (TEEB), which aims to demonstrate to national and local policy makers the economic value of the benefits of ecosystems through the concept of ecosystem services, has further increased the demand for transfer values, especially in European member states (Brouwer et al. 2013).

Although unit value and other rudimentary benefit transfer methods remain commonplace, there have been multiple high-profile, collaborative efforts to develop more valid and flexible mechanisms to support benefit transfer for European policy analysis. Unlike most efforts to develop benefit transfer methods in the U.S. and Australia, many of the efforts in Europe have involved major international, inter-agency collaborations. These efforts have sought to bridge the gap between the sophisticated methods proposed in the academic literature and the data and expertise available for applied policy analysis. Although these efforts have advanced benefit transfer methods available for application in Europe, they still face challenges. These include a shortage of available valuation studies (particularly for some areas and types of environmental changes) and the assumptions required to generate broadly applicable benefit functions. Other challenges include the need to regularly update benefit functions to account for the temporal instability of values over longer periods of time.

This chapter provides a more or less chronological overview of benefit transfer in Europe. It does so by linking demand for transfer values to important pieces of European regulation and legislation. We also discuss national and European projects that have further developed benefit transfer guidelines and applied the resulting benefit estimates to specific environmental issues. We do not pretend that the overview is complete and covers all efforts. The goal is to provide a general perspective on the use and continual improvement of benefit transfer within European policy making, focusing on applications within the last two decades.

4.2 Air Pollution and Mortality Risk Valuation in Europe

One of the major areas in which benefit transfer has been used is in the valuation of health impacts. Value of Statistical Life (VSL) unit values were established and used in the early 1990s as part of the European Commission ExternE project (www.externe.info) introduced above. Based on a review of existing valuation studies, mainly hedonic wage (HW) studies from the U.S. and expert assessment, external health damage costs (the basis of VSL estimates) were estimated from air pollution due to the combustion of fossil fuels. Many European countries had already advocated VSL estimates in general CBA guidelines, and CBA guidelines for transportation projects in particular. However, most of these VSL estimates were based on the human capital method (HCM), which accounts for productivity losses only. In contrast, HW studies and stated preference (SP) studies of mortality risk provide estimates of the total loss in welfare (Jones-Lee et al. 1985).

In response to this situation, the Directorate-General (DG) for the Environment of the European Commission organized an expert workshop in 2000 in order to establish a unit value for VSL to be used in CBA of new EU directives and programs with impacts on environmentally related mortality risks, such as the CBA of the Clean Air For Europe (CAFE) program (Holland et al. 2005). An expert assessment based mostly on valuation studies of transport-related mortality risks led to adjustments of these estimates. These adjustments accounted primarily for differences in age at death in car accidents versus air pollution (i.e., people on average lose less than one year of their life expectancy due to air pollution as opposed to more than thirty years in traffic accidents).

The identified lack of SP studies for environmentally related mortality risk motivated the EU-project NewExt within ExternE. Here, identical SP studies, based on state-of-the art contingent valuation (CV) studies of mortality risks, conducted initially in North America (Krupnick et al. 2002), were performed in three European countries (France, Italy and the U.K.). The resulting VSL estimates supported an expert VSL estimate of about one million euro, which was about one-third of the estimate used in the initial ExternE projects and about twice the estimates based on the HCM. The initial ExternE estimates were based mainly on HW studies focusing on workers' acceptance of compensation for mortality risks rather than willingness to pay (WTP) of the general population for risk reductions.

During the last decade several subsequent SP studies have been carried out in Europe in this field, and in 2011 the OECD initiated a global meta-analysis of SP studies of mortality risks related to the environment, transport and health (Lindhjem et al. 2011; OECD 2012). This project suggested the use of meta-regression functions for value transfer of VSL estimates. Contrary to Dekker et al. (2011), who found limited overlap in the set of context-specific predictive VSL distributions from road safety, air pollution and general mortality risk, these latter benefit transfer procedures suggested simple unit value transfer with adjustments for differences in purchasing power parity (PPP) and adjusted gross domestic product (GDP) per capita across countries. These adjusted values have recently been used, together with World Health Organization (WHO) estimates of the number of premature deaths due to air pollution, to assess the social costs of air pollution from road traffic in European countries as well as other OECD countries such as China and India (OECD 2014).

4.3 Transportation Noise Valuation in Europe

Another area in which benefit transfer has played a significant role in policy analysis is the valuation of transportation noise impacts. The European Commission's DG Environment initiated work in 2001 to provide a state-of-the-art review of all valuation studies of transportation noise in order to establish unit values per A-weighted decibels (dBA) for amenity loss due to noise from road traffic, railways and aircrafts. The resulting review (Navrud 2002) identified many hedonic pricing (HP) studies, mainly from the U.S., carried out in the 1970s and 1980s, but also some conducted in Europe in the early 1990s. In addition, the review found a few European SP studies. Navrud (2002) provided a preliminary unit value for WTP per decibel per household per year based on the SP studies of amenity loss from road traffic noise. SP studies were preferred to HP studies as the latter provide estimates of loss in house prices from noise, which capture welfare losses from *all* disamenities related to road traffic, measured through noise as an indicator of all these disamenities. In contrast, SP studies could target WTP for noise reductions alone.

For aircraft and railroad noise, there were not enough SP studies to determine similar unit values that could be recommended for use in CBA. The unit value for road traffic noise was adopted by the DG Environment for use in CBA, and also provided the basis for unit value estimates in national CBA guidelines for road transportation projects (e.g., in Norway). These preliminary unit values were later updated based on new SP studies in other European countries financed by national authorities, and a six-country study using the same CV survey to estimate WTP for avoiding different levels of road traffic and railroad noise annoyance as part of the EU-project HEATCO (Developing Harmonized European Approaches for

Transport Costing and Project Assessment) provided another update. Detailed results for the Norwegian study are presented in Navrud (2010), while Navrud et al. (2005) present a summary of the results for all six countries.

4.4 The Economic Benefits of Natura 2000 in Europe

Natura 2000 is the most important European legislation related to nature and biodiversity protection. It is a network of nature protection areas designed to conserve Europe's most valuable and threatened species and habitats. It comprises 26,000 sites, including Special Areas of Conservation designated by European member states under the Habitats Directive and Special Protection Areas designated under the Birds Directive. Together these cover almost 20 % of the EU territory. Natura 2000 also includes an increasing number of marine protected areas. It is not a system of nature reserves in which all human activities are excluded. Most of the land continues to be privately owned and the emphasis is on ensuring that future management is sustainable, both ecologically and economically. In addition to its biodiversity benefits, the Natura 2000 network provides a range of co-benefits to society and the economy through the flow of ecosystem services associated with protected areas. This includes provisioning services such as timber, fish and crops, regulating services such as water purification, and cultural services such as recreation.

In order to help policy makers understand the benefits of protected areas and the important role of ecosystem services these protected areas provide, ten Brink et al. (2011) applied unit values from the existing literature to estimate the total economic value of implementing Natura 2000 in the whole of Europe. Their assessment of the network's economic value from the flow of terrestrial ecosystem services in Europe amounts to 200–300 billion euros annually. This value was derived by scaling up from a limited number (34 values from 20 studies) and limited geographic focus of site-based assessments of the value of Natura 2000, with most valuations coming from the EU-15, in particular the United Kingdom, Netherlands, and Belgium. The study recommends that more values from Natura 2000 case studies be developed under a comparable valuation protocol. Despite the potentially serious concerns with scaling up ecosystem service benefit estimates in this way (see Chaps. 2 and 12), this example shows yet another way in which benefit transfer has been used to help influence European policy decisions.

4.5 The UK National Ecosystem Assessment

One of the most detailed examples of the application of benefit transfer at a national level is the UK National Ecosystem Assessment (UK NEA) (Watson et al. 2011). The UK NEA was funded over several years by the Department for Environment,

Food and Rural Affairs (Defra), the Natural Environment Research Council (NERC), the Economic and Social Research Council (ESRC), Northern Ireland Environment Agency (NIEA), the Scottish Government, the Countryside Council for Wales (CCW), and the Welsh Assembly Government (WAG), and resulted in one of the most extensive benefit transfer valuation exercises of natural capital and ecosystem services in Europe in recent years.

As part of the UK NEA, a wide variety of market and non-market valuation methods were applied to estimate the economic value of the identified flows of ecosystem services. The study focused both on past trends in ecosystem services provision and associated economic values and future flows of ecosystem services based on policy scenario simulations. The authors of the study acknowledge that there are limits to the ability of economists to capture all values associated with ecosystem services and argue that this applies in particular to certain shared social values, especially those which are not evident in observable behavior, such as the spiritual value of the environment. Moreover, the ability to derive robust monetary estimates for the nonuse value of biodiversity may be debatable. Nonetheless, the study represents one of the most ambitious efforts to quantify ecosystem service values over a large scale, using existing studies.

The main valuation methodology is outlined in Bateman et al. (2011). Where possible, use was made of available benefit estimates; these included existing values of multi-purpose woodland, maintenance of agricultural productivity, recreation, peace and quiet, and water quality. In cases where no benefit assessments were available, avoided cost estimates were used, for instance to estimate the avoided damage costs by not allowing ecosystems to degrade or the costs of clean water supply. Market prices were adjusted for market distortions such as taxes and subsidies, while production function approaches tried to adjust for the costs of other inputs in order to isolate the marginal effect of ecosystem services as inputs into production processes. For less tangible benefits, such as biodiversity, both available use and nonuse values were used to estimate the economic value of U.K. biodiversity. In the latter case, use was also made of available information about fundraising and legacy income of environmental charities as a revealed preference proxy for nonuse values.

In some cases the results from prior meta-analyses, synthesizing the economic values of ecosystem services, were applied. Examples include meta-analyses of ecosystem service values provided by wetlands and urban green space. Where possible the valuation accounted for spatial characteristics of ecosystem services delivery (supply) and beneficiaries (demand), such as recreational values based on travel cost models. In the case of local green urban space, the geographical distribution of environmental amenity values captured in house prices was mapped for the whole of England, based on HP models. However, in many cases the authors had to rely upon a transfer of simple average point values.

In the case of water quality, for example, figures compiled by the Environment Agency were used to estimate the benefits of improvements in water quality per kilometer for the main river basins in England and Wales. Average benefit estimates were £15.6/km, £18.6/km and £34.2/km for water quality improvements from low to medium, medium to high and low to high, respectively. Similar unit values are supplied in the study by Morris and Camino (2010). An example is the value of water quality improvements provided by inland and coastal wetlands in the U.K., measured in £ per hectare per year. Aggregated across the U.K., economic values range between just under £100 million per year for timber or marine-based raw biotic materials to £430 million per year for biodiversity pollination services, £600 million per year for fish landings, £680 million per year for carbon sequestration by U.K. woodlands, to £1.3 billion per year for the economic amenity value of all wetlands in the U.K. Planned river quality improvements may generate values up to £1.1 billion per year.

4.6 Benefit Transfer Guidelines for Non-market Forest Benefits

Concerned by the lack of common protocols for primary valuation studies and benefit transfer of non-market forest benefits, a group of European scientists with diverse disciplinary backgrounds took the initiative of discussing, and eventually agreeing upon good practice protocols for revealed preference (RP) and SP methods, as well as benefit transfer procedures for use and nonuse values of non-market forest benefits. The work was performed within the framework of the Cooperation in Science and Technology (COST) action, a European Union framework program instrument supporting cooperation among scientists and practitioners across Europe, called EUROpean FORest EXternalities (EUROFOREX).

Prior to EUROFOREX, there was no equivalent in Europe to the list of unit values of the U.S. Forest Service of the Department of Agriculture for recreational values per activity day for recreational activities, which have been used in CBA of measures which improve accessibility and quality of recreational sites. The EUROFOREX project showed that for forest recreation in Europe there are now sufficient valuation studies available to establish similar preliminary unit values. The same is true for nonuse values of forests (Lindhjem 2007; Elsasser et al. 2009). However, current valuation databases, including EVRI (Environmental Value Reference Inventory—now containing 3800 studies), currently provide insufficient coverage of these studies to enable valid benefit transfer. Hence, additional information on these studies must be added to EVRI to enable benefit transfers of this type. Alternatively, a new, more detailed database of forest recreational use and nonuse values must be developed to support benefit transfer on a wider scale in Europe.

4.7 Non-market Values Related to the Water Framework Directive

The development and application of similar valuation protocols for ecosystem services was also used within the European project AquaMoney. This project was funded by the European Commission to support implementation of the European Water Framework Directive (WFD), in particular the valuation of water resources and the services they provide across European member states. The WFD is the first European Directive in the domain of water, which explicitly recognizes the role of economics in reaching environmental water quality objectives. The Directive calls for the application of economic principles (e.g., polluter pays principle), methods and tools (e.g., cost-effectiveness analysis) and for the consideration of economic instruments (e.g., water-pricing methods) for achieving good chemical and ecological water status for all European water bodies. Although water resources are often unpriced or underpriced due to their public good characteristics, they generate important socioeconomic costs and benefits. Recognition of these costs and benefits within policy analysis is required to ensure policies are developed and implemented that maximize social welfare.

The project AquaMoney developed practical guidelines for the assessment of non-market values of water resources (Brouwer et al. 2009a). It did so by focusing on some of the key water policy issues in EU member states. Case studies were grouped around some of the main water management issues in Europe, such as the ecological restoration of heavily modified water bodies in the international Danube river basin in central and eastern Europe (Austria, Hungary, Romania) (Brouwer et al. 2009b), chemical and ecological water quality improvement in northern Europe (U.K., Denmark, Norway, Belgium, Netherlands, Lithuania) (Bateman et al. 2011), and water allocation and conservation in southern Europe (Spain, Italy, Greece) (Brouwer et al. 2015). By developing standardized water quality scales or "ladders" and employing identical valuation procedures, the transferability of the estimated non-market values was tested across member states. Guidance was also offered on the appropriate specification of transferable value functions, based on theoretical considerations such as the role of income (ability to pay) and distance decay, replacing previous ad hoc approaches. Special attention was paid to spatial heterogeneity in value transfer functions in view of the fact that many valuation studies involve spatial choices among environmental improvements at different locations within a confined geographical area, such as a watershed or river basin (e.g., Schaafsma and Brouwer 2013; Schaafsma et al. 2013).

4.8 Online Benefit Transfer Tool for Ecosystem Services in Flanders

We end our overview with a discussion of a practical online benefit transfer tool developed for the Flemish Government. In 2011 the Department of Environment, Nature and Energy (LNE) of the Flemish Government launched an online benefit transfer tool to support the economic valuation of ecosystem services. The goal of this effort was to support spatial planning related to the creation, restoration and design of natural areas. The online tool was developed to estimate, among others, the economic value of cultural services such as landscape amenity, biodiversity and recreation. The tool was officially launched in December 2011 by the Flemish Minister for the Environment in Brussels during a one-day workshop in which about thirty Flemish policy advisors participated as potential users. The tool is available at http://natuurwaardeverkenner.be/nwv2/.

Among other components, the tool includes a value function for the non-market benefits associated with the conversion of agricultural land use into natural areas. This value function is based on a choice experiment conducted among a representative sample of 3000 Flemish households in 2010 (Liekens et al. 2013). Although hypothetical, the choice experiment has several advantages compared to other stated preference methods, including the fact that it allows for the inclusion of ecosystem service and site characteristics and accounts for important trade-offs and substitution effects between alternative policy scenarios. The policy scenarios in this case concerned land use change, in which existing agricultural land is converted into nature areas, such as natural grasslands, forests, wetlands and marshes. The different nature types included in the choice experiment are based on the Flemish Biological Value Map. Important spatial characteristics include area size, accessibility, species richness, adjacent land use of the area, and the distance from a household's place of residence to the location of the proposed land use change. The distance measure is included to account for distance-decay effects in demarcating the size of the economic market of beneficiaries, i.e., that values typically decline with distance from an affected area.

The value function provides a value estimate for any additional hectare of nature area or restoration of lost habitat, and is used in combination with available GIS data on population density, population characteristics, and surrounding land use. Application of the function demonstrates that the average value per hectare of land for specific nature types differs significantly depending on size, distance and other site and population characteristics. Not controlling for these influencing factors, which is common practice in many benefit transfer exercises in practice, can result in severe under- or over-estimation of the non-market values of the proposed land use changes, leading to misguided policy and decision making.

The economic valuation tool aims to support decision making in local and regional spatial planning, including the creation, restoration and compensation of nature areas. Specifically, it enables cost-benefit analysis in which the ecosystem service values generated by land use plans can be compared with the financial costs of the plans. In densely populated countries such as Belgium and the Netherlands, nature areas are under increasing pressure from urban and infrastructure development. In order to compensate for these developments, the Government of Flanders has designated almost 10 % of its total land cover as protected areas such as Special Protection Areas and Special Areas of Conservation, comprising around 105 thousand ha. However, the Flemish Decree for Nature Conservation requires that the government delineates an area of 125 thousand ha of natural area as part of the Flemish Ecological Network and an additional 150 thousand ha as part of the Integral Interrelation and Support Network. Hence, further expansion of nature areas and buffer areas is required.

To evaluate the robustness of this value function over time, the estimated transfer parameters were tested for temporal stability using a test-retest study. As part of this study, the same choice experiment used to estimate the original value function was implemented one year after the original choice experiment, using the same sample of households (Schaafsma et al. 2014). The results were then compared to those of the original choice experiment. The results of the retest study show that the estimated value function one year later is slightly different, but does not result in significantly different WTP values, suggesting that the originally estimated value function is robust over the one-year time period. The value function will need regular testing to evaluate the robustness of the results over a longer period of time, and to enable updating as necessary.

4.9 Conclusions

In Europe, work to improve benefit transfer procedures and develop benefit transfer guidelines has been initiated by both national environmental protection agencies (EPAs) like the UK Defra (Bateman et al. 2010) and the Danish EPA (Navrud 2007), national research councils, and European agencies such as the European Environment Agency (Brander et al. 2012) and DG Environment and DG Research of the European Commission (see above). Some European countries, mainly the United Kingdom and France, have focused on the development of an updated web-based database of valuation studies, the Environmental Valuation Reference Inventory (EVRI). The information on primary valuation studies provided by such databases, together with the benefit transfer guidelines and methods and guidelines for evaluating the quality of available primary valuation studies (e.g. Söderquist and Soutukorva 2006 for the Swedish EPA), are prerequisites for valid benefit transfer.

Work is currently progressing to improve the basis for benefit transfers used to inform policy in various European countries and the continent as a whole. For example, the online tool presented in the previous section is an important step forward compared to existing attempts to inform environmental policy and decisionmaking on the basis of so-called indicator values for different ecosystem services. These indicator values are included in CBA of environmental policy interventions as constants, without consideration of important spatial characteristics of the ecosystem services or population of beneficiaries. The scaling up of such constant values was considered one of the primary flaws of Costanza et al. (1997), which sought to quantify the value of the world's ecosystem services and natural capital. Revisiting the estimated economic value of wetland ecosystem services in Costanza et al. (1997), Brander et al. (2013) use a meta-regression model instead of a constant unit value per hectare for wetland ecosystem services, accounting for spatial characteristics related to the service and population of beneficiaries in GIS. Predicting the global value of one particular wetland ecosystem service (regulating services) based on this more sophisticated approach, the economic value is only 10 % of the value originally estimated by Costanza et al. (1997). This result suggests that the function approach to benefit transfer may not only produce better verifiable and validated results, but also produce more conservative and hence acceptable values for decisionmakers.

More than fifteen years later, European research projects such as AquaMoney provide improved guidelines for more reliable and valid benefits transfer, based on spatially sensitive (GIS-based) value functions. Similar functions provide the basis for the Flemish Government online tool summarized above. In these and other cases, ongoing research is seeking to replace simple unit value transfers and indicator values with more sophisticated benefit function transfers that are able to better account for differences in natural resources, populations and policy contexts (e.g., spatial characteristics, availability of substitutes, baseline environmental quality levels). At the same time, new work is seeking to identify cases in which unit value transfers provide acceptable approximations of true underlying values (e.g., Bateman et al. 2011). These and other efforts are helping to provide the cost and benefit estimates that are increasingly requested by European government agencies as a basis for policy decisions.

References

Bateman, I. J., Brouwer, R., Cranford, M., Hime, S., Ozdemiroglu, E., Phang Z., & Provins, A. (2010). *Valuing environmental impacts: Practical guidelines for the use of value transfer in policy and project appraisal.* London: Defra. https://www.gov.uk/government/publications/valuing-environmental-impacts-guidelines-for-the-use-of-value-transfer.

Bateman, I. J., Brouwer, R., Ferrini, S., Schaafsma, M., Barton, D. N., Dubgaard, A., et al. (2011). Making benefit transfers work: Deriving and testing principles for value transfers for similar and dissimilar sites using a case study of the non-market benefits of water quality improvements across Europe. *Environmental and Resource Economics, 50*, 365–387.

Bickel, P., & Friedrich, R. (2005). *ExternE—externalities of energy. Methodology 2005 update.* Brussels: Directorate-General for Research and Innovation, Sustainable Energy Systems, Report No. EUR 21951.

Brander, L., Bräuer, I., Gerdes, H., Ghermandi, A., Kuik, O., Markandya, A., et al. (2012). Using meta-analysis and GIS for value transfer and scaling up: Valuing climate induced losses of European wetlands. *Environmental and Resource Economics, 52*, 395–413.

Brander, L., Brouwer, R., & Wagtendonk, A. (2013). Economic valuation of regulating services provided by wetlands in agricultural landscapes: A meta-analysis. *Ecological Engineering, 56*, 89–96.

Brouwer, R. (2000). Environmental value transfer: State of the art and future prospects. *Ecological Economics, 32*, 137–152.

Brouwer, R., Barton, D., Bateman, I. J., Brander, L., Georgiou, S., & Martín-Ortega, J., et al. (2009a). *Economic valuation of environmental and resource costs and benefits in the water framework directive: Technical guidelines for practitioners.* AquaMoney (SSPI-022723). Deliverable D23 for DG Research of the European Commission, Institute for Environmental Studies, Vrije Universiteit Amsterdam, The Netherlands.

Brouwer, R., Bliem, M., Flachner, S., Getzner, M., Kerekes, S., Milton, S., et al. (2009b). *Ecosystem service valuation from floodplain restoration in the Danube River Basin: An international choice experiment application.* Paper presented at the Annual Conference of the Association of Environmental and Resource Economists, Amsterdam, June 2009.

Brouwer, R., Brander, L., Kuik, O., Papyrakis, E., & Bateman, I. (2013). *A synthesis of approaches to assess and value ecosystem services in the EU in the context of TEEB*. Brussels: Report for the Directorate-General for the Environment.

Brouwer, R., Martin-Ortega, J., Dekker, T., Sardonini, L., Andreu, J., Kontogianni, A., Skourtos, M., Raggi, M., Viaggi, D., Pulido-Velazquez, M., Rolfe, J., & Windle, J. (2015). Improving value transfer through socio-economic adjustments in a multi-country choice experiment of water conservation alternatives. *Australian Journal of Agricultural and Resource Economics, 59*, 1–21.

Costanza, R., d'Arge, R., de Groot, R., Farber, S., Grasso, M., Hannon, B., et al. (1997). The value of the world's ecosystem services and natural capital. *Nature, 387*, 253–260.

Dekker, T., Brouwer, R., Hofkes, M., & Moeltner, K. (2011). The effect of risk context on the statistical value of life: A Bayesian meta-model. *Environmental and Resource Economics, 49*, 597–624.

Elsasser, P., Meyerhoff, J., Montagné, C., & Stenger, A. (2009). A bibliography and database on forest benefit valuation studies from Austria, France, Germany and Switzerland: A possible base for a concerted European approach. *Journal of Forest Economics, 15*, 93–107.

Holland, M, Hunt, A., Hurley, F., Navrud, S., & Watkiss, P. (2005). *Methodology for cost-benefit analysis of CAFÉ (Clean Air for Europe).* Volume 1: *Overview and methodology*. Report to DG for the Environment. Brussels: European Commission. http://ec.europa.eu/environment/archives/cafe/pdf/cba_methodology_vol1.pdf.

Jones-Lee, M. W., Hammerton, W. M., & Philips, P. R. (1985). The value of safety: Results of a national sample survey. *Economic Journal, 95*, 49–72.

Krupnick, A., Alberini, A., Cropper, M., Simon, N., O'Brien, B., Goeree, R., & Heintzelman, M. (2002). Age, health and the willingness to pay for mortality risk reductions: A contingent valuation survey of Ontario residents. *Journal of Risk and Uncertainty, 24*, 161–186.

Liekens, I., Schaafsma, M., De Nocker, L., Broekx, S., Staes, J., Aertsens, J., & Brouwer, R. (2013). Developing a value function for nature development and land use policy in Flanders. *Land Use Policy, 30*, 549–559.

Lindhjem, H. (2007). 20 years of stated preference valuation of non-timber benefits from Fennoscandian forests: A meta-analysis. *Journal of Forest Economics, 12*, 251–277.

Lindhjem, H., Navrud, S., Braathen, N. A., & Biausque, V. (2011). Valuing lives saved from environment, transport and health policies: A meta analysis. *Risk Analysis, 31*, 1381–1407.

Morris, J., & Camino, M. (2010). *Economic assessment of freshwater, wetland and floodplain (FWF) ecosystem services*. UK National Ecosystem Assessment, Working Paper. Cranfield, UK: Cranfield University, School of Applied Sciences.

Navrud, S. (2002). *The state-of-the-art of the economic valuation of noise*. Final Report to European Commission DG for the Environment. Brussels. http://ec.europa.eu/environment/noise/pdf/020414noisereport.pdf, April 14, 2002.

Navrud, S. (2007). *Practical tools for value transfer in Denmark: Guidelines and an example*. Working Report No. 28, Miljøstyrelsen (Danish Environmental Protection Agency). At http://www2.mst.dk/common/Udgivramme/Frame.asp? http://www2.mst.dk/Udgiv/publications/2007/978-87-7052-656-2/html/default_eng.htm.

Navrud, S. (2010). Economic valuation of transportation noise in Europe. *Rivista italiana di acustica, 34*, 15–25.

Navrud, S., & Ready, R. (Eds.). (2007). *Environmental value transfer: Issues and methods*. Dordrect: Springer.

Navrud, S., Trædal, Y., Hunt, A., Longo, A., Greßmann, A., Leon, C., Espino, R., Markovits-Somogyi, R., & Meszaros, F. (2005). Economic values for key impacts valued in the Stated Preference surveys. Deliverable 4. Brussels: EU-project HEATCO (Developing Harmonized European Approaches for Transport Costing and Project Assessment). 62 pp and appendices. http://heatco.ier.uni-stuttgart.de/internal/hd4704.pdf.

OECD. (2012). *Mortality risk valuation in environment, health and transport policies*. Paris: OECD Publishing. http://dx.doi.org/10.1787/9789264130807-en.

OECD. (2014). *The cost of air pollution: Health impacts of road transport*. Paris: OECD Publishing. http://dx.doi.org/10.1787/9789264210448-en.

Schaafsma, M., & Brouwer, R. (2013). Testing geographical framing and substitution effects in spatial choice experiments. *Journal of Choice Modelling, 8*, 32–48.

Schaafsma, M., Brouwer, R., Gilbert, A., van den Bergh, J. C. J. M., & Wagtendonk, A. (2013). Estimation of distance-decay functions to account for substitution and spatial heterogeneity in stated preference research. *Land Economics, 89*, 514–537.

Schaafsma, M., Brouwer, R., Liekens, I., & De Nocker, L. (2014). Temporal stability of preferences and willingness to pay for natural areas in choice experiments: A test-retest. *Resource and Energy Economics, 38*, 243–260.

Söderquist, T., & Soutukorva, A. (2006). *An instrument for assessing the quality of environmental valuation studies*. Report. Stockholm: Swedish Environmental Protection Agency. http://www.naturvardsverket.se/Nerladdningssida/?fileType=pdf&downloadUrl=/Documents/publikationer/620-1252-5.pdf.

ten Brink, P., Badura, T., Bassi, S., Daly, E., Dickie, I., Ding, H., Gantioler, S., Gerdes, H., Kettunen, M., Lago, M., Lang, S., Markandya, A., Nunes, P. A. L. D., Pieterse, M., Rayment, M. & Tinch, R. (2011). *Estimating the overall economic value of the benefits provided by the Natura 2000 network*. Final Synthesis Report to the European Commission, DG Environment, on Contract ENV.B.2/SER/2008/0038. Brussels: Institute for European Environmental Policy / GHK / Ecologic.

Watson, R., Albon, S., Aspinall, R. et al. (2011). *U.K. National Ecosystem Assessment: Understanding nature's value to society. Synthesis of key findings*. Oxford: Information Press.

Chapter 5
Applied Benefit Transfer: An Australian and New Zealand Policy Perspective

John Rolfe, Jeffrey Bennett and Geoffrey Kerr

Abstract This chapter provides an introduction to and review of the use of benefit transfer approaches and data within Australasian policy making. The focus is on applications within the last two decades and the role of transfer methods within legal, policy and institutional structures. While there has been substantial interest in benefit transfer, the number of practical applications remains limited in both Australia and New Zealand. The limited pool of primary valuation studies and challenges in value transfer has meant that to date, understanding about the validity and reliability of benefit transfer and the development of protocols to guide its use are still limited. Nonetheless, recent major policy issues and controversies such as conservation of the Great Barrier Reef and management of water in the Murray-Darling Basin have led to an increase in applications of benefit transfer, and also to the potential for misuse. Included in this chapter is a discussion of the acceptance of benefit transfer approaches for various applications, the prevalence of benefit transfer, and the legal role of benefit transfers within Australasian policy analysis. The chapter will also highlight the potential for benefit transfer to make benefit- cost analysis more useful to policy makers and more easily evaluated within Australasian policy contexts. The need for more work to provide confidence around processes and results is assessed.

Keywords Benefit transfer · Cost-benefit analysis · Policy · Australia · New Zealand

J. Rolfe (✉)
School of Business and Law, CQUniversity, Rockhampton, QLD, Australia
e-mail: j.rolfe@cqu.edu.au

J. Bennett
Crawford School of Public Policy, Australian National University, Canberra, ACT, Australia
e-mail: Jeff.Bennett@anu.edu.au

G. Kerr
Faculty of Environment, Society and Design, Lincoln University, Christchurch, New Zealand
e-mail: Geoffrey.Kerr@lincoln.ac.nz

R.J. Johnston et al. (eds.), *Benefit Transfer of Environmental and Resource Values*,
The Economics of Non-Market Goods and Resources 14,
DOI 10.1007/978-94-017-9930-0_5

5.1 Introduction

There has long been interest in the development of non-market valuation and benefit transfer approaches in Australia and New Zealand. A number of environmental issues have become prominent in national debates in both countries since the 1960s, with subsequent interest in assessing the net benefits of protection or restoration. Turning points in environmental awareness occurred in both countries with campaigns to stop new dams being built in wilderness areas for generating hydroelectric power, including campaigns to stop the raising of Lake Manapouri in New Zealand and Lake Pedder in Tasmania. In the subsequent four decades, issues such as the loss of environmental systems to dams and mining, logging of old growth forests, water allocations to irrigation, and broad-scale tree clearing for pasture development have all been flashpoint issues in major public debates in which an assessment of the public benefits of development or protection have been required in policy analysis.

The development of non-market valuation and benefit transfer approaches has been driven in part by a range of "demand" factors. Growing public interest in environmental issues, increased requirements for, and scrutiny of, assessment and decision processes, and the proliferation of government regulations and other requirements have been key factors contributing to the demand for environmental values to be included in decisions about environmental issues. Although major debates have been played out through the political process, a general trend has emerged in government towards more systematic evaluation of changes in policy and proposed developments. This has been aimed at demonstrating competence and thoroughness in decision processes, as well as having a process to reject proposals that do not meet the public interest test. Requirements for systematic evaluation have largely focused on environmental impact assessment (EIA). However, there has been varying (and sometimes sporadic) inclusion of economic criteria in the decision-making process. Some landmark controversies, such as the proposal to make Fraser Island a national park and to mine Coronation Hill (at Kakadu in the Northern Territory) were accompanied by economic analysis to assess the public values of protection measures (Bennett 1996, 2005). The different cases and requirements for economic assessment have stimulated varying demands for environmental values to be generated through primary data collection or benefit transfer processes.

A range of supply factors has also contributed to non-market valuation and benefit transfer approaches in Australia and New Zealand. These include an active academic and research community involved in developing and applying non-market valuation techniques to provide "source" values, a pool of researchers, consultants and policy makers advancing benefit transfer approaches to meet policy needs, the development of data bases of environmental value estimates such as ENVALUE in New South Wales, and the development of case study examples and guides. The intersection of supply and demand factors has led to exploratory and development activity around the transfer of environmental values, even though rates of take-up and application at policy level remain limited.

Some of the developments in BT have been tied to advances in non-market valuation techniques, where BT outcomes have been implicit or explicit outcomes of the improvements. The development of the Choice Modeling (CM) technique in environmental application by Jeff Bennett and colleagues during the 1990s (e.g. Blamey et al. 1997) is the key area of focus in relation to BT outcomes. The initial focus of establishing choice experiments in order to value environmental changes was to provide an alternative to the contingent valuation method. However, a further outcome of CM's development was that environmental values could be disaggregated by attributes. This provided insights into how values could be transferred within the frame of a relevant issue (Bennett and Blamey 2001; Rolfe et al. 2002). Over time, some further developments of the CM technique were more explicitly focused on framing the experiments in ways that allowed subsequent value estimates to be used for wider benefit transfer applications. This included the identification of "adjustment factors" that allowed values to be transferred across contexts and variations in frames (van Bueren and Bennett 2004; Rolfe and Windle 2008). Rolfe and Bennett (2006) provide a key summary of how advances and applications in CM have provided both implicit and explicit inputs into BT.

Some developments in BT have been more independent of specific non-market valuation techniques. One area of focus has been the development of databases of values for BT use, such as the ENVALUE database in New South Wales and the New Zealand database (Kerr n.d.). A second area has been the establishment of sets of values over national or regional areas that can be extrapolated to case studies of interest. Examples of this approach include the Land and Water Audit project in Australia (van Bueren and Bennett 2004), river valuation projects in New South Wales and Victoria (Bennett et al. 2008a, b), the regional soil, water and land condition project in Queensland (Rolfe and Windle 2008), and the biosecurity (Bell et al. 2009) and stream mitigation projects (Kerr and Sharp 2006) in New Zealand. A third area has been the demonstration of integrated applications of BT in specific case study exercises. Examples include the assessment of the net values of protecting ground water reserves of the Great Artesian Basin (Rolfe 2010), environmental values associated with water availability in the Murray Darling Basin (Morrison and Hatton McDonald 2010), and values for different ecosystems in the Lake Macquarie region of New South Wales (Windle and Rolfe 2012).

The development of BT in Australia and New Zealand has not been without challenges. A major limitation is the small pool of primary studies available to provide source values. Another key problem has been the difficulty of applying different source study values without any adjustment for scale and population differences, together with limited understanding about the rationale and application of those adjustment factors. A more recent issue has been the use of BT applications by environmental interest groups, where the accuracy of value estimates may be secondary to political purposes.

5.2 The Demand for Values to Be Transferred

The use of benefit transfer in Australia and New Zealand has largely been driven by direct and indirect demand from governments for cost-benefit and other economic assessments of policy and development proposals. Economic criteria and economic evaluation processes requiring benefit transfer form a major part of at least three important policy settings:

- scrutiny of new legislation and regulation,
- environmental impact assessment, and
- government policy analysis.

Each of these important areas is considered in turn.

5.2.1 *Scrutiny of Legislation and Regulation*

The most specific area where primary studies or benefit transfer are used to provide environmental values is when new legislation or regulations need to be evaluated for possible adverse economic impacts. In Australia these are performed through Regulatory Impact Statements (RIS) at both Australian and State Government levels, while in New Zealand these are performed through Regulatory Impact Analysis requirements at a national level.

The Australian Government established the Office of Best Practice Regulation in 2006 and released the *Best Practice Regulation Handbook* to guide the assessment process (Australian Government 2010). The Office of Best Practice Regulation oversees a regulatory impact assessment process in which the risks and net economic impacts of new proposals are assessed. Parallel arrangements exist at the State level, coordinated through a Council of Australian Government (COAG) agreement. Where new legislation or regulation meets some size of potential impacts threshold, then relevant departments or Ministerial Councils (in the case of inter-governmental bodies) must prepare a Regulatory Impact Assessment. These RIS evaluations are fundamentally structured around the principles and practice of cost-benefit analysis, but many proposals are either exempted from preparing a RIS or are poorly executed in practice (Productivity Commission 2012).

The New Zealand Government requires a Regulatory Impact Analysis for policy initiatives or reviews that would involve creating, amending or repealing primary or delegated legislation where the changes are sufficient to involve a paper being submitted to Cabinet (New Zealand Treasury 2009). The New Zealand guidelines for a Regulatory Impact Analysis are broadly structured in a cost-benefit analysis framework, similar to the RIS process in Australia. For the central government agencies that may need to prepare a Regulatory Impact Statement, the Act broadly identifies that impacts should be quantified and expressed in dollar terms, and that net benefits should be assessed (New Zealand Treasury 2009, Sect. 2.7.2).

However the application of a Regulatory Impact Analysis is left very flexible, with expectations that smaller changes require less rigorous assessment, and no binding requirements for costs and benefits to be valued. Similarly, Sect. 32 of the Resource Management Act 1991 requires management agents to evaluate "whether, having regard to their efficiency and effectiveness, the policies, rules, or other methods are the most appropriate for achieving the objectives … an evaluation must take into account … the benefits and costs of policies, rules, or other methods." In practice, the lack of binding requirements under the relevant legislation means that the use of cost-benefit analysis or monetary evaluation of non-marketed environmental benefits is rarely undertaken.

5.2.2 Environmental Impact Assessment

The second key area in which economic analysis is required is where governments specify that some form of environmental impact assessment needs to be performed as a part of the evaluation and decision process for major projects where environmental losses are involved. In these cases the analysis is generated and reported by the private sector proponent.

In Australia, requirements for impact assessments are largely set by the States, with variations in the legislative framework and process across them. At the national level, the Australian Government also assesses projects that meet the criteria for major environmental impacts under the Environmental Protection and Biodiversity Conservation (EPBC) Act 1999, requiring that an EIS prepared under state legislation is also assessed under the EPBC 1999. In New Zealand the Resource Management Act 1991 specifies that an Assessment of Environmental Effects be conducted, and although there is no specific requirement to assess benefits and costs, under Sect. 7 of the Act applicants are required to "have particular regard to [inter alia]… efficient use and development of natural and physical resources, … maintenance and enhancement of amenity values, and … intrinsic values of ecosystems." While the framework exists in both countries to use cost-benefit analysis as a part of the impact assessment process, in practice there is limited assessment and use of non-market values.

In both Australia and New Zealand the thoroughness of assessment typically required increases in the scale of the potential impacts, from self assessment at the lowest level through to a major EIS or Assessment of Environmental Effects at the most complex level. While lower level assessments may require only a business justification or some prediction of economic impacts, such as employment changes, higher level assessments such as the EPBC 1999 in Australia require that a comprehensive cost-benefit analysis be performed. In New South Wales the guidelines provided by James and Gillespie (2002) identify how a cost-benefit analysis can be conducted in an Environmental Impact Assessment framework.

Associated with the variations in scale of requirements are differences in process and requirements for impact assessment across states (Thomas and Elliot 2005), and variations in the enthusiasm for the thoroughness of economic analysis. Even where there are formal requirements for cost-benefit analysis or the assessment of environmental values, these are not always enforced or are interpreted loosely by policy makers (Bennett 2005; Dobes 2008; Productivity Commission 2012). Governments in both Australia and New Zealand appear to treat non-market values as useful but not essential components of cost-benefit analysis needed to meet legislative requirements. Many jurisdictions do not identify benefit transfer as a mechanism for sourcing values, although the Queensland Government (2003) identifies this option as a means to provide values into environmental impact assessment. However' there have been primary valuation studies undertaken for use within Environmental Impact Assessments in Australia (e.g. Gillespie and Bennett 2013), and benefit transfer has been applied to evaluate resource consent implications in New Zealand (e.g. Kerr 2009 assesses a hydro-electric case).

Although the network of environmental impact assessment laws and regulations in Australia and New Zealand generates the largest requirements for environmental values to be supplied, there are few examples of rigorous primary valuation studies or benefit transfer applications. Many impact assessments are completed to meet a regulatory requirement in which economic analysis is a minor criterion and variations in results or accuracy have limited impact on outcomes. The assessments are typically compiled by consultants and reviewed by public officials, with neither group necessarily having particular expertise in specialized environmental economics. The outcome is that while requirements for economic analysis to be included in impact assessment cast a wide net, there are variations in the quantity and accuracy of environmental values needed across jurisdictions and time, and the use, application and accuracy of any benefit transfer approaches are patchy.

5.2.3 Government Policy Analysis

The third key area where economic evaluation and non-market values are required is in more general policy analysis. Cost-benefit analysis has a core role in public policy evaluation in Australia and New Zealand, and its application is mandated in some circumstances. The use and depth of cost-benefit analysis studies varies, with studies prepared to examine different policy, development, expenditure and revenue options by government. National and state governments have handbooks to guide applications (e.g. Commonwealth of Australia 2006a; New Zealand Treasury 2005). While the key focus of application has been on major capital works (New South Wales Treasury 2007), more formal evaluations have also been applied to issues such as pest management and health programs (Commonwealth of Australia 2006b).

However, the use of cost-benefit analysis has not been institutionalized in Australasia to the same extent that it has in other countries, such as the United States, and while there are many examples of rigorous work being performed, there are also many examples of analysis being ignored by government (Dobes 2008).

Many past developments, carried out in the guise of nation building, such as the Ord River Dam, would have benefited from more formal applications of cost-benefit analysis (Davidson 1965). Even in more recent years, the application and thoroughness of cost-benefit analysis by government is limited and patchy (Bennett 2005; Dobes 2008; Dobes and Bennett 2009). Reasons include treatment of cost-benefit analysis as a bureaucratic exercise, resistance at the political level, lack of acceptance by the public service, costs involved, a paucity of "plug-in" values, few standardized approaches, and limited skills and expertise.

New Zealand environmental legislation, such as the Resource Management Act 1991 and the Biosecurity Act 1993, calls for assessments of economic impacts, efficiency, and comparison of costs and benefits to help evaluate the merits of proposed policies, plans and regulations. Some, often significant, elements of these evaluations are typically non-marketed. The most prevalent application of benefit transfer in New Zealand has been the assessment of the value of a statistical life. Values originally developed for road transport purposes (Miller and Guria 1991) have been applied to evaluate fire safety, general accident prevention, family violence, aviation crashes, drowning, fire regulations (Wren and Barrell 2010 categorize these applications) and the National Environmental Standard for Air Quality (Clough et al. 2009). In practice, however, the use of cost-benefit analysis is rare.

While the use of cost-benefit analysis is formally required within governments for some processes, informal use is also important, although inconsistently applied. The latter occurs when analyses are developed, at varying levels of sophistication, in order to advise ministers and senior public servants about the pros and cons of different policy options. Little information is available to judge the extent of use of cost-benefit analysis and benefit transfer. Many large consulting firms are known to regularly use benefit transfer for government, quasi-government and private clients, but their reports are rarely published and so the work goes largely unrecognized.

There is little formal recognition or guide to the use of benefit transfer for environmental values at either formal or informal levels. For example, the New Zealand Treasury (2005) makes no mention of benefit transfer as a mechanism to source values into cost-benefit analysis. It is similar in Australia, although there is one passing reference to benefit transfer at the end of an appendix in the Cost Benefit Analysis Handbook, with little guide to application or quality (Commonwealth of Australia 2006a, p. 133).

Of course, it is often not feasible to conduct primary research for an economic evaluation. Analysts must adopt and modify benefit values found in other studies, especially research studies, rather than undertake a large amount of primary data collection and analysis. The process of benefit transfers involves the transfer of existing estimates of non-market values to the present study, which invariably differs in some features from the original studies. Ideally, a meta study would have analyzed the reasons for the differences between studies, so that the most relevant values can be selected. However, it is common practice to adopt mean estimated values from studies that are considered broadly similar. In some cases it may be appropriate to adopt a higher or lower value to reflect special local conditions.

5.3 The Supply of Environmental Values for Benefit Transfer

Interest and activity in benefit transfer has also been stimulated by a number of supply side factors. There is a long history of non-market environmental valuation applications in Australia and New Zealand, using both revealed preference and stated preference techniques, with the first contingent valuation study reported in New Zealand by Gluck (1974) and in Australia by Bennett (1982). Non-market valuation has been a key focus of activity in the environmental economics field in both countries, with a number of researchers and students specializing in the application of different techniques (Bennett 2005).

The interest in non-market valuation has generated a number of source studies, most of which have been associated with the travel cost method for recreation studies and the contingent valuation and choice modeling techniques for environmental protection studies (Bennett 2005). Acceptance of the contingent valuation technique in Australia has been limited since a controversy erupted over values estimated for protecting Coronation Hill (adjacent to Kakadu National Park) from mining (Bennett 1996, 2005). There have been a large number of studies published around some topics; for example, Rolfe and Brouwer (2013) document 154 different value estimates generated from 19 separate choice experiment case studies valuing river protection in Australia.

Other factors contributing to the supply of benefit transfer has been the development of data bases to help find appropriate source studies, as well as relevant guides to benefit transfer applications. Australian policy makers were among the early starters in the field with the establishment of the ENVALUE data base by the New South Wales Government in 1995 (available at http://www.environment.nsw.gov.au/envalueapp/). The site was regularly updated until 2004, but activity has since lapsed in favor of the Environmental Valuation Reference Inventory (EVRI) in Canada (http://www.evri.ca/). In New Zealand Geoff Kerr has maintained a database of non-market valuation studies for that country at: http://www2.lincoln.ac.nz/nonmarketvaluation/. By February 2011, that site referenced 135 studies, including 41 on recreation issues and 21 on environmental preservation or enhancement studies. New Zealand studies have also been migrated to the EVRI database, which is funded by Australia, Canada, France, New Zealand, United Kingdom and the United States of America.

By September 2012, there were 311 records in the EVRI database for Oceania, including 194 records focusing on Australia, with the balance largely from New Zealand. The Oceania records comprise about 8.5 % of the data base. Usage is limited but consistent, with approximately fifty website visits per month from Australia in the two years to March 2012 and 79 active user accounts, indicating ongoing use by academics, consultants and policy makers.

Several environmental valuation studies have been designed specifically to provide valuation functions for subsequent benefit transfer purposes. Morrison and Bennett (2004) and Bennett et al. (2008a, b) assessed river health values for representative samples of rivers in New South Wales and Victoria, respectively, with the aim of providing values for a benefit transfer function in each state. van Bueren and Bennett (2004) generated a similar outcome with their study of land and water assets in Australia, identifying different adjustments depending on whether assets were valued in a regional or national context, or by a regional or national population. Rolfe and Windle (2008) identified how values for soil, land and waterway condition varied across key regional areas of Queensland with the aim of providing source values for regional transfer purposes.

Kerr and Sharp (2008a) undertook a research program designed for the purpose of evaluating mitigation of riparian impacts of development in New Zealand's Auckland Region. The work was funded by the Auckland Regional Council and Transit NZ, with the intent of publishing a guideline for evaluating riparian mitigation throughout the Auckland region. Also in New Zealand, Bell et al. (2009) assembled a series of non-market values into a benefit transfer framework to be used to assess response strategies for new exotic pests entering the country. Separate choice modeling studies were used to estimate protection values for four key ecosystem types: high country, marine ecosystems, beech forests and freshwater streams. It is difficult, however, to apply the results in a consistent framework because of issues such as representativeness for specialized assets and the difficulty of accounting for diminishing marginal benefits of control as pest incursions consolidate.

Rolfe and Bennett (2006) consolidated a number of Australian and New Zealand choice modeling studies together with other international examples into an edited book on benefit transfer. Contributions to the volume summarized some of the major choice modeling studies on tests of benefit transfer conducted in Australia and New Zealand. One chapter in the text provided a simple guide to the use of benefit transfer aimed at practitioners, while another provided a theoretical and technical analysis of the process in an effort to provide a stronger base for benefit transfer to be performed.

5.4 Advances in Non-Market Valuation Techniques

The development of the choice modeling technique after its formative stages has been closely associated with benefit transfer applications. One key topic area has been to identify whether there are differences in the ways that populations value the same good. Morrison and Bennett (2004) explicitly tested how values for river health differed between populations inside and outside catchments in New South Wales so that adjustment factors could be identified. van Bueren and Bennett (2004) identified that values for land and water assets in Australia varied according to whether regional or national populations were assessed. Kerr and Sharp (2006, 2008a) explored how different populations in the Auckland area valued better stream protection, Zander et al. (2010) tested how different capital city populations in southern Australia valued river catchment protection in northern Australia, while Rolfe and Windle (2012a) tested for value equivalency and distance effects with protection values for the Great Barrier Reef in Australia.

A second benefit transfer issue associated with the development of the choice modeling technique has been to test how the same population valued the same good in different locations. Rolfe and Bennett (2002) explored values held by the Brisbane population for rainforest protection at different national and international locations. In a similar way, Morrison et al. (2002) and Morrison and Bennett (2004) tested whether Sydney residents held similar values for wetlands protection and river systems in good health, respectively, in varying locations across New South Wales. Rolfe et al. (2006) reported valuation experiments with Brisbane residents for different river catchments in central Queensland, and Rolfe and Windle (2012b) report values held by Brisbane populations for protection of three different local areas of the Great Barrier Reef.

A third area of particular focus for the development of benefit transfer approaches has been to identify how a population valued the same environmental assets presented in different contexts, particularly when the scope of the environmental asset being presented varies. van Bueren and Bennett (2004) first explored this issue in their study of national and regional values for protection of land and water assets in Australia. They found that significant value differences existed according to whether the assets were presented in a national or regional context and they estimated adjustment factors to transfer values between contexts. Similar results have been reported by Mazur and Bennett (2009) and Rolfe and Windle (2012b) in the contexts of catchment protection in New South Wales and the health of the Great Barrier Reef, respectively, and by Kerr and Sharp (2008b) in tests of spatial differences in values for protection of endangered species.

A fourth area of interest has been to understand how the context of a resource trade-off can influence value estimates. Rogers and Cleland (2010, 2011) report tests of values held by scientists and the public for the same environmental good (wetlands and waterways in the Kimberly and the reserve system in southwest Australia, respectively), and demonstrate that preferences diverge. Distance decay effects have been noted by Concu (2007) and Rolfe and Windle (2012a). Rolfe and Bennett (2002) identify a type of responsibility effect across jurisdiction boundaries, where respondents exhibiting higher values for rainforests in their home State. Rolfe and Windle (2012b) find that while values for protection of the Great Barrier Reef were robust to various site and population differences, they did appear to diverge according to whether losses or gains are involved.

5.5 Performing Benefit Transfer from Independent Source Studies

Benefit transfer applications in both Australia and New Zealand can mostly be characterized as either research tests or case study applications. In the research tests, values for the source and target studies were assessed and compared in the one project. In the case studies, values were assessed to meet a particular reporting or assessment need, typically focused around a narrowly defined issue. Source and target valuation studies for the first group are rarely independent, although the benefit-transfer tests that have been conducted are rigorous. Independent source studies are typically "harvested" for inputs into the case study applications, but these rarely face strong review or independent scrutiny. There are a very limited number of major technical reports and published papers where the primary focus has been on the application of benefit transfer to assess environmental values or to perform a cost-benefit analysis.

Morrison and Hatton McDonald (2010) used benefit transfer from 15 source studies to help assess protection values for the Murray Darling system in southern Australia. They restricted the pool of potential source values to studies that had been conducted within the basin in the past 15 years, and assessed values for direct use (recreation), indirect use (water filtration) and nonuse purposes. Categories of nonuse values that were assessed included native vegetation, fish species and populations, waterbird breeding, waterbirds and other species, and other values of interest. The relevance of each attribute and the quantity of potential improvements were assessed for 19 separate sub-catchments, and then values were assessed and summed across the catchments. While the analysis was comprehensive, questions remain about issues of coverage and overlap of attributes, and the failure to account for marginal effects when summing values from individual studies into values for a larger whole.

Rolfe (2010) reported a benefit transfer exercise to assess values for groundwater protection in the Great Artesian Basin (GAB) in Australia. In that study, values were assessed for direct uses (recreation use and maintaining water supply for regional communities), indirect use (contributing to the reduction of greenhouse gases), and non-uses (maintaining ecological and biodiversity assets, cultural heritage and the options for future use and conservation). Values were drawn from eight separate source studies and then adjusted for each estimate of component value so as to reflect accurately the different impacts involved. Potential limitations of the case study approach include differences between source studies and the GAB, difficulties in identifying the scale of change to be valued, and assumptions of linearity made in extrapolating value estimates.

A development in recent years has been the application of benefit transfer by special interest groups to mount economic arguments for political purposes. Oxford Economics (2009) (commissioned by the Great Barrier Reef Foundation) used benefit transfer to estimate the value of the Great Barrier Reef in Australia across five key categories of values: consumer surplus use values for tourism and recreation, producer surplus use values for tourism and commercial fishing, indirect use values, and nonuse values for both Australian and international communities. The Oxford report concluded that the present value of the whole Great Barrier Reef was $51.4 billion, and that the cost of its total and permanent coral bleaching from climate change would be $37.7 billion.

The Australian Conservation Foundation (ACF) released a brief report titled *What's a Healthy Murray-Darling Basin Worth to Australians?* (ACF 2011). The analysis involved an extrapolation of the values reported in the benefit transfer exercise of Morrison and Hatton McDonald (2010). The ACF estimated that the protection values that Australians hold for improving the health of the Murray-Darling system and the Coorong at the mouth of the Murray at $9.8 billion, and that these values substantially outweighed the costs of water reform in the basin.

A number of weaknesses can be identified with the ACF (2011) and Oxford Economics (2010) reports. A key conceptual problem involves attempts to identify total values for an environmental asset instead of the policy-relevant question of marginal values for changes in protection. Other methodological problems relate to the difficulties of finding appropriate source studies, problems of overlap or inadequate coverage of the key value elements, and problems in value transfer because of variations among sites, populations, geographic scale and iconic status. Of particular note is the scale effect when benefit transfers based on source studies involving smaller scales are used to "extrapolate" across to larger scale target cases. This practice results in upwardly exaggerated estimates of value in the target setting. While these issues are not restricted to applications by special interest groups, there is a risk that groups wanting input into political debates will have incentives to use the large values estimated with a "total asset" approach and to present estimated values as more certain than can be justified.

5.6 Conclusions

In Australia and New Zealand the practice and performance of benefit transfer remains patchy, and the full extent of use remains difficult to assess accurately. On the positive side, an active focus on non-market valuation, particularly the development of the choice modeling technique, has generated pockets of expertise, case studies and interest in both countries. However, a number of the benefit transfer test studies have shown that while transfers are possible, they also demonstrate a number of limitations to accuracy, even in cases where variations among source and target sites and populations are small. Applications are further limited by the small pool of source studies to draw on, the difficulties of establishing environmental protection functions that enable estimates of changes in attribute levels to be linked to management and policy actions, and the complexity of performing accurate benefit transfer.

Three observations can be made about the complexity in benefit transfer applications. The first is that benefit transfer is becoming more technically complex, aiming to provide better precision of transferred value estimates by statistically accounting for methodological, quality and statistical attributes of source studies. The second is that benefit transfer is becoming more specialized, with expert knowledge needed to understand where benefit transfer may be appropriate, to select appropriate source studies, and to apply benefit transfer with appropriate adjustments. The third is that there is scope for benefit transfer applications to be inaccurate or misleading, and there are limited processes to guard against this. The increased sophistication of benefit transfer methods risks creating the impression of apparent scientific validity that can hide inaccuracies when inadequate source studies are available.

A key problem in Australia and New Zealand is that while the case and need for benefit transfer remain very strong, policy support is weak and varied. In part this is because of the failure by state and national governments to have a rigorous approach to policy evaluation, with limited use of cost-benefit analysis, thus restricting the demand for benefit transfer. It is also because of the very small set of primary source studies (across all non-market valuation techniques), and the limited number of people in government and policy circles with the appropriate training and understanding of the use of economic assessment techniques. The lack of specialized knowledge and processes to guide benefit transfer generates some risks that inaccurate studies will be input into policy settings. While the use of benefit transfer remains promising in policy applications, better guidelines are needed to ensure appropriate and accurate application.

References

ACF (Australian Conservation Foundation). (2011). *What's a healthy Murray-Darling Basin worth to Australians?* Analysis released on the 2nd of February 2011. Carlton, Victoria.

ACF (Australian Government). (2010). *Best Practice Regulation Handbook.* Australian Government, Canberra, at http://www.finance.gov.au.obpr/proposal/gov-requirements.html.

Bell, B., Cudby, C. & Yap, M. (2009, February). Establishing a benefits transfer database for biosecurity decision making on indigenous biodiversity. *Paper presented at the 53rd Annual Conference of the Australian Agricultural and Resource Economics Society*, Cairns.

Bennett, J. (1982). Valuing the existence of a natural ecosystem. *Search, 13*, 232–235.

Bennett, J. (1996). The contingent valuation method: A Post-Kakadu assessment. *Agenda, 3*, 185–194.

Bennett, J. (2005). Australasian environmental economics: Contributions, conflicts and 'cop-outs'. *Australian Journal of Agricultural and Resource Economics, 49*, 243–261.

Bennett, J., & Blamey, R. (Eds.). (2001). *The Choice Modelling Approach to Environmental Valuation.* Cheltenham, UK and Northhampton, MA, USA: Edward Elgar.

Bennett, J., Dumsday, R. & Gillespie, R. (2008a). Analysing options for the River Red Gum Forests along the Murray River. *Paper presented at the 52nd Australian Agricultural and Resource Economics Society Conference*, Queenstown, NZ.

Bennett, J., Dumsday, R., Howell, G., Lloyd, C., Sturgess, N., & Van Raalte, L. (2008b). The economic value of improved environmental health in Victorian Rivers. *Australasian Journal of Environmental Management, 15*, 138–148.

Blamey, R. K., Rolfe, J. C., Bennett, J. W. & Morrison, M. D. (1997). Environmental choice modelling: Issues and qualitative insights. *Choice Modelling Research Report No 4*, University College, University of New South Wales, Canberra.

Blamey, R., Rolfe, J., Bennett, J., & Morrison, M. (2000). Valuing remnant vegetation in Central Queensland using choice modelling. *Australian Journal of Agricultural and Resource Economics, 44*, 439–456.

Clough, P., Guria, J., & Ballingall, J. (2009). *The Value of Air Quality Standards: Review And Update of Cost Benefit Analysis of National Environmental Standards on Air Quality.* Wellington: Ministry for the Environment, New Zealand Institute of Economic Research.

Commonwealth of Australia. (2006a). *Handbook of Cost-Benefit Analysis.* Australian Government, Canberra, at: http://www.finance.gov.au/publications/finance-circulars/2006/docs/Handbook_of_CB_analysis.pdf.

Commonwealth of Australia. (2006b) *Introduction to cost-benefit analysis and alternative methodologies*. Australian Government, Canberra, at: http://www.finance.gov.au/publications/finance-circulars/2006/docs/Intro_to_CB_analysis.pdf.

Concu, G. (2007). Investigating distance effects on environmental values: A choice modelling approach. *Australian Journal of Agricultural and Resource Economics, 51*, 175–194.

Davidson, B. (1965). *The Northern Myth: A Study of the Physical and Economic Limits to Agricultural and Pastoral Development in Northern Australia*. Melbourne: Melbourne University Press.

Dobes, L. (2008). *A century of Australian cost benefit analysis*. Working paper, Department of Finance and Deregulation, Canberra, at: http://www.finance.gov.au/obpr/docs/Working-paper-1-Leo-Dobes.pdf (accessed July 24, 2012).

Dobes, L., & Bennett, J. W. (2009). Multi-criteria analysis: Good enough for government work? *Agenda, 16*, 7–30.

Gillespie, R., & Bennett, J. (2013). Willingness to pay for kerbside recycling in Brisbane, Australia. *Journal of Environmental Planning and Management, 56*, 362–377.

Gluck, R. J. (1974). *An economic evaluation of the Rakaia fishery as a recreational resource*. MAgSci thesis. Christchurch: University of Canterbury (Lincoln College).

James, D., & Gillespie, R. (2002). *Guidelines for Economic Effects and Evaluation for Eia*. Sydney: Prepared for New South Wales Department of Urban Affairs and Planning.

Kerr, G. N. (n.d.). *New Zealand Non-Market Valuation Database*. http://www2.lincoln.ac.nz/nonmarketvaluation/.

Kerr, G. N. (2009). Statement of Evidence in Chief of Dr Geoffrey Neville Kerr in the Environment Court at Christchurch under the Resource Management Act 1991 in the matter of Appeals to the Environment Court under Section 120 of the Act between Garth William Dovey, Lower Waitaki River Management Society Incorporated, Te Runanga O Ngai Tahu, Ngai Tahu-Mamoe Fisher People Incorporated, Waitaki Protection Trust, Appellants and Canterbury Regional Council, Respondent and Meridian Energy Limited, Applicant.

Kerr, G. N., & Sharp, B. M. H. (2006). Transferring mitigation values for small streams. In J. Rolfe & J. Bennett (Eds.), *Transferring Environmental Values from Choice Modelling Applications* (pp. 136–163). Cheltenham, UK: Edward Elgar.

Kerr, G. N., & Sharp, B. M. H. (2008a). Evaluating off-site mitigation using choice modelling. *Australian Journal of Agricultural and Resource Economics, 52*, 381–399.

Kerr, G. N. & Sharp, B. M. H. (2008b). *The impact of wilding trees on indigenous biodiversity: A choice modelling study*. AERU Research Report No. 303. Agribusiness and Economics Research Unit, Lincoln University, New Zealand.

Mazur, K. & Bennett, J. (2009). *Scale and scope effects on communities' values for environmental improvements in the Namoi Catchment: A choice modelling approach*. Environmental Economics Research Hub Research Report No. 42, at: http://www.crawford.anu.edu.au/research_units/eerh/publications/.

Miller, T. R & Guria, J. (1991). *The value of statistical life in New Zealand market research on road safety*. Land Transport Division, Ministry of Transport, Wellington, at: http://nzier.org.nz/publications/fix-flawed-values-of-statistical-life-and-life-years-to-get-better-policy-outcomes#sthash.Dbej18f2.dpuf.

Morrison, M., & Bennett, J. (2004). Valuing New South Wales rivers for use in benefit transfer. *Australian Journal of Agricultural and Resource Economics, 48*, 591–611.

Morrison, M., Bennett, J., Blamey, R., & Louviere, J. (2002). Choice modelling and tests of benefit transfer. *American Journal of Agricultural Economics, 84*, 161–170.

Morrison, M., & Hatton McDonald, D. (2010). *Economic Valuation of Environmental Benefits in the Murray-Darling Basin*. Commonwealth of Australia, Canberra: Report prepared for the Murray-Darling Basin Authority.

New South Wales Treasury. (2007). *NSW Government Guidelines for Economic Appraisal*. Sydney: New South Wales Government.

New Zealand Treasury. (2005). *Cost benefit analysis primer*. New Zealand Government, Wellington, at: http://www.treasury.govt.nz/publications/guidance/planning/costbenefitanalysis/primer/cba-primer-v12.pdf (accessed 24th of July, 2012).

New Zealand Treasury. (2009). *Regulatory Impact Analysis Handbook*. New Zealand Government, Wellington, at: http://www.treasury.govt.nz/publications/guidance/regulatory/impactanalysis/ria-handbk-nov09.pdf (accessed 9th of September, 2012).

Oxford Economics. (2009). *Valuing the effects of Great Barrier Reef bleaching*. Report prepared for the Great Barrier Reef Foundation, Brisbane, Australia.

Productivity Commission. (2012). *Regulatory impact analysis: Benchmarking*. Draft Report. Canberra: Australian Government.

Queensland Government (Environmental Protection Agency). (2003). *Techniques for environmental economic valuations*. Information sheet, Environmental Impact Assessment. Brisbane: Queensland Government.

Rogers, A. & Cleland, J. (2010). *Comparing scientist and public preferences for conserving environmental systems: A case of the Kimberley's Tropical waterways and wetlands*. Environmental Economics Research Hub Research Report No. 80, at: http://www.crawford.anu.edu.au/research_units/eerh/publications/.

Rogers, A. & Cleland, J. (2011). *Conserving biodiversity in the Southwest Australia ecoregion: The policy implications of scientist and community values*. Environmental Economics Research Hub Research Report No. 91, at: http://www.crawford.anu.edu.au/research_units/eerh/publications/.

Rolfe, J. (2010). Valuing deductions in water extractions from ground water basins with benefit transfer: The Great Artesian Basin in Australia. *Water Resources Research, 46*, W06301. doi:10.1029/2009WR008458.

Rolfe, J., & Bennett, J. (2002). Assessing rainforest conservation demands. *Economic Analysis and Policy, 32*, 51–67.

Rolfe, J., & Bennett, J. (Eds.). (2006). *Choice Modelling and the Transfer of Environmental Values*. Cheltenham, UK: Edward Elgar.

Rolfe, J., Bennett, J., & Louviere, J. (2002). Stated values and reminders of substitute goods: Testing for framing effects with choice modelling. *Australian Journal of Agricultural and Resource Economics, 46*, 1–20.

Rolfe, J., & Brouwer, R. (2013). Design effects in a meta-analysis of river health choice experiments in Australia. *Journal of Choice Modelling, 5*, 81–97.

Rolfe, J., Loch, A., & Bennett, J. (2006). Testing benefit transfer with water resources in central Queensland, Australia. In J. Rolfe & J. Bennett (Eds.), *Choice Modelling and the Transfer of Environmental Values* (pp. 112–135). Cheltenham, UK: Edward Elgar.

Rolfe, J., & Windle, J. (2008). Testing for differences in benefit transfer values between state and regional frameworks. *Australian Journal of Agricultural and Resource Economics, 52*, 149–168.

Rolfe, J., & Windle, J. (2012a). Distance decay functions for iconic assets: Assessing national values to protect the health of the Great Barrier Reef in Australia. *Environmental and Resource Economics, 53*, 347–365.

Rolfe, J., & Windle, J. (2012b). Testing benefit transfer of reef protection values between local case studies: The Great Barrier Reef in Australia. *Ecological Economics, 81*, 60–69.

Thomas, I. G., & Elliot, M. (2005). *Environmental Impact Assessment In Australia: Theory and Practice* (4th ed.). Sydney: The Federation Press.

van Bueren, M., & Bennett, J. W. (2004). Towards the development of a transferable set of value estimates for environmental attributes. *Australian Journal of Agricultural and Resource Economics, 48*, 1–32.

Windle, J., & Rolfe, J. C. (2012). *Assessing the Values of Natural Capital and Ecosystem Services in the Lake Macquarie City Council Area*. New South Wales, February: Report prepared for the Lake Macquarie City Council.

Wren, J. & Barrell, K. (2010). *The costs of injury in New Zealand and methods for prioritizing resource allocation: A background the briefing paper to inform evaluation of the New Zealand Injury Prevention Strategy*. New Zealand Injury Prevention Secretariat, Accident Compensation Corporation, Wellington.

Zander, K. K., Garnett, S. T., & Straton, A. (2010). Trade-offs between development, culture and conservation: Willingness to pay for tropical river management among urban Australians. *Journal of Environmental Management, 91*, 2519–2528.

Chapter 6
Benefit Transfer for Water Quality Regulatory Rulemaking in the United States

William J. Wheeler

Abstract This chapter describes the use of benefit transfer in Regulatory Impact Analyses for surface water quality regulations at the U.S. Environmental Protection Agency (EPA). It begins by explaining the regulatory framework at the EPA and existing guidance on the use of benefit transfer. It then describes how benefit transfer has been applied in practice for water quality RIAs and how that practice has changed over time: from the use of unit value transfers to more sophisticated function and meta-analytic transfers. The chapter concludes with observations about the practical application of benefit transfer methods at the EPA and how this practice diverges from the academic literature.

Keywords Water quality valuation · Regulatory impact analysis · Benefit transfer

6.1 Introduction and Background[1]

6.1.1 The Clean Water Act

The Environmental Protection Agency (EPA) is the primary agency in the United States responsible for protecting the nation's water resources. This involves protecting lakes, rivers, and streams; and ensuring that ecosystems can sustain plants, fish, and wildlife. Starting in 1972, in response to rising public concern about water quality, the U.S. Congress amended the Federal Water Pollution Control Act to shift the authority for pollution control from the states to the federal government

[1]For a more detailed history of the Clean Water Act, see Freeman (2000). For a more specific description of the regulations EPA is required to put forth under the CWA, see Griffiths and Wheeler (2005).

W.J. Wheeler (✉)
U.S. Environmental Protection Agency, National Center for Environmental Economics, Washington, DC, USA
e-mail: Wheeler.william@epa.gov

R.J. Johnston et al. (eds.), *Benefit Transfer of Environmental and Resource Values*, The Economics of Non-Market Goods and Resources 14,
DOI 10.1007/978-94-017-9930-0_6

and also changed the way in which pollution was controlled by requiring federal limits on sources rather than state-level ambient standards. These amendments became known as the Clean Water Act (CWA), and provide most of the authority to regulate surface water quality at EPA.

With a few exceptions, the regulation of surface water quality under the CWA is achieved by using two types of controls: technology-based effluent limits on point sources holding discharge permits coupled with use-based water quality designations. All point-source dischargers are required to have a National Pollutant Discharge Elimination System (NPDES) permit issued by states and EPA regional offices. These permits specify pollutants that a point source must control, limits on those pollutants, and the required frequency of monitoring. EPA is required by the CWA to promulgate pollutant limitations known as effluent limitation guidelines (ELGs), which are intended to represent the greatest pollutant reductions that are economically achievable for that industry. A point source that is covered by an ELG must have pollutant limits that are at least as strict as those described by the ELG.

In addition to technology-based permit limits, facilities may have additional—or more stringent—limits based on ambient water quality standards. States are required to assign a "designated use" (such as fishing or swimming) to all water bodies and then assign water quality criteria with numeric pollutant concentrations that designate the maximum allowable in-stream pollutant levels to support that designated use. If any criteria are exceeded then facilities can have more stringent (technically and economically feasible) limits written into their NPDES permit.

All states are required to periodically assess waters to see if they are meeting their water quality standards and designated uses. If the water quality standards are not being met then a Total Maximum Daily Load (TMDL) must be developed. TMDLs determine what level of pollutant load would be consistent with meeting the violated standard. TMDLs also allocate acceptable loads among sources of the relevant pollutants for both point and nonpoint sources. Once a load allocation is determined for point sources, limits reflecting these allocations can be written into NPDES permits. In general, EPA has no regulatory authority over nonpoint sources, but states may also require nonpoint sources to implement best management practices (BMPs) to meet a load allocation or may institute a water quality trading program to meet the TMDL.

While the CWA does not require cost-benefit analysis, President Reagan instituted such a requirement in 1981 with the publication of Executive Order 12291. According to EO 12291, all "major rules" (those costing $100 million per year[2]) had to be accompanied by a Regulatory Impact Analysis (RIA) that presented the cost and benefits of the rule. Although EO 12291 was replaced by EO 12866 in 1993, essentially the same requirements for cost-benefit analysis were maintained, so this requirement has been in force at EPA since 1981.

---

[2]Rules that caused cost or price increases or had other adverse effects on the economy also had to have an RIA.

Prior to the publication of EO 12291, EPA had published a large number of effluent guidelines to comply with the requirements of the CWA and this pace continued for some time.[3] However, the EPA's Office of Water had not performed a cost-benefit analysis for any of their regulations before EO 12291; after the EO, EPA published only two RIAs prior to 1998: one in 1982 covering the Iron and Steel Plants ELG, and one in 1995 for the Great Lakes Water Quality Guidance.[4] Starting in 1998, all effluent guidelines (even those under the $100 million threshold) were accompanied by an RIA and two water quality standards in addition to the Guidance had RIAs.

6.1.2 EPA Guidance on Benefit Transfer

In response to these requirements for RIAs and cost-benefit analysis, EPA published guidance on performing these evaluations. The first version (U.S. EPA 1983) predated the 1992 *Water Resources Research* volume that popularized the general concept of benefit transfer, and so does not discuss the technique. Prior to 1992 a variety of simple benefit transfer approaches had been used in Agency analyses. These techniques had not been formally grouped or recognized under the heading of benefit transfer, but the Agency quickly adopted the methods of benefit transfer in its analyses. The second version of official Agency guidance (U.S. EPA 2000g) discusses benefit transfer throughout a chapter on benefits estimation; emphasizing that time and expense often preclude original analyses. It casts the discussion of benefits estimation in terms of evaluating studies, primary estimation methods (e.g. stated preference), and results for use in benefit transfer. The 2000 guidance does not recommend the use of mean unit value transfers (but notes that "analysts will often adjust point estimates based on judged differences between the study and policy cases"), describes function transfer as "more refined but also more complex," and states that meta-analysis is the "most rigorous" transfer method.

In the 2010 update to the guidance (U.S. EPA 2010a), the chapter on benefits was again written with benefit transfer in mind. In addition, the chapter discusses primary valuation methods with recommendations for evaluating (and conducting) original valuation studies for transfer. The update also notes "the reality is that benefit transfer is one of the most common approaches for completing a BCA at EPA" while also stating that the cost and time advantages of the final estimates

[3]See http://water.epa.gov/scitech/wastetech/guide/industry.cfm#exist for a list of existing effluent guidelines. Because the CWA calls for zero discharge by industrial facilities, EPA is sometimes sued by outside groups to enforce this provision; the frequency of the publication of effluent guidelines is usually determined by the outcome of these suits.

[4]Although titled "guidance," this set minimum water quality standards, specified numeric criteria, and described procedures to implement these criteria in permits. It is clear from the RIA that the benefit analysis was conducted to comply with EO requirements.

benefit transfer must be weighed against the potential reduced accuracy.[5] In terms of evaluating specific methods, the 2010 update argues that benefit transfer from a single study case is unlikely to be as accurate as a tailored primary study. The update discusses the pros and cons of each of the benefit transfer methods, concluding that, in general, function transfer is preferable to unit value transfer. The update discusses meta-analysis and structural benefit transfer (preference calibration) but does not evaluate the relative accuracy of these methods. The 2010 guidance does state that "as a general rule, the more related case study estimates involved in a benefit transfer, the more reliable the estimate."

The next section discusses the benefit transfer approaches EPA has used in its rules enacting water quality regulations. To keep the discussion manageable, the section focuses on the transfer technique(s) used to estimate primary use value estimates (such recreational fishing) and nonuse values (which may or may not be estimated separately from use values). Information on other benefit categories is provided by Griffiths et al. (2005). The final section offers several conclusions regarding the history of benefit transfer applied to EPA water quality regulations.

6.2 EPA Water Quality Regulations with RIAs

6.2.1 Early Rules Using Shares to Transfer

The Iron and Steel effluent guideline was the first EPA water quality regulation to be accompanied by an RIA (U.S. EPA 1982). Within this RIA the EPA Office of Water (OW) performed a relatively simple transfer exercise that does not conform to more recent benefit transfer typologies and recommended approaches; essentially EPA adjusted and scaled an existing unit value transfer exercise. EPA calculated benefits based on Freeman's (1979) estimate of the national benefits of water pollution control. Freeman's original estimates reflected the goals laid out in the CWA to achieve by 1985[6] compared to a 1978 baseline. These estimates were calculated by transferring estimates of water quality for the following categories: recreation (fishing, swimming, boating, and water fowl hunting), nonuser benefits,[7] diversionary uses (drinking water and health; avoided treatment costs for municipal water supplies, avoided household costs of water hardness, and industrial treatment costs), and commercial fisheries. Identifying existing, relevant studies, Freeman

[5]The 2010 guidance also quotes the Office of Management and Budget's guidance on RIAs, which states that benefit transfer "be treated as a last-resort option and not used without explicit justification" after noting that original studies might not be feasible due to time or expense.

[6]Many of his results were based on two reviews performed under contract to EPA: Unger (1975) and Heintz et al. (1976). Since both of these reviews were aimed at estimating the benefits of the CWA goals, they were directly relevant to Freeman's objective.

[7]Based on results from early stated preference studies, his central estimate for nonusers was based on scaling the (use) benefit to recreational anglers by 50 %.

used his best judgment to weigh and synthesize results, thereby calculating central estimates and ranges for values within each of the benefit categories. EPA made several adjustments to these estimates,[8] updated the baseline from 1972, and calculated the share of remaining pollutants that would be controlled by the regulation. This share was then applied to the adjusted and updated benefit estimates to calculate the estimated benefits of the effluent guideline.

The Organic Chemicals, Plastics, and Synthetic Fibers (OCPSF) rule (U.S. EPA 1987; Caulkins and Sessions 1997) used a very similar approach to the "shares" approach used in the Iron and Steel ELG, but replaced most of the Freeman estimates with the national aggregate estimates from the Mitchell and Carson (1984) willingness to pay survey for freshwater quality as the primary source of their benefit transfers (EPA augmented the benefits estimates with additional categories from Freeman that were not included in the Mitchell and Carson estimates). The survey used a national, in-person stated preference survey to ask respondents to value changes in a water quality index anchored to achievement of the goals of the Clean Water Act (that is, moving from boatable to fishable and swimmable water and maintaining boatable water). The focus of the survey was a national change in water quality. EPA adjusted the baseline from Mitchell and Carson to be contemporaneous with the rule and scaled the benefits to be commensurate with expected water quality improvements from the OCPSF regulation.

6.2.2 Toxic Controls Valued Using Transfers of Percentage Improvements

The 1995 Great Lakes Water Quality Guidance (GLWG) did not contain a full benefits analysis, but did include three case studies (U.S. EPA 1993, 1995; Castillo et al. 1997). EPA estimated benefits using two different benefit transfer approaches. One approach included a transfer of a benefit percentage that is again not easily described using the standard benefit transfer terminology (and is not discussed in the academic literature); this approach is probably most notable for the influence it had on subsequent RIAs. The approach was based on a dissertation by Lyke (1993), who applied a random utility model (RUM) using survey data to estimate the value of the Wisconsin open water fishery to anglers. Within the same survey used to gather the data for the RUM, Lyke (1993) asked a contingent valuation (CV) question regarding willingness to pay (WTP) for a policy that would make the same fishery "completely free" of contaminants that "may threaten human health." Because Lyke used multiple RUM and CV models, she obtained a range of values for both the baseline value of the fishery and the WTP for the improvement.

[8]Adjustments were made to account for instances in which they disagreed with Freeman's assumptions and to account for newer research results (i.e., draft versions of Mitchell and Carson (1984)).

Comparing these ranges, EPA determined that the value of a contaminant-free fishery was 11.1–31.3 % of the baseline value of the fishery. EPA then scaled the baseline value calculated by Lyke (1993) downward to account for the smaller size of the case studies (compared to the entire Wisconsin open water fishery) and to allow for the fact that the GLWG achieved only a proportion of a completely contaminant-free fishery (EPA assumed 50 % in the proposal but lowered this to 1.41–2.82 % in the Final RIA).

The second approach used in the GLWG combined an estimate of the number of angler days in the fishery, based on state Department of Natural Resources data, with average consumer surplus values for fishing days from Milliman et al. (1992) and Walsh et al. (1988, 1990). These fishing day values were multiplied by the 11.1–31.3 % increase used in the previous approach to calculate the value to anglers of a contaminant-free fishery; these values were again scaled to account for the incompleteness of the guidance in achieving a contaminant-free fishery.

To estimate nonuse values for the GLWG, EPA relied on the results of Fisher and Raucher (1984), who reviewed the available literature on water quality valuation, focusing on studies that estimated use and nonuse values for recreational fishing. They found that nonuse values are "roughly half (or more)" of use values in these studies. Based on this conclusion, EPA estimated nonuse values as 50 % of the use values estimated using the Lyke (1993) percentage increase approach.

Following the GLWG, EPA estimated the benefits of several rules using essentially the same methodology. In the Pulp and Paper "Cluster Rule" (U.S. EPA 1997), EPA used dilution models to estimate in-stream pollutant concentrations as well as the Dioxin Reassessment Evaluation model to estimate concentrations of dioxins and furans, the two major pollutants of concern for the rule, both of which are known to cause fish consumption advisories. EPA's models predicted that these advisories could be lifted after the rule. The value of lifted advisories was calculated by multiplying the average consumer surplus per day of fishing (Walsh et al. 1990) by the percentage increase in value (11.1–31.3 %) estimated by Lyke (1993); that is, EPA assumed that the contaminant-free fishery described by Lyke for the Great Lakes "may be equated by anglers with the lifting of consumption advisories" downstream of pulp and paper mills. EPA also estimated that lifting fish consumption advisories would increase recreational fishing participation by 20 % at affected reaches and included an estimate of the value of these additional fishing days, but did not add it to the benefit estimate of removing advisories because this increase in participation may simply have reflected substitution from other sites. EPA did not include an estimate of nonuse benefits in its analysis of the benefits of the Cluster Rule.

In 2000, EPA OW promulgated four effluent guidelines—Centralized Waste Treatment (U.S. EPA 2000b), Waste Combustors (U.S. EPA 2000c), Landfills (U.S. EPA 2000d), and Transportation Equipment Cleaning Industry (U.S. EPA 2000e)—and one water quality standard—the California Toxics Rule (U.S. EPA 2000f). In all of these cases, benefit transfers were implemented following an approach that was parallel to that within the Guidance and Cluster Rule analysis. Each replicated the transfer approach using the Lyke (1993) percentages and

Walsh et al. (1990) values to calculate recreational benefits; however, instead of using fish consumption advisories, these analyses used removal of all exceedances of acute (short-term) water quality criteria as the policy change equivalent to Lyke's contaminant-free fishery. For the Waste Combustors, Landfills, TECI, and California Toxics regulations,[9] nonuse benefits were estimated as 50 % of the recreational benefits calculated using the water quality criteria exceedance approach, based on the survey of Fisher and Raucher (1984).

The final effluent guideline centered on toxic pollutants, Metal Products & Machinery (MP&M) (U.S. EPA 2003c), expanded the Lyke-based approach by using eight studies to estimate the percentage increase in value; these studies are listed in Table 6.1. The RIA also added wildlife viewing and boating as categories of benefits,[10] based on an analogous approach of transferring percentage increases to day use values. For recreational fishing benefits, EPA assumed that eliminating water quality criteria exceedances (the policy change valued in the rule) was roughly comparable to the following discrete water quality changes: achieving a contaminant free fishery, reducing the level of toxins in fish tissue, removing fish consumption advisories; and improving water quality from "boatable" to "fishable," from "fair" to "good," and from "moderately polluted" to "unpolluted" (see Table 6.1). That is, these statements all describe the scenarios transferred from published studies and represent a fairly large range of policy cases. For the MP&M rule, EPA estimated nonuse values to be 50 % of the use values, again based on Fisher and Raucher (1984) but also citing Brown (1993).

In all cases summarized above, EPA analysis applied benefit transfer approaches that relied on expert judgment regarding the equivalence of different types of environmental and policy changes for purposes of benefit estimation, along with transfers based on percentage changes in values rather than on benefit functions and other approaches more commonly recommended in the academic literature.

This review also shows a pattern in which similar methods are used within sequential RIAs conducted by the Agency; these methods make repeated use of findings from a few key studies in the literature (e.g., the analysis of Fisher and Raucher (1984) used to justify a 2:1 ratio of transferred use to nonuse values).

6.2.3 Function and Meta-Analytic Transfers for Conventional Pollutants and Nutrients

After the MP&M guideline, the focus of EPA's water program turned to conventional pollutants, especially from animal agriculture. The Concentrated Animal Feeding Operations (CAFO) rule introduced the National Water Pollution Control

[9]The Centralized Waste Treatment Rule did not include an estimate of nonuse benefits.

[10]All three approaches included a RUM model for Ohio, estimated specifically for the rule, and estimated nonuse values as 50 % of use values.

Table 6.1 Percentage changes in value used for metal products and machinery analysis

Study Citation	Scenario and geographic scope	Methodology	Value of change as percent of baseline (%)
Lyke (1993)	Contaminant—free Wisconsin open water fishery	RUM + CV	11.1–31.3
Jakus et al. (1997)	Lifting fish consumption advisories in Tennessee reservoirs	RUM	6.0–8.0
Montgomery and Needleman (1997)	Elimination of toxic impairment for fishing in New York State lakes and ponds	RUM	13.7
Phaneuf et al. (1998)	20 % reduction in toxins, reanalysis of Lyke data	RUM, demand system, Kuhn-Tucker model	27.5–34.3
Desvousges et al. (1987)	Moving the Monongahela River from "boatable" to "fishable" waters	CV	5.9–7.9
Lant and Roberts (1990)	Moving Illinois and Iowa river basins from "fair" to "good" water quality	CV	9.7–13.1
Farber and Griner (2000)	Moving from "moderately polluted" to "unpolluted" water	CE	3.9–9.0
Tudor et al. (1999)	Elimination of AWQC exceedances	RUM	0.77

Source List of studies and percentage changes taken from U.S. EPA (2003c), Table 15.3. Scenario, scope, and methodology identified from original studies

Assessment Model (NWPCAM) into RIAs. NWPCAM was designed by OW for use in water quality benefits estimation and was first used to estimate the retrospective benefits of the CWA controls on conventional pollutants (U.S. EPA 2000a). NWPCAM was later expanded to include other pollutants, including nutrients and pathogens.

For the CAFO Rule (U.S. EPA 2003a, b), EPA estimated benefits with NWPCAM using two approaches applying the results from the Mitchell and Carson (1989)[11] national contingent valuation study of freshwater quality (and so encompassed both use and nonuse values). The first approach used by EPA in the CAFO rule was an aggregate unit value transfer; with this approach, EPA estimated the number of river and stream segments that were achieving each level of water quality

[11]A note on citations: Mitchell and Carson (1989) refers to their book, *Using Surveys to Value Public Goods*, which is by far the most commonly-cited reference to their survey. However, it does not present as many analytic details and results as the final report for the EPA cooperative agreement that funded the survey, Mitchell and Carson (1986). The latter was the primary reference for EPA (2000a), which describes the integration of the NWPCAM model and the survey results. Carson and Mitchell (1993) is a summary of the results in the report.

(boatable, fishable, swimmable) without the CAFO rule and then with the CAFO regulatory options. EPA then transferred the values for meeting each of these water quality levels from the Mitchell and Carson (1986) detailed survey results. The second approach utilized a function transfer with the Vaughn (1981) water quality ladder, which was also prominent in the Mitchell and Carson survey instrument. The ladder relied on a water quality index that transforms each pollutant in the model into a 0–100 subindex that reflects the different contributions of these pollutants to water quality. The ladder approach then weights and aggregates those subindex values into a 100-point index. The index values were linked to WTP using benefit transfer function by applying the equation estimated by Mitchell and Carson (1986; see also Carson and Mitchell 1993):

$$\Delta TOTWTP = \exp[0.8341 + 0.819 * \log(WQI_1/10) + 0.959 * \log(Y)] - \exp[0.8341 + 0.819 * \log(WQI_0/10) + 0.959 * \log(Y)].$$

where $\Delta TOTWTP$ is the change in total household willingness to pay for a change in water quality, WQI_1 is projected WQI, WQI_0 is baseline WQI, and Y is statewide annual household income.[12] The state values are aggregated to obtain the national estimate of benefits. For both approaches, the analysis of the benefits of the revised CAFO regulations examined water-quality improvements on a state-by-state basis and separately calculated the benefits of in-state and out-of-state improvements, assuming that households will allocate two-thirds of their willingness to pay values to the improvement of in-state waters (Mitchell and Carson 1986).

The NWPCAM model with Mitchell and Carson benefits estimates was also used for the Meat and Poultry Products effluent guideline (U.S. EPA 2004a) although only using the second, function transfer approach. In the Concentrated Aquatic Animal Production rule (U.S. EPA 2004b), OW continued to apply the function approach to Mitchell-Carson benefits estimates (Carson ad Mitchell 1993) but without using NWPCAM.[13]

In the Construction and Development ELG (U.S. EPA 2009), EPA used the SPARROW water-quality model (Smith et al. 1997) and replaced the Carson and Mitchell (1993) estimates with values from a meta-analysis of studies that use the water-quality index approach or a scenario that can be mapped onto the index. This meta-analysis was described as following Johnston et al. (2005). The dependent variable in the meta-analysis was ln (*WTP*) (the natural log of WTP for water quality improvements in each study in the meta-analysis). The explanatory variables in the meta-analysis included specific details of the valued water bodies, the extent of water-quality improvements, whether the improvements occur in estuarine

[12]The published version of this equation includes covariates for household use and the importance of controlling pollution, as expressed by respondents. In the CAFO analysis, EPA used the Carson and Mitchell sample averages as a scalar value for the entire sample and incorporated the scalar value times the coefficient (for each variable) into the constant term.

[13]EPA used the QUAL2E model because of resource and data constraints.

or fresh water, the geographic region and scale of water-quality improvements, baseline conditions, the extent of water-quality change, characteristics of surveyed populations, methodological variables, and other specific details of each study. Most importantly, two of the dependent variables were baseline and post-regulatory water-quality index levels; some the studies in the meta-analysis used a water-quality index directly and others were mapped onto the index by EPA. EPA compared semi-log and trans-log functional forms in their econometric estimation, but used the trans-log functional form for benefits estimation because it was less sensitive to very small changes in water quality and most of the changes in the rule were also small; thus, the choice of the trans-log would avoid overstating benefits.

To estimate WTP for the changes in the rule, EPA plugged appropriate values into each variable. For example, the baseline and modeled levels of water quality were used in the water quality index variables, household income values were assigned for each state based on data from the American Community Survey, the variable *nonusers* was set to zero so WTP was total value for both users and nonusers, and methodological variables were set to methodologically-preferred approaches. The same meta-analysis was used for the Florida Nutrients Rule (U.S. EPA 2010b), with no changes to the meta-analysis or its application.

In contrast to earlier approaches discussed in Sect. 6.2.2, these benefit function and meta-analysis transfers correspond much more closely to typologies and recommended approaches for benefit transfer in the more recent academic literature (Johnston and Rosenberger 2010). This suggests a gradual improvement in the sophistication of benefit transfers applied by EPA, and increasing correspondence between Agency practice and recommendations of the academic literature in this area.

6.3 Conclusions

This review of the use of benefit transfer in RIAs for water quality at EPA yields several conclusions. First, it seems that EPA has made steady progress in improving the sophistication of benefit transfers. EPA has moved from the use of unit value transfers (or methods that are related to unit value transfers), which are generally considered the least accurate form of benefit transfer, to function transfers. At the same time, this progress has been slower than might be expected based on trends in the literature (a point also noted by Wilson and Hoehn (2006) and Johnston and Rosenberger (2010)): EPA did not make the transition to function transfer until the development of NWPCAM in 2000 (U.S. EPA 2000a) and the CAFO Rule (which was developed starting in 1998, proposed in 2001, and finalized in 2003), whereas the potential increased accuracy of function transfer was first proposed by Loomis (1992). EPA first adopted a meta-analysis in 2009 with the Construction and Development Rule, again much later than the technique was suggest for benefit transfer (Bergstrom and DeCivita 1999; Rosenberger and Loomis 2000).

Second, EPA appears to have been slow to update the studies it has used repeatedly for water-quality RIAs. EPA relied on the Lyke (1993) study, an unpublished dissertation, as a primary component of its water-quality benefit transfers until 2003, when it added a number of studies to the transfer underlying the benefits analysis for the MP&M rule, in spite of the fact that most of these studies were published well before (see Table 6.1). Similarly, the use of Fisher and Raucher (1984) as the basis for transfers of a 2:1 use/nonuse value ratio continued until at least 2003 with the MP&M rule.

Third, it appears that EPA's progress has been driven by economically significant rules. For example, the introduction of the Mitchell and Carson (1986) benefit function transfer accompanied the CAFO rule, the additional studies (to Lyke 1993) for toxics impairment were introduced in the MP&M rule, and the meta-analysis was first applied in the Construction and Development Rule. Each of these rules was economically significant[14] and thus high-profile (and likely had a larger budget that enabled methodological improvements). Less expensive rules have typically applied the same transfer methods used in prior rules with similar pollutants.

Fourth, outside of EPA's guidance documents, there is surprisingly little discussion of the academic literature on benefit transfer. While there are frequent citations in RIAs to a handful of studies discussing the criteria for selecting study sites (e.g., Boyle and Bergstrom 1992; Desvousges et al. 1987, 1992), there are few (and sometimes no) citations to the relatively large literature that compares the accuracy of methods. None of the RIAs using unit value transfers cite this literature, and neither the documentation for the development of NWPCAM nor the CAFO rule cite the literature on the potential increased accuracy of benefit function transfer, even though the CAFO rule uses this approach. The Construction and Development rule is an exception, and does discuss the pros and cons of using a meta-analysis for transfer.

A fifth conclusion is that the relatively slow (but laudable) improvements in the sophistication of EPA's benefit transfer methods of surface water quality RIAs are in contrast to the expressed need for improved benefit transfer methods in EPA's research programs. EPA named benefit transfer as a research priority in 2005 because of the impracticality of its funding enough primary valuation studies to cover all of the health and ecological endpoints requiring value estimates across EPA (U.S. EPA 2005). EPA's Office of Research and Development's funding opportunities highlighted this research need for several years.

While these conclusions suggest room for improvement in EPA's benefit transfer for water quality, these issues are recognized within the agency and EPA's current practices are continuing to change. A current driver is the analysis of the Chesapeake Bay TMDL; EPA is implementing a number of both original and transfer analyses to value cleaning up the Bay with an additional objective of

[14]Although the final MP&M regulation set limits only for a relatively small number of facilities, the 2001 proposal set limits for a large number of toxic pollutants for 89,000 facilities. It was estimated to cost $1.98 billion per year, so EPA put forth considerable effort to try to estimate benefits comprehensively.

improving the practice of water quality valuation and benefit transfer (Griffiths 2011 and related presentations). Further, EPA is working with other agencies to extend existing water quality models to support benefit assessments (Wells et al. 2011) and EPA will have to adapt transfer methods to the outputs of these newer modeling efforts. This will be a continuing effort as both benefit transfer methods and water quality models evolve.

Acknowledgments This chapter is an expanded and revised version of a section originally written for Griffiths et al. (2012). The views expressed in this article are those of the author and do not necessarily represent those of the U.S. Environmental Protection Agency. No official agency endorsement should be inferred.

I would like to thank Tess Sison for her help in obtaining many of the Regulatory Impact Analyses discussed here and Charles Griffiths for helpful suggestions.

References

Bergstrom, J. C., & DeCivita, P. (1999). Status of benefits transfer in the United States and Canada: A review. *Canadian Journal of Agricultural Economics, 47*, 79–87.

Boyle, K. J., & Bergstrom, J. C. (1992). Benefit transfer studies: Myths, pragmatism and idealism. *Water Resources Research, 28*, 657–663.

Brown, T. (1993). Measuring nonuse value: A comparison of recent contingent valuation studies. In J. C. Bergstrom (Ed.), *Benefits and costs transfer in natural resource planning*. Sixth Interim Report. Athens, GA: University of Georgia, Department of Agriculture and Applied Economics.

Carson, R. T., & Mitchell, R. C. (1993). The value of clean water: The public's willingness to pay for boatable, fishable, and swimmable quality water. *Water Resources Research, 29*, 2445–2454.

Castillo, E. T., Morris, M. L., & Raucher, R. S. (1997). Great Lakes water quality guidance. In R. D. Morgenstern (Ed.), *Economic analyses at EPA* (pp. 419–454). Washington, DC: Resources for the Future.

Caulkins, P., & Sessions, S. (1997). Water pollution and the organic chemicals industry. In R. D. Morgenstern (Ed.), *Economic analyses at EPA* (pp. 87–130). Washington, DC: Resources for the Future.

Desvousges, W. H. (1992). Benefit transfer: Conceptual problems in estimating water quality benefits using existing studies. *Water Resources Research, 28*, 675–683.

Desvousges, W. H., Smith, V. K., & Fisher, A. (1987). Option price estimates for water quality improvements: A contingent valuation study for the Monongahela River. *Journal of Environmental Economics and Management, 14*, 248–267.

Farber, S., & Griner, B. (2000). Valuing watershed quality improvements using conjoint analysis. *Ecological Economics, 34*, 63–76.

Fisher, A., & Raucher, R. (1984). Intrinsic benefits of improved water quality: Conceptual and empirical perspectives. In V. K. Smith (Ed.), *Advances in applied micro-economics: A research annual* (pp. 37–66). Greenwich, CT: JAI Press.

Freeman, A. M., III. (1979). *The benefits of air and water pollution control: A review and synthesis of recent estimates*. Washington, DC: Council on Environmental Quality.

Freeman, A. M., III. (2000). Water pollution policy. In P. R. Portney & R. N. Stavins (Eds.), *Public policies for environmental protection* (pp. 169–213), 2nd edition. Washington, DC: RFF Press.

Griffiths, C. & Wheeler, W. (2005). Benefit-cost analysis of U.S. regulations affecting surface water quality. In R. Brouwer & D. Pearce (Eds.), *Cost-benefit analysis and water resources management* (pp. 223–250). Cheltenham, UK: Edward Elgar.

Griffiths, C. (October 31–November 1, 2011). *Considerations for a benefit-cost analysis of the Chesapeake Bay TMDL*. (Presentation at Chesapeake Bay Economic Researchers' Workshop, Washington, DC). http://abtassociates.com/projects/2011/assessment-of-economic-effects-of-the-chesapeake-b.aspx.

Griffiths, C., Klemick, H., Massey, D. M., Moore, C., Newbold, S., Simpson, R. D., et al. (2012). EPA valuation of surface water quality improvements. *Review of Environmental Economics and Policy, 6*, 130–146.

Heintz, H. T., A. Hershaft & G. Horak. (1976). National damages of air and water pollution. Report submitted to the U.S. Environmental Protection Agency.

Jakus, P. M., Downing, M., Bevelhimer, M. S., & Fly, J. M. (1997). Do fish consumption advisories affect reservoir anglers' site choice? *Agricultural and Resource Economics Review, 26*, 198–204.

Johnston, R. J., Besedin, E. Y., Iovanna, R., Miller, C. J., Wardwell, R. F., & Ranson, M. H. (2005). Systematic variation in willingness to pay for aquatic resource improvements and implications for benefit transfer: A meta-analysis. *Canadian Journal of Agricultural Economics, 53*, 221–248.

Johnston, R. J. & Rosenberger, R. S. (2010). Methods, trends and controversies in contemporary benefit transfer. *Journal of Economic Surveys, 24*, 479–510. http://dx.doi.org/10.1111/j.1467-6419.2009.00592.x.

Lant, C. L., & Roberts, R. S. (1990). Greenbelts in the corn belt: Riparian wetlands, intrinsic values, and market failure. *Environment and Planning A, 22*, 1375–1388.

Loomis, J. B. (1992). The evolution of a more rigorous approach to benefit transfer: Benefit function transfer. *Water Resources Research, 28*, 701–705. doi:10.1029/91WR02596.

Lyke, A. J. (1993). *Discrete choice models to value changes in environmental quality: A Great Lakes case study*. Dissertation, University of Wisconsin, Department of Agricultural Economics.

Milliman, S. R., Johnson, B.L., Bishop, R.C. & Boyle, K.J. (1992). The bioeconomics of resource rehabilitation: A commercial-sport analysis for a Great Lakes fishery. *Land Economics, 68*, 191–210. http://www.jstor.org/stable/3146774.

Mitchell, R. C. & Carson, R. T. (1984). *Willingness to pay for freshwater quality improvements*. Report prepared for U.S. Environmental Protection Agency, Office of Policy Analysis, under cooperative agreement CR810224-01-4. Washington, DC. http://yosemite.epa.gov/ee/epa/eerm.nsf/vwGA/21E2CA5C213441AD8525644D0053BEA2.

Mitchell, R. C., & Carson, R. T. (1986). *The use of contingent valuation data for benefit/cost analysis in water pollution control*. CR-810224-02. Report prepared for U.S. Environmental Protection Agency, Office of Policy Analysis, under cooperative agreement CR810224-01-4, Washington, DC. http://yosemite.epa.gov/ee/epa/eerm.nsf/vwGA/2B37DFD67EE8870A85256 44D0053BE87.

Mitchell, R. C., & Carson, R. T. (1989). *Using surveys to value public goods: The contingent valuation method*. Washington, DC: Resources for the Future.

Montgomery, M., & Needelman, M. (1997). The welfare effects of toxic contamination in freshwater fish. *Land Economics, 73*, 211–223.

Phaneuf, D. J., Kling, C. L., & Herriges, J. A. (1998). Valuing water quality improvements using revealed preference methods when corner solutions are present. *American Journal of Agricultural Economics, 80*, 1025–1031.

Rosenberger, R. S., & Loomis, J. B. (2000). Using meta-analysis for benefit transfer: In-sample convergent validity tests of an outdoor recreation database. *Water Resources Research, 36*, 1097–1107.

Smith, R. A., Schwarz, G. E., & Alexander, R. B. (1997). Regional interpretation of water-quality monitoring data. *Water Resources Research, 33*, 2781–2798.

Tudor, L., Besedin, E., Smith, S. & Snyder, L. (1999). *What pollutants matter for consumers of water-based recreation?* Paper presented at the annual Northeastern Agricultural and Resource Economics Association meeting. Morgantown, WV, June.

Unger, S. G. (1975). *National benefits of achieving the 1977, 1983, and 1985 water quality goals*. Report submitted to the U.S: Environmental Protection Agency.

U.S. EPA. (1982). Regulatory impact analysis of the effluent limitation guidelines regulation for the iron and steel industry.

U.S. EPA. (1983). *Guidelines for performing regulatory Impact analysis*. (Reprinted in 1991). http://yosemite.epa.gov/ee/epa/eerm.nsf/vwAN/EE-0228A-1.pdf/%24file/EE-0228A-1.pdf.

U.S. EPA. (1987). Regulatory impact analysis of the effluent guidelines regulation for the organic chemicals, plastics, and synthetic fibers industry.

U.S. EPA. (1993). Regulatory impact analysis of the proposed Great Lakes water quality guidance: Final report. EPA 820-B95-011. Downloaded from nepis.epa.gov, last checked 7/19/2012.

U.S. EPA. (1995). Regulatory impact analysis of the final Great Lakes water quality guidance: Final report. EPA 820-B95-011. Downloaded from nepis.epa.gov, last checked 7/19/2012.

U.S. EPA. (1997). *Economic analysis for the national emission standards for hazardous air pollutants for source category: Pulp and paper production, effluent limitations guidelines, pretreatment standards, and new source performance standards: Pulp, paper, and paperboard category—Phase 1*. http://water.epa.gov/scitech/wastetech/guide/pulppaper/upload/1997_11_13_guide_pulppaper_jd_pulp.pdf, last accessed 7/20/12.

U.S. EPA. (2000a). *A benefits assessment of water pollution control programs since 1972: Part 1, the benefits of point source controls for conventional pollutants in rivers and streams*. http://water.epa.gov/lawsregs/lawsguidance/cwa/316b/upload/2000_04_17_economics_assessment.pdf.

U.S. EPA. (2000b). *Economic analysis of effluent limitation guidelines and standards for the centralized waste treatment industry*. http://water.epa.gov/scitech/wastetech/guide/treatment/upload/2006_12_28_guide_cwt_final_econanal.pdf.

U.S. EPA. (2000c). *Economic analysis of final effluent limitations guidelines and standards for commercial hazardous waste combustors*. EPA 821-B-99-008. http://water.epa.gov/scitech/wastetech/guide/chwc/upload/2000_05_23_guide_chwc_final_economic.pdf.

U.S. EPA. (2000d). *Economic analysis of final effluent limitations guidelines and standards for the landfills point source category*. EPA-821-B-99-005. http://water.epa.gov/scitech/wastetech/guide/landfills/upload/economics.pdf.

U.S. EPA. (2000e). *Economic analysis of final effluent limitations guidelines and standards for the transportation equipment cleaning category*. EPA-821-R-00-013. http://water.epa.gov/scitech/wastetech/guide/tec/upload/TEC_Economic-Analysis_2000.pdf.

U.S. EPA. (2000f). *Economic analysis of the California Toxics Rule*. http://water.epa.gov/lawsregs/rulesregs/ctr/upload/2000_04_28_standards_rules_ctr_rule.pdf.

U.S. EPA. (2000g). *Guidelines for preparing economic analyses*. EPA 240-R-00-003. Washington, DC: Office of the Administrator. http://yosemite.epa.gov/ee/epa/eerm.nsf/vwRepNumLookup/EE-0516?OpenDocument.

U.S. EPA. (2003a). *Environmental and economic benefit analysis of the final revisions to the national pollutant discharge elimination system regulation and the effluent guidelines for concentrated animal feeding operations*. EPA-821-R-03-003. Washington, DC: Office of Water. http://cfpub.epa.gov/npdes/docs.cfm?view=allprog&program_id=7&sort=name#cafofinalrule_envecon_2003.

U.S. EPA. (2003b). *Estimation of national economic benefits using the national water pollution control assessment model to evaluate regulatory options for concentrated animal feeding operations (CAFOs)*. EPA-821-R-03-009. Washington, DC: Office of Water. http://cfpub.epa.gov/npdes/docs.cfm?view=allprog&program_id=7&sort=name#cafofinalrule_nationaleconbenefits_2003.

U.S. EPA. (2003c). *Economic, environmental, and benefits analysis of the Final Metal Products & Machinery Rule*. EPA-821-B-03-002. http://water.epa.gov/scitech/wastetech/guide/mpm/eeba_index.cfm, last accessed 7/20/12.

U.S. EPA. (2004a), *Economic and environmental benefits analysis of the final meat and poultry products rule*. EPA-821-R-04-010. Washington, DC: Office of Water. http://water.epa.gov/scitech/wastetech/guide/mpp/upload/2004_08_02_guide_mpp_eeba_part1.pdf and http://water.epa.gov/scitech/wastetech/guide/mpp/upload/2004_08_02_guide_mpp_eeba_part2.pdf .

U.S. EPA. (2004b). *Economic and environmental benefits analysis of the final effluent limitations guidelines and new source performance standards for the concentrated aquatic animal production industry point source category*. EPA-821-R-04-13. Washington, DC: Office of Water. http://water.epa.gov/scitech/wastetech/guide/aquaculture/EEBA_index.cfm.

U.S. EPA. (2005). *Environmental economics research strategy*. EPA/600/R-04/195. http://www.epa.gov/research/htm/documents/econresearch.pdf.

U.S. EPA. (2009). *Environmental impact and benefits assessment for final effluent guidelines and standards for the construction and development category*. EPA-821-R-09-012. http://water.epa.gov/scitech/wastetech/guide/construction/upload/2008_12_8_guide_construction_files_environment.pdf.

U.S. EPA. (2010a). *Guidelines for preparing economic analyses*. EPA 240-R-10-001. Washington, DC: Office of the Administrator. http://yosemite.epa.gov/ee/epa/eed.nsf/webpages/Guidelines.html.

U.S. EPA. (2010b). *Economic analysis of final water quality standards for nutrients for lakeand flowing waters in Florida*. http://water.epa.gov/lawsregs/rulesregs/upload/florida_econ.pdf.

Vaughan, W. J. (1981). The water quality ladder, Appendix II. In: R. C. Mitchell &. T. Carson (Eds.), *An experiment in determining willingness to pay for national water quality improvement, draft report*. http://yosemite.epa.gov/ee/epa/eerm.nsf/vwAN/EE-0011-04.pdf/%24file/EE-0011-04.pdf.

Walsh. R. G., D. M. Johnson & J. R. McKean. (1988). *Review of outdoor recreation demand studies with nonmarket benefit estimates*. Colorado Water Resources Research Institute Technical Report No. 54. Fort Collins: Colorado State University.

Walsh, R. G., Johnson, D. M., & McKean, J. R. (1990). Nonmarket values from two decades of research on recreation demand. *Advances in Applied Micro-Economics, 5*, 167–193.

Wells, D., Srinivasan, R. & Arnold, J. (2011). *Hydrologic and Water Quality System (HAWQS) Project*. (Presentation at the 47th Annual American Water Resources Association Conference, Albuquerque, NM). http://www.awra.org/annual2011/doc/pres/S7-Wells.pdf.

Wilson, M. A., & Hoehn, J. P. (2006). Valuing environmental goods and services using benefit transfer: The state-of-the art and science. *Ecological Economics, 60*, 335–342.

Part II
Methods and Applications

Chapter 7
Benefit Transfers with the Contingent Valuation Method

John C. Whitehead, O. Ashton Morgan and William L. Huth

Abstract The results of contingent valuation analyses are often used for benefit transfer. The contingent valuation method is a stated preference approach to the valuation of non-market goods in which survey respondents are asked hypothetical questions directly about their total economic values. The advantages of the method include flexibility, ability to estimate nonuse values and an ability to incorporate ex-ante uncertainty. Previous benefit transfer research with contingent valuation is difficult to assess since each study uses different forms of the valuation question and benefit transfer tests are not uniform. Nevertheless, there is some evidence that dichotomous choice valuation questions may produce lower transfer errors relative to other question formats. We present a case study using the dichotomous choice referendum question format with key tests for theoretical validity and find evidence that these study features may improve benefit transfer reliability.

Keywords Benefit transfer · Contingent valuation method · Willingness to pay · Accuracy · Validity · Reliability · Oyster safety

---

J.C. Whitehead (✉) · O.A. Morgan
Department of Economics, Appalachian State University, Boone, NC, USA
e-mail: whiteheadjc@appstate.edu

O.A. Morgan
e-mail: morganoa@appstate.edu

W.L. Huth
Department of Marketing and Economics, University of West Florida, Pensacola, FL, USA
e-mail: whuth@uwf.edu

R.J. Johnston et al. (eds.), *Benefit Transfer of Environmental and Resource Values*,
The Economics of Non-Market Goods and Resources 14,
DOI 10.1007/978-94-017-9930-0_7

7.1 Introduction

The contingent valuation method (CVM) is a stated preference approach to the valuation of non-market goods (Mitchell and Carson 1989).[1] Preferences are "stated" in the sense that survey respondents are asked hypothetical questions directly about their total economic values. This contrasts to revealed preference methods, such as the travel cost method and the hedonic pricing method, in which use values are revealed by observable behavior. The CVM has been used for major policy analyses associated with the U.S. Clean Water Act, the U.S. Clean Air Act and the Natural Resource Damage Assessment associated with the Exxon Valdez oil spill. It is also a major source of economic values for benefit transfer-based policy analysis in recreation, health, environmental quality and other public goods. Adamowicz (2004) argues that the environmental valuation literature was dominated by the CVM during the 1990s, with hedonic pricing a distant second and travel cost, choice experiment, and other stated preference methods even further behind.[2] Carson (2011) finds that the CVM publication trend continues through 2007 with more than 2000 CVM articles in the ISI Web of Science database. Carson's (2011) CVM bibliography includes more than 7000 entries.

The CVM typically asks respondents about their willingness to pay for a policy proposal that leads to a change in a non-market good relative to the status quo. Empirical estimation tends to focus on the total value of the non-market good. This differs to varying extents from other stated preference approaches such as choice modeling and contingent behavior (Carson and Louviere 2011). The choice modeling approach (i.e., discrete choice experiment), for example, typically asks respondents about their preferences for multi-attribute policy proposals with two or more choice alternatives in addition to the status quo. Empirical estimation focuses on the marginal valuation of attributes of the non-market goods, although total values may also be estimated. The CVM and choice modeling approach can both estimate use and nonuse values. In contrast, the contingent behavior approach asks questions about hypothetical behavior with discrete choice or frequency of choice questions (Whitehead et al. 2008). For example, a contingent behavior question might ask respondents how the number of trips to a particular beach might change as a result of possible changes in beach width (see Chap. 9). Use values are then estimated using the empirical methods of revealed preference approaches (Bockstael and McConnell 2010). Of the alternative methods for stated preference value elicitation and benefit transfer, this chapter emphasizes the CVM. We do so for two reasons: (1) benefit transfer with choice modeling and contingent behavior are addressed elsewhere in this volume and (2) to our knowledge, there are

[1]See Whitehead (2006) for a discussion of how to conduct CVM studies and Whitehead and Blomquist (2006) for a discussion of CVM and benefit-cost analysis.

[2]This is in part due to the controversy surrounding the Exxon Valdez oil spill (Portney 1994).

no benefit transfer studies using the contingent behavior method in the literature outside of this volume.[3]

Like many forms of primary data collection for non-market valuation, original contingent valuation surveys can be time-consuming and expensive to conduct. The benefit transfer approach to environmental valuation was developed for situations in which these time and/or money costs are prohibitive (Boyle et al. 2010; Johnston and Rosenberger 2010). With benefit transfer, benefit estimates from existing case studies (i.e., the study sites) are spatially and/or temporally transferred to a policy case study (i.e., the policy site). Benefit transfer has been widely used to inform policy analysis since the 1950s (Bergstrom and DeCivita 1999). Yet it was not until a special section in volume 28(3) of *Water Resources Research* (*WRR*) that attention was given to benefit transfer as a distinct field of research, with emphasis on both theory and practice (Brookshire and Neill 1992).

Brookshire and Neill (1992) state that the special section was motivated, in part, by Luken et al. (1992) and Desvousges et al. (1992) who conduct cost-benefit analyses of Clean Water Act standards, with funding from the U.S. Environmental Protection Agency. Both of these efforts transfer benefit estimates to the policy sites using CVM benefit estimates from different study sites and highlight a number of issues and concerns related to the transfer exercise. Smith (1992) compares the Luken et al. (1992) and Desvousges et al. (1992) studies and highlights the major role played by researcher judgment. One primary conclusion from the special section was that benefit transfer should become a research topic and not a "back-room exercise" conducted by policy analysts (Brookshire and Neil 1992).

Many of the approaches now used to implement and evaluate benefit transfers were first introduced in the original 1992 *WRR* articles. Luken et al. (1992) and Desvousges et al. (1992) employed unit value transfer, in which an estimate of willingness to pay developed for the study site is directly applied at the policy site. This was a standard approach up to that time. Within the *WRR* special section, however, Loomis (1992) introduced benefit function transfer in which an empirical model of benefits and determinants developed at a study site is used to estimate benefits at the policy site. Characteristics of the current policy situation or case study (e.g., population demographics, site characteristics) are substituted into the statistical model from the study site to develop benefit estimates that are tailored to the current policy situation. It is now accepted that benefit function transfer generally outperforms unit value transfer (Johnston and Rosenberger 2010). Also in the special section, Boyle and Bergstrom (1992) suggest the now-common convergent validity (i.e., percent transfer error) test for unit value and benefit function transfers. Benefit function transfer can also be assessed using tests of coefficient equality. These and other *WRR* articles recommended components of a systematic research agenda that has been adhered to by most benefit transfer studies since then.

[3]This is a much needed area for future research since there are a large number of contingent behavior studies which can serve as a source of environmental values for benefit transfer-based policy analysis.

Two early benefit transfer validity studies illustrate these tests for cases of CVM benefit function transfer. Downing and Ozuna (1996) use the CVM to estimate the benefits of marine recreational fishing in Texas bays. They transfer benefits across time and space and find few benefit functions with coefficient equality. Few benefit estimates generated from the benefit functions transfer accurately. Similarly, in a study of recreation sites in Arizona and New Mexico using the CVM, Kirchhoff et al. (1997) find that between 55 and 90 % of the Arizona benefit function transfer tests and 90 % of the New Mexico tests reject convergent validity. In a review of the convergent validity literature, Kaul et al. (2013) identify 31 benefit transfer studies that provide enough information for meta-analysis. Twelve of the 31 studies use the CVM. For the entire sample of valuation studies, Kaul et al. (2013) find that benefit function transfer outperforms unit value transfer and the CVM performs (1) no worse than other valuation methods and (2) better than meta-analysis transfer.

The remainder of this chapter discusses a number of CVM issues in the context of the benefit transfer literature. We describe the advantages of the CVM for benefit transfer relative to other valuation methods; these include flexibility, ability to estimate nonuse values and incorporation of ex-ante uncertainty. Next we describe standard convergent validity tests for benefit transfer in the context of the CVM benefit transfer literature. We then review the various forms of the willingness to pay valuation question used in benefit transfer studies. Various tests for validity and reliability of the CVM are described and their absence in the benefit transfer literature noted. Finally, we provide a case study of a typical benefit transfer exercise. Among the contributions of this chapter is the first example of convergent validity testing with the preferred CVM question format, a discrete choice referendum, with associated validity tests for price and quantity (i.e., scope effects).

7.2 Advantages of the CVM for Benefit Transfer

Stated preference methods, including the CVM, are more flexible relative to revealed preference methods. All of the revealed preference methods are constrained to quasi-public goods—those for which the non-market good may be modeled as a characteristic of a market good. Revealed preference methods require that the observed demand for the market good first be estimated and then that the effect of changes in the non-market good on the market good be isolated. If these two empirical conditions are satisfied the revealed preference method can be used to estimate use value resulting from the change in the quasi-public good.

In contrast, valuation of nearly all quasi-public and pure public goods is within the domain of two of the stated preference methods—CVM and choice modeling (i.e., discrete choice experiments). The only constraint that application of the CVM and choice modeling imposes is that a realistic valuation scenario must be constructed around payment and delivery of the change in the non-market good. Contingent behavior, in turn, is limited only by the necessity of framing the hypothetical question in the appropriate behavioral context. As a result, compared to revealed preference methods, stated preference methods including the CVM enable valuation of a much greater range of non-market resources.

In addition, realistic policy analysis often requires valuation beyond the observable range of historic behavior. The stated preference methods introduce the flexibility to value wide ranges of future quality changes. This flexibility extends to valuation of projects of different scope. For example, multiple valuation questions can be used to estimate the value of the incremental benefits of a project to determine the scope at which the net benefits are maximized. Most applications of the revealed preference methods are limited to simulated changes in scope. The validity of these simulations, particularly for large changes, is often tenuous due to potential non-linearities and other complications.

Beyond flexibility, stated preference methods offer advantages over revealed preference methods in the types of values that can be measured. Willingness to pay is a measure of the total value of a policy change to an individual, and can be decomposed into use and nonuse values. Stated preference methods are the only broadly accepted valuation methods capable of capturing nonuse values within total value estimates. For some policies, nonuse values may exist but their contribution to total value is not substantial. In these cases revealed preference methods may be sufficient. For other policies, however, ignoring the measurement of nonuse values would lead to significant errors in policy analysis.

Finally, stated preference methods can have advantages when evaluating willingness to pay for policies whose outcomes are subject to considerable uncertainty. Ex post measures of willingness to pay sometimes incorporate uncertainty by assigning probabilities to different outcomes. The sum of the probability weighted ex post willingness to pay amounts from revealed preference methods yields expected surplus. For policies and projects that involve significant uncertainty, however, the appropriate measure of the impacts of policy is an ex ante measure. Stated preference scenarios can be designed with uncertainty as part of the experimental design, thereby providing these more theoretically appropriate ex ante measures (e.g., respondents can be asked their ex ante willingness to pay for policies that incorporate explicit uncertainty).

7.3 Benefit Transfer Tests

Three major benefit transfer tests have appeared in the CVM convergent validity literature, designed to evaluate unit value transfer, benefit function transfer and benefit function estimate transfer. These tests can only be conducted for cases in which a primary valuation study has been conducted at both the study and policy sites. Hence, they are not typically possible as a part of policy-motivated (i.e., real world) benefit transfers, because transfers are used only when a primary study at the policy site is infeasible. However, they can be used to evaluate the accuracy of benefit transfers for illustrative cases in which comparative policy site values are available.

The validity of unit value transfer is tested with a comparison of unadjusted willingness to pay estimates. The null hypothesis is:

$$H0\colon \Delta WTP_{s,p} = 0 \tag{7.1}$$

where $\Delta WTP_{s,p} = WTP_s - WTP_p$ is the difference in study, s, and policy, p, site willingness to pay, WTP. If benefit functions are estimated, $WTP_j = \beta_j' x_j$, where β is a coefficient vector, x includes policy relevant and socioeconomic variables, and $j = p, s$, then a statistical test of benefit function transfer is one for equality of the coefficient vectors. The null hypothesis is:

$$H0\colon \beta_s = \beta_p. \tag{7.2}$$

A statistical test for benefit function value transfer is

$$H0\colon \Delta WTP_{sp,pp} = 0 \tag{7.3}$$

where $\Delta WTP_{sp,pp} = WTP_{sp} - WTP_{pp}$, $WTP_{sp} = \beta_s' x_p$ and $WTP_{pp} = \beta_p' x_p$.

These tests illustrate some of the necessary conditions proposed by Brouwer (2000) for accurate benefit transfer: (1) that willingness to pay estimates are accurate (i.e., valid and reliable), (2) that the populations in the study and policy sites must be similar, (3) that the difference between pre-policy and post-policy quality (or quantity) levels must be similar and (4) that the study and policy sites must be similar in terms of baseline environmental characteristics. The accuracy of benefit transfer will often suffer if these conditions are violated. Other problematic issues identified in the benefit transfer literature involve preference heterogeneity and functional form. Unobserved preference heterogeneity, such as differences in environmental attitudes, may decrease the accuracy of benefit transfers (Brouwer and Spaninks 1999) and can lead to rejection of each of the three hypotheses. In addition, differences between the underlying functional form of preferences or willingness to pay at the study and policy sites can diminish the accuracy of benefit transfer.[4]

Of the 12 CVM studies considered by Kaul et al. (2013),[5] including Scarpa et al. (2010), which uses the same data as Matthews et al. (2009), nine conduct unit value transfer tests. Of these, three reject equality in all comparisons (Barton 2002; Bergland et al. 2002; Rozan 2004), and one rejects the hypothesis in 89 % of the comparisons (Kristofersson and Navrud 2007). Five studies reject the hypothesis of unit value transfer in 42–50 % of the comparisons (Brouwer and Bateman 2005; Brouwer and Spaninks 1999; Scarpa et al. 2010; Ready et al. 2004; Vandenberg et al. 2001).

---

[4]Rosenberger and Stanley (2006) raise additional (and related) issues that may reduce benefit transfer accuracy.

[5]See Bateman et al. (2011) for a more recent CVM benefit transfer study.

Eight studies test for benefit function transfer accuracy across policy and study sites. Three of these reject coefficient equality in all comparisons (Barton 2002; Bergland et al. 2002; Ready et al. 2004), and two reject the hypothesis in more than 75 % of the comparisons (Brouwer and Bateman 2005; Brouwer and Spanicks 1999). Only two failed to reject the hypothesis of coefficient equality (Matthews et al. 2009; Rozan 2004).

Six studies conduct tests of benefit function value transfer. Of these, two reject equality in all comparisons (Barton 2002; Bergland et al. 2002), and three reject the hypothesis in 57–78 % of the comparisons (Brouwer and Bateman 2005; Kristofersson and Navrud 2007; Vandenberg et al. 2001). Only Groothuis (2005) finds that benefit function values transfer in 80 % of the comparisons.

One conclusion from this review is that Brouwer's (2000) stringent conditions for benefit transfer accuracy are typically not met under the best of conditions, and are even less likely to be met when conducting real world policy analysis. In this context the concept of reliability is important. Reliability is the empirical accuracy of a benefit transfer measured by the magnitude of transfer error, and is quantified with convergent validity tests. Convergent validity is a measure of benefit transfer accuracy in which transfer error is calculated based on the difference between a transferred value estimate and an alternative value estimate for the same site. For unit value (*UV*) transfers and benefit function value (*BFV*) transfers, the percentage transfer error (*TE*) is

$$\%TE_{UV} = 100 \times \left(WTP_s - WTP_p\right)/WTP_p \tag{7.4}$$

$$\%TE_{BFV} = 100 \times \left(WTP_{sp} - WTP_{pp}\right)/WTP_{pp} \tag{7.5}$$

The weighted mean transfer error of the CVM convergent validity studies in Kaul et al. (2013) is 36 %. The range of average transfer across study is 20 % (Barton 2002) to 125 % (Kristofersson and Navrud 2007). The range of transfer error within studies with more than two comparisons is from 13 % (Barton 2002) to 312 % (Kristofersson and Navrud 2007). An open question is the acceptable level of transfer error for policy analysis.

A conclusion from Kaul et al. (2013) is that benefit function value transfer tends to outperform unit value transfer in the benefit transfer literature. Only two of the eight CVM studies that clearly provide this comparison find that benefit function value transfer improves accuracy (Barton 2002; Groothuis 2005), with three studies providing mixed results (Brouwer and Bateman 2005; Kristofersson and Navrud 2007; Vandenberg et al. 2001). Bergland et al. (2002), Brouwer and Spaninks (1999) and Ready and Navrud (2007) conclude that unit value transfer outperforms benefit function value transfer. More recently, Bateman et al. (2011) conclude that unit value transfer is accurate for similar study and policy sites. Benefit function transfer is more accurate for dissimilar study and policy sites.

7.4 Question Format and Convergent Validity

Although there has been significant research on transfer errors emerging from CVM unit and function transfer, this research generally overlooks whether the format of the valuation question influences transfer reliability. This is a relatively surprising omission, given that past research has shown that benefit estimates, and therefore benefit transfer, can be affected by the question format (Carson and Groves 2007). The CVM benefit transfer literature reflects the same differences in forms of valuation questions that exist in the larger CVM literature, yet the impact of these differences on transfer accuracy remains unknown.

Early applications of the CVM often asked respondents iterative bidding questions, in which respondents answer yes or no to payment of increasing or decreasing bids. Upon finding that iterative bidding values were sensitive to the starting bid, open-ended questions about willingness to pay were introduced (Piper and Martin 2001; Brouwer and Bateman 2005). Rozan (2004) employs open-ended questions that follow an iterative bidding exercise. Open-ended questions are prone to incentive incompatibility, outliers and item nonresponse. The payment card question, in contrast, asks an open-ended question but provides dollar interval response categories to respondents (Vandenberg et al. 2001; Brouwer and Spaninks 1999). Payment card questions are prone to range bias.

In part due to limitations in the elicitation formats described above, the dichotomous choice question has now become the dominant form of the valuation question (Groothuis 2005; Scarpa et al. 2010). The single-bound dichotomous choice question is similar to the initial iterative bidding question with two differences: (1) the starting point is varied across survey respondents and (2) there are no follow-up willingness to pay questions. The advantage of the dichotomous choice question is that each respondent is asked a valuation question that is relatively easy to answer. The major disadvantage is that the researcher learns only whether each respondent's willingness to pay is above or below the dollar amount threshold included in the question. More sophisticated econometric methods are necessary to estimate average willingness to pay. Moreover, the variance on these estimates tends to be larger relative to valuation questions that produce continuous willingness to pay distributions, for any given sample size. As a result, larger samples are necessary to implement the dichotomous choice approach.

The double-bounded question was introduced to reduce the variance of single-bound willingness to pay. The double-bounded format adds one follow-up question to the single-bound dichotomous choice question. If the respondent answers "yes" to the first question then the dollar amount is increased and the question is asked again. If the respondent initially answers "no," then the dollar amount is reduced and the question is asked again. Different econometric methods may be used to analyze the resulting data. Barton (2002), Scarpa et al. (2010) and Matthews et al. (2009) analyze the double-bounded data with discrete choice methods. Bergland et al. (2002) analyze the double-bounded data with interval data methods. Although these approaches can increase the efficiency of willingness to pay estimation, they are subject to starting point bias and incentive incompatibility. Hence, they are less common than single-bounded questions.

Another form of multiple-bounded valuation question is a combination of the payment card and polychotomous choice question. Polychotomous choice questions add categories such as "definitely yes," "probably yes," "probably no," and "definitely no" to the dichotomous choice question format. In this visual question format a wide range of dollar amounts are offered in the first column of a table and respondents are asked to state their willingness to pay each amount by checking boxes in columns expressing various levels of certainty (Kristofersson and Navrud 2007). In a single column "payment ladder" variation respondents may indicate all of the lower values that they would certainly pay and all of the higher values they certainly would not pay (e.g., Ready et al. 2004; Ready and Navrud 2007). A benefit of the multiple-bounded approach is the expression of uncertainty and the collection of multiple observations for each respondent, which increases econometric efficiency. A potential cost is the unfamiliarity of the valuation task, which can lead to various response biases.

Despite past work comparing the performance and properties of alternative willingness to pay elicitation methods, the influence of question format on transfer error is currently unknown. There is some evidence that studies that use some form of dichotomous choice question might have lower transfer errors. Considering the studies presented by Kaul et al. (2013), the lone study to use single-bound dichotomous choice has an average transfer error of 30 % (Groothuis 2005). Studies that use double-bounded dichotomous choice questions have transfer errors of 20 % (Barton 2002), 21 % (Bergland et al. 2002) and 27 % (Matthews et al. 2009). Studies that use some form of an open-ended, payment card or payment ladder question have average transfer errors of 25 % (Rozan 2004); 29 % (Vandenberg et al. 2001); 34 % (Brouwer and Bateman 2005); 37 % (Ready et al. 2004); 39 % (Piper and Martin 2001); 42 % (Brouwer and Spaninks 1999) and 125 % (Kristofersson and Navrud 2007). The weighted mean transfer error is 28 % for the dichotomous choice studies and 47 % for the open-ended studies.

While the difference is suggestive, it is based on a very small sample of studies. Moreover, none of the studies cited above conducts a direct test of transfer error under multiple elicitation formats for the same non-market good. Hence, differences that appear to be caused by question format may instead be related to other, unobserved differences across studies. Given the lack of systematic, controlled evidence, it is not yet possible to conclude that dichotomous choice question formats are preferred for benefit transfer.

7.5 Accuracy of the CVM and Benefit Transfer

Accuracy of the CVM is a necessary condition for accuracy of the benefit transfers that rely on this method (Brouwer 2000). Accuracy is comprised of validity and reliability. Validity is the extent to which the CVM generates unbiased willingness to pay.[6] Reliability is the extent to which the CVM consistently generates the same estimate of willingness to pay.[7]

There are several types of validity that have been considered in the CVM literature. Criterion validity is the extent to which hypothetical willingness to pay compares favorably to actual willingness to pay. Often, CVM survey respondents may state that they will pay for a good when in fact they will not, or they will actually pay less, when placed in a similar purchase decision. A difference between hypothetical and actual willingness to pay has been found in a variety of private and public good applications (Murphy et al. 2005). Two approaches to "hypothetical bias" mitigation dominate the literature (Loomis 2011). Several researchers find that the divergence between hypothetical and actual willingness pay is mitigated or eliminated by providing additional instructions to respondents encouraging them to carefully consider their budget constraints and substitutes and to treat the hypothetical scenario as if an actual monetary transaction were taking place. Other researchers find that hypothetical willingness to pay is similar to actual willingness to pay when the level of certainty respondents have about making payment is taken into account. Only one of the CVM benefit transfer studies (Ready et al. 2004) explicitly considers the effects of hypothetical bias, using a certainty scale correction when evaluating transfer accuracy.

Content validity is the extent to which a stated preference survey presents information and questions in a way that enables valid willingness to pay elicitation. Among the components of content validity is the incentive compatibility of different question formats. Carson and Groves (2007) argue that scenarios that involve the provision of public goods with a voluntary contribution format and the purchase of private goods could be expected to lead to overstatements of hypothetical willingness to pay (e.g., Ready et al. 2004). On the other hand, Carson and Groves (2007) argue that respondents, when considering a public good with individual policy costs and a referendum vote, will tend to truthfully reveal their willingness to pay if they believe their votes will ultimately impact a decision to implement the project. If respondents believe that there is some probability that the project will be implemented if the hypothetical referendum passes, incentives exist for truthful revelation of willingness to pay. None of the CVM benefit transfer studies adopt such an incentive-compatible referendum format.

Convergent validity, in the non-benefit transfer context, is the extent that willingness to pay from the CVM is correlated with similar measures of willingness to pay estimated using another valuation method. Estimates that are statistically similar achieve convergent validity, increasing the confidence in both valuation estimates. There is some consensus that CVM can achieve convergent validity with other methods for the estimation of use values (Carson et al. 1996). A more rigorous approach to convergent validity is the combination and joint estimation of stated and

[6]Measurement error refers to a transfer error caused by bias in the original valuation study.

[7]This is distinct from the use of the term "reliability" in the benefit transfer literature, where it refers to the size of transfer errors (Johnston and Rosenberger 2010).

revealed preference data. Joint estimation raises the possibility that convergent validity tests might be used to recover nonuse values. Gonzalez-Sepulveda and Loomis (2011) consider whether joint estimation can improve benefit transfer for recreation trip values. They estimate individual and joint dichotomous choice CVM and travel cost method models and find that transfers from both individually estimated travel cost and jointly estimated models pass statistical validity tests, in part due to large confidence intervals. Unit value transfer errors are 15 % for the CVM and 37 % for the travel cost method. Jointly estimated willingness to pay is virtually identical to the CVM estimates. Since the joint model results in more accurate estimates of seasonal benefits, they recommend its use for benefit transfer.

Expectation-based (or theoretical) validity is the extent to which willingness to pay changes in response to the changes in conditions under which it is evaluated, as predicted by theory. For example, do willingness to pay estimates decrease with own-price, increase or decrease with price of substitutes or complements and increase with income (for normal goods)? Only one of the twelve CVM benefit transfer studies conducts formal validity tests based on economic theory. Barton (2002) finds that agreement to pay in a dichotomous choice question declines with the price of the policy. Following the NOAA guidelines (Arrow et al. 1993), sensitivity to the scope of the resource allocation change exists if willingness to pay is non-decreasing in quality or quantity. A number of meta-analyses based solely on stated preference data demonstrate sensitivity to scope (e.g., Johnston et al. 2005). Barton (2002) is the only CVM benefit transfer study that provides a scope test, finding only mixed evidence that the willingness to pay estimates are sensitive to scope. Reliability tests in the CVM literature focus on the within and across study variation in estimates rather than the ability of studies to accurately measure unbiased value. There are several tests for reliability of the CVM including econometric and test-retest reliability. Econometric reliability is the ability to explain the variation in willingness to pay through observable variables included in the econometric specification. Econometric reliability can be tested with statistical measures of overall fit of a regression model. All of the CVM benefit transfer studies exhibit some degree of econometric reliability.

Temporal reliability is the stability of welfare estimates over time, and is tested by comparing results from CVM surveys conducted at different times. If the magnitude of willingness to pay is consistent across time then willingness to pay is considered temporally reliable (e.g., Brouwer 2006). However, a temporal difference in willingness to pay does not necessarily indicate unreliable results. If willingness to pay changes over time in response to changing factors that affect willingness to pay then the researcher may conclude that the CVM is temporally reliable and that benefit estimates transfer accurately over time (but not necessarily across sites). Whitehead and Hoban (1999) find that benefit functions transfer accurately over time only after adjustment for changing environmental attitudes. Benefit transfers typically vary across both space and time since the willingness to pay for the study site policy is determined in the past and willingness to pay at the policy site is for a future policy. Considering this result in a policy context with no policy site willingness to pay estimate, the benefit transfer analyst will have the same problem as the macroeconomic forecaster who must deal with structural breaks in time series data.

7.6 A Benefit Transfer Case Study

As described above, while the CVM is a common source of values used for benefit transfer, existing research provides only limited information on the expected performance of these transfers. For example, no benefit transfer convergent validity study to date has used the most recommended CVM question format, the referendum, with a successful validity test for quantity (i.e., scope) effects.[8] This section addresses this important but currently unexplored question in CVM content validity, as applied to benefit transfer. Specifically, do benefit transfers from referendum CVM questions exhibit expected sensitivity to scope?

7.6.1 Methods and Data

The data are drawn from a recent CVM study that considers an oyster safety policy.[9] For this study we developed an internet survey of oyster consumers (aged 18 and over) sampled from Georgia, Florida, Alabama, Mississippi, Louisiana, Texas, and California. The sample was drawn from an online panel and the survey was administered between March and April, 2010. The response rate was 53 %. In total, there were 1849 completed responses from oyster consumers across the seven states.

The willingness to pay question is in a referendum format: "Suppose that in order to minimize the risks from eating raw oysters, the U.S. Food and Drug Administration (FDA) proposes a federal law to ensure that all oysters are post-harvest processed (PHP) before going to market. It is believed that this will reduce the average annual number of deaths in the U.S. from eating raw oysters from the current 16 to 20 people to [*d*] people. However, because of the additional costs incurred by oyster producers to process their product, the program will result in an increase in the price of an average oyster meal for all consumers. Imagine that you have the opportunity to vote on this proposed law. If more than 50 % of those vote for the federal law, the FDA would put it into practice. If you could vote today and you knew that the price of your average oyster meal would go up by [Δp] but the price of all other food would stay the same, would you vote for or against the proposed law?" There are three randomly assigned versions of the annual number of deaths [*d*], 1–5, 6–10 and 11–15, and four randomly assigned bid levels, \$1, \$3,

[8]Bateman et al. (2011) test for scope effects with the payment card question format.

[9]See Whitehead et al. (2012) for more details.

Table 7.1 Variable descriptions

Variable	Descriptions	Values
FOR	One if vote "for" the proposal, zero if "against"	0, 1
Δp	Change in the price per oyster meal of the proposal	1, 3, 5, 7
Δq	Change in the number of lives saved by the proposal	5, 10, 15
ATRISK	One if consumer is at risk from *V. Vulnificus*, zero otherwise	0, 1
RAW	One if consumer eats raw oysters, zero otherwise	0, 1
HOUSE	Household size	1–9
CHILD	Number of children in the household	0–7
WHITE	Equal to one if consumer is white, zero otherwise	0, 1
MALE	Equal to one if consumer is male, zero otherwise	0, 1
INCOME	Household income (in $1000s)	8–150

$5 and $7. Respondents were given three choice options, for, against and undecided (would not vote).

For this empirical exercise we consider only those states that have (1) theoretically valid referendum responses in terms of both price and quantity effects and (2) positive willingness to pay for the program.[10] This leads to the deletion of California, Georgia and Louisiana responses, leaving 1267 respondents in four states. In order to keep the econometric models straightforward, we also discard n = 365 undecided voters leaving a sample of n = 659.[11] Forty-nine percent of the remaining sub-sample would vote for the proposal.

Variable descriptions are presented in Table 7.1 and a data summary is presented in Table 7.2. The data suggest some relevant differences across states. The proportion of "for" votes differs across state with a low of 44 % in Mississippi to a high of 53 % in Texas. The average change in oyster price does not vary significantly across states since this variable was randomly assigned. We convert the range of deaths avoided to a lives saved variable, Δq, equal to the change in the number of deaths, 5, 10 and 15. This variable mean is also similar across states since it was randomly assigned. The number of consumers who are at risk from tainted oysters due to autoimmunity issues (e.g., liver disease) varies from 16 % in Florida to 22 % in Texas. Only 31 % of Texas consumers eat oysters raw while 40 and 41 % of Florida and Mississippi consumers eat raw oysters. Household size, number of children and gender are similar across states. The Texas sample has a lower proportion of white oyster consumers, whereas average household income is lower in Florida. In addition to the variables presented in Table 7.2, the number of oyster meals consumed varies across state with averages of 16 in Texas, 25 in Mississippi and 16 in Florida.

[10]Louisiana willingness to pay is negative at the mean number of lives saved. Willingness to pay is positive and statistically significant at $p = 0.20$ when lives saved are equal to 26.

[11]Alternatives include recoding undecided voters to "against" votes or estimating multinomial or ordered models (Groothuis and Whitehead 2002).

Table 7.2 Data summary

Variable	Texas		Mississippi		Florida	
	Mean	SD	Mean	SD	Mean	SD
FOR	0.53	0.50	0.44	0.50	0.49	0.50
Δp	3.91	2.22	3.80	2.23	3.96	2.25
Δq	9.96	4.14	9.92	4.18	9.91	4.18
ATRISK	0.22	0.41	0.21	0.41	0.16	0.37
RAW	0.31	0.46	0.41	0.49	0.40	0.49
HOUSE	2.51	1.30	2.51	1.33	2.64	1.35
CHILD	0.46	0.84	0.55	1.06	0.62	1.02
WHITE	0.74	0.44	0.86	0.34	0.84	0.37
MALE	0.54	0.50	0.48	0.50	0.52	0.50
INCOME	72.64	38.95	68.60	38.75	62.81	38.09
Cases	249		177		233	

7.6.2 Benefit Function Transfer

If willingness to pay is given by $WTP = \alpha + \gamma\Delta q + \beta' x + \varepsilon$, $\varepsilon \sim N(0, \sigma^2)$ the probability of a "yes" response to the referendum question may be modeled using a probit model as the probability that willingness to pay is greater than or equal to the price change, Δp. This is given by

$$\begin{aligned}\Pr(for) &= Pr(\alpha + \gamma\Delta q + \beta' x + \varepsilon \geq \Delta p) \\ &= Pr\left(\frac{\alpha + \gamma\Delta q + \beta' x - \Delta p}{\sigma} \geq \frac{\varepsilon}{\sigma}\right)\end{aligned} \tag{7.6}$$

in which σ is the scale parameter, α/σ, γ/σ, β/σ is a probit coefficient vector and $-1/\sigma$ is the probit coefficient on the change in price variable (Cameron and James 1987).

As an initial step in the benefit transfer evaluation, Table 7.3 presents probit model results for each state. These models omit covariates that could otherwise be used to adapt estimates across states (e.g., for demographic differences in oyster consumers). Each of the change in price coefficients is negative and statistically significant at, at least, the $p = 0.05$ level lending validity to the transfer exercise. Each of the constant terms is positive and statistically significant at, at least, the $p = 0.05$ level. This result leads to a positive willingness to pay estimate. We conduct likelihood ratio tests to determine if the vectors of coefficients are equal in pairwise comparisons. The tests indicate that coefficient vectors for (i) Texas and Florida and (ii) Mississippi and Florida are not statistically different. The coefficient vectors for Texas and Mississippi are statistically different.

Table 7.4 presents probit models with covariates which are used for "fitted" unit value transfer and benefit function value transfer. Each of the price coefficients is negative and statistically significant at the $p = 0.01$ level. The quantity (i.e., scope)

Table 7.3 Unit value transfer probit models: dependent variable = *FOR*

	Texas		Mississippi		Florida	
	Coeff	SE	Coeff	SE	Coeff	SE
Constant	0.419**	0.164	0.421**	0.190	0.542***	0.171
Δp	−0.085**	0.036	−0.153***	0.044	−0.144***	0.038
Model χ^2	5.49*	12.19***	8.49**			
LL function	−169.27	−115.34	−154.00			
Cases	249	177	233			
LR test (χ^2[2 d.f.])						
TX and MS	5.58*					
TX and FL	2.18					
MS and FL	1.48					

$^*p \leq 0.10$, $^{**}p \leq 0.05$, $^{***}p \leq 0.01$

Table 7.4 Benefit function transfer probit models: dependent variable = *FOR*

	Texas		Mississippi		Florida	
	Coeff	SE	Coeff	SE	Coeff	SE
Constant	0.032	0.406	−0.466	0.879	−0.552	0.773
Δp	−0.130***	0.040	−0.134***	0.049	−0.164***	0.041
Δq	0.078***	0.021	0.334*	0.177	0.377**	0.156
Δq^2			−0.018**	0.009	−0.018**	0.007
ATRISK	0.074	0.201	0.690***	0.259	0.647***	0.247
RAW	−0.371**	0.188	−0.389*	0.219	−0.233	0.186
HOUSE	0.191*	0.105	0.198	0.159	0.284**	0.117
CHILD	−0.151	0.155	−0.142	0.198	−0.353**	0.152
WHITE	−0.249	0.198	−0.768**	0.310	−0.540**	0.251
MALE	−0.466***	0.177	0.099	0.223	−0.531***	0.190
INCOME	−0.001	0.002	−0.004	0.003	−0.003	0.002
Model χ^2	34.63***	31.88***	49.80***			
LL function	−154.70	−105.50	−136.55			
Cases	249	177	233			
LR test (χ^2[11 d.f.])						
TX and MS	24.20**					
TX and FL	78.26***					
MS and FL	10.70					

$^*p \leq 0.10$, $^{**}p \leq 0.05$, $^{***}p \leq 0.01$

coefficients are statistically significant at various levels. Both of these results lend validity to the transfer exercise, a necessary condition for valid transfer (Brouwer 2000). No other independent variable has a consistent effect on the "for" votes across model. Mississippi and Florida consumers who are at risk are more likely to vote for the policy. Texas and Mississippi raw oyster consumers are more likely to vote against the policy. The probability of a "for" vote increases with the size of Texas and Florida households. An increase in the number of children decreases the probability of a "for" vote for Florida consumers. White consumers are less likely to vote "for" the policy in Mississippi and Florida and male consumers are less likely to vote "for" the policy in Texas and Florida. The best-fit Texas model scope test indicates that willingness to pay increases linearly with each life saved by the policy. The marginal willingness to pay per meal for each life saved is $\frac{\partial WTP}{\partial \Delta Q} = 0.60(0.22)$, with the standard error in parentheses. The best-fit Mississippi and Florida models are quadratic with willingness to pay increasing at a decreasing rate with each life saved by the policy. The marginal willingness to pay per meal for each life saved is $\frac{\partial WTP}{\partial \Delta q} = 2.50(1.69) - 0.26(0.17) \times \Delta q$ for Mississippi and $\frac{\partial WTP}{\partial \Delta Q} = 2.27(1.07) - 0.22(0.11) \times \Delta q$ for Florida. In a joint Mississippi and Florida probit model a likelihood ratio test for equality of these coefficients indicates that there is no statistical difference across models ($\chi^2 = 2.10[2df]$). These results indicate a valid transfer of marginal willingness to pay between Mississippi and Florida but not between Texas and the other two states.

We conduct likelihood ratio tests to determine if the vectors of coefficients are equal in pairwise comparisons. Since functional forms are different we use the log-likelihood function from a quadratic model for Texas. The tests indicate that the coefficient vectors for (1) Texas and Mississippi and (2) Texas and Florida are statistically different. In spite of several differences in individual coefficients, the coefficient vectors for Mississippi and Florida are not statistically different.

7.6.3 Value Transfer

Unit value willingness to pay estimated from a model without covariates is $WTP = \hat{\alpha}$. Fitted unit value willingness to pay is estimated with same site coefficients and variables with constant quantity across sites, $WTP_j = \hat{\alpha}_j + \gamma_j \Delta q + \hat{\beta}_j' \bar{x}_j$, where $\Delta q = 10$. Benefit function value transfer willingness to pay is evaluated at the means of the site specific variables and the estimated coefficients, $WTP_{ps} = \hat{\alpha}_p + \gamma_p \Delta q + \hat{\beta}_p' \bar{x}_s$, where $\Delta q = 10$. Standard errors are constructed using the Delta Method (Cameron 1991). The difference between "fitted" unit value transfer and benefit function transfer is that the latter adjusts for differences in respondent characteristics.

In Table 7.5 we present unit value transfer tests based on the models in Table 7.3. The willingness to pay per meal estimates are \$5 for Texas, \$3 for Mississippi and \$4 for Florida (with rounding). Each of these are significantly different from zero at the $p = 0.01$ level. The bottom portion of Table 7.5 is the difference in willingness to pay and percentage transfer errors. Each row presents two study site values transferred to a policy site. In this way we present two sets of benefit transfers with each pair of states considered a policy site and a study site.

Table 7.5 Unit value transfer

Study sites						
Policy Sites	Texas		Mississippi		Florida	
	WTP	SE	WTP	SE	WTP	SE
Texas	4.94	1.04				
Mississippi			2.76	0.69		
Florida					3.77	0.583
	ΔWTP	%TE[a]	ΔWTP	%TE	ΔWTP	%TE
Texas[b]			−2.18	44.13	−1.17	23.68
Mississippi[c]	2.18	78.99			1.01	36.59
Florida[d]	1.17	31.03	−1.01	26.79		

[a]Transfer error
[b]Transferring MS and FL study sites to TX policy site
[c]Transferring TX and FL study sites to MS policy site
[d]Transferring TX and MS study sites to FL policy site

While the willingness to pay estimates are highly significant, their confidence intervals are wide so that the differences in willingness to pay are not statistically significant.[12] This indicates that the six unit value transfers are valid. The average transfer error is 40 %.

In Table 7.6 we present "fitted" unit value and benefit function value transfer tests based on the models in Table 7.4. The willingness to pay estimates in the diagonal of the top portion of the table are the benefit function unit values with the means of the coefficients from the same state and $\Delta q = 10$. The fitted values are $5 for Texas, $5 for Mississippi and $6 for Florida (with rounding). Each of these are significantly different from zero at the $p = 0.01$ level. The off-diagonal willingness to pay estimates are benefit function values with means from the row policy sites substituted into the column study site models.

The bottom portion of Table 7.6 presents the difference in willingness to pay and percentage transfer errors. The first set of comparisons presents fitted unit value transfer where only the willingness to pay estimates from the diagonal are compared. As in Table 7.4, the confidence intervals on each of the differences in willingness to pay overlap and the differences are not statistically significant. The average transfer error is 15 %. The second set of comparisons presents benefit function value transfer where the study site willingness to pay estimates from the off-diagonal are compared to the policy site diagonal estimates. The differences are not statistically significant. The average transfer error is 22 %.

[12]When willingness to pay values have wide confidence intervals the standard tests of differences in means are relatively weak. Equivalence testing may be more important in the benefit transfer context (Kristofferson and Navrud 2005). Equivalence tests specify a range of acceptable transfer errors. Johnston and Duke (2008) suggest a range of 40–60 % for acceptable transfer errors in an equivalence test.

Table 7.6 Fitted unit value and benefit function value transfer (policy site variable means ($\Delta q = 10$) and study site coefficient estimates)

Study sites						
Policy sites	Texas		Mississippi		Florida	
	WTP	SE	WTP	SE	WTP	SE
Texas	4.74	0.68	5.96	1.69	6.41	1.14
Mississippi	4.35	0.73	4.94	1.47	5.96	1.07
Florida	4.41	0.77	5.16	1.52	5.93	1.07
Fitted unit value transfer[a]						
	ΔWTP	%TE[b]	ΔWTP	%TE	ΔWTP	%TE
Texas[c]			0.20	4.22	1.19	25.11
Mississippi[d]	−0.20	4.05			0.99	20.04
Florida[e]	−1.19	20.07	−0.99	16.69		
Benefit function value transfer[f]						
	ΔWTP	%TE[b]	ΔWTP	%TE	ΔWTP	%TE
Texas[c]			−1.22	25.74	−1.67	35.23
Mississippi[d]	−0.59	11.94			1.02	20.65
Florida[e]	−1.52	25.63	−0.77	12.98		

[a]Comparisons along the diagonal "true" study site values
[b]Transfer error
[c]Transferring MS and FL study sites to TX policy site
[d]Transferring TX and FL study sites to MS policy site
[e]Transferring TX and MS study sites to FL policy site
[f]Comparisons of off-diagonal transfer values with diagonal "true" study site values

Comparing the transfer errors across modeling approach, it appears that (1) "fitted" unit value transfer is more accurate than benefit function estimate transfer and that (2) both are more accurate than unit value transfer. However, considering the transfer errors as data (n = 18) and conducting nonparametric tests, only (2) can be stated with confidence. Considering the samples as independent, the Mann-Whitney U test finds that the distribution of transfer errors across fitted unit value and benefit function value are not statistically different ($p = 0.174$). The distribution of transfer errors across unit value and (1) fitted unit value and (2) benefit function value methods are statistically different (p = 0.008 and p = 0.045).

Considering the samples as dependent, there is some evidence that the transfer errors from the fitted unit values are lower than the benefit estimate values, which are lower than the unit values. The Wilcoxon signed rank test finds that the differences between fitted unit value and benefit function value transfer errors are statistically significant at the p = 0.094 level. The Wilcoxon signed rank test for differences between unit value and fitted unit value finds statistically significant differences at the p = 0.063 level. The Wilcoxon signed rank test for differences between unit value and benefit function value finds statistically significant differences at the p = 0.094 level.

7.7 Conclusions

The contingent valuation method is a highly flexible method for the estimation of willingness to pay for non-market goods and services, including nonuse values and values under uncertainty. It is also frequently used for benefit transfer. Yet, a review of the literature finds a number of gaps in past evaluations of CVM benefit transfer. For example, no study to date has thoroughly evaluated the properties of benefit transfers conducted using a CVM referendum format, including an analysis of scope sensitivity and tests of unit versus function transfer. This chapter provides a systematic valuation of these issues using a case study of programs to enhance oyster safety.

In our empirical example we focus on study and policy sites with data that pass basic theoretical validity tests, such as sensitivity to scope. In each comparison, unit value estimates are not statistically different, indicating the potential for valid unit value transfer. While only one of the three benefit function transfer tests suggests valid transferability, benefit function value estimates are not statistically different, indicating that benefit transfers are valid. The average benefit transfer error across each of the comparison methods is 25 % (n = 18) which is lower than all but three of the CVM convergent validity studies in the literature. These results suggest that the use of the referendum format and thorough validity testing (e.g., eliminating studies or samples that fail basic validity tests) may improve the performance of benefit transfers with the CVM. We also find some evidence that benefit function value transfer improves accuracy relative to unit value transfer.

We have raised a number of issues that provide fodder for future CVM and benefit transfer research. None of the convergent validity studies have considered the potential for hypothetical bias nor assessed whether implementation of the mitigation approaches improves benefit transfer. There is also a need for further conduct of validity tests, including joint estimation with revealed preference methods, and reliability tests across time to assess their effect on benefit transfer. Also, present analysis does not explicitly address the issue of use versus nonuse value and implications for benefit transfer. Since one of the advantages of the CVM is that it can be used to estimate nonuse values, it is surprising that most studies have conducted convergent validity tests with goods that generate only use values. The relative accuracy of benefit transfer for goods that generate only use, only (or primarily) nonuse, and a combination of use and nonuse values is needed to evaluate the potential performance of CVM benefit transfer over a range of different use versus nonuse applications.

Given that numerous conditions are necessary for benefit transfer accuracy with the CVM, it is not surprising that many studies reject convergent validity. Yet statistical validity is not always required for benefit transfers to be relevant. In some cases expected transfer errors may fall in an acceptable range for a particular policy application, despite a failure to achieve a statistically valid transfer. Bergstrom and DeCivita (1999) provide a discussion on the degree of accuracy required for benefit transfer in various contexts. The key determinant is the role of the benefit estimate in the policy process and the costs of a wrong decision. For example, lower accuracy is sufficient for benefit estimates that are used to set policy agendas. Benefit transfers may be ideal in this situation. When used to inform policy and court decisions, however, the required accuracy of benefit transfer is increased, and the costs of an incorrect decision may be large. In these cases, primary data CVM studies are strongly preferred. Additional work is required to help identify when and how different types of CVM analyses might be suitable for particular transfer applications. Findings of this chapter suggest that this work should incorporate greater attention to the validity of the underlying primary data.

Acknowledgments The data used in this chapter were collected under research supported by Gulf Oyster Industry Program Grant No. R/LR-Q-32. The authors thank Greg Martin and Richard Sjolander for their contributions to the data collection effort and Rob Johnston for a number of helpful comments and suggestions.

References

Adamowicz, W. L. (2004). What's it worth? An examination of historical trends and future directions in environmental valuation. *Australian Journal of Agricultural and Resource Economics, 48*, 419–443.

Arrow, K., Solow, R., Portney, P. R., Leamer, E. E., Radner, R., & Schuman, H. (1993). Report of the NOAA panel on contingent valuation. *Federal Register 58*, 4601–46124.

Barton, D. N. (2002). The transferability of benefit transfer: Contingent valuation of water quality improvements in Costa Rica. *Ecological Economics, 42*, 147–164.

Bateman, I. J., Brouwer, R., Ferrini, S., Schaafsma, M., Barton, D. N., Dubgaard, A., et al. (2011). Making benefit transfer work: Deriving and testing principles for value transfers for similar and dissimilar sites using a case study of the non-market benefits of water quality improvements across Europe. *Environmental and Resource Economics, 50*, 365–387.

Bergland, O., Magnussen, K., & Navrud, S. (2002). Benefit transfer: Testing for accuracy and reliability. In R. J. G. M. Florax, P. Nijkamp, & K. G. Willis (Eds.), *Comparative environmental economic assessment* (pp. 117–132). Cheltenham, UK: Edward Elgar.

Bergstrom, J. C., & DeCivita, P. (1999). Status of benefit transfer in the United States and Canada: A review. *Canadian Journal of Agricultural Economics, 47*, 79–87.

Bockstael, N. E., & McConnell, K. E. (2010). *Environmental and resource valuation with revealed preferences: A theoretical guide to empirical models*. Dordrecht, The Netherlands: Springer.

Boyle, K. J., & Bergstrom, J. C. (1992). Benefit transfer studies: Myths, pragmatism, and idealism. *Water Resources Research, 28*, 657–663.

Boyle, K. J., Kuminoff, N. V., Parmeter, C. F., & Pope, J. C. (2010). The benefit-transfer challenges. *Annual Review of Resource Economics, 2*, 161–182.

Brookshire, D. S., & Neill, H. (1992). Benefit transfers: Conceptual and empirical issues. *Water Resources Research, 28*, 651–655.

Brouwer, R. (2000). Environmental value transfer: State of the art and future prospects. *Ecological Economics, 32*, 137–152.

Brouwer, R. (2006). Do stated preference methods stand the test of time? A test of the stability of contingent values and models for health risks when facing an extreme event. *Ecological Economics, 60*, 399–406.

Brouwer, R., & Bateman, I. J. (2005). Benefits transfer of willingness to pay estimates and functions for health-risk reductions: A cross-country study. *Journal of Health Economics, 24*, 591–611.

Brouwer, R., & Spaninks, F. A. (1999). The validity of environmental benefits transfer: Further empirical testing. *Environmental and Resource Economics, 14*, 95–117.

Cameron, T. A. (1991). Interval estimates of non-market resource values from referendum contingent valuation surveys. *Land Economics, 67*, 413–421.

Cameron, T. A., & James, M. D. (1987). Efficient estimation methods for 'closed-ended' contingent valuation surveys. *Review of Economics and Statistics, 68*, 269–276.

Carson, R. T. (2011). *Contingent valuation: A comprehensive bibliography and history*. Cheltenham, UK: Edward Elgar.

Carson, R. T., & Groves, T. (2007). Incentive and informational properties of preference questions. *Environmental and Resource Economics, 37*, 181–210.

Carson, R. T., & Louviere, J. J. (2011). A common nomenclature for stated preference elicitation approaches. *Environmental and Resource Economics, 49*, 539–559.

Carson, R. T., Flores, N. E., Martin, K. M., & Wright, J. L. (1996). Contingent valuation and revealed preference methodologies: Comparing the estimates for quasi-public goods. *Land Economics, 72*, 80–99.

Desvousges, W. H., Naughton, M. C., & Parsons, G. R. (1992). Benefit transfer: Conceptual problems in estimating water quality benefits using existing studies. *Water Resources Research, 28*, 675–683.

Downing, M. E., & Ozuna, T. (1996). Testing the reliability of the benefit function transfer approach. *Journal of Environmental Economics and Management, 30*, 316–322.

Gonzalez-Sepulveda, J. M., & Loomis, J. B. (2011). Are benefit transfers using a joint revealed and stated preference model more accurate than revealed and stated preference data alone? In J. C. Whitehead, T. Haab, & J.-C. Huang (Eds.), *Preference data for environmental valuation: Combining revealed and stated approaches* (pp. 289–302). New York: Routledge.

Groothuis, P. A. (2005). Benefit transfer: A comparison of approaches. *Growth and Change, 36*, 551–564.

Groothuis, P. A., & Whitehead, J. C. (2002). Does don't know mean no? Analysis of 'don't know' responses in dichotomous choice contingent valuation questions. *Applied Economics, 34*, 1935–1940.

Johnston, R. J., & Duke, J. M. (2008). Benefit transfer equivalence tests with non-normal distributions. *Environmental and Resource Economics, 41*, 1–23.

Johnston, R. J., Elena Y. B., Richard, I., Christopher J. M., Ryan F. W., & Matthew H. R. (2005). Systematic variation in willingness to pay for aquatic resource improvements and implications for benefit transfer: A meta-analysis. *Canadian Journal of Agricultural Economics/Revue canadienne d'agroeconomie, 53*(2-3), 221–248.

Johnston, R. J., & Rosenberger, R. S. (2010). Methods, trends and controversies in contemporary benefit transfer. *Journal of Economic Surveys, 24*, 479–510.

Kaul, S., Boyle, K. J., Kuminoff, N., Parmeter, C., & Pope, J. (2013). What can we learn from benefit transfer errors? Evidence from 20 years of research on convergent validity. *Journal of Environmental Economics and Management, 66*, 90–104.

Kirchhoff, S., Colby, B. G., & LaFrance, J. T. (1997). Evaluating the performance of benefit transfer: An empirical inquiry. *Journal of Environmental Economics and Management, 33*, 75–93.

Kristofersson, D., & Navrud, S. (2005). Validity tests of benefit transfer–are we performing the wrong tests? *Environmental and Resource Economics, 30*, 279–286.

Kristofersson, D., & Navrud, S. (2007). Can use and non-use value be transferred across countries? In S. Navrud & R. Ready (Eds.), *Environmental value transfer: Issues and methods* (pp. 207–226). Dordrecht: Springer.

Loomis, J. B. (1992). The evolution of a more rigorous approach to benefit transfer: Benefit function transfer. *Water Resources Research, 28*, 701–705.

Loomis, J. (2011). What's to know about hypothetical bias in stated preference valuation studies? *Journal of Economic Surveys, 25*, 363–370.
Luken, R. A., Johnson, F. R., & Kliber, V. (1992). Benefits and costs of pulp and paper effluent controls under the Clean Water Act. *Water Resources Research, 28*, 665–674.
Matthews, D. I., Hutchinson, W. G., & Scarpa, R. (2009). Testing the stability of the benefit transfer function for discrete choice contingent valuation data. *Journal of Forest Economics, 15*, 131–146.
Mitchell, R. C., & Carson, R. T. (1989). *Using surveys to value public goods: The contingent valuation method*. Washington, DC: Resources for the Future.
Murphy, J. J., Allen, P. G., Stevens, T. H., & Weatherhead, D. (2005). A meta-analysis of hypothetical bias in stated preference valuation. *Environmental and Resource Economics, 30*, 313–325.
Piper, S., & Martin, W. E. (2001). Evaluating the accuracy of the benefit transfer method: A rural water supply application in the USA. *Journal of Environmental Management, 63*, 223–235.
Portney, P. R. (1994). The contingent valuation debate: Why economists should care. *Journal of Economic Perspectives, 8*, 3–17.
Ready, R., & Navrud, S. (2007). Morbidity value transfer. In S. Navrud & R. Ready (Eds.), *Environmental value transfer: Issues and methods* (pp. 77–88). Dordrecht: Springer.
Ready, R., Navrud, S., Day, B., Dubourg, R., Machado, F., Mourato, S., et al. (2004). Benefit transfer in Europe: How reliable are transfers between countries? *Environmental and Resource Economics, 29*, 67–82.
Rosenberger, R. S., & Stanley, T. D. (2006). Measurement, generalization, and publication: Sources of error in benefit transfers and their management. *Ecological Economics, 60*, 372–378.
Rozan, A. (2004). Benefit transfer: A comparison of WTP for air quality between France and Germany. *Environmental and Resource Economics, 29*, 295–306.
Scarpa, R., Hutchinson, G., Chilton, S., & Buongiorno, J. (2010). Reliability of benefit value transfers from contingent valuation data with forest-specific attributes. In S. Navrud & R. Ready (Eds.), *Environmental value transfer: Issues and methods* (pp. 179–205). New York: Springer.
Smith, V. K. (1992). On separating defensible benefit transfers from "smoke and mirrors". *Water Resources Research, 28*, 685–694.
Vandenberg, T. P., Poe, G. L., & Powell, J. R. (2001). Assessing the accuracy of benefits transfers: Evidence from a multi-site contingent valuation study of groundwater quality. In J. C. Bergstrom, K. J. Boyle, & G. L. Poe (Eds.), *The economic value of water quality* (pp. 100–120). Northampton: Edward Elgar.
Whitehead, J. C. (2006). A practitioner's primer on the contingent valuation method. In A. Alberini & J. R. Kahn (Eds.), *Handbook on contingent valuation* (pp. 66–91). Northampton: Edward Elgar.
Whitehead, J. C., & Blomquist, G. C. (2006). The use of contingent valuation in benefit-cost analysis. In A. Alberini & J. R. Kahn (Eds.), *Handbook on contingent valuation* (pp. 92–115). Northampton: Edward Elgar.
Whitehead, J. C., & Hoban, T. J. (1999). Testing for temporal reliability in contingent valuation with time for changes in factors affecting demand. *Land Economics, 75*, 453–465.
Whitehead, J. C., Christopher, F. D., Jeffery, H., & Bob B. (2008). Valuing beach access and width with revealed and stated preference data.*Marine Resource Economics, 23*(2), 119–135.
Whitehead, J. C., Morgan, O. A., Huth, W. L., Martin, G. S., & Sjolander, R. (2012). *Willingness-to-pay for oyster consumption mortality risk reductions*. Paper presented at the 2012 AERE Summer Meeting, Asheville, June.

Chapter 8
Applying Benefit Transfer with Limited Data: Unit Value Transfers in Practice

John Rolfe, Jill Windle and Robert J. Johnston

Abstract Unit value or point value transfers from individual source studies remain the oldest and most common form of benefit transfer. Although practitioners generally recommend benefit function transfers, these are not always possible. Where unit value transfers are to be performed, appropriate protocols must be followed to select source studies, transfer values, and perform necessary value adjustments. This chapter demonstrates the processes and challenges involved in the implementation of unit value transfers, using case studies of environmental values in a peri-urban community on the east coast of Australia where key ecosystems ranged from coastal beaches to inland forests. Key issues in evaluating the potential for benefit transfer included the availability and quality of source studies, the extent of overlap between source studies and the target site, the need for different forms of adjustment to account for variations in scope and scale, and the limitations to unit value transfers.

Keywords Benefit transfer · Unit values · Value adjustment · Australia · Peri-urban

8.1 Introduction

There is a growing demand for environmental valuations to support cost-benefit assessments and improve environmental policy and management decisions. In many situations (e.g., cases in which no primary valuations have been conducted

J. Rolfe (✉) · J. Windle
School of Business and Law, CQUniversity, Rockhampton, QLD, Australia
e-mail: j.rolfe@cqu.edu.au

J. Windle
e-mail: j.windle@cqu.edu.au

R.J. Johnston
Department of Economics and George Perkins Marsh Institute, Clark University, Worcester, MA, USA
e-mail: rjohnston@clarku.edu

R.J. Johnston et al. (eds.), *Benefit Transfer of Environmental and Resource Values*, The Economics of Non-Market Goods and Resources 14,
DOI 10.1007/978-94-017-9930-0_8

and resources are restricted), it is necessary to use research results from pre-existing primary studies at one or more sites or policy contexts (often called study sites) to predict welfare estimates or related information at target policy sites; this process is known as benefit transfer (Brouwer 2000; Johnston and Rosenberger 2010; Navrud and Ready 2007; Rolfe and Bennett 2006).

Benefit transfer methods are generally classified into two primary categories: unit value transfers and benefit function transfers (Johnston and Rosenberger 2010; Navrud and Ready 2007; Rolfe 2006). A unit value transfer involves the transfer of a single value estimate or set of value estimates from source studies as point source estimates, while a benefit function transfer involves the transfer of a parameterized benefit function. This function is combined with at least some information (on independent variables) drawn from the policy site to generate an adjusted welfare estimate for that site. One key difference between these two approaches is that unit values from source studies are typically transferred with limited or no adjustments. Adjustments, where they occur, are generally performed ex post and ad hoc. In contrast, adjustments within benefit function transfers are primarily based on the underlying function(s) estimated at the study site, combined with information on independent variables observed at the policy site. As such, adjustments within benefit function transfer may be viewed as a natural extension of the originally estimated function, albeit at a new site. Another difference is that unit value transfers are often transferred from single source studies,[1] while benefit function transfers can be generated from single source studies or a meta-analysis that synthesizes information from a number of source studies.

Many authors have recommended the transfer of benefit functions because these allow for adjustments to be made according to a variety of factors that can influence values, including site and population differences (Johnston and Rosenberger 2010; Morrison and Bergland 2006; Rolfe 2006; Rosenberger and Stanley 2006). Yet despite the limitations of unit value transfer, it is still commonly applied (see Chaps. 3–6). Unit value transfers are simple to use, are often the only approach available when source studies are limited or when benefit functions are not reported, and can provide relatively accurate results under certain conditions (Bateman et al. 2011; Colombo and Hanley 2008). Moreover, although the literature suggests that function transfers generally outperform unit value transfers (e.g. Kaul et al. 2013), the evidence on this issue is somewhat mixed (Colombo and Hanley 2008; Johnston and Rosenberger 2010). Bateman et al. (2011), for example, suggest that unit value transfers may be more appropriate and generate lower transfer errors where source and target sites are similar, but that benefit function transfers will outperform unit value transfers as differences across sites increase.[2] In addition, the concepts and

[1]Although sometimes values from several source studies are assessed before choosing a single unit value for transfer. Unit value transfers may also transfer a mean or median value from prior studies.

[2]They note that value functions explicitly incorporate differences between sites and hence are appropriate where differences between source and target sites are involved, but may be over-parameterized when limited differences between sites exist.

applications within unit value transfers are often easier to communicate (due to the simplicity of the methods), and hence the results may be more acceptable to policy makers.

At the same time, the apparent simplicity of unit value transfer masks a number of issues that can lead to unacceptably large transfer errors. For example, the accuracy of point source estimates that are transferred with unit values are very dependent on the quality of the source study, the extent of alignment between source and target sites, and the appropriateness of any ex post adjustments to unit values that are made. The selection of source studies, the performance of the benefit transfer and any adjustment to values all require expert judgment, as variations in any step of the process can lead to large differences in prediction. The experience and skills of the analyst can also be critical, in part because the process remains a combination of art and science. Not all unit value transfers are conducted accurately or appropriately, and a major concern is that the individual judgments involved in the process make it difficult for outside analysts to assess the quality of transfers.

The key challenge in any benefit transfer exercise, including unit value transfers, is to avoid errors that lead to improper inferences regarding welfare effects and misguided policy. These would include errors transferred from the original primary studies (*measurement errors*) and errors generated by the transfer process itself (*generalization errors*). Compared to other types of transfer, unit value transfers often face greater challenges related to both types of error. The standard reliance on individual source studies for point source estimates potentially magnifies risks related to the possible selection of a single inaccurate or inappropriate source study for transfer, thereby increasing the risk of measurement errors. In addition, the lack of a benefit function to support value adjustments can lead to greater generalization errors, particularly when the policy and study sites are not closely matched. These challenges imply that an analyst applying unit value transfers must give particular attention to the quality and appropriateness of source studies and to the similarity between study and policy sites (or valuation contexts).

This chapter reviews and illustrates the basic steps and protocols involved in unit value transfers. The goal is to provide clear methodological guidance and illustrations for this type of transfer when more sophisticated types of transfer are infeasible or otherwise considered inappropriate. We give particular attention to the decisions and assumptions involved in selecting primary studies and values that best match policy sites and needs. Steps and protocols are illustrated using a case study addressing ecosystem changes in a peri-urban community on the eastern seaboard of Australia. The chapter begins with a review of methodology for unit value transfers. This is followed by the case study application, illustrating the practical steps involved and discussing some of the caveats and concerns that must be considered when conducting unit value transfers.

8.2 Methodology for Unit Value Transfers

Unit value estimates can be transferred from a single study or multiple studies, and either transferred directly or adjusted in some way to account for variations between the source and target sites and contexts. The value of changes in an environmental asset may be influenced by factors such as the change in provision (i.e., the extent of quantity or quality change), characteristics of the site (including special features), the availability of substitutes and the characteristics of the population (Bateman et al. 2011). Care must be taken to assure that the economic framework is appropriate and that the concepts being measured are consistent between source and target applications (e.g., that willingness to pay (WTP) measures are not being used to predict willingness to accept (WTA) values).

A number of different steps are involved in conducting a benefit transfer (see Chap. 2). These can be grouped into three broad stages (Table 8.1). The first step is to establish the context and framework for a benefit transfer exercise. The second is to identify and evaluate the source studies that are available and to select the benefit transfer approach to be used. The third is performing the benefit transfer. This final

Table 8.1 The key tasks and objectives to conduct a benefit transfer

Task	Objective
Stage 1: establishing the context and framework	
1. Define the benefit transfer context	Scope the valuation and policy context
2. Establish the need for benefit transfer	Evaluate whether benefit transfer is preferred over a primary study
3. Define the policy, good and population	Establish the characteristics of the "target" study
4. Define and quantify policy options and changes in goods	Quantify the marginal changes to be valued
Stage 2: selecting source studies and transfer methods	
5. Gather and evaluate valuation data and evidence	Identify, screen and evaluate source studies, and any additional data requirements
6. Determine benefit transfer methods	Select appropriate method(s), given the policy site characteristics and availability of source studies
Stage 3: performing the benefit transfers	
7. Design and implement transfers	Select source studies to be used, perform benefit transfer and adjust value estimates where necessary and appropriate
8. Aggregate values over populations, area and time	Extrapolate values from unit or benefit function transfers to the population, area and time frame relevant to the target study
9. Conduct sensitivity analysis and test reliability	Test sensitivity of transfer estimates to changes in assumptions and treatment of the data; identify limiting factors
10. Report results	Detail the procedures, data and testing involved

step includes implementing adjustments and extrapolation of values as appropriate, along with appropriate sensitivity testing and reporting.

8.2.1 Stage 1: Establishing the Context and Framework for a Benefit Transfer Exercise

The key steps here are largely uniform across both unit value and benefit function approaches. However, information from step 1 (the benefit transfer context) can identify key factors that influence the choice of unit value versus benefit function transfer. As discussed in Chap. 2, the context in which the information will be used, the level of accuracy that is needed, and the limitations over time and resources are important guides to the selection of the benefit transfer method. Unit value transfers tend to be more appropriate for applications in which the pragmatic need for welfare estimates outweigh the need for accuracy, and in which available time and resources are limited (Brookshire and Neill 1992).

8.2.2 Stage 2: Selecting Source Studies and Transfer Methods

The initial focus of the second stage is to identify and evaluate the availability and quality of potential source studies and other relevant information. The type of material available will determine whether it is possible to conduct a benefit transfer, and if so, the type of transfer that might be applied. It is recommended that some protocol be applied to restrict the extent of the initial literature search and to ensure some form of quality control on the selected studies (Johnston and Rosenberger 2010; Smith and Pattanayak 2002).

In situations with limited data sources, there is less opportunity to identify source studies that closely match the target site (site similarity). Consequently, a key issue is to identify limitations in source data that can be remediated during the transfer process, as well as limitations that make benefit transfer unsuitable. In some cases, ex post adjustments can be applied to the transferred values (e.g. to account for site differences), while in other cases, there may be underlying disparities in values for the amenity that cannot be remediated or adjusted in some way. However, in some situations an admittedly imprecise value might be satisfactory (or better than no value at all)—for example when the primary goal of valuation is environmental advocacy rather than policy analysis (Kline and Mazzotta 2012). In such cases it is important to clearly identify and outline the limitations of the transferred value estimate.

When evaluating unit values in this way, analysts must consider both the representativeness and quality of source studies. *Representativeness* relates to the similarity of the source and target case studies. It is largely focused on site similarity (including populations), although methodological differences may also be important. The analyst must identify source studies that correspond to the target site across all relevant dimensions, as well as evaluate the extent of divergence. One way of judging representativeness is to consider the extent to which the definition of the good and the context in which it is valued can be transferred accurately with or without adjustments to the resulting unit values. A related consideration is the availability of information required to inform any related value adjustments. If the definition and/or contexts differ to the extent that accurate transfers are unlikely regardless of any possible adjustments, then the transfer fails the representativeness criteria. For example, if the underlying commodity valued by potential source studies does not correspond closely to that for which values are required at the target (or policy) site, then available studies are not sufficiently representative.[3]

It is also important that source studies are of appropriate *quality*, otherwise measurement errors from the source study can be transferred to the target site. Relevant dimensions of primary study quality include both accuracy and robustness. These are influenced by the appropriateness of the methodology used to generate value estimates. It is often difficult to identify the quality of source studies; Bateman et al. (2011), Nelson and Kennedy (2009), and Rosenberger and Johnston (2009) note this in relation to meta-analysis, but similar issues are relevant to all benefit transfer techniques. Although the analyst can make a case-by-case evaluation of potential source studies for unit value transfers, many issues of incomplete information about studies often remain.

The second key aim of stage 2 involves identifying whether a unit value transfer or a benefit function transfer is to be performed. In most cases the choice is determined by the availability and suitability of source studies as discussed above. Also relevant to this decision are the time and resources available, the precision of estimates that are required and the level of expertise available (Johnston and Rosenberger 2010; also see Chap. 2).

8.2.3 Stage 3: Performing the Benefit Transfers

The third stage involves implementing the benefit transfer. While most unit value transfers involve point estimates from single studies, transfers are sometimes made from multiple source studies. Typical approaches include the transfer of a weighted or unweighted mean of welfare estimates provided by the set of available studies. In other cases the protocols outlined in stage 2 are used to select a preferred study from available options to implement the transfer.

As described above, the implementation of unit value transfers often includes ex post adjustments to the source data with the aim of improving the accuracy of the

[3]See discussions of commodity consistency in Johnston and Rosenberger (2010).

benefit transfer, and to meet with the requirements of value aggregation in transfer step 8 (Table 8.1). Most unit value transfers involve some relatively standard adjustments. For example, source study welfare estimates are generally adjusted to current prices. Household or individual values from the source study might also be extrapolated to the total population at the target site. The extent of the population expected to hold values for a given environmental change will depend on the relative importance of the environmental asset, the distance between the asset and the relevant population, and the proportion of the population likely to have values for the asset in question.[4] These values will also frequently decline with distance (distance decay)—a pattern that is often difficult to accommodate within unit value transfers (Bateman et al. 2006). In practice, unit value transfers typically require subjective decisions about (a) the extent of the relevant population; (b) the proportion of the population that is deemed to hold similar values as that of the population sample; and (c) any adjustments to account for variations in the values held by the population of interest (for example, decays in values with increasing distance of the population from the asset of interest).

With a few exceptions (e.g., adjustments to account for changes in currency value over time), these and other adjustments for unit value transfers, such as those used to account for variations in site or population characteristics between source and target situations, typically occur on an ad hoc basis. The rationale and performance of these adjustments should be detailed in step 10 (reporting), and may also be tested in step 9 (sensitivity testing).

8.3 Case Study Details and Application of Benefit Transfer

This section presents a practical example of a unit value benefit transfer. To illustrate the process (see Table 8.1), we use a case study application to a peri-urban community on the east coast of Australia, where the local government wanted to assess the value of key ecosystems that ranged from coastal beaches to inland forests. This exemplifies a typical situation in which unit value transfers are put to use. Specifically, an initial literature review failed to identify any suitable source studies for combined ecosystem values in peri-urban communities that would allow a benefit transfer function to be applied. Thus options were limited for benefit transfer to a series of unit value transfers for each ecosystem of interest.

[4]This is related to the concept of the economic jurisdiction, or the size of the population that holds value for a given environmental change (Loomis 2000).

8.3.1 Establishing the Context and Framework

The protection of local natural assets in peri-urban communities is a major issue in Australia, as the majority of the population lives on the coast and rapid population growth generates resource tradeoffs and pressure on natural resources (DSEWPC 2011). Nonmarket value estimates are needed to evaluate the tradeoffs involved in cases where increased housing, infrastructure, industry and services generate environmental losses (e.g., vegetation clearing) at the same time that communities desire improved environmental protection and attractive natural amenities. However, as limited financial resources typically restrict the ability of government agencies and other groups to requisition primary valuation studies, benefit transfer must often be used to provide needed value estimates. In this study, details of the target site were generalized to represent a wide range of coastal areas, so that the benefit transfers could potentially be used in different coastal regions in Australia.

The target site for the illustrative case study was a town in a peri-urban environment (i.e., not a major urban city) that included urban residential and rural residential areas. The local council jurisdiction covered an area of between 500 and 800 km^2. At the time of the study the rural residential areas were surrounded by native vegetation, while smaller vegetation patches still existed in urban residential areas. The study area also included some wetland areas. There were several waterways in the catchment but no large rivers (river order four). Only river orders 2 and 3 were considered (not the smallest streams—river order 1). See Fig. 8.1 for an illustrative example of the four river order classifications.

The town had an expanding population of approximately 100,000. The main pressure on the environmental assets came from population growth and increased human activity. There was growing demand for new land to be released for housing

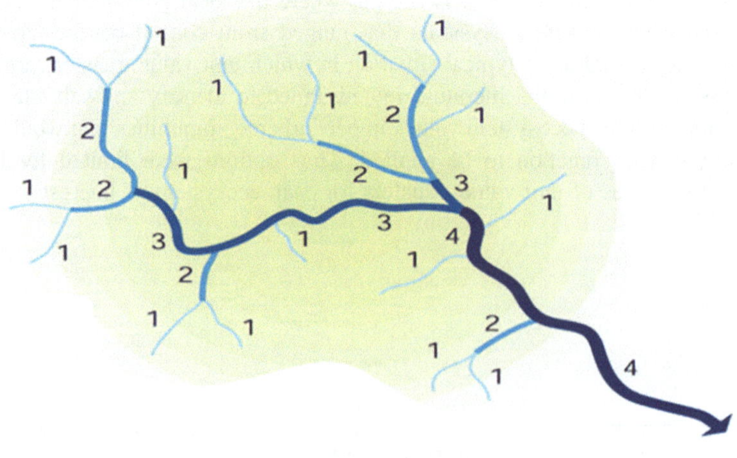

Fig. 8.1 Illustrative example of the four river order classifications

Table 8.2 Source study site selection: desirable and undesirable characteristics

Key criteria	Characteristic	Desirable	Less desirable
Scope similarity	Physical characteristics	Peri-urban contexts	Rural, urban iconic/famous sites
Measurement similarity	Quantitative/qualitative changes	Loss in area/length/extent/quality	Gain in area/length/extent/quality
Framing similarity	Type of policy changes	Urban development (population pressures)	Rural development (agricultural development)
	Types of impact (attributes in the valuation)	General descriptions for remnant vegetation, wetlands, waterways	Specific or specialized ecosystem types
Scale similarity	Site size/valuation range	Similar to target	Much larger or smaller than target
Population similarity	Population sample	Local	External

Note The relative importance of the different criteria will be case study-specific; expert opinion is often required to identify the significance of variations

development (causing remnant native vegetation to be cleared), and more people were accessing natural areas to detrimental effects (i.e., rubbish dumping, trail bike-riding). This meant that a comprehensive valuation of environmental changes would need to account for both quantitative changes (loss in area) as well as qualitative changes (both loss and improvement in condition). The following environment assets were included in the valuation:

- *Remnant native vegetation*: Up to 50,000 ha with condition ranging from good in some areas to degraded and fragmented in other areas.
- *Wetlands*: Up to 5000 ha with condition ranging from good in some areas to degraded and fragmented in others.
- *Waterways*: Up to 1000 km of waterways (limited use for recreation)
- *Beaches*: Up to 50 km of beaches used for recreation

The key selection criteria to assess the contexts (or study sites) addressed by source studies for site similarity are outlined in Table 8.2.

8.3.2 Selecting Source Studies and Transfer Methods

Two threshold criteria were specified to establish a literature search protocol and to maintain a degree of quality control. These criteria are designed to minimize concerns about transferability over space and time; valuation studies conducted overseas and any Australian studies that were more than 15 years old were excluded. International studies were excluded because of potential differences among populations (e.g., income, exchange rates, attitudes, knowledge, culture) and sites (e.g., potential differences in priorities for conservation). The time limit on studies was imposed for a several reasons: the accuracy of valuation studies has improved over the past 30 years; people's preferences may have changed over a 15-year period; population dynamics may mean source communities are more likely to have changed over longer time periods; and missing data on population characteristics make it difficult to perform adjustments (cf. Rosenberger and Johnston 2009).

Relatively few source studies were identified in the initial literature search. Studies were mainly sourced from the peer-reviewed published literature, with studies in publically available reports also considered. Where results appeared in multiple outlets, the results of only a single publication were considered, preferably from a peer-reviewed journal article.

Potential source studies were then evaluated for relevance using the key factors identified in Table 8.2. There is no inherent ordering or priority of issues, so both the issues and their relative importance had to be determined and evaluated on a case-by-case basis. In this case, the key issues related to factors expected to generate significant differences in welfare estimates, based on theory and prior evidence in the literature. The issues identified in Table 8.2 are relatively simplistic, but as is shown in the subsequent analysis, the small sample of available source studies in Australia meant that even these limited criteria narrowed the choices of studies available for unit value transfers.

The selection of benefit transfer studies and unit values for each environmental asset is outlined below. In the interests of brevity, full case study information is provided for the remnant vegetation asset and summary details for the remaining assets.

8.3.2.1 Remnant Vegetation (Up to 50,000 ha)

Although several potential source studies were identified, none closely matched the conditions of the target site, raising the potential for generalization errors. There was only one source study for which the amount of the asset involved (henceforth referred to as scale) and the general valuation context matched the target site.

Mallawaarachchi et al. (2006) valued vegetation change in a peri-urban area under pressure from population growth and new housing development. However, the native vegetation attribute in this study was not well-suited to transfer in the present case, as it related only to rare or unique vegetation, there was no payment period specified for the annual payments, and the study was conducted in 1999. As a result, this study was considered unsuitable for benefit transfer in this case. Another study by Concu (2007) was deemed potentially suitable for transfer, as the site was in an urban area (Perth), and one of the valuation attributes related to public access to bush land. However, after further review this study was disqualified, as the public access attribute was not significant in the data analysis and no time-frame details were reported for the cost attribute (an annual payment). Other studies were considered, but were excluded if only specific vegetation types (e.g., river red gums) or land types (e.g., riparian vegetation) were used, as these were unlikely to be representative.

After discarding Concu (2007) and Mallawaarachchi et al. (2006) as potential sources for the transfer, seven source (choice modeling) studies involving 10 value estimates were given further consideration. Value estimates from these studies were converted into 2012 values (AUD) and to lump sum [present] values if annual payments were used (Table 8.3).[5] Present value estimates ranged from $0.06/ha to $4.79/ha, using a 15 % discount rate to reflect the relatively high social discount rate that has been found in valuation studies (e.g. Kovacs and Larson 2008). Given this range in the value estimate, the challenge was to select the study/studies that best matched the target site.

Comparisons between the target site and the source studies revealed that none of the latter focused entirely on native vegetation, with only one relevant attribute in each study that would be suitable for transfer. None of these studies matched the scope of the target site, as all related to large rural areas. Neither did any match the context of the target site (pressure from urban development), as the source studies were focused on tradeoffs between agricultural development and environmental protection (studies RV1-RV6) or tradeoffs around wetland management (study RV7). In addition, the target site involved very fragmented vegetation communities, including small patches in urban settings that were not replicated in the source studies. All of these reflect differences relevant to generalization errors in unit value transfers.

There were also differences in the quantities involved. Based on theory and empirical evidence, economists generally expect that the larger the scale or size of the valued asset (i.e., the good or service being valued), the lower will be the expected value for marginal changes, all else held constant. Corresponding to this general intuition, the present values per household, for a marginal improvement of 1000 ha of remnant vegetation (in good condition) were the lowest for the Fitzroy Basin in Queensland (studies RV2-RV5), the largest area under valuation. The relatively low value in the South Australian study (study RV7), given the smaller scale involved, was possibly because it was primarily focused on wetlands, which may have diminished the relative importance of other remnant vegetation. The influence of scale on valuation estimates was most clearly represented in the New South Wales study (study RV1). The area of the 10 % case study most closely aligns with the target site (i.e., 50,000 ha of remnant vegetation). This is the only value ($4.79/household/1000 ha) identified that would be suitable for benefit transfer in terms of scale, although the study does not meet other desirable conditions for transfer.

A further challenge was to adapt transferred unit values to represent marginal changes in quantity and quality dimensions. The target site involved both losses in the area of vegetation as well as declines in the amount of vegetation in good condition. In contrast, the source studies all described the relevant attribute in terms

[5]Note: in this chapter all dollar values refer to Australian dollars.

Table 8.3 Source Study details for remnant vegetation (target: up to 50,000 ha)

#	Source	Site/year	Attributes in CM valuation[a]	Current size at time of valuation (*1000 ha)	Size range under valuation (*1000 ha)	Sample	Value in 2012 $[b]	Present value
RV1	Mazur and Bennett (2009)	Namoi catchment NSW 2008	• **Native vegetation in good condition** • Native species • Healthy waterways • People working in agriculture	1. 10 %: 18 2. 50 %: 90 3. 100 %; 180	1. 10 % = 18–60 2. 50 % = 90–300 3. 100 % = 180–600	Local External Local	1. $1.43 2. $0.66 3. $0.11	1. $4.79 2. $2.21 3. $0.37
RV2	Rolfe et al. (2002)	Fitzroy River Basin Qld 2000	• **Healthy remnant vegetation** • Healthy waterways • People leaving country areas • Amount of water left unallocated	>50 % of area cleared (total area = 142,000 km^2)	20–50 % (2800–7000)	Local	$0.01	$0.06
RV3	Rolfe and Windle (2003)	Fitzroy River Basin Qld 2001	• **Healthy remnant vegetation** • Healthy waterways • Protection of Aboriginal cultural sites • Unallocated water	>50 % of area cleared (total area = 142,000 km^2)	20–50 % (2800–7000)	Local	$0.02	$0.14
RV4	Rolfe and Bennett (2009)	Fitzroy River Basin Qld 2002	• **Healthy remnant vegetation** • Healthy waterways • People leaving country areas • Amount of water left in reserve	>50 % of area cleared (total area = 142,000 km^2)	20–30 % (2800–4200)	External	$0.03	$0.17

(continued)

Table 8.3 (continued)

#	Source	Site/year	Attributes in CM valuation[a]	Current size at time of valuation (*1000 ha)	Size range under valuation (*1000 ha)	Sample	Value in 2012 $[b]	Present value
RV5	Windle and Rolfe (2005)	Fitzroy River Basin Qld 2003	• **Healthy remnant vegetation** • Healthy waterways • Protection of Aboriginal cultural sites • River estuary in good health	>50 % of area cleared (total area = 142,000 km^2)	20–50 % (2800–7000)	External	$0.03	$0.20
RV6	Rolfe and Windle (2008)	1. S.E Qld 2. Cent Coast Qld 2005	• **Healthy remnant vegetation** • Soils in good condition • Healthy waterways	1. 45 % = 1035 2. 65 % = 585	1 = 25–40 % (575–920) 2 = 45–65 % (405–585)	Local Local	1. $0.16 2. $0.33	1. $0.93 2. $1.92
RV7	Whitten and Bennett (2006)	Upper SE, SA 2001	• **Healthy remnant vegetation** • Healthy wetlands • Threatened species that benefit • Ducks hunted	Not reported	50–100	Local and external	$1.25	$1.25

[a]Attributes in bold are those given further consideration for benefit transfer
[b]Willingness to pay per household for a marginal change of 1000 ha

of changes in the area in good condition, without distinguishing between quantity and quality changes. This limited the potential for benefit transfer values to be tailored to different scenarios for vegetation protection, including the potential for improvements to vegetation condition.

All the potential source studies identified for remnant vegetation were from choice modelling experiments[6] used to estimate WTP for increased protection, which meant that there were no important methodological variations to consider. Not all source studies provided a local population sample (Table 8.3), which was required at the target site. However, these differences were not considered as important as those outlined above, because some of the studies reported no significant difference in the value estimates between local and non-local samples.

In summary, after the initial review, none of the source studies was considered suitable for direct unit value transfer. Reasons included:

- *Scope differences*: All source studies were set in a rural context, rather than a peri-urban setting. It is not currently known how these differences affect value estimates for either quantitative or qualitative changes;
- *Measurement differences*: All source studies referred to vegetation in good condition, which limited their potential application;
- *Framing differences*: No source studies were framed in the context of increased population pressure;
- *Scale differences*: Only one study matched the scale of the target site (study RV.1), but adjustments could be applied; and
- *Population differences*: Not all studies assessed values from a local population.

In this case, value adjustments could potentially be used to accommodate some but not all differences between the source study and a target application. For example, adjustments for scale differences could be estimated from the split-sample experiments in study RV1 (as reported in Rolfe et al. (2013), while one could follow methods of Morrison et al. (2002) and Morrison and Bennett (2004) to adjust for differences between the values of local and non-local populations. However, no information was available to allow adjustments for scope, measurement and framing differences. Given these limitations, the recommendation was to not continue with a benefit transfer for remnant vegetation.

8.3.2.2 Wetlands (Up to 5000 ha)

Seven potential source studies were identified for wetlands, but two were removed due to attribute misspecification.[7] Of the remaining five studies (Table 8.4), two

[6]Choice modelling is a stated preference technique capable of estimating both use and nonuse values (Bateman et al. 2002; Rolfe and Bennett 2006).

[7]Both Mallawaarachchi et al. (2001) and Morrison and Bennett (2004) included a wetlands attribute in their choice modelling studies, but in both cases riparian vegetation was also included in the description, making it unsuitable to transfer to the target site.

used measurement units that did not relate to any specified ecological quality (studies WL1 and WL2). Only Study WL1 related to a loss in area, although the study did not match the scope or policy frame of the target site. None of the remaining studies was considered suitable for benefit transfer. The main issues of concern (summarized in Table 8.4) matched those outlined for the remnant vegetation asset.

8.3.2.3 Waterways (Up to 1000 km: River Orders 2 and 3)

Ten potential source studies were identified for waterways. Since many studies overlapped with those identified for other environmental assets as described above, the key issues and lack of suitability remained the same (Table 8.5). In this case, the river order (the main river versus smaller tributaries) added another level of complexity to the issue of site similarity. The large, catchment-scale source studies referred to the length of rivers and did not distinguish among river orders. There was an implicit assumption that details of river length referred to large (main) rivers (river order 4 and possibly river order 3) and not to smaller tributaries (river order 2). In contrast, although there were several hundred kilometers of waterways at the target site, they were all small rivers with limited options for recreation.

Nine choice modeling studies were identified as potential sources (Table 8.5). The first three studies were rejected because the key attribute did not align well with the target study. These are included in Table 8.5 to illustrate the scope differences in terms of river order. An additional meta-analysis (WW10) was also identified and included for comparative purposes.

The scale of the target site was within the valuation range of two source studies (WW8 and WW9), although the total size of the environmental asset at both sites was much larger than the target. The southeast Queensland site (study WW9) was in a peri-urban area and more closely matched the scope of the target. However, in both studies the associated attribute related only to waterways in good condition, which limited the potential extent of application at the target site. The present value for a 1 km improvement (of waterways in good condition) was estimated at $1.07 per household for study WW8 and at $1.21 for study WW9.[8] These values can be compared to an estimate generated from the meta-analysis function (study WW10), where allowing for 800 km of waterways generates value predictions of $1.24/km.

There were two limitations to consider, however. First, the source studies involved higher order (larger) rivers than the rivers in the target area, creating the potential for amenity mis-specification (or lack of commodity consistency) in the benefit transfer. Unfortunately there is no evidence in the current literature to help understand how values might vary between larger and smaller rivers. The second

[8]The higher value for study WW9 might be a reflection of the peri-urban context and/or higher values to avoid a loss than for a gain.

Table 8.4 Source study review for wetlands (target: up to 5000 ha)

ID	Source	Year	Valuation/policy scenario	Method	Comment
WL1	Hatton MacDonald and Morrison (2010)	2003	Land use/agricultural development	CM	**Scope and framing**: rural valuation context
					Measurement: *loss* in area—no specific quality
					Scale differences: 73,000–99,000 ha
WL2	Morrison et al. (2002)	1997	Water management for wetlands	CM	**Scope and framing**: rural valuation context
					Measurement: *increase* in area—no specific quality
					Scale differences: site 1 = 1000 km^2; site 2 = 400 km^2
WL3	Rolfe and Dyack (2010)	2006	Recreational use of iconic wetlands	TCM + CVM	**Scope and population differences**: iconic wetland site. External (visitors) sample rather than local sample
WL4	Tapsuwan et al. (2009)	2005/2006	Value of urban wetlands/lakes	HP	**Scope and framing**: urban (capital city) valuation context
					Transfer challenges: additional data requirements from target site
WL5	Whitten and Bennett (2006)	2001	Wetland management	CM	**Scope, framing and population issues**: unsuitable valuation context; external sample
					Measurement: related only to area in good condition

CM choice modelling; *TCM* travel cost method; *CVM* contingent valuation method; *HP* hedonic pricing

limitation was that the values did not allow for differentiation between gains and losses or between varying changes in quality.

8.3.2.4 Beaches (Up to 50 km)

Beaches were assessed in terms of the transferability of recreational values. Only three source studies were identified that estimated beach recreation values, with one (B2) providing a value considered suitable for transfer (Table 8.6). This study was broad-scale, encompassing major regional urban centers as well as smaller population centers in regional areas of Queensland. There was no underlying reason to

Table 8.5 Source study review for waterways (target: <1000 km: river orders 2 and 3)

ID	Source	Year	Valuation/ policy scenario	Method	Comment
WW1	Bennett et al. (2008)	2005	Rivers and water quality/ development	CM	**Attribute misalignment**: percentage of waterways suitable for primary contact (n/a for target site)
WW2	Morrison and Bennett (2004)	2000	Rivers and water quality/ development	CM	**Attribute misalignment**: categorical not metric: suitable for either fishing or fishing and swimming for whole of river change (n/a for target site)
WW3	Van Bueren and Bennett (2004)	2000	Natural resource management in a rural area	CM	**Attribute misalignment**: per km waterways restored for fishing or swimming (n/a for target site) **Scale**: details not reported
WW4	Rolfe et al. (2002)	2000	Water resource and agricultural development	CM	**Scope and framing**: Rural valuation context **Measurement**: km in good condition **Scale difference**: 1500–2400 km
WW5	Rolfe and Windle (2003)	2001	Water resource and agricultural development	CM	Same site: comments as for Rolfe et al. (2002)
WW6	Rolfe and Bennett (2009)	2002	Water resource and agricultural development	CM	Same site: comments as for Rolfe et al. (2002)
WW7	Windle and Rolfe (2005)	2003	Water resource and agricultural development	CM	Same site: comments as for Rolfe et al. (2002)
WW8	Mazur and Bennett (2009)	2008	Natural resource management in a rural area	CM	**Scope and framing**: rural valuation context **Measurement**: km in good condition **Scale**: within range: 950–1500 km
WW9	Rolfe and Windle (2008)	2005	Natural resource management	CM	**Some scope and scale similarity**: 1 site included a more urbanised area with overlapping scale (700–1100 km) **Measurement**: km in good condition
WW10	Rolfe and Brouwer (2012)	Various	Various	Meta-analysis	Incorporated studies outlined above

Table 8.6 Source study review for beaches (target: up to 50 km)

ID	Source	Year	Valuation/policy scenario	Method	Comment
B1	Blackwell (2007)	1999/2000	Recreational use of 4 beaches	TCM	Small scale study (n = 243) with unknown proportion of local visitors. Study (and analytical methods) dated
B2	Rolfe and Gregg (2012)	2010	Recreational use of 1400 km beaches	TCM	Broad scale study (n = 1049) of local residents. Best sample match
B3	Windle and Rolfe (2013)	2012	Recreational use of 250 km beaches in an urbanized area	TCM	Broad scale study (n = 1001) of nearby capital city residents. Average travel distance (80–100 km) greater than local users at target site

believe the estimated trip value for a beach visit would be significantly different at the target site, making this suitable for unit value transfer.

8.3.3 Performing the Benefit Transfers

Based on the assessments described above, unit value transfers were defensible only for beaches and to a lesser extent, waterways. Two suitable source studies and a meta-analysis were identified for the waterways asset, creating three possible options for transferring values.[9] The first was to apply the value of the \$1.21/km from the southeast Queensland site (study WW9) because the site represented the "best fit" with the target. The second was to use the average value from the two single studies, WW8 and WW9, i.e. (\$1.07 + \$1.21)/2 = \$1.14/km, while the third option was to apply the value of \$1.24/km from the meta-analysis. The first option is the most common with unit value transfers.

Once source study values had been selected, adjustments were made to tailor and extrapolate values at the target site. Beginning with a present value of \$1.21 per household for a one kilometer improvement in the length of waterways in good condition, the primary adjustment was to extrapolate this value to the population at the target site (assumed to be 40,000 households). The transferred value of \$48,400 per kilometer should be applied only to assess the value of a marginal change in quality (not the total value of the asset), but could be applied to either gains or losses.[10]

[9]Note that the use of meta-analysis in this way is generally considered a type of benefit function transfer.

[10]In the source study, the status quo was set as a future base, with a lower level of provision than the current situation. All alternative levels represented an improvement on the future base, but

For beaches, the Rolfe and Gregg (2012) study provided a value estimate of \$35.09 per visit. As there was no information about the frequency of beach trips at the target site, the visit rate information was also transferred from the source study (20 trips per adult per year). If the adult population at the target is assumed to be 50,000 adults, the annual value of beach recreation would be approximately \$35 million. This is an estimate of total recreation value. Further information would be required to estimate values for marginal changes in quality. For example, the source study also provided information from a contingent behavior experiment about changes in visitation rates if water quality at beaches declined, reporting that a one percent decline in water quality reduced the value of a beach visit by \$1.40.

The final steps in the benefit transfer involves sensitivity analysis, where standard procedures were applied (these are suppressed here for conciseness). It is also important in the final evaluation and reporting to outline the limitations of any selected source study estimates. These are outlined in the sections above.

8.4 Conclusion

This chapter outlines the process of conducting a unit value transfer and describes key issues that must be considered when conducting this type of transfer. To illustrate these concepts, it presents a case study example from Australia. Because unit value transfers provide little flexibility to adjust value estimates for differences between study and policy valuation contexts, the correspondence between these contexts is critical to accuracy.

Here, three main limiting factors constrained the pool of information available for transfer. The first was the limited pool of potential source site valuations. Having so few studies to draw upon not only limited the potential of finding a suitable match, but also provided little information to evaluate how values might differ across different types of contexts. Second, most source study valuations had been conducted in large scale rural catchments. These were not well matched to the small scale peri-urban coastal towns for which values were required (scale limitation). Third, the broad scale and rural focus of many source studies did not match the features of environmental assets in the targeted peri-urban residential areas (scope and context differences). While there was potential to make adjustments for scale differences, no information existed to help account for rural/urban context differences.

Based on this evaluation, unit value transfer was considered to be defensible only for two assets, beaches and waterways. The accuracy of these transfers was supported indirectly—at least for the waterways attribute—by the similarity of

(Footnote 10 continued)
were lower than the current level. Consequently, it is not clear if respondents were indicating their WTP to avoid a loss or achieve a gain.

value estimates from different source studies (from \$1.07 to \$1.24 per km). For other attributes, available source studies did not meet at least one of the necessary criteria for accurate unit value transfer (see Table 8.2). Empirical results such as these support the general contention of Bateman et al. (2011) that unit value transfers can be defensible in some cases. However, they also highlight the extremely restrictive conditions under which such transfers are expected to provide accurate results.

References

Bateman, I. J., Carson, R. T., Day, B., Hanemann, M., Hanley, N., Hett, T., et al. (2002). *Economic valuation with stated preference techniques: A manual*. Cheltenham: Edward Elgar.

Bateman, I. J., Day, B. H., Georgiou, S., & Lake, I. (2006). The aggregation of environmental benefit values: Welfare measures, distance decay and total WTP. *Ecological Economics, 60*, 450–460.

Bateman, I., Brouwer, R., Ferrini, S., Schaafsma, M., Barton, D., Dubgaard, A., et al. (2011). Making benefit transfers work: Deriving and testing principles for value transfers for similar and dissimilar sites using a case study of the non-market benefits of water quality improvements across Europe. *Environmental and Resource Economics, 50*, 365–387.

Bennett, J., Dumsday, R., Howell, G., Lloyd, C., Sturgess, N., & Van Raalte, L. (2008). The economic value of improved environmental health in Victorian Rivers. *Australasian Journal of Environmental Management, 15*, 138–148.

Blackwell, B. (2007). The value of a beach recreational visit: An application to Mooloolaba beach and comparisons with other outdoor recreational sites. *Economic Analysis and Policy, 37*, 77–97.

Brookshire, D., & Neill, H. (1992). Benefit transfers: Conceptual and ethical issues. *Water Resources Research, 28*, 651–655.

Brouwer, R. (2000). Environmental value transfer: State of the art and future prospects. *Ecological Economics, 32*, 137–152.

Colombo, S., & Hanley, N. (2008). How can we reduce the errors from benefits transfer? An investigation using the choice experiment method. *Land Economics, 84*, 128–147.

Concu, G. (2007). Investigating distance effects on environmental values: A choice modelling approach. *The Australian Journal of Agricultural and Resource Economics, 51*, 175–194.

DSEWPC (Department of Sustainability, Environment, Water, Population & Communities). (2011). *State of the Environment Report 2011*. Chapter 11 Coasts, Section 2.7 Population growth and urban development. Canberra: Australian Government.

Hatton MacDonald, D., & Morrison, M. (2010). Valuing biodiversity using habitat types. *Australasian Journal of Environmental Management, 17*, 235–243.

Johnston, R. J., & Rosenberger, R. S. (2010). Methods, trends and controversies in contemporary benefit transfer. *Journal of Economic Surveys, 24*, 479–510.

Kaul, S., Boyle, K. J., Kuminoff, N. V., Parmeter, C. F., & Pope, J. C. (2013). What can we learn from benefit transfer errors? Evidence from 20 years of research on convergent validity. *Journal of Environmental Economics and Management, 66*, 90–104.

Kline, J. D., & Mazzotta, M. J. (2012). *Evaluating tradeoffs among ecosystem services in the management of public lands*. General Technical Report. PNW-GTR-865. Portland, OR: U.S. Department of Agriculture, Forest Service, Pacific Northwest Research Station. 48 p. http://www.fs.fed.us/pnw/pubs/pnw_gtr865.pdf.

Kovacs, K. F., & Larson, D. M. (2008). Identifying individual discount rates and valuing public open space with stated-preference models. *Land Economics, 84*, 209–224.

Loomis, J. B. (2000). Vertically summing public good demand curves: An empirical comparison of economic versus political jurisdictions. *Land Economics, 76*, 312–321.

Mallawaarachchi, T., Blamey, R., Morrison, M., Johnson, A., & Bennett, J. W. (2001). Community values for environmental protection in a cane farming catchment in northern Australia. *Journal of Environmental Management, 62*, 301–316.

Mallawaarachchi, T., Morrison, M. D., & Blamey, R. K. (2006). Choice modelling to determine the significance of environmental amenity and production alternatives in the community value of peri-urban land: Sunshine Coast, Australia. *Land Use Policy, 23*, 323–332.

Mazur, K., & Bennett, J. (2009). Scope and scale effects on communities; values for environmental improvements in the Namoi catchment: A choice modelling approach. *Environmental Economics Research Hub*. Research Report #42. http://www.crawford.anu.edu.au/research_units/eerh/publications/ (accessed 11/13/2012).

Morrison, M., & Bennett, J. (2004). Valuing New South Wales rivers for use in benefits transfer. *Australian Journal of Agricultural and Resource Economics, 48*, 591–611.

Morrison, M., & Bergland, O. (2006). Prospects for the use of choice modelling for benefit transfer. *Ecological Economics, 60*, 420–428.

Morrison, M., Bennett, J., Blamey, R., & Louviere, J. (2002). Choice modelling and tests of benefit transfer. *American Journal of Agricultural Economics, 84*, 161–170.

Navrud, S., & Ready, R. (Eds.). (2007). *Environmental value transfer: Issues and methods*. Dordrecht: Springer.

Nelson, J. P., & Kennedy, P. E. (2009). The use (and abuse) of meta-analysis in environmental and resource economics: An assessment. *Environmental and Resource Economics, 42*, 345–377.

Rolfe, J. (2006). A simple guide to choice modelling and benefit transfer. In J. Rolfe & J. Bennett (Eds.), *Choice modelling and the transfer of environmental values* (pp. 10–27). Cheltenham: Edward Elgar.

Rolfe, J., & Bennett, J. (Eds.). (2006). *Choice modelling and the transfer of environmental values*. Cheltenham: Edward Elgar.

Rolfe, J., & Bennett, J. (2009). The impact of offering two versus three alternatives in choice modelling experiments. *Ecological Economics, 68*, 1140–1148.

Rolfe, J., & Brouwer, R. (2012). Design effects in a meta-analysis of river health choice experiments in Australia. *Journal of Choice Modelling, 5*, 81–97.

Rolfe, J., & Dyack, B. (2010). Testing for convergent validity between travel cost and contingent valuation estimates of recreation values in the Coorong, Australia. *Australian Journal of Agricultural and Resource Economics, 54*, 583–599.

Rolfe, J., & Gregg, D. (2012). Valuing beach recreation across a regional area: The Great Barrier Reef in Australia. *Ocean and Coastal Management, 69*, 282–290.

Rolfe, J., & Windle, J. (2003). Valuing the protection of Aboriginal cultural heritage sites. *Economic Record, 79* (Special Issue), 85–95.

Rolfe, J., & Windle, J. (2008). Testing for differences in benefit transfer values between state and regional frameworks. *Australian Journal of Agricultural and Resource Economics, 52*, 149–168.

Rolfe, J., Loch, A., & Bennett, J. (2002). *Tests of benefit transfer across sites and populations in the Fitzroy Basin*. Valuing Floodplain Development in the Fitzroy Basin. Research Report #4, Central Queensland University. http://resourceeconomics.cqu.edu.au/. Accessed October 29, 2012.

Rolfe, J., Windle, J., Bennett, J., & Mazur K. (2013). *Calibration of values in benefit transfer to account for variations in geographic scale and scope: Comparing two choice modelling experiments*. Paper presented at the 57th Annual conference of the Australian Agricultural and Resource Economics Society, Sydney, February.

Rosenberger, R. S., & Johnston, R. J. (2009). Selection effects in meta-analysis and benefit transfer: Avoiding unintended consequences. *Land Economics, 85*, 410–428.

Rosenberger, R. S., & Stanley, T. D. (2006). Measurement, generalization, and publication: Sources of error in benefit transfers and their management. *Ecological Economics, 60*, 372–378.

Smith, V. K., & Pattanayak, S. K. (2002). Is meta-analysis a Noah's Ark for non-market valuation? *Environmental and Resource Economics, 22*, 271–296.

Tapsuwan, S., Ingram, G., Burton, M., & Brennan, D. (2009). Capitalized amenity value of urban wetlands: A hedonic property price approach to urban wetlands in Perth, Western Australia. *The Australian Journal of Agricultural and Resource Economics, 53*, 527–545.

Van Bueren, M., & Bennett, J. (2004). Towards the development of a transferable set of value estimates for environmental attributes. *The Australian Journal of Agricultural and Resource Economics, 48*, 1–32.

Whitten, S., & Bennett, J. (2006). Transferring the environmental values of wetlands. In J. Rolfe & J. Bennett (Eds.), *Choice modelling and the transfer of environmental values* (pp. 164–190). Cheltenham: Edward Elgar.

Windle, J., & Rolfe, J. (2005). Assessing non-use values for environmental protection of an estuary in a Great Barrier Reef catchment. *Australasian Journal of Environmental Management, 12*, 147–156.

Windle, J., & Rolfe, J. (2013). Estimating nonmarket values of Brisbane (state capital) residents for state based beach recreation. *Ocean and Coastal Management, 85*, 103–111.

Chapter 9
Benefit Transfer Combining Revealed and Stated Preference Data

Robert J. Johnston, Mahesh Ramachandran and George R. Parsons

Abstract Benefit transfers often combine data from revealed preference (RP) and stated preference (SP) analyses. RP/SP estimation allows a researcher to generate more broadly-applicable benefit functions, leading to potential improvements in transfer reliability and validity. The appropriateness of various types of RP/SP data combinations within benefit transfer has also been subject to disagreement, for example with regard to the potential pooling of theoretically inconsistent welfare measures within meta-analysis. This chapter provides a summary and case study illustration of RP/SP modeling within benefit transfer. The chapter begins with an introduction to the use of these techniques for benefit transfer and typology of applicable methods. This is followed by an illustration that uses RP/SP micro-data to quantify recreational benefit changes under Delaware Department of Natural Resources and Environmental Control beach nourishment and retreat scenarios. Unlike most benefit transfers in the academic literature implemented in artificial and idealized circumstances, the present case study represents an actual, policy-driven benefit transfer used within agency cost benefit analysis (CBA). The chapter concludes with a discussion of the role of RP/SP data within benefit transfer.

Keywords Beach erosion · Beach nourishment · Coastal adaptation · Willingness to pay · Recreation demand · Travel cost method · Contingent behavior · Joint estimation

---

R.J. Johnston (✉)
Department of Economics and George Perkins Marsh Institute, Clark University, Worcester, MA, USA
e-mail: rjohnston@clarku.edu

M. Ramachandran
Cape Cod Commission, Barnstable, MA, USA
e-mail: mramachandran@capecodcommission.org

G.R. Parsons
School of Marine Science and Policy, Department of Economics and Department of Applied Economics and Statistics, University of Delaware, Newark, DE, USA
e-mail: gparsons@udel.edu

R.J. Johnston et al. (eds.), *Benefit Transfer of Environmental and Resource Values*, The Economics of Non-Market Goods and Resources 14,
DOI 10.1007/978-94-017-9930-0_9

9.1 Introduction

Benefit transfers often combine data from revealed and stated preference analyses. Revealed preference (RP) valuation techniques use data on observed behaviors to estimate monetary measures of welfare change ex post. Stated preference (SP) valuation techniques estimate these measures ex ante using responses to survey questions that ask how individuals would behave if faced with hypothetical scenarios describing a resource or policy change to be evaluated. Despite a historical tendency among economists to view RP and SP techniques as substitutes—often with RP methods preferred due to their reliance on observed behavior—there is increasing recognition that the two approaches offer complementary strengths and offsetting weaknesses (Whitehead et al. 2008b, 2011a, b). For example, the grounding of RP models in observed behavior implies that valuation is limited to use values under observable conditions. In contrast, the use of hypothetical data by SP models allows estimation of use and nonuse values for both observable and potential future conditions, including hypothetical resource or policy scenarios that do not yet exist. These complementary strengths and offsetting weaknesses have led to a rapidly expanding literature on combined revealed and stated preference (RP/SP) techniques for environmental valuation (Whitehead et al. 2008b, 2011a, b).[1,2]

Most of the advantages of RP/SP techniques for primary valuation apply to benefit transfer. Benefit transfer has also been one of the areas in which methods for RP/SP estimation have seen significant recent advances, including work in traditional single-study joint estimation and meta-analysis (e.g., González-Sepúlveda and Loomis 2011; Johnston and Rosenberger 2010; Smith et al. 2002; Van Houtven et al. 2011; Bergstrom and Taylor 2006; Rosenberger and Johnston 2009). In addition, there is at least one RP/SP method—structural benefit transfer or preference calibration—that has been developed primarily for transfer (e.g., Smith et al. 2002, 2006; Van Houtven et al. 2011). Advantages of RP/SP estimation relevant for benefit transfer include an ability to capitalize on data from one or more primary studies to generate broadly-applicable benefit functions. This can lead to improvements in reliability, validity, and efficiency (González-Sepúlveda and Loomis 2011; Van Houtven et al. 2011; Johnston and Rosenberger 2010).

The appropriateness of various types of RP/SP data combinations within benefit transfer has also been subject to disagreement, for example with regard to the potential pooling of theoretically inconsistent welfare measures from RP and SP studies within meta-regression models (Bergstrom and Taylor 2006; Londoño and Johnston 2012;

[1]A few examples include Adamowicz et al. (1997), Cameron et al. (1996), Huang et al. (1997), Haener et al. (2001), Kling (1997), Boxall et al. (2003), Englin and Cameron (1996) and Parsons et al. (1999).

[2]Borrowing the terminology of Whitehead et al. (2011a, b, p. 3), this chapter emphasizes joint estimation RP/SP analysis, in which "relationships between the independent … and … dependent variables are estimated in a single model." We do not address data comparison or other analyses in which RP and SP data are not jointly estimated, including the large literatures on convergent validity and hypothetical bias in SP estimation (Carson et al. 1996; Murphy et al. 2005).

Johnston and Moeltner 2014; Johnston and Rosenberger 2010; Nelson and Kennedy 2009; Smith and Pattanayak 2002). Hence, while benefit transfer is an area in which RP/SP techniques offer great promise, it is also an area in which these techniques have been subject to controversy. Moreover, unlike the mainstream valuation literature in which RP/SP models are generally seen as a distinct methodological class with associated summary publications (Whitehead et al. 2008b, 2011a, b), parallel approaches within benefit transfer tend to be scattered across multiple areas of the literature (e.g., meta-analysis, structural benefit transfer, RP/SP micro-data models), often with little methodological cohesiveness. This reflects a broader disorganization in the benefit transfer literature (Johnston and Rosenberger 2010).

This chapter provides a summary perspective and case study illustration of RP/SP modeling within benefit transfer. To our knowledge, this is the first review of this kind. The chapter begins with an introduction to the use of RP/SP techniques for benefit transfer and typology of applicable methods. This is followed by an illustration of benefit transfer that uses RP/SP data to quantify recreational benefit changes under Delaware Department of Natural Resources and Environmental Control (DNREC) beach nourishment and retreat scenarios. Unlike most benefit transfers in the academic literature implemented in artificial and idealized circumstances, the present case study represents an actual, policy-driven benefit transfer used within agency cost benefit analysis (CBA). The illustrated benefit transfer is also differentiated from most in the literature by identical study and policy sites; the reason that the application is a benefit transfer is a difference between the specific policy outcomes addressed by the original study and those for which benefits were required. This illustrates a type of benefit transfer not often highlighted in the academic literature; one in which the sites are the same but the projected policy outcomes differ. The chapter concludes with a discussion of the role of RP/SP data within benefit transfer, both in the present case study and in transfer applications more broadly. We also discuss future prospects and research needs.

9.2 Benefit Transfer Combining Revealed and Stated Preference Data

As noted above, the ways that RP/SP techniques appear in the primary valuation and benefit transfer literatures differ. First, while RP/SP data are frequently applied for benefit transfer, most applications do not emphasize this fact. This stands in contrast to the primary valuation literature, in which the use of RP/SP data is often given primary emphasis. Second, some RP/SP techniques are solely or primarily found within benefit transfer. Examples include meta-analysis and structural benefit transfer. Despite these and other differences, similar general principles apply to the use of RP/SP data in primary valuation and benefit transfer.

This chapter does not include a comprehensive review of all RP/SP techniques in environmental valuation; such a review is provided elsewhere (Whitehead et al. 2008b, 2011b). Rather, we emphasize those methods with particular relevance for benefit transfer, most typically for the estimation of transferable benefit functions. As a precursor to this discussion, we introduce the primary methods for joint RP/SP estimation that may be applied to benefit transfer. We then present a simple typology of the ways in which these methods are applied.

9.2.1 Applicable Methods

Whitehead et al. (2011a, b) propose a typology in which RP/SP approaches in the valuation literature are divided into three categories: (1) frequency data models, (2) mixed data models and (3) discrete data models. The first category involves analysis of stacked RP/SP data on the frequency of some activity, most often trips within a recreation demand framework. Examples include Englin and Cameron (1996) and Whitehead et al. (2000). Within the second category, the form of dependent variables from RP and SP methods differ. Because the data cannot be easily stacked or pooled—for example these models often combine discrete and continuous data—utility theoretic specifications or related assumptions are required for joint estimation. Early examples include Cameron (1992), Kling (1997) and Huang et al. (1997). Finally, discrete data models include cases in which both RP and SP data are drawn from comparable random utility models, so that the data may be stacked and jointly estimated. The first published example was Adamowicz et al. (1994); more recent examples include Von Haefen and Phaneuf (2008).

While such distinctions are informative for RP/SP models in general, they are not necessarily the most useful for characterizing benefit transfer methods. Rather, within benefit transfer, four broad approaches have emerged that use RP/SP data in different ways. These include: (1) stacked micro-data models, (2) jointly estimated mixed data models, (3) meta-analysis combining RP and SP data, and (4) structural benefit transfer or preference calibration.

The first two categories overlap those of Whitehead et al. (2011a), with frequency data and discrete data models combined within our broader category of stacked micro-data models, and jointly estimated mixed data models the same in both groupings. Within these first two categories, an additional distinguishing factor for benefit transfer is whether the transfer itself combines RP and SP data, or whether the transfer employs the results of one or more primary valuation studies that have *already* combined these data. For example, preference functions or unit values generated by a previously published mixed data model might be used directly for benefit transfer; here the benefit transfer practitioner does not directly combine the RP/SP data, but rather transfers a value or function that has already been generated using data combination. Alternatively, a similar mixed data model might be estimated expressly for transfer (e.g., González-Sepúlveda and Loomis 2011). There are similar examples of stacked micro-data models developed expressly to evaluate benefit transfer (e.g., Haener et al. 2001).

In addition to the methods covered in Whitehead et al. (2011a), benefit transfer includes two additional data combination approaches: meta-analysis and structural benefit transfer. Unlike other methods that pool *raw data* from RP and SP research, these latter two approaches typically combine the *final empirical results* of prior RP and SP studies. Both of these areas of research are the subject of a growing research literature (Johnston and Rosenberger 2010).

9.2.1.1 Stacked Micro-Data Models

This category includes the transfer of welfare results (either unit values or benefit functions) from a study that pools RP and SP micro-data; as noted above the RP/SP estimation may be conducted as part of an existing primary study or directly for transfer. Such models may stack either frequency or discrete data, as detailed by Whitehead et al. (2008b, 2011a); applicable empirical models are determined by the type of data employed. A common example is the transfer of models that pool parallel revealed and contingent data on a continuous behavior of interest, most commonly recreation behavior such as the number of trips to a site (e.g., Englin and Cameron 1996; Whitehead et al. 2008a). Haener et al. (2001) provide an example of this type of benefit transfer. The underlying methods used to combine RP/SP data within contingent behavior and other stacked micro-data models do not typically differ between primary valuation and benefit transfer applications. Whitehead et al. (2008b) provides an extensive review of these underlying methods, to which readers are referred for additional details.

9.2.1.2 Jointly Estimated Mixed Data Models

This category parallels the eponymous category of Whitehead et al. (2008b, 2011a). It is distinguished by the joint estimation of models including RP and SP data that resist direct stacking. Typical examples involve discrete choice contingent valuation data paired with a continuous demand relationship estimated using RP data, often applied to recreational behavior. Although such data cannot be stacked directly, the RP and SP models may be jointly estimated by assuming correlated errors (Whitehead et al. 2008b). While many examples exist within the broader environmental valuation literature, few jointly estimated mixed data models have been developed explicitly for benefit transfer. An example of this type of benefit transfer is provided by González-Sepúlveda and Loomis (2011). As above, the empirical data combination methods applied for benefit transfer do not differ fundamentally from those applied in parallel primary valuation applications, and are hence not reviewed in detail here.

9.2.1.3 Meta-Analysis Combining RP and SP Data

Meta-analysis (MA) may be characterized as "the statistical analysis of a large collection of results for individual studies for the purposes of integrating the findings" (Glass 1976, p. 3). When applied to benefit transfer, MA is often used to "identify and test systematic influences of study, economic, and resource attributes on WTP, characterize results of the literature addressing certain classes of non-market values, and generate reduced form functions for direct transfer applications" (Johnston and Rosenberger 2010, pp. 483–484). Most examples involve use of a meta-regression model (MRM) with a dependent variable reflecting a summary welfare statistic drawn from comparable primary studies (often a WTP estimate for a particular type of marginal resource change). Independent moderator variables characterize resource, policy, context and population attributes hypothesized to explain observed variation in the chosen welfare statistic across primary study observations (Johnston and Rosenberger 2010; Nelson and Kennedy 2009).

In order to obtain metadata of sufficient size and variation, MRMs in the benefit transfer literature often pool data from prior RP and SP analyses. This is distinct from typical RP/SP models, as MRM dependent variables typically reflect *final welfare estimates* derived from primary studies using either RP or SP data. That is, the MRMs pool the welfare estimates provided by distinct RP and SP studies rather than relying on raw RP/SP data. Past MRMs have pooled stated preference (SP) and recreation demand model (RP) estimates of WTP for such commodities as per fish catch by recreational anglers (Johnston et al. 2006), recreational visitor/days to coral reefs (Londoño and Johnston 2012), and visitor/days of outdoor recreation (Rosenberger and Loomis 2000), among many others. The review of Nelson and Kennedy (2009) includes multiple MRMs that pool RP and SP data.

The pooling of RP/SP data within MRMs, however, has been the subject of controversy (Johnston and Rosenberger 2010; Johnston and Moeltner 2014). Within valuation metadata, welfare consistency requires that pooled welfare measures represent the same theoretical construct (Bergstrom and Taylor 2006; Nelson and Kennedy 2009; Smith et al. 2002), for example a well-defined measure such as Hicksian compensating surplus. Only observations that satisfy a minimum degree of welfare consistency should be pooled within an MRM (Nelson and Kennedy 2009; Smith and Pattanayak 2002). However, there is a lack of consensus regarding the minimum level of consistency that should be required (Londoño and Johnston 2012). For example, some authors caution against pooling otherwise equivalent Marshallian and Hicksian welfare measures within the same MRM (Nelson and Kennedy 2009; Smith and Pattanayak 2002); this would preclude most RP/SP pooling found within the meta-analysis literature. Others, however, allow for greater flexibility in pooling Hicksian versus Marshallian welfare measures, arguing that the ability of pooled RP/SP MRMs to generate reliable (i.e., low error) benefit transfers is an empirical question (Londoño and Johnston 2012). With only a few exceptions (e.g., Johnston and Moeltner 2014; Londoño and Johnston 2012), systematic empirical evidence regarding effects of welfare inconsistency on transfer reliability is sparse, and results are mixed. Hence, it is currently unclear whether the combination of strictly welfare inconsistent RP/SP data within MRMs is able to promote more reliable benefit transfers.

9.2.1.4 Structural Benefit Transfer or Preference Calibration

The final area in which benefit transfers combine RP and SP data is within structural benefit transfer or preference calibration. Structural benefit transfer is distinguished by a formal basis in an explicit utility function, designed to impose theoretical rigor on the combination of preference data. First proposed by Smith et al. (2002), structural benefit transfer requires the analyst to specify a utility function that describes a representative individual's choices over a set of market and non-market goods modeled in prior valuation studies, presuming standard budget-constrained utility maximization (Johnston and Rosenberger 2010). The specified function enables WTP or another welfare measure to be expressed as a function of arguments including the resource change in question, income, prices and other factors indicated by economic theory. These variables are observed or inferred from the available prior studies or gathered from supplemental sources. The prior studies used in preference calibration typically include RP and SP analyses implemented over similar populations, so that one can argue that the same umbrella utility (or preference) function underpins the results of each prior study.

Based on the specified utility function, the analyst derives parameterized analytical expressions that determine relationships between each available RP and SP welfare estimate and other observable factors. These expressions must "assure the variables assumed to enter the preference function are consistently measured across each study and linked to preference parameters" (Smith et al. 2002, p. 136). Empirical methods are then used to calibrate parameters to the specified utility-theoretic structure. This calibration solves for unknown parameters implied by analytical expressions, ensuring the consistency of RP and SP data within the specified utility structure.

Like many valuation MRMs, preference calibration combines the results of previously implemented RP and SP models to estimate an umbrella preference function. This differs from commonly encountered RP/SP stacked micro-data and jointly estimated mixed data models that combine raw RP/SP data prior to preference estimation. Unlike most MRMs, however, the data combination is implemented within a strong structural utility theoretic (SSUT) framework that imposes theoretical consistency on model results (Bergstrom and Taylor 2006). As noted by Johnston and Rosenberger (2010, p. 485), "[p]roponents of preference calibration argue that such approaches provide advantages over other transfer methods; these advantages include a model that imposes theoretical consistency on the use of prior information. The method, however, is not without limitations, not the least of which is a requirement of strong a priori assumptions regarding the underlying structural model…" Recent illustrations of structural benefit transfer are provided by Pattanayak et al. (2007), Smith et al. (2002, 2006), and Van Houtven et al. (2011), among others. Chapter 23 also provides an example.

9.3 Benefit Transfer with Identical Sites and Different Resource Changes: An Application to Delaware Bay Beaches

The remainder of this chapter illustrates a benefit transfer using a stacked, RP/SP micro-data model, drawing on results of Parsons et al. (2013). Unlike transfers in the published literature implemented over artificial and often idealized test cases, results of this analysis were used directly for agency CBA. The purpose is this illustration is twofold. First, it demonstrates the use of RP/SP micro-data for benefit transfer. Second, it illustrates empirical challenges and potential solutions encountered in actual benefit transfer situations. In doing so, it provides a practical illustration of transfer suitable for practitioners; this contrasts to works in the academic literature that emphasize methodological advances. Because the case study addresses an actual rather than hypothetical benefit transfer, true underlying values are unknown. Hence, transfer reliability and validity cannot be assessed.

Within the application, RP data are drawn from a count data recreation demand (travel cost) analysis of seven Delaware Bay beaches under existing beach widths. Corresponding SP data are drawn from contingent behavior questions enabling demand estimation under hypothetical alternative widths. RP/SP micro-data pooling enables estimation of consumer surplus (CS, reflecting recreational access value) at different widths. The benefit transfer adapts information from the original model to project beyond the scope of originally analyzed effects, providing transferred estimates for a new set of projected policy outcomes.

The application estimates recreational benefit changes under four proposed DNREC beach nourishment and retreat scenarios designed to address ongoing erosion projected through 2041. These estimates reflect the projected economic benefits of beach recreation gained or lost at seven Delaware Bay beaches: (1) Pickering, (2) Kitts Hummock, (3) Bowers, (4) South Bowers, (5) Slaughter, (6) Primehook, and (7) Broadkill. The transfers quantify changes in recreational access values due to projected changes in beach width and associated losses of housing structures under the DNREC management scenarios.

As noted above, a distinguishing feature of this transfer is that it occurs *across different policy outcomes* rather than *across different sites*. Although benefit transfer is most often associated with a geographical transfer of benefits from a study site to a policy site, transfers can occur with identical sites if affected populations or policy outcomes differ (cf., Morrison et al. 2002). Here the original RP/SP analysis was conducted for a different scope of policy effects than those for which benefit estimates were required. Specifically, compared to scenarios evaluated by the primary studies, the proposed policy effects involve larger beach width changes and an additional loss of housing structures on affected beaches. The latter is relevant because it reduces lodging available for those engaged in multi-day beach visits. Because the CBA required estimation of benefits outside the scope of those evaluated by the original study, it was necessary to conduct a transfer across different policy outcomes and time frames at the same sites to obtain the needed estimates.

9.3.1 The Policy Context

Beach erosion is a continuing concern along the Delaware Bay, threatening recreational activities, coastal homes and other resources that depend on the width or existence of beaches. To address this problem Delaware has traditionally adopted a strategy of beach nourishment, replenishing sand 'as needed' on ocean and bay beaches to maintain a tolerable width for recreation and storm protection (Parsons et al. 2013).[3] Recently, however, in the face of concerns regarding sea level rise, increased storm severity and the costs of indefinite beach nourishment, the state is reconsidering this practice. This mirrors similar debates elsewhere (Whitehead et al. 2008a; Pendleton et al. 2012; Smith et al. 2009). Among the primary benefits affected are those related to beach recreation. The illustrated transfer of recreational benefits presented herein contributed to a larger CBA of beach nourishment and retreat alternatives for the Delaware Bay beaches identified above.[4]

Four beach nourishment and retreat scenarios were considered for the benefit transfer, each applied to the seven Bay beaches. These are (1) beach nourishment, (2) strategic retreat, (3) basic retreat, and (4) do nothing. Key variables of interest for the recreational benefit models are the average beach width and number of housing structures lost due to beach erosion and flooding; data for both were provided by Johnson, Mirmiran and Thompson, Inc. (2012). Tables 9.1, 9.2, 9.3 and 9.4 illustrate projected beach widths and housing losses in ten-year increments beginning in 2011 (the analysis was conducted during 2010 and 2011).

Summarizing these scenarios, scenario one (beach nourishment) would replenish sand on each beach to reach new widths between 176 and 563 % of current widths, depending on the beach. Sand would then be added as needed to maintain these average widths. Scenario two (strategic retreat), removes houses to provide a beach profile identical in width to scenario one, yet without nourishment (i.e., the beach is extended landward). Beaches then migrate naturally, maintaining the same average width as they move landward. Scenario three is a basic retreat option, in which the beach migrates naturally and houses are removed as needed to accommodate the landward retreat. The current width of the beach is maintained as the beach retreats. Scenario four is the "do nothing" scenario, in which beaches retreat but houses are not removed by the State. Because structures are not removed, the natural migration of the beach is interrupted, and beach widths decline. Many of these average widths reach zero before 2041.

[3]As described by Whitehead et al. (2008a), "[b]each … nourishment is the placement of sand on beaches to increase … width for the purposes of protecting property and maintaining recreation opportunities (Jones and Mangun 2001)."

[4]As discussed by Parsons et al. (2013), the Delaware Bay beaches under consideration for the present analysis are smaller and less populated than the better-known ocean beaches located further south. Current average beach widths above mean high tide range from approximately 27 to 70 ft, with some areas as narrow as 13 ft (Johnson, Mirmiran and Thompson, Inc. 2012). Parsons et al. (2013) estimated a total of approximately 49,000 adult visitors per year to the seven beaches combined in 2010.

Table 9.1 Projected beach widths and housing structure losses, scenario one (nourishment)

Beach community	Number of 2011 community structures	Current mean width[a]	2011 width	2011 structures lost	2021 width	2021 structures lost	2031 width	2031 structures lost	2041 width	2041 structures lost
Pickering	44	38	174	0	174	0	174	0	174	0
Kitts Hummock	122	29	163.5	0	163.5	0	163.5	0	163.5	0
Bowers	354	54	150	0	150	0	150	0	150	0
South Bowers	84	34	125	0	125	0	125	0	125	0
Slaughter	372	43	109	0	109	0	109	0	109	0
Primehook	195	27	108	0	108	0	108	0	108	0
Broadkill	592	70	123	0	123	0	123	0	123	0

[a]All widths measured in feet

Table 9.2 Projected beach widths and housing structure losses, scenario two (strategic retreat)

Beach community	Number of 2011 community structures	Current Mean width[a]	2011 width	2011 structures lost[b]	2021 width	2021 structures lost	2031 width	2031 structures lost	2041 width	2041 structures lost
Pickering	44	38	174	38	174	1	174	0	174	0
Kitts Hummock	122	29	163.5	52	163.5	10	163.5	9	163.5	1
Bowers	354	54	150	35	150	4	150	2	150	2
South Bowers	84	34	125	8	125	2	125	1	125	1
Slaughter	372	43	109	5	109	5	109	14	109	21
Primehook	195	27	108	63	108	0	108	0	108	0
Broadkill	592	70	123	92	123	24	123	39	123	24

[a]All widths measured in feet

[b]Housing structure losses reflect losses realized by the year indicated, in addition to previously indicated losses

Table 9.3 Projected beach widths and housing structure losses, scenario three (basic retreat)

Beach community	Number of 2011 community structures	Current mean width[a]	2011 width	2011 structures lost[b]	2021 width	2021 structures lost	2031 width	2031 structures lost	2041 width	2041 structures lost
Pickering	44	38	38	0	38	10	38	27	38	1
Kitts Hummock	122	29	29	0	29	9	29	18	29	24
Bowers	354	54	54	0	54	4	54	5	54	8
South Bowers	84	54	54	0	54	1	54	4	54	2
Slaughter	372	43	43	0	43	0	43	0	43	4
Primehook	195	27	27	0	27	1	27	6	27	5
Broadkill	592	70	70	2	70	53	70	36	70	25

[a]All widths measured in feet

[b]Housing structure losses reflect losses realized by the year indicated, in addition to previously indicated losses

Table 9.4 Projected beach widths and housing structure losses, scenario four (do nothing)

Beach community	Number of 2011 community structures	Current mean width[a]	2011 width	2011 structures lost[b]	2021 width	2021 structures lost	2031 width	2031 structures lost	2041 width	2041 structures lost
Pickering	44	38	38	0	0	2	0	14	0	22
Kitts Hummock	122	29	29	0	0	0	0	13	0	18
Bowers	354	54	54	0	34	0	14	2	3	2
South Bowers	84	34	34	0	21	0	15	1	13	2
Slaughter	372	43	43	0	24	0	6	0	0	0
Primehook	195	27	27	0	22	0	21	0	21	4
Broadkill	592	70	70	0	40	4	12	12	0	33

[a] All widths measured in feet

[b] Housing structure losses reflect losses realized by the year indicated, in addition to previously indicated losses

9.4 Data and Primary RP/SP Study

The original RP/SP model of Parsons et al. (2013) estimates both total recreational access values and changes in access values under alternative widths for the seven Bay beaches. Within this study, total access value refers to the total annual CS from recreation at each beach. Widths are modeled as "dry beach," or the average distance between mean high tide and the dune edge. The original study estimates recreational access values for each beach under three scenarios: (1) existing 2010 widths; (2) if the beach were 25 % of its current average width, and (3) if the beach were 200 % of its current average width. These values are estimated using a two-component model. The first component is a pooled travel cost/contingent behavior model. The second component is a visitor count model. Benefit estimates are obtained by multiplying average CS per visitor by the estimated number of visitors to each beach. The benefit transfer described in this chapter uses prepublication data and results that are similar but not always identical to those reported in Parsons et al. (2013); minor differences are due to modeling and other changes in the original study made after the benefit transfer was completed.

9.4.1 Pooled Travel Cost/Contingent Behavior Model

The first component of Parsons et al. (2013) is an individual, count data travel cost model that pools RP/SP data from all seven beaches to estimate average CS per day under the different beach widths noted above. Distinct values are estimated for different visitor types (owners vs. non-owners of beach homes), and visit lengths (day trips, overnight trips). The underlying approach is similar to that of Whitehead et al. (2008a), who estimate recreational benefits related to beach nourishment and parking improvements on selected North Carolina beaches.

Both RP and SP data were collected from on-site surveys of visitors at each beach. RP data included: (1) the location of the respondent's permanent residence and whether he or she owned a secondary residence in one of the seven beach communities; (2) attributes of the current trip, such as how long he or she would they be on the beach that day, types of activities engaged in, and number of nights spent at the beach; (3) the number of trips taken to the beach since the beginning of the year and the number of trips expected over the balance of the year; (4) a breakdown of past and expected visits into day trips, short overnight trips,[5] and long overnight trips; (5) the number of trips to the six other Delaware Bay beaches in the study.[6]

These data were supplemented with responses to two contingent behavior (SP) questions. The first pertained to whether or not the quality of the person's recreation experience would be affected by the beach being one-quarter (25 %) of its current

[5]A short overnight trip was defined as three or fewer nights.

[6]The overwhelming majority of the respondents made no visits to other beaches in the set.

Table 9.5 Per-trip values for beach access at alternative beach widths

Type of trip	Width	Mean per day value
Day trip	25 % of current width	$28.15
Day trip	Current width	$32.87
Day trip	200 % of current width	$35.47
Overnight trip	25 % of current width	$31.53
Overnight trip	Current width	$36.82
Overnight trip	200 % of current width	$39.73

average width and, if so, if this would have altered the number of trips taken over the year. The second was a parallel question addressing a situation in which the beach was double (200 %) its current width.

The original analysis is based on 572 survey responses, with roughly 43 % of respondents intercepted during a day trip, 34 % on a short overnight trip (three or fewer nights), and 23 % on a long overnight trip (greater than three nights). These data were used to estimate a pooled single-site recreation demand model over the seven beaches. The count data model was estimated using a multivariate poisson gamma specification that corrected for on-site sampling (Parsons et al. 2013, cf. Landry and Liu 2011; Whitehead et al. 2008a). The model stacked the three demand equations for each individual, with discrete explanatory (demand shift) variables distinguishing observations at current widths (RP), 25 % of current width (SP), and 200 % of current width (SP). The area between these demand curves is interpreted as the change in CS due to the changes in beach width. From these results, one may calculate average CS per day, per person under different width conditions (Table 9.5).

9.4.2 Visitor Count Model

The second component of Parsons et al. (2013) predicts annual visitation (the number of annual days) at each of the seven Delaware Bay beaches, using an RP visitor count prediction model. The tally data model is estimated based on a stratified on-site sample of visitors counted at all seven beaches during 214 observation periods from June 2010 to July 2011. Sampling was stratified to provide representative coverage across different times-of-day, weekend versus weekdays, and months across the sample period. Using these data, a Hurdle-Poisson Model is estimated to predict the probable count of visitors on each sampling occasion (Cameron and Trivedi 1998, p. 124), with probability assumed to depend on the beach, time-of-day, month, and weekend versus weekday. The estimated model accounts for the high concentration of zero visits (times during which zero visitors were observed) by partitioning the model into two components. The first component predicts the probability that a sampling occasion will have zero visitors. The second component predicts the number of visitors conditional on there being at least one visitor. Parameters were estimated using maximum likelihood.

Table 9.6 Predicted annual days spent visiting Delaware Bay beaches

Beach and visitor type	Predicted days during single-day trips	Predicted days during short overnight trips	Predicted days during long overnight trips
Bowers (non-owners)	3604	697	232
Bowers (owners)	1162	1162	0
Broadkill (non-owners)	4542	6704	4433
Broadkill (owners)	2812	2703	541
Primehook (non-owners)	586	1759	1173
Primehook (owners)	1320	1026	293
Slaughter (non-owners)	5188	2136	813
Slaughter (owners)	1525	1322	102
Pickering (non-owners)	563	241	80
Pickering (owners)	241	80	0
Kitts Hummock (non-owners)	231	347	0
Kitts Hummock (owners)	347	115	0
South Bowers (non-owners)	1146	264	528
South Bowers (owners)	176	176	0

Note The average length of a short overnight trip is 2.197 days. The average length of a long overnight trip is 8.621 days

These results were used to forecast annual visitation days to each beach by owners and non-owners of beach community homes (Table 9.6). This forecast required an additional transformation to adjust predicted *instantaneous counts* of visitors at different sampling periods during each day to a *total estimate of unique beach visits* per day, based on the average length of stay (per day) observed at each beach. Predictions are disaggregated by beach, owners versus non-owners and trip length. These results, combined with per day CS estimates in Table 9.5, provide the central results that are adapted for benefit transfer.

9.5 Benefit Transfer Methods

The goal of the benefit transfer was to estimate recreational benefit changes from 2011 to 2041 under the four DNREC policy scenarios in Table 9.1 through Table 9.4. Under these scenarios, beach widths vary from 0 to 564 % of current widths; this is outside the scope of changes modeled by Parsons et al. (2013). In addition, the DNREC scenarios involve a loss of housing structures (due to flooding or removal by the State) that was not incorporated into the models of Parsons et al. (2013). This is relevant to recreational benefits because nearly all visitors to the seven Bay beaches stay in private residences (most of these communities are entirely residential, with no hotel or motel lodging). As houses disappear there is reduced lodging for these beach visitors. Due to these differences benefit transfer is required to adapt the results of Parsons et al. (2013) to the projected policy scenarios.

A sequence of steps is required to transfer these benefit estimates. First, drawing from width and housing loss forecasts available for 2011, 2021, 2031 and 2041 (Tables 9.1, 9.2, 9.3 and 9.4), we interpolate widths and housing losses for each intervening year. Second, drawing from results in Tables 9.5 and 9.6, we estimate a piecewise-linear value surface that provides a unique estimate of annual aggregate CS for each beach and visitor type, at any given width. These estimates, combined with interpolated width forecasts, enable calculation of preliminary estimates of annual recreational access value for each year, at each beach. An additional adjustment then accounts for the projected loss of housing structures, which is assumed to affect the number of beach visits. Finally, results are discounted and aggregated to obtain a final benefit estimate for each policy scenario. Each of these steps is described below.

9.5.1 Predicting Beach Widths and Housing Losses

Piecewise linear interpolation is first used to estimate predicted annual average dry beach widths for 2011–2041, based on values provided by the engineering firm Johnson, Mirmiran and Thompson, Inc. (2012) for 2011, 2021, 2031 and 2041 (Tables 9.1, 9.2, 9.3 and 9.4). This provides a forecast of average width for each beach, during each year, under each management scenario. Figure 9.1 illustrates this interpolation for Slaughter Beach under scenario four. Bold points represent predicted widths in 2011, 2021, 2031 and 2041. Other widths are interpolated between these

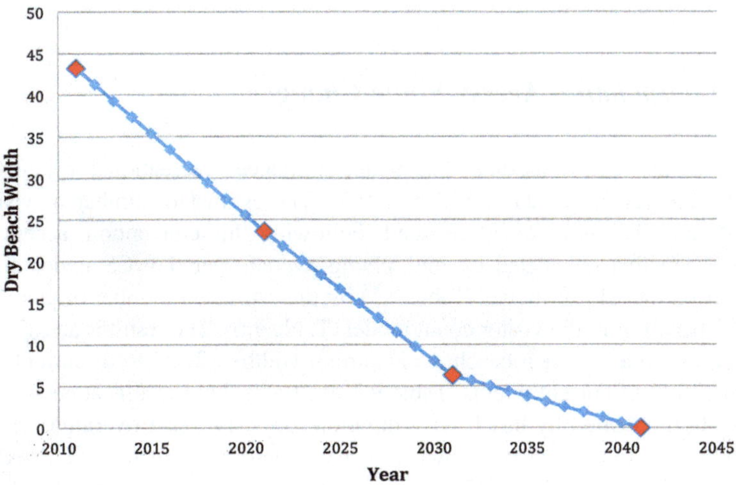

Fig. 9.1 Predicted average beach width: Slaughter Beach, scenario 4

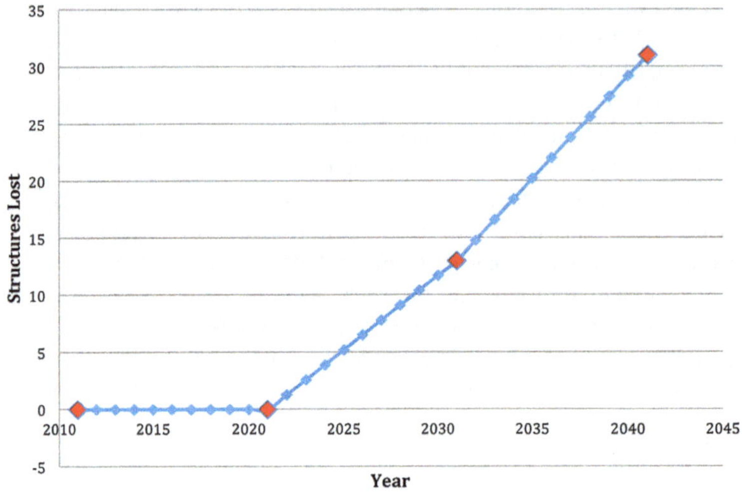

Fig. 9.2 Predicted total housing structures lost: Kitts Hummock, scenario 4

anchor points. A parallel piecewise linear interpolation is used to project housing structure losses for each year, also based on data from Tables 9.1, 9.2, 9.3 and 9.4. Structures are lost both due to flooding and purposeful removal by the State, depending on the scenario. Figure 9.2 illustrates this interpolation for Kitts Hummock under scenario four. Bold points represent predicted total structures lost by 2011, 2021, 2031 and 2041. Losses between these points are interpolated. The combination of these two steps provides an estimate of average beach width and remaining housing structures for each beach, under each management scenario, during each year. These results provide the biophysical basis for benefit adjustments described below.

9.5.2 Interpolating Access Value Changes

Parsons et al. (2013) assume that width influences recreational benefits only through changes in per day CS (Table 9.5). The estimated number of visits per beach (Table 9.6) is assumed constant. Following this convention, access value estimates are first calculated by multiplying average per day CS under different widths from the RP/SP model (Table 9.5) by the estimated number of annual visit days per beach from the visitor count model (Table 9.6). The result is an estimate of total access value for each beach at (1) current widths, (2) 25 % of current widths and (3) 200 % of current widths. Table 9.7 illustrates these results at current beach widths, disaggregated by beach, visit duration (day trip, short overnight trip, long overnight trip) and owner type (owners versus non-owners of beach homes).

From these initial data points, piecewise linear interpolation is used to forecast values for all possible beach widths, at each beach. The slope between the current

Table 9.7 Predicted recreational access value of Delaware Bay Beaches at current widths

Beach and visitor type	Access value day trips	Access value short overnight trips	Access value long overnight trips
Bowers (non-owners)	$118,488	$25,690	$8563
Bowers (owners)	$38,225	$42,817	$0
Broadkill (non-owners)	$149,308	$246,865	$163,243
Broadkill (owners)	$92,442	$99,546	$19,925
Primehook (non-owners)	$19,276	$64,797	$43,205
Primehook (owners)	$43,391	$37,807	$10,795
Slaughter (non-owners)	$170,532	$78,668	$29,964
Slaughter (owners)	$50,150	$48,703	$3755
Pickering (non-owners)	$18,515	$8887	$2964
Pickering (owners)	$7934	$2964	$0
Kitts Hummock (non-owners)	$7617	$12,799	$0
Kitts Hummock (owners)	$11,426	$4266	$0
South Bowers (non-owners)	$37,669	$9738	$19,476
South Bowers (owners)	$5793	$6489	$0

value (100 %) and the value at 200 % width is assumed to hold for values beyond 200 %, imposing piecewise linearity in our transfer.[7] Total access value at zero width is assumed to be zero for all beaches. Figure 9.3 provides an example of this interpolation for Bowers Beach. The figure shows the change in total day trip value for non-owners at different proportional widths. By definition, at 100 % (the current width), there is no change in value from the current situation. At widths greater than 100 % of current width, there is a positive change in value. At widths less than 100 % of current width, there is a negative change in value. At 0 % width (no beach), all access value is lost. For example, as shown by Fig. 9.3 the current access value for non-owner visits to Bowers Beach is $118,489, all of which is lost at zero width. At 200 % of current width this value increases by $12,083. Analogous value surfaces are estimated for all beaches.

These value surfaces provide a transfer function mapping widths to recreational values for each beach, based on combined RP/SP data. Combined with interpolated beach widths, the results provide an initial, transferrable estimate of recreational value for all beaches, under each management scenario, for each year between 2011 and 2041.

9.5.3 Accounting for Housing Loss

From these initial estimates, a final adjustment is made to account for the projected loss of housing structures due to flooding or State removal. The true relationship

[7]We assume that for all practical purposes widths in 2010 and 2011 are identical.

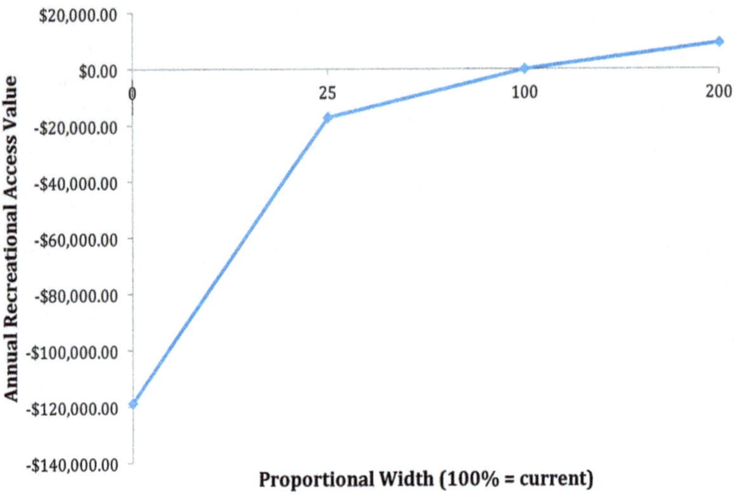

Fig. 9.3 Predicted change in access value: day trips to Bowers Beach, non-owners

between the number of housing structures on each beach and the number of recreational visits is unknown. Accordingly, we make the simplifying assumption that all visits involving the use of beach housing will vary in direct proportion with the number of housing structures still standing. We further assume that all visits (day and overnight) by current beach homeowners and all overnight visits by non-owners require the use of beach houses.[8] For these visit types, we assume that an X % loss of beach community houses will lead to an X % decline in visits, and therefore an X % decline in recreational value. All other visit types are assumed to be unaffected. Beyond this, the overall rate of recreation use from 2011 to 2041 is assumed to be unchanged (i.e., no growth or decline in participation due to population, new housing in the bay beach communities, or other exogenous influences). These assumptions are used to project recreational values. While these assumptions risk the introduction of additional error into final benefit estimates, protocols such as this are required in virtually all benefit transfers to account for differences between the contexts under which primary studies were conducted and those for which values are to be transferred. Here, we make these assumptions transparent.

[8]This is likely a reasonable assumption for these communities, given the lack of hotel accommodations.

9.5.4 Results

Transferred benefit estimates for each year are discounted at a 4 % annual rate and aggregated over time to generate a total net present value for each scenario. Benefits are calculated relative to a hypothetical baseline in which 2011 widths are maintained indefinitely. Results are shown in Table 9.8. Compared to this baseline scenario, the present value of recreational benefit changes for the seven beaches are $3,888,068.10 under scenario one (nourishment), −$1,415,683 under scenario two (strategic retreat), −$1,394,400 under scenario three (basic retreat), and −$12,177,926 under

Table 9.8 Benefit transfer results: net present recreational value of beach management alternatives, 2011–2041[a]

Beach and visitor type	Scenario 1 (nourishment)	Scenario 2 (strategic retreat)	Scenario 3 (basic retreat)	Scenario 4 (do nothing)
Bowers (non-owners)	$370,764	$296,206	−$9825	−$402,061
Bowers (owners)	$190,609	$19,707	−$22,523	−$209,614
Bowers (total)	$561,373	$315,914	−$32,349	−$611,675
Broadkill (non-owners)	$585,236	−$1,032,852	−$695,509	−$8,062,889
Broadkill (owners)	$251,304	−$591,174	−$359,389	−$829,680
Broadkill (total)	$836,541	−$1,624,027	−$1,054,898	−$8,892,570
Primehook (non-owners)	$526,972	−$230,370	−$34,446	−$73,432
Primehook (owners)	$438,707	−$262,533	−$29,341	−$53,591
Primehook (total)	$965,679	−$492,903	−$63,787	−$127,023
Slaughter (non-owners)	$579,431	$482,224	−$2238	−$1,161,274
Slaughter (owners)	$224,058	$123,911	−$2114	−$426,840
Slaughter (total)	$803,490	$606,136	−$4353	−$1,588,114
Pickering (non-owners)	$148,673	−$126,302	−$84,112	−$336,675
Pickering (owners)	$53,356	−$162,330	−$67,121	−$121,125
Pickering (total)	$202,030	−$288,633	−$151,234	−$457,801
Kitts Hummock (non-owners)	$129,540	−$26,687	−$29,806	−$224,392
Kitts Hummock (owners)	$99,559	−$91,980	−$36,545	−$172,473
Kitts Hummock (total)	$229,100	−$118,668	−$66,352	−$396,866
South Bowers (non-owners)	$244,880	$172,117	−$15,083	−$87,115
South Bowers (owners)	$44,972	$14,380	−$6341	−$16,758
South Bowers (total)	$289,852	$186,497	−$21,424	−$103,874
Total all beaches	$3,888,068	−$1,415,683	−$1,394,400	−$12,177,926

[a]Assumes a 4 % discount rate; values are relative to a baseline of constant 2011 beach widths

scenario 4 (do nothing).[9] That is, the nourishment option provides the greatest net benefits, while the "do nothing" scenario generates the greatest projected losses. More modest losses are experienced under the two retreat options.

While the benefit estimates presented in Table 9.8 are useful to envision changes relative to current conditions (a hypothetical but unrealistic baseline in which present conditions are maintained indefinitely), this is not the most relevant perspective for policy analysis. A more traditional policy analysis perspective would normalize net benefits relative to scenario 4 (no action). Normalized in this way, the present value of recreational benefit changes for the seven beaches would be $16,065,994 under scenario one (nourishment), $10,762,242 under scenario two (strategic retreat), $10,783,525 under scenario three (basic retreat), and $0 under scenario 4 (do nothing). This alternative normalization of the summary results in Table 9.8 highlights the additional net benefits generated by the nourishment and retreat options, compared to a default case in which the State does nothing.

As expected, the estimated change in recreational benefits varies over beaches and scenarios. The largest gains and losses occur at Broadkill and Slaughter beaches; these are the beaches supporting the greatest number of current visits (Table 9.6). Smaller beaches such as Kitts Hummock support a much smaller number of visits, and changes in recreational values are similarly small. Value patterns also vary according to relationships between management scenario effects and beach characteristics. For example, the large number of housing structure losses under scenario three at Broadkill Beach compared to 2011 levels (Table 9.3) lead to significant recreational benefit losses despite no change in beach width (Table 9.8); this is due to the proximity of houses to the shoreline (i.e., in the beach retreat zone). The same scenario at Slaughter Beach leads to negligible benefit change, because few houses are in the beach retreat zone. Benefit changes also vary between owners and non-owners of beach homes in the affected areas, with value changes for non-owners typically (but not always) exceeding those for owners.

The benefit transfer also reveals a pattern wherein losses due to State inaction (scenario 4) tend to outweigh gains due to beach nourishment (scenario 1). That is, potential gains due to nourishment are relatively small compared to the potential losses caused by doing nothing. Again, these estimates are compared to a hypothetical baseline of current beach widths maintained indefinitely. The two exceptions to this pattern are Primehook and South Bowers; these are beaches characterized by unusually large width gains due to nourishment (Table 9.1) and few potential structure losses due to erosion (Table 9.4). This mirrors underlying RP/SP results of Parsons et al. (2013), which suggest relatively small increases in CS due to beach width increases, but relatively large losses due to width decreases. Housing loss patterns not considered by Parsons et al. (2013) yet modeled for benefit transfer can either mitigate or amplify these patterns.

[9]These estimates only reflect changes in recreational consumer surplus (benefits). The costs of each scenario (e.g., to nourish a beach, remove homes, etc.) are not included.

Another relevant pattern in benefit transfer results is the proportional allocation of benefit changes across beaches under different scenarios. In many of the scenarios—and particularly scenarios 3 and 4—a large proportion of benefit losses can be prevented through policy actions at only one or two beaches. This suggests the existence of cost-effective solutions that maintain the majority of recreational benefits. For example, under scenario 4, over 73 % of all benefit losses occur at Broadkill Beach. Hence, were DNREC to adopt scenario four (do nothing) on all beaches except Broadkill, where current widths would be maintained, the projected loss of net benefits would be reduced by 73 %. Hence, the benefit transfer not only characterizes overall benefits, but suggests cost-effective policy alternatives that could potentially minimize the loss of recreational benefits.

9.6 Conclusions

The use RP/SP data enrichment is a common feature of benefit transfer, with multiple approaches available to practitioners. These include the transfer of results from primary studies that already combine RP/SP micro-data, along with methods such as meta-analysis and structural benefit transfer that combine the results of independent primary studies using either RP or SP data. In the former case the primary study combines RP/SP micro-data. In the latter case the transfer practitioner combines the results of prior RP and SP studies; in most cases these primary studies use either RP or SP data, but not both. Each of these approaches has distinct uses and advantages, reflecting potential data enrichment benefits reported in the primary valuation literature (Whitehead et al. 2008b, 2011a, b).

The case study application detailed above illustrates a transfer of results from a primary study that combines RP/SP micro-data, applied to the management of erosion on Delaware Bay beaches. The presented results employ (1) RP/SP data from Parsons et al. (2013) to generate underlying benefit estimates and (2) subsequent benefit transfer to adapt these prior estimates to policy scenarios. Specifically, the original micro-data RP/SP analysis is first used to estimate relationships between recreational access values and a limited number of beach widths. Because these results apply to a range of beach widths that is much smaller than those projected under the DNREC beach management scenarios, and do not account for a projected loss of housing structures, benefit transfer is required to adapt these results to predicted policy outcomes.

The illustrated application demonstrates ways in which a micro-data RP/SP benefit transfer can tailor results to specific policy scenarios when a targeted primary study is not possible. As implied above, however, such transfers are subject to a number of concerns. In addition to challenges facing all benefit transfers,[10]

[10]For example, underlying assumptions and errors in primary studies will carry through to subsequent benefit transfers.

these include a potential compounding of data consistency problems that may be encountered with all RP/SP data (Whitehead et al. 2008b) with additional sources of error inherent in benefit transfer (Rosenberger and Stanley 2006). In the present case study, the consistency of RP and SP data is a maintained assumption; we do not (and cannot given available data) test the consistency of RP and SP results. Within meta-analysis, the pooling of RP/SP data typically requires similar assumptions related to the comparability of Marshallian and Hicksian welfare measures (Bergstrom and Taylor 2006). Structural benefit transfers seek to avoid such problems through the imposition of a formal utility structure through which RP and SP results are combined (Smith et al. 2002), but even these approaches require strong and often influential assumptions regarding functional forms.

Given the potential advantages and disadvantages of RP/SP data enrichment within different types of benefit transfer, the net effect on transfer validity and reliability remains largely unknown; this is a relatively recent area of research. Unlike the broader valuation literature (cf. Whitehead et al. 2011b), benefit transfer research includes a relatively small body of work that explicitly quantifies the empirical advantages or disadvantages (e.g., in terms of increased/decreased transfer reliability or validity) of RP/SP data enrichment (e.g., González-Sepúlveda and Loomis 2011; Johnston and Moeltner 2014; Londoño and Johnston 2012). Results from this work are mixed. For example, while González-Sepúlveda and Loomis (2011) find that jointly estimated RP/SP benefit transfers are more accurate than parallel transfers using either RP or SP data alone, Londoño and Johnston (2012) find that an MRM including only SP observations generates lower transfer errors than a parallel MRM that combines RP and SP observations. Johnston and Moeltner (2014) find that combining Hicksian and Marshallian measures can often (but not always) improve the efficiency of benefit transfer. In summary, despite theoretical concerns related to certain types of data combinations (e.g., pooling Marshallian and Hicksian welfare measures within meta-analysis; Smith and Pattanayak 2002), and theoretical advantages of others (e.g., structural benefit transfer; Smith et al. 2002), the empirical consequences of these practices for transfer reliability and validity remains subject to uncertainty (Johnston and Rosenberger 2010). At the same time, the success of these methods in the primary valuation literature suggests that similar advantages may be possible with benefit transfer. The likelihood that benefit transfers employing RP/SP data will continue to be a central part of agency cost benefit analyses underscores the need for future research in this area.

Acknowledgments This research is partially supported by the Delaware Department of Natural Resources and Environmental Control in partnership with Johnson, Mirmiran and Thompson, Inc. We acknowledge helpful comments of Ari Michelsen. Opinions are those of the authors and do not imply endorsement of the funding agency. Any remaining errors are the sole responsibility of the authors.

References

Adamowicz, W., Louviere, J., & Williams, M. (1994). Combining revealed and stated preference methods for valuing environmental amenities. *Journal of Environmental Economics and Management, 26*, 271–292.

Adamowicz, W., Swait, J., Boxall, P., Louviere, J., & Williams, M. (1997). Perceptions versus objective measures of environmental quality in combined revealed and stated preference models of environmental valuation. *Journal of Environmental Economics and Management, 32*, 65–84.

Bergstrom, J. C., & Taylor, L. O. (2006). Using meta-analysis for benefits transfer: Theory and practice. *Ecological Economics, 60*, 351–360.

Boxall, P. C., Englin, J., & Adamowicz, W. L. (2003). Valuing aboriginal artifacts: A combined revealed–stated preference approach. *Journal of Environmental Economics and Management, 45*, 213–230.

Cameron, T. A. (1992). Combining contingent valuation and travel cost data for the valuation of nonmarket goods. *Land Economics, 68*, 302–317.

Cameron, A. C., & Trivedi, P. R. (1998). *Regression analysis of count data. Econometric society monograph* (Vol. 30). Cambridge, UK: Cambridge University Press.

Cameron, T. A., Shaw, W. D., Ragland, S. E., Callaway, J. M., & Keefe, S. (1996). Using actual and contingent behavior data with differing levels of time aggregation to model recreation demand. *Journal of Agricultural and Resource Economics, 21*, 130–149.

Carson, R. T., Flores, N. E., Martin, K. M., & Wright, J. L. (1996). Contingent valuation and revealed preference methodologies: Comparing the estimates for quasi-public goods. *Land Economics, 72*, 80–99.

Englin, J., & Cameron, T. A. (1996). Augmenting travel cost models with contingent behavior data. *Environmental & Resource Economics, 7*, 133–147.

Glass, G. V. (1976). Primary, secondary, and meta-analysis of research. *Educational Researcher, 5*, 3–8.

González-Sepúlveda, J. M., & Loomis, J. B. (2011). Are benefit transfers using a joint revealed and stated preference model more accurate than revealed and stated preference data alone? In J. C. Whitehead, T. C. Haab, & J. C. Huang (Eds.), *Preference data for environmental valuation: Combining revealed and stated approaches* (pp. 289–302). London: Routledge.

Haener, M. K., Boxall, P. C., & Adamowicz, W. L. (2001). Modeling recreation site choice: Do hypothetical choices reflect actual behavior. *American Journal of Agricultural Economics, 83*, 629–642.

Huang, J. C., Haab, T. C., & Whitehead, J. C. (1997). Willingness to pay for quality improvements: Should revealed and stated preference data be combined? *Journal of Environmental Economics and Management, 34*, 240–255.

Johnson, Mirmiran, & Thompson, Inc. (2012). *Shoreline and housing loss projections, Delaware Department of Natural Resources and Environmental Control scenarios for Delaware Bay beaches*. Dover, DE: Delaware Department of Natural Resources and Environmental Control.

Johnston, R. J., Ranson, M. H., Besedin, E. Y., & Helm, E. C. (2006). What determines willingness to pay per fish? A meta-analysis of recreational fishing values. *Marine Resource Economics, 21*, 1–32.

Johnston, R. J., & Moeltner, K. (2014). Meta-modeling and benefit transfer: The empirical relevance of source-consistency in welfare measures. *Environmental and Resource Economics, 59*, 337–361.

Johnston, R. J., & Rosenberger, R. S. (2010). Methods, trends and controversies in contemporary benefit transfer. *Journal of Economic Surveys, 24*, 479–510.

Jones, S. R., & Mangun, W. R. (2001). Beach nourishment and public policy after Hurricane Floyd: Where do we go from here? *Ocean and Coastal Management, 44*, 207–220.

Kling, C. L. (1997). The gains from combining travel cost and contingent valuation data to value nonmarket goods. *Land Economics, 73*, 428–439.

Landry, C. E., & Liu, H. (2011). Econometric models for joint estimation of revealed and stated preference site-frequency recreation demand models. In J. C. Whitehead, T. C. Haab, & J. C. Huang (Eds.), *Preference data for environmental valuation: Combining revealed and stated approaches* (pp. 87–100). London: Routledge.

Londoño, L. M., & Johnston, R. J. (2012). Enhancing the reliability of benefit transfer over heterogeneous sites: A meta-analysis of international coral reef values. *Ecological Economics, 78*, 80–89.

Morrison, M., Bennett, J., Blamey, R., & Louviere, J. (2002). Choice modeling and tests of benefit transfer. *American Journal of Agricultural Economics, 84*, 161–170.

Murphy, J. J., Allen, P. G., Stevens, T. H., & Weatherhead, D. (2005). A meta-analysis of hypothetical bias in stated preference valuation. *Environmental & Resource Economics, 30*, 313–325.

Nelson, J. P., & Kennedy, P. E. (2009). The use (and abuse) of meta-analysis in environmental and resource economics: An assessment. *Environmental & Resource Economics, 42*, 345–377.

Parsons, G. R., Jakus, P. M., & Tomasi, T. (1999). A comparison of welfare estimates from four models for linking seasonal recreational trips to multinomial logit models of site choice. *Journal of Environmental Economics and Management, 38*, 143–157.

Parsons, G. R., Chen, Z., Hidrue, M. K., Standing, N., & Lilley, J. (2013). Valuing beach width for recreational use: Combining revealed and stated preference data. *Marine Resource Economics, 28*, 221–241.

Pattanayak, S. K., Smith, V. K., & Van Houtven, G. (2007). Improving the practice of benefits transfer: A preference calibration approach. In S. Navrud & R. Ready (Eds.), *Environmental value transfer: Issues and methods* (pp. 241–260). Dordrecht, The Netherlands: Springer.

Pendleton, L., Mohn, C., Vaughn, R. K., King, P., & Zoulas, J. G. (2012). Size matters: The economic value of beach erosion and nourishment in Southern California. *Contemporary Economic Policy, 30*, 223–237.

Rosenberger, R. S., & Johnston, R. J. (2009). Selection effects in meta-analysis and benefit transfer: Avoiding unintended consequences. *Land Economics, 85*, 410–428.

Rosenberger, R. S., & Loomis, J. B. (2000). Using meta-analysis for benefit transfer: In-sample convergent validity tests of an outdoor recreation database. *Water Resources Research, 36*, 1097–1107.

Rosenberger, R. S., & Stanley, T. D. (2006). Measurement, generalization and publication: Sources of error in benefit transfers and their management. *Ecological Economics, 60*, 372–378.

Smith, V. K., & Pattanayak, S. K. (2002). Is meta-analysis a Noah's Ark for non-market valuation? *Environmental & Resource Economics, 22*, 271–296.

Smith, V. K., Pattanayak, S. K., & Van Houtven, G. (2006). Structural benefit transfer: An example using VSL estimates. *Ecological Economics, 60*, 361–371.

Smith, M. D., Slott, J. M., McNamara, D., & Murray, A. B. (2009). Beach nourishment as a dynamic capital accumulation problem. *Journal of Environmental Economics and Management, 58*, 58–71.

Smith, V. K., Van Houtven, G., & Pattanayak, S. K. (2002). Benefit transfer via preference calibration: "Prudential algebra" for policy. *Land Economics, 78*, 132–152.

Van Houtven, G. L., Pattanayak, S. K., Patil, S., & Depro, B. (2011). Benefits transfer of a third kind: An examination of structural benefits transfer. In J. C. Whitehead, T. C. Haab, & J. C. Huang (Eds.), *Preference data for environmental valuation: Combining revealed and stated approaches* (pp. 303–321). London: Routledge.

Von Haefen, R. F., & Phaneuf, D. J. (2008). Identifying demand parameters in the presence of unobservables: A combined revealed and stated preference approach. *Journal of Environmental Economics and Management, 56*, 19–32.

Whitehead, J. C., Haab, T. C., & Huang, J. C. (2000). Measuring recreation benefits of quality improvements with revealed and stated behavior data. *Resource and Energy Economics, 22*, 339–354.

Whitehead, J. C., Dumas, C. F., Herstine, J., Hill, J., & Buerger, B. (2008a). Valuing beach access and width with revealed and stated preference data. *Marine Resource Economics, 23*, 119–135.

Whitehead, J. C., Pattanayak, S. K., Van Houtven, G. L., & Gelso, B. R. (2008b). Combining revealed and stated preference data to estimate the nonmarket value of ecological services: An assessment of the state of the science. *Journal of Economic Surveys, 22*, 872–908.

Whitehead, J. C., Haab, T. C., & Huang, J. C. (2011a). Introduction. In J. C. Whitehead, T. C. Haab, & J. C. Huang (Eds.), *Preference data for environmental valuation: Combining revealed and stated approaches* (pp. 1–11). London: Routledge.

Whitehead, J. C., Haab, T. C., & Huang, J. C. (Eds.). (2011b). *Preference data for environmental valuation: Combining revealed and stated approaches*. London: Routledge.

Chapter 10
Benefit Transfer: Insights from Choice Experiments

John Rolfe, Jill Windle and Jeffrey Bennett

Abstract In this chapter we explore six key reasons for the close alignment of choice modeling (CM) experiments with benefit transfer applications. Of these six, some relate to the richness of value estimate output that is generated in CM applications, whereas others involve the insights into choice behavior and the nature of preferences that are gained through the use of the technique. These outcomes improve the accuracy of the benefit transfer process and also provide more verification and confidence in the results. An additional focus of the chapter is to explore the tension between improving the accuracy and insights from CM on the one hand against, on the other, the need to make benefit transfer practical and operational. Although there is an extensive literature on the development and operation of the CM technique, it is not practical to cover this in a single chapter; instead the focus here is on the aspects of CMs that offer the most insight into benefit transfer processes.

Keywords Benefit transfer · Choice modeling · Methodology

J. Rolfe (✉) · J. Windle
School of Business and Law, CQUniversity, Rockhampton, QLD, Australia
e-mail: j.rolfe@cqu.edu.au

J. Windle
e-mail: j.windle@cqu.edu.au

J. Bennett
Crawford School of Public Policy, Australian National University, Canberra, ACT, Australia
e-mail: Jeff.Bennett@anu.edu.au

R.J. Johnston et al. (eds.), *Benefit Transfer of Environmental and Resource Values*, The Economics of Non-Market Goods and Resources 14,
DOI 10.1007/978-94-017-9930-0_10

10.1 Introduction

Choice modeling (CM), also known as choice experiments (CE) or discrete choice experiments (DCE), was developed as a valuation technique in the early-to mid-1990s (Carson et al. 1994), after the establishment of choice-based experimental methods for application in marketing and transport (Louviere and Hensher 1982; Louviere and Woodworth 1983). Within a few years, the technique was developed for application to environmental contexts (Adamowicz et al. 1998; Blamey et al. 1999; Rolfe et al. 2000). While the formulation of CM for the analysis of environmental tradeoffs was driven in part by the controversy over the contingent valuation method following the Exxon Valdez case (see Arrow et al. 1993), it was accompanied by an interest in the use of CM for potential application as a source of value estimates for benefit transfer. Some of the foundation studies in environmental CM were conducted, at least in part, with the expressed aim of generating benefit transfer functions (Jiang et al. 2005; Johnston 2007; Morrison and Bennett 2000; Morrison et al. 2002; van Bueren and Bennett 2004). Many reviews of benefit transfer (e.g. Brouwer 2000; Johnston and Rosenberger 2010; Morrison and Bergland 2006; Navrud and Ready 2007; Rolfe and Bennett 2006) have drawn on insights from the use of CM studies as generators of source estimates of value.

The reasons for the expansion in use of the CM valuation technique being so closely associated with benefit transfer can be grouped into six areas. Of these six, some relate to the richness of value estimate output that is generated in CM applications, whereas others involve the insights into choice behavior and the nature of preferences that are gained through the use of the technique.

The first reason relates to the hedonic characteristics of a CM, where describing the issue of interest in terms of component attributes, labels and levels generates a more disaggregated output compared to other non-market valuation techniques. The second is that CM is capable of generating estimates of compensating surplus that are related to both site and respondent characteristics in the case study of interest. These estimates can then be used to arrive at values for any combination of attributes, labels and levels through a benefit transfer function. A related third reason is that the richness of predictive outputs allows greater opportunities for testing the equivalence and convergent validity of value estimates for use in benefit transfer.

The fourth reason relates to the insights that the analysis of CM data offers into the decision processes that respondents employ. A variety of models and analytical techniques is available to allow analysts to identify and test a number of potential biases and effects that relate to both choice behavior and methodological factors. The fifth reason relates to the framing of choice experiments. Tests have been conducted to identify how the context of choice decisions can affect value estimates through factors such as payment vehicles, geographic proximity of respondents to the study site, feelings of responsibility, types of management actions, and outcome likelihoods. The sixth reason relates to the ability of CM to provide insights into preference structures. Together with framing advantages and the potential use of benefit transfer functions, these insights can make subsequent assemblages of values more consistent with utility functions and address concerns about limited theoretical foundations.

These methodological advantages come at a cost for benefit transfer applications. Researchers have focused on explaining choice tradeoffs at finer and finer levels, with more attention being given both to the number of different factors that may influence respondents' choices and ways of modeling those choices. These attempts have largely been successful, as they have demonstrated with increasing levels of precision that values are sensitive to a large number of factors and influences. Paradoxically, these efforts to increase the precision of value estimates make benefit transfer more complex and problematic. Each split sample choice experiment that successfully demonstrates value sensitivity to site, population, framing or methodological factors generates another predictive or adjustment factor that can be incorporated into benefit transfer functions.

The six key reasons why CM applications are closely aligned with benefit transfer are explored in the following sections of this chapter. An additional focus is to explore the tension between improving the accuracy and insights from CM on the one hand against, on the other hand, the need to validate and make operational benefit transfer. This tension is discussed in the final section. An important point to note is that there is an extensive literature on the development of the CM technique, methodological issues and valuation case studies that is not practical to cover in a single chapter; instead the focus here is on the aspects of CMs that offer most insight into benefit transfer processes.

10.2 Describing the Tradeoffs

10.2.1 Representing Issues with Attributes

A defining aspect of CM is the decomposition of a case study issue into component attributes, labels and levels. At an operational level, this "unpacking" of the elements that comprise values as specified by Kelvin Lancaster's demand analysis (Lancaster 1966) helps respondents to comprehend and construct the choice tasks, identify and remember the key factors that might be significant, and shows different scenarios in more comprehensive and realistic ways (Adamowicz et al. 1998; Bennett and Blamey 2001; Rolfe et al. 2002). At an analytical level, the disaggregation of values into sub-components allows for a richer set of predictor variables to be generated and allows decision structures to be better modeled. Furthermore, the set of value component estimates creates opportunities for a wider spectrum of point value transfers from a single CM data collection exercise (Holmes and Adamowicz 2003; Louviere et al. 2000; Rolfe and Bennett 2006).

A particular advantage of CM is that it allows the description of resource management contexts to be presented in much broader terms in comparison to other non-market valuation techniques. This can be illustrated in environmental applications, where as well as describing effects on the environmental assets involved, tradeoffs can also reflect the impacts of changing resource management on social and economic conditions (Morrison and Bennett 2004; Rolfe et al. 2000; van Bueren and Bennett 2004). Environmental impacts can also be described using attributes that focus on ecological processes (e.g. Johnston et al. 2012; Liekens et al. 2013). Other ways of extending the context of resource tradeoffs have been to include risk outcomes (e.g. Glenk and Colombo 2011; Wielgus et al. 2009) and management options (e.g. Czajkowski and Hanley 2009; Hanley et al. 2010; Johnston and Duke 2007).

The analyst designing a CM application typically has discretion over the selection and description of attributes, the number of choice alternatives and choice sets that are presented to survey respondents, and the way that choices are framed, including those against opportunity costs (Hensher 2006; Louviere et al. 2000). Researchers have paid some attention to issues of choice set dimensions and application, concerned that presentational differences could affect subsequent value estimates. There is some evidence that the structure of choice sets in terms of the number of attributes and choice alternatives can impact on value estimates (Caussade et al. 2005; Hensher 2006, 2008; Rolfe and Bennett 2009). In most cases the analyst balances the desire to make choice sets realistic (including more choice alternatives, attributes, levels and labels) against the desire to contain choice complexity (reducing the number of alternatives, attributes, levels and labels) within respondents' cognitive capacities. The decision about what attributes to include in an environmental valuation exercise and how to describe them is likely to continue to be a multifarious task, with tradeoffs being made between respondent comprehension, ecological validity, policy relevance and content validity (Johnston et al. 2012).

The use of values from CM applications for benefit transfer can be limited by variations in the selection of attributes and attribute ambiguity (Johnston et al. 2012). There is no standard approach to the selection of attributes to be included in any particular valuation, even for relatively common environmental contexts such as forests, rivers, or wetlands. For example, two Scandinavia valuations for marine water quality (Eggert and Olsson 2009; Kosenius 2010) applied two quite different sets of attributes. A related problem occurs when indicator or iconic species are used to represent a species group or ecosystem. Care must be taken with the interpretation of values for those species, as people are known to have higher willingness to pay (WTP) for some types of species, such as mammals, over others (Loomis and White 1996; Tisdell et al. 2006); for more charismatic species (White et al. 1997); and for rarer species (Christie et al. 2006). Jacobsen et al. (2008) have shown that simply naming and hence "iconizing" only a few species can attract higher value estimates than using a quantitative description.

Issues can also emerge in the way that attributes are described. Attribute levels are often described in CM applications in subjective terms; for example, quantitative changes may be described as "small/medium/large" while qualitative changes may be described as "high/medium/low" or "good/medium/poor" (Eggert and

Olsson 2009; Kosenius 2010). Such subjective descriptions make it very difficult to estimate specific outcome benefits, and this reduces the potential for application in benefit transfer. Despite these complexities and potential limitations, the ability to present issues as a set of attributes within a CM and to generate individual part-worths by attribute is a key reason why the development of CM techniques has been closely associated with benefit transfer.

10.2.2 The Cost Attribute

The framing of the cost attribute and associated payment vehicle has received considerable attention in the wider stated preference literature because of the potential for starting point bias, anchoring effects, and different forms of protest bids that bias value estimates. Choice modeling experiments have allowed these issues to be tested more thoroughly. Both Hanley et al. (2005) and Kragt (2012) found that higher cost levels did not lead to significantly higher value estimates. In contrast, Carlsson and Martinsson (2008) found that higher cost levels resulted in higher WTP estimates, but that the design of the first choice set (starting point bias) did not have a significant impact on WTP estimates. The latter contrasted with the results of Ladenburg and Olsen (2008) who found that varying price levels in the first choice set did influence the WTP estimates, but only for females and not males. They also found that impact of the starting point bias diminished as the number of choice sets increased, in line with the "discovered preferences hypothesis" (Braga and Starmer 2005) or learning effects (Bateman et al. 2008). The concept of a choke price (Kragt 2012; Mørkbak et al. 2010) has been a useful contribution to the design of CM and the need to include high enough cost levels to invoke an income effect.

10.2.3 Labeling the Alternatives

The use of labeled alternatives in choice sets allows more nuanced descriptions of tradeoffs as well as better opportunities to test for the influences of potentially relevant factors. Alternative labels can also help to communicate key issues of importance, and to distinguish policy dimensions in the available options. There are two main approaches to the use of labels. The first is to use a label to capture other factors that may be important to choices, holding attributes and levels constant across choice alternatives. For example, Carlsson et al. (2011), Morse-Jones et al. (2012) and Rolfe et al. (2000) used geographical or country labels to help communicate to respondents that other factors such as institutional settings, cost, control and responsibility may differentiate alternatives.

In the second approach, labels can be used to signal that the policy options vary between choice alternatives, with the levels for each attribute tailored to the relevant label, helping to represent case study scenarios more accurately. For example, both

Czajkowski and Hanley (2009) and Rolfe and Windle (2013) found that using management policy labels provided respondents with relevant information about the way in which the environmental good is provided, leading to a significant increase in the scope sensitivity of welfare measures.

On the other hand, labeled alternatives may increase the cognitive burden faced by respondents, leading them to use a form of choice heuristic by which choices are based primarily on the labels, with less attention being paid to variations in the levels of the attributes. Blamey et al. (2000) reported that the inclusion of policy labels appeared to shift respondents' attention from the attributes to the labels, but they found no significant differences in the welfare estimates.

10.3 Extrapolating to Benefit Transfer Functions

Since the focus on benefit transfer in the 1992 special issue of *Water Resources Research*, there has been a preference in the literature away from the transfer of point values toward transfer of benefit functions (Johnston and Rosenberger 2010; Morrison and Bergland 2006). The arguments in favor of using benefit functions are that more detailed information is involved and that adjustments for different site and population characteristics between source and target case studies can be more easily applied (Rolfe and Bennett 2006). There is also the argument that benefit transfer functions are likely to be more consistent estimators of value than an amalgam of point source estimates. This point is explored further in Sect. 10.6.

Benefit functions can be derived in different ways, including those from single studies and meta-analyses (Johnston and Rosenberger 2010). A key strength of the CM valuation technique is that both the site and respondent characteristics in the source case study can be used to estimate compensating surpluses for any combinations of attributes, labels and levels. The same function can then be used for benefit transfer to target case studies with differing levels for site and population characteristics, so long as those levels lie within the respective ranges used in the source study. Some CM studies have been explicitly focused on framing the applications in ways that allowed subsequent value estimates to be used for wider benefit transfer applications, or to identify adjustment factors that facilitated values to be transferred across variations in contexts and frames (Morrison and Bennett 2004; Rolfe and Windle 2008; van Bueren and Bennett 2004). This approach essentially internalizes the potential for transferring benefit functions into the design of the application.

Although there is strong support in the literature to move away from point source transfers to benefit function transfers, the evidence from case study examples remains mixed (Johnston and Rosenberger 2010). Some researchers (e.g. Kerr and Sharp 2006; Morrison and Bennett 2004; van Bueren and Bennett 2004) have reported that many benefit transfer functions derived from CM applications do not satisfy convergent validity tests. Others (e.g. Rolfe and Windle 2008, 2012a) report benefit transfer functions which are robust to site and population differences. One conclusion drawn from these findings is that adjustment factors can be developed that account for differences in components such as scope (e.g. van Bueren and Bennett 2004), population types (e.g. Morrison and Bennett 2004), or distance effects (e.g. Concu 2007, Rolfe and Windle 2012b). Another conclusion is that although the results of CM applications are suited for benefit function transfer, there is some support for results to be harvested for point source estimates (such as when only part-worth values are transferred) rather than for benefit functions only.

10.4 Testing Equivalence and Convergent Validity of Value Estimates

Much of the literature relating to benefit transfer has focused on identifying the accuracy and validity of transferred values (Johnston and Rosenberger 2010). Two key foci of these approaches are the identification of measurement errors within a source study and the transfer errors associated with the application of source study values to a target site (Johnston and Rosenberger 2010; Rosenberger and Stanley 2006). Tests for measurement errors are typically assessed with split-sample experiments, whereas transfer errors are assessed by comparing source study value estimates against estimates derived from a primary study of values for the target site. In both cases these are typically performed as convergent validity or reliability tests, with welfare estimates assumed to be equal unless testing reveals otherwise (Johnston and Rosenberger 2010). However, some value differences between source and target sites can be expected because of site and population differences, complicating convergent validity tests (Chap. 18; Rosenberger and Stanley 2006). Alternative approaches are to set the null hypothesis that environmental values differ, and then use equivalence testing (e.g. Johnston and Duke 2008; Kristofersson and Navrud 2005) or to compare transfer errors against a benchmark (e.g. Brouwer 2000).

Choice modeling applications have provided insights into both measurement and transfer errors. In relation to the measurement errors, the ability to test for and incorporate site and population differences, deal with heterogeneity, specify functional relationships more accurately, and predict values by particular sub-groups has both improved the accuracy of CM estimates and helped to identify where remaining prediction variances and errors exist. In relation to transfer errors, the richness of predictive values provided by CM models for part-worths, compensating surplus estimates, benefit functions and error terms means that multiple comparisons are possible. Reasons for the satisfaction or failure of convergent validity tests can thus be forensically identified. Rolfe and Windle (2012a, b) demonstrate that transfer errors vary by attributes and labels, as well as between use and nonuse values and by the iconic nature of assets.

Many tests for convergent validity remain difficult to satisfy, with substantial transfer errors in some applications (Brouwer 2000; Johnston and Rosenberger

2010; Rolfe and Bennett 2006; Rosenberger and Stanley 2006). In some, but not all, cases, failures appear to be linked with larger differences between sites and populations, or because of unincorporated factors such as scope differences (Rolfe and Bennett 2006). However, failures may also be driven by increasing accuracy and requiring tighter specifications of primary studies, making it more difficult to transfer values to other sites that do not have identical characteristics. In these cases the use of equivalence testing (Johnston and Duke 2008) or a move towards preference calibration (e.g. Smith et al. 2002) may be required.

10.5 Respondent Behavior

In CM, the analyst is faced with the challenge of explaining the link between respondents' choices and their preferences for different attributes and their levels, in order to elicit meaningful welfare estimates. The use of benefit transfer has been enhanced by the insights that CM studies have allowed into respondent behavior, helping to identify key factors that influence choice decisions as well as to understand how choices may be influenced by methodological design. This has occurred in two main ways:

1. through refinements in statistical methods; and
2. through analysis of choice patterns.

10.5.1 Refinements in Statistical Methods

Refinements in statistical methods have occurred through a move away from use of the standard multinomial logit (MNL) model of CM respondent choices. More advanced models have allowed improved analysis of choice behavior by better representation of respondent heterogeneity in responses, more precise identification of random error components, and accommodation of variations in the ways that alternatives are considered (Adamowicz et al. 2008; Louviere et al. 2000). Researchers have dealt with preference heterogeneity by including attitudinal and behavioral variables (e.g. Brouwer and Spaninks 1999) or using random parameter or error component logit models to capture functional forms (e.g. Colombo et al. 2005), with improvements in the accuracy of benefit transfer to different population groups.

One area of focus has been to capture choice variation across respondents through the estimation of latent class models. These models, through their identification of sub-groups of respondents that share similar preferences, allow benefit transfer to be directed according to those sub-groups (Boxall and Adamowicz 2002). The accuracy of benefit transfer functions have been further developed through the estimation of utility in willingness to pay space to minimize confounding effects of heterogeneity in preference construction (Scarpa et al. 2008, 2009). The combination of CM predictive models with geographic information system (GIS) data has identified where values might need to be adjusted by location or other geographic factors (Tait et al. 2012), while van Bueren and Bennett (2004) identify that adjustments to benefit transfer functions may be required where scope differences exist, such as those between regional and national contexts.

10.5.2 Analysis of Choice Patterns

One strength of the CM technique is that it allows more detailed analysis of choice behavior through more comprehensive and accurate models. The testing of methodological issues is also facilitated. An example of the former is the use of nested logit models to identify path-dependent choices (Louviere et al. 2000). Other tests have identified respondents who had made choices representing lexicographic preferences (Rulleau and Dachary-Bernard 2012), or who have used different patterns of decision heuristics (Leong and Hensher 2012). Tests for incentive compatibility (e.g. Lusk and Schroeder 2004) have identified how elements of choice behavior have varied between hypothetical and real purchase settings.

Another area of focus in understanding choice behavior has been “attribute non-attendance” (Alemu et al. 2013; Campbell et al. 2011, 2012; Carlsson et al. 2010; Scarpa et al. 2009, 2010). While some studies have established that respondents do ignore some attributes, including the cost attribute (Campbell et al. 2012), the results of other studies produce ambiguous results. One of the sources of ambiguity is that the exact nature of non-attendance and the reasons for the behavior are not clear. Some evidence suggests that respondents place less weight on some attributes rather than ignoring them (Carlsson et al. 2010). Alemu et al. (2013) distinguished non-attendance responses into three separate categories (discontinuous preferences, zero preferences, and possible low preferences), allowing separate adjustments to be made. Herein lies the difficulty: Attribute non-attendance due to low or zero respondent preferences would appear to pose no challenge to value estimates. However, attribute non-attendance caused by respondents ignoring attributes because of the particular formulation of the choice task is problematic. Distinguishing between these two types of behaviour poses a particular challenge to CM practitioners.

Data from CM applications have also been used to explore methodological issues around the structure and complexity of choice experiments. A number of studies have identified sequencing or ordering effects where systematic changes in expressed preferences are observed along the sequence of valuation tasks, potentially related to learning and fatigue effects (e.g. Day et al. 2012; Day and Prades 2010; McNair et al. 2011; Rulleau and Dachary-Bernard 2012; Scheufele and Bennett 2012). One argument is that these effects indicate a lack of respondent familiarity and experience with changes in environmental quality and that resultant choices are not stable or coherent (Brouwer et al. 2010). Other researchers argue that choices may be strongly anchored to some initial starting point (Ariely et al. 2003), and that more experience (gained through undertaking repeated choice tasks) helps to reduce inconsistencies and stabilize preferences (List 2003).

Increasing complexity has been shown to increase choice inconsistency (DeShazo and Fermo 2002), the use of simplifying heuristics (Dhar 1997; Dhar and Simpson 2003; Hensher 2008; Swait and Adamowicz 2001) or the avoidance of choices (Dhar 1997). Boxall et al. (2009) found that respondents were more likely to select the status quo alternative as task complexity increased. (Complexity was defined by multiple attribute level changes occurring across all alternatives in a choice set as compared to single level changes.) There is mixed evidence about the influence on respondent behavior of the structure and dimensions of choice tasks. Some evidence suggests that the structure of choice sets in terms of the number of attributes and alternatives can impact on value estimates (Boyle and Özdemir 2009; Caussade et al. 2005; Hensher 2006, 2008) or serial non-participation (Rolfe and Bennett 2009; Von Haefen et al. 2005).

10.6 Framing Choice Tradeoffs

A key advantage of CM is that its rich statistical output allows insights into whether factors additional to the description of the scenarios and the socio-economic characteristics of respondents affect value estimates. CM has advantages in being able to present complex scenarios to respondents. Elements of complexity, such as the presence of complementary and substitute goods, can be incorporated into component attributes or tested through split-sample experiments (Rolfe and Bennett 2006). Framing problems occur when the respondent to a survey is sensitive to the context in which a particular tradeoff is offered in ways that are fundamentally different from the context of the actual policy issue being investigated. The presence of differential sensitivity creates risks that any subsequent benefit transfer process may be inaccurate if the frame varies between source and target sites. Three areas of focus for framing effects in benefit transfer relate to:

1. adjustments for scope factors;
2. variations in management policy; and
3. treatment of risk and uncertainty.

Each of these is discussed in turn.

10.6.1 Scope Adjustments

A particular area of interest for calibration in benefit transfer studies is the potential for scope effects, where unit values vary according to the amount of the amenity being valued and the extent of the context in which the amenity is being offered (Czajkowski and Hanley 2009; Johnston and Rosenberger 2010). Where the size of the tradeoff and its context are different between source and target studies, then sensitivity to unit value differences makes the benefit transfer process problematic without calibration (Rolfe and Wang 2011).

Many of the theoretical arguments and earlier tests with the contingent valuation method have focused on only one dimension at a time; however, emerging applications of CM (e.g., Lew and Wallmo 2011) allow both dimensions of scope to be tested. For the purposes of this study we distinguish between two types of scope effects: one where there are only changes in one attribute (a quantity or quality effect), and one where the dimensions of the tradeoffs occur (i.e. there are changes in the number or framing of the attributes). This is similar to the distinction made by Bateman et al. (2002), in which they identify changes in only one argument in the utility function as a scope effect, and changes in multiple arguments in the utility function as an embedding effect. Here we refer to them as quantity and dimension scope effects.

Economic theory predicts that larger amounts of a good are expected to have higher values than a lesser amount of the same good, but values for marginal changes are expected to be smaller for larger sized goods compared to smaller sized goods as a consequence of diminishing marginal utility (Hoehn 1991; Hoehn and Randall 1989). There may also be effects when there are changes in the frame or context of the amenity of interest as the dimensions of a good change, and hence the pool of substitute and complement goods that may be considered. The default assumption in the transfer of stated preference values is that quantity scope effects have little impact on marginal value estimates. This allows analysts to transfer unit values estimated, for instance, at one level of scope (e.g. a local river catchment) to target sites at different scope levels (e.g. a regional river catchment). If this default assumption does not hold, then benefit transfers across scopes should also involve some application of adjustment factors to take account of the impacts on unit value estimates (Johnston and Rosenberger 2010; Rolfe and Wang 2011; van Bueren and Bennett 2004).

The estimation of calibration factors for scope changes is complex. While there have been some case study calibrations (e.g. van Bueren and Bennett 2004) there has been no systematic approach to develop calibration factors that can be applied more widely. Rolfe et al. (2013) compiled the results of two case studies in Australia to develop a calibration factor that can be applied in BT related to the ratios of scope amounts. The authors found statistically significant correlation between the ratios of the quantities involved and the WTP estimates (expressed in log form) for each of the 41 different scope tests that were examined across two case studies.

10.6.2 Policy Options

Information about the policy used to achieve environmental protection outcomes is rarely included as a variable in CM. Some policy situations can be addressed with very different management strategies, and these may generate a range of other impacts (such as, restrictions of property rights, individual benefits and localized outcomes) independent of a cost variable. In such cases, people may have different preferences for environmental protection options that achieve the same outcome arising from different management strategies. In welfare terms, the utility of environmental protection options may be sensitive to the choice of inputs used to achieve the protection because those inputs may signal the presence of other positive and negative impacts on individual welfare. A number of studies have demonstrated that including information about management policy has a significant impact on values for environmental assets (Czajkowski and Hanley 2009; Hanley et al. 2010; Johnston and Duke 2007; Rolfe and Windle 2013).

In some situations, labeled alternatives may be a more appropriate mechanism for incorporating management policy scope into choice sets than the use of a separate policy attributes. A label is different from other attributes because it is independent from all the elements of the good, with responses depending on participant perceptions (Czajkowski and Hanley 2009) or emotional connection (Blamey et al. 2000) with the label. The use of labeled alternatives also means that levels for each attribute can be tailored to the relevant label, helping to represent case study scenarios more accurately (Rolfe and Windle 2013).

10.6.3 Risk and Uncertainty

There have been some attempts to incorporate information about outcome certainty into the design of CM. The goal has been to generate a more accurate depiction of choice alternative outcomes, particularly for scenarios with different likelihoods of occurrence, and to help make scenarios more realistic to respondents. There are two broad approaches to including information about output certainty into CM. The first is to provide general framing statements in the questionnaire that inform respondents that predictions about future environmental conditions are not necessarily certain. Studies that have tested this approach (e.g. Macmillan et al. 1996; Wielgus et al. 2009) have shown that WTP estimates for environmental attribute improvements are lower when the chance of occurrence is reduced.

The second broad approach is to include outcome certainty directly into choice experiments by incorporating certainty information into labels, attributes and levels. For example, Roberts et al. (2008) included different levels of uncertainty in the description of each of the two outcome attributes (algal blooms and water levels) and found that respondents' WTP was higher when information about outcome uncertainty was provided. Glenk and Colombo (2011) included outcome certainty as a separate stand-alone attribute focused on a valuation of the benefits of soil carbon sequestration in Scotland, with results showing that WTP estimates increased when an outcome certainty attribute was included.

10.7 Consistency of Values from Different Sources

Benefit transfer applications typically have poor theoretical foundations (Bergstrom and Taylor 2006; Johnston and Rosenberger 2010; Smith et al. 2002; Smith and Pattanayak 2002), particularly when values from different studies are combined, either as a compilation of point source transfers or in a meta-analysis. A key weakness is that values for commodities may not be consistent between case studies as a consequence of variations in the frame of the tradeoffs involved or where there are methodological differences between studies. This means that an assemblage of non-market values may not be consistent with individual utility functions (Johnston and Rosenberger 2010; Johnston and Thomassin 2010), particularly when point source values are amalgamated from different studies into a benefit transfer function. Smith et al. (2002, 2006) suggest initially setting a structured utility function as a framework, with transferred values then calibrated into that framework. This would minimize risks that assembled values are inconsistent.

The use of CM applications for BT can improve the consistency of values in two important ways. First, there is potential for CM applications to inform the setting of an initial structural utility function as suggested by Smith et al. (2002, 2006), essentially identifying the broad architecture of preference structures. Second, the multi-attribute nature of a CM means that values for labels and attributes are assessed in the context of the other elements of the choice set and background information, so that the frame for value discovery is more explicit. Further, the benefit transfer function can be wholly or partially transferred to the case study of interest with the potential to make some framing adjustments by accounting for site, population, and other differences. This means that the values generated in a CM are already consistent within the framework that has been established, and limits the amount of calibration required for values to be transferred into a structural utility function.

10.8 Conclusions

The richness of data available from CM applications has impacted on benefit transfer in a number of ways. Some of the impacts are in terms of precision, where the hedonic description of issues in terms of attributes, labels and levels provides a greater number of value estimates, while the benefit function derived from a CM application allows those values to be set in a more consistent framework. The substantial advances in statistical analysis have also helped to generate primary values that are more accurate and reflective of a wider array of causal factors.

Another key advantage of using CM results for BT is that they provide better insights into the validity and complexity of benefit transfer approaches. Transfer errors can be specified by attribute or population characteristics, by the choice processes or the error terms involved. Results from studies that have assessed the nature and extent of errors are helping to identify when benefit transfer works well, or where some form of adjustment is required. They also assist by providing estimates of adjustment coefficients. Insights into preference structures and the ability to transfer values or value functions that have already been framed in relevant settings help to minimize risks that an assembled transfer function will be inconsistent with utility preference structures.

Advances in the estimation of non-market value estimates come at a cost for BT applications. Improvements in precision or increases in the array of explanatory factors make it more difficult to transfer values between source and target sites, and make differences between study and policy site values more evident. Studies have clearly demonstrated that "one size does not fit all" when it comes to the use of CM-generated source values for benefit transfer across multiple target sites. There are ongoing challenges to identify how to vary the precision of BT estimates according to need, and where values need to be calibrated for BT purposes.

References

Adamowicz, W., Boxall, P., Williams, M., & Louviere, J. (1998). Stated preference approaches for measuring passive use values: Choice experiments and contingent valuation. *American Journal of Agricultural Economics, 80*, 64–75.

Adamowicz, W., Bunch, D., Cameron, T. A., Dellaert, B. G. C., Hanneman, M., Keane, M., et al. (2008). Behavioral frontiers in choice modelling. *Marketing Letters, 19*, 215–228.

Alemu, M. H., Mørkbak, M. R., Olsen, S. B., & Jensen, C. L. (2013). Attending to the reasons for attribute non-attendance in choice experiments. *Environmental & Resource Economics, 54*, 333–359.

Ariely, D., Loewenstein, G., & Prelec, D. (2003). Coherent arbitrariness: Stable demand curves without stable preferences. *Quarterly Journal of Economics, 118*, 73–105.

Arrow, K., Solow, R., Portney, P., Learner, E., Radner, R., & Schuman, H. (1993). Report of the NOAA panel on contingent valuation. *Federal Register, 58*, 4601–4614.

Bateman, I. J., Burgess, D., Hutchinson, W. G., & Matthews, D. I. (2008). Learning design contingent valuation (LDCV): NOAA guidelines, preference learning and coherent arbitrariness. *Journal of Environmental Economics and Management, 55*, 127–141.

Bateman, I., Carson, R., Day, B., Hanemann, W., Hanley, N., Hett, T., et al. (2002). *Environmental valuation with stated preference techniques: A manual*. Cheltenham, UK: Edward Elgar.

Bennett, J., & Blamey, R. (Eds.). (2001). *Choice modelling approach to environmental valuation*. Cheltenham, UK: Edward Elgar.

Bergstrom, J.C., & Taylor, L.O. (2006). Using meta-analysis for benefits transfer: Theory and practice. *Ecological Economics, 60*, 351–360.

Blamey, R. K., Bennett, J. W., Louviere, J. J., Morrison, M. D., & Rolfe, J. (2000). A test of policy labels in environmental choice modelling studies. *Ecological Economics, 32*, 269–286.

Blamey, R., Gordon, J., & Chapman, R. (1999). Choice modeling: Assessing the environmental values of water supply options. *Australian Journal of Agricultural and Resource Economics, 43*, 337–357.

Boxall, P., & Adamowicz, W. (2002). Understanding heterogeneous preferences in random utility models: A latent class approach. *Environmental & Resource Economics, 23*, 421–446.

Boxall, P., Adamowicz, W. L., & Moon, A. (2009). Complexity in choice experiments: Choice of the status quo alternative and implications for welfare measurement. *Australian Journal of Agricultural and Resource Economics, 53*, 503–519.

Boyle, K. J., & Özdemir, S. (2009). Convergent validity of attribute-based, choice questions in stated-preference studies. *Environmental & Resource Economics, 42*, 247–264.

Braga, J., & Starmer, C. (2005). Preference anomalies, preference elicitation and the discovered preference hypothesis. *Environmental & Resource Economics, 32*, 55–89.

Brouwer, R. (2000). Environmental value transfer: State of the art and future prospects. *Ecological Economics, 32*, 137–152.

Brouwer, R., Dekker, T., Rolfe, J., & Windle, J. (2010). Choice certainty and consistency in repeated choice experiments. *Environmental & Resource Economics, 46*, 93–109.

Brouwer, R., & Spaninks, F. A. (1999). The validity of environmental benefits transfer: Further empirical testing. *Environmental & Resource Economics, 14*, 95–117.

Campbell, D., Hensher, D. A., & Scarpa, R. (2011). Non-atttendance to attributes in environmental choice analysis: A latent class specification. *Journal of Environmental Planning and Management, 54*, 1061–1076.

Campbell, D., Hensher, D. A., & Scarpa, R. (2012). Cost thresholds, cut-offs and sensitivities in stated choice analysis: Identification and implications. *Resource and Energy Economics, 34*, 396–411.

Carlsson, F., Kataria, M., & Lampi, E. (2010). Dealing with ignored attributes in choice experiments on valuation of Sweden's environmental quality objectives. *Environmental & Resource Economics, 47*, 65–89.

Carlsson, F., Kataria, M., Lampi, E., Löfgren, A., & Sterner, T. (2011). Is fairness blind?—The effect of framing on preferences for effort-sharing rules. *Ecological Economics, 70*, 1529–1535.

Carlsson, F., & Martinsson, P. (2008). How much is too much? An investigation of the effect of the number of choice sets, context dependence and the choice of bid vectors in choice experiments. *Environmental & Resource Economics, 40*, 165–176.

Carson, R. T., Louviere, J., Anderson, D., Arabie, P., Bunch, D., Hensher, D., et al. (1994). Experimental analysis of choice. *Marketing Letters, 5*, 351–368.

Caussade, S., Ortuzar, J., Rizzi, L., & Hensher, D. (2005). Assessing the influence of design dimensions on stated choice experiment estimates. *Transportation Research Part B, 39*, 621–640.

Christie, M., Hanley, N., Warren, J., Murphy, K., Wright, R., & Hyde, T. (2006). Valuing the diversity of biodiversity. *Ecological Economics, 58*, 304–317.

Colombo, S., Hanley, N., & Calatrava-Requena, J. (2005). Designing policy for reducing the off-farm effects of soil erosion using choice experiments. *Journal of Agricultural Economics, 56*, 81–95.

Concu, G. (2007). Investigating distance effects on environmental values: A choice modelling approach. *Australian Journal of Agricultural and Resource Economics, 51*, 175–194.

Czajkowski, M., & Hanley, N. (2009). Using labels to investigate scope effects in stated preference methods. *Environmental & Resource Economics, 44*, 521–535.

Day, B., Bateman, I. J., Carson, R. T., Dupont, D., Louviere, J. J., Morimoto, S., et al. (2012). Ordering effects and choice set awareness in repeat-response stated preference studies. *Journal of Environmental Economics and Management, 63*, 73–91.

Day, B., & Prades, P. J.-L. (2010). Ordering anomalies in choice experiments. *Journal of Environmental Economics and Management, 59*, 271–285.

DeShazo, J. R., & Fermo, G. (2002). Designing choice sets for stated preference methods: The effects of complexity on choice consistency. *Journal of Environmental Economics and Management, 44*, 123–143.

Dhar, R. (1997). Context and task effect on choice deferral. *Marketing Letters, 8*, 119–130.

Dhar, R., & Simpson, I. (2003). The effect of forced choice on choice. *Journal of Marketing Research, 40*, 146–160.

Eggert, H., & Olsson, B. (2009). Valuing multi-attribute marine water quality. *Marine Policy, 33*, 201–206.

Glenk, K., & Colombo, S. (2011). How sure can you be? A framework for considering delivery uncertainty in benefit assessments based on stated preference methods. *Journal of Agricultural Economics, 62*, 25–46.

Hanley, N., Adamowicz, W., & Wright, R. E. (2005). Price vector effects in choice experiments: An empirical test. *Resource and Energy Economics, 27*, 227–234.

Hanley, N., Czajkowski, M., Hanley-Nickolls, R., & Redpath, S. (2010). Economic values of species management options in human–wildlife conflicts: Hen harriers in Scotland. *Ecological Economics, 70*, 107–111.

Hensher, D. A. (2006). Revealing differences in willingness to pay due to the dimensionalities of stated choice designs: An initial assessment. *Environmental & Resource Economics, 34*, 7–44.

Hensher, D. A. (2008). Joint estimation of process and outcome in choice experiments and implications for willingness to pay. *Journal of Transport Economics and Policy, 42*, 297–322.

Hoehn, J. P. (1991). Valuing the multidimensional impacts of environmental policy: Theory and methods. *American Journal of Agricultural Economics, 73*, 289–299.

Hoehn, J. P., & Randall, A. (1989). Too many proposals pass the benefit cost test. *American Economic Review, 79*, 544–551.

Holmes, T. P., & Adamowicz, W. L. (2003). Attribute-based methods. In P. Champ, K. J. Boyle, & T. C. Brown (Eds.), *A primer on nonmarket valuation* (pp. 171–219). Dordrecht: Kluwer Academic Publishers.

Jacobsen, J. B., Boisen, J. H., Thorsen, B. J., & Strange, N. (2008). What's in a name? The use of quantitative measures versus 'iconised' species when valuing biodiversity. *Environmental & Resource Economics, 39*, 247–263.

Jiang, Y., Swallow, S. K., & McGonagle, M. (2005). Context-sensitive benefit transfer using stated choice models: Specification and convergent validity for policy analysis. *Environmental & Resource Economics, 31*, 477–499.

Johnston, R. J. (2007). Choice experiments, site similarity and benefits transfer. *Environmental & Resource Economics, 38*, 331–351.

Johnston, R. J., & Duke, J. M. (2007). Willingness to pay and policy process attributes. *American Journal of Agricultural Economics, 89*, 1098–1115.

Johnston, R. J., & Duke, J. M. (2008). Benefit transfer equivalence tests with non-normal distributions. *Environmental & Resource Economics, 41*, 1–23.

Johnston, R. J., & Rosenberger, R. S. (2010). Methods, trends and controversies in contemporary benefit transfer. *Journal of Economic Surveys, 24*, 479–510.

Johnston, R. J., Schultz, E. T., Segerson, K., Besedin, E. Y., & Ramachandran, M. (2012). Enhancing the content validity of stated preference valuation: The structure and function of ecological indicators. *Land Economics, 88*, 102–120.

Johnston, R.J., & Thomassin, P.J. (2010). Willingness to pay for water quality improvements in the United States and Canada: Considering possibilities for international meta-analysis and benefit transfer. *Agricultural and Resource Economics Review, 39*, 114–131.

Kerr, G.N., & Sharp, B.M.H. (2006). Transferring mitigation values for small streams. In J. Rolfe & J. Bennett (Eds.), *Choice Modelling and the Transfer of Environmental Values*. Cheltenham, UK: Edward Elgar.

Kosenius, A. (2010). Heterogeneous preferences for water quality attributes: The case of eutrophication of the Gulf of Finland, Baltic Sea. *Ecological Economics, 69*, 528–538.

Kragt, M. E. (2012). The effects of changing cost vectors on choices and scale heterogeneity. *Environmental & Resource Economics, 54*, 201–221.

Kristofersson, D., & Navrud, S. (2005). Validity tests of benefit transfer: Are we performing the wrong tests? *Environmental & Resource Economics, 30*, 279–286.

Ladenburg, J., & Olsen, S. B. (2008). Gender-specific starting point bias in choice experiments: Evidence from an empirical study. *Journal of Environmental Economics and Management, 56*, 275–285.

Lancaster, K. (1966). A new approach to consumer theory. *Journal of Political Economy, 74*, 132–157.

Leong, W., & Hensher, D. A. (2012). Embedding decision heuristics in discrete choice models: A review. *Transport Reviews, 32*, 313–331.

Lew, D. K., & Wallmo, K. (2011). External tests of scope and embedding in stated preference choice experiments: An application for endangered species valuation. *Environmental & Resource Economics, 48*, 1–23.

Liekens, I., Schaafsma, M., De Nocker, L., Broekx, S., Staes, J., Aertsens, J., & Brouwer, R. (2013). Developing a value function for nature development and land use policy in Flanders, Belgium. *Land Use Policy, 30*, 549–559.

List, J. (2003). Does market experience eliminate market anomalies? *Quarterly Journal of Economics, 118*, 41–72.

Loomis, J. B., & White, D. S. (1996). Economic benefits of rare and endangered species: Summary and meta-analysis. *Ecological Economics, 18*, 197–206.

Louviere, J., & Hensher, D. (1982). Design and analysis of simulated choice or allocation experiments in travel choice modeling. *Transportation Research Record, 890*, 11–17.

Louviere, J. J., Hensher, D. A., & Swait, J. D. (2000). *Stated choice methods: Analysis and applications*. New York: Cambridge University Press.

Louviere, J. J., & Woodworth, G. (1983). Design and analysis of simulated consumer choice of allocation experiments. *Journal of Marketing Research, 20*, 350–367.

Lusk, J. L., & Schroeder, T. C. (2004). Are choice experiments incentive compatible? A test with quality differentiated steaks. *American Journal of Agricultural Economics, 86*, 467–482.

Macmillan, D., Hanley, N., & Buckland, S. (1996). A contingent valuation study of uncertain environmental gains. *Scottish Journal of Political Economy, 43*, 519–533.

McNair, B., Bennett, J., & Hensher, D. (2011). A comparison of responses to single and repeated discrete choice questions. *Resource and Energy Economics, 33*, 554–557.

Mørkbak, M., Christensen, T., & Gyrd-Hansen, D. (2010). Choke price bias in choice experiments. *Environmental & Resource Economics, 45*, 537–551.

Morrison, M., & Bennett, J. (2000). Choice modelling, non-use values and benefit transfer. *Economic Analysis & Policy, 30*, 13–32.

Morrison, M., & Bennett, J. (2004). Valuing NSW rivers using benefit transfer. *Australian Journal of Agricultural and Resource Economics, 48*, 591–612.

Morrison, M., Bennett, J., Blamey, R., & Louviere, J. (2002). Choice modeling and tests of benefit transfer. *American Journal of Agricultural Economics, 84*, 161–170.

Morrison, M., & Bergland, O. (2006). Prospects for the use of choice modelling for benefit transfer. *Ecological Economics, 60*, 420–428.

Morse-Jones, S., Bateman, I., Kontoleon, A., Ferrini, S., Burgess, N., & Turner, K. (2012). Stated preferences for tropical wildlife conservation amongst distant beneficiaries: Charisma, endemism, scope and substitution effects. *Ecological Economics, 78*, 9–18.

Navrud, S., & Ready, R. (Eds.). (2007). *Environmental value transfer: Issues and methods*. Dordrecht: Springer.

Roberts, D. C., Boyer, T. A., & Lusk, J. L. (2008). Preferences for environmental quality under uncertainty. *Ecological Economics, 66*, 584–593.

Rolfe, J., & Bennett, J. (Eds.). (2006). *Choice modelling and the transfer of environmental values*. Cheltenham, UK: Edward Elgar.

Rolfe, J., & Bennett, J. (2009). The impact of offering two versus three alternatives in choice modelling experiments. *Ecological Economics, 68*, 1140–1148.

Rolfe, J., Bennett, J., & Louviere, J. (2000). Choice modelling and its potential application to tropical rainforest preservation. *Ecological Economics, 35*, 289–302.

Rolfe, J., Bennett, J., & Louviere, J. (2002). Stated values and reminders of substitute goods: Testing for framing effects with choice modelling. *Australian Journal of Agricultural and Resource Economics, 46*, 1–20.

Rolfe, J., & Wang, X. (2011). Dealing with scale and scope issues in stated preference experiments. In J. Bennett (Ed.), *The international handbook on non-market environmental valuation* (pp. 254–272). Cheltenham, UK: Edward Elgar.

Rolfe, J., & Windle, J. (2008). Testing for differences in benefit transfer values between state and regional frameworks. *Australian Journal of Agricultural and Resource Economics, 52*, 149–168.

Rolfe, J., & Windle, J. (2012a). Testing benefit transfer of reef protection values between local case studies: The Great Barrier Reef in Australia. *Ecological Economics, 81*, 60–69.

Rolfe, J., & Windle, J. (2012b). Distance decay functions for iconic assets: Assessing national values to protect the health of the Great Barrier Reef in Australia. *Environmental & Resource Economics, 53*, 347–365.

Rolfe, J., & Windle, J. (2013). Including management policy options in discrete choice experiments: A case study of the Great Barrier Reef. *Canadian Journal of Agricultural Economics, 61*, 197–215.

Rolfe, J., Windle, J., Bennett, J., & Mazur K. (2013). Calibration of values in benefit transfer to account for variations in geographic scale and scope: comparing two choice modelling experiments. Paper presented at the 57th Annual conference of the Australian Agricultural and Resource Economics Society, Sydney, February.

Rosenberger, R.S., & Stanley, T.D. (2006). Measurement, generalization and publication: Sources of error in benefit transfers and their management. *Ecological Economics, 60*, 372–378.

Rulleau, B., & Dachary-Bernard, J. (2012). Preferences, rational choices and economic valuation: Some empirical tests. *The Journal of Socio-Economics, 41*, 198–206.

Scarpa, R., Gilbride, T., Campbell, D., & Hensher, D. A. (2009). Modelling attribute non-attendance in choice experiments for rural landscape valuation. *European Review of Agricultural Economics, 36*, 151–174.

Scarpa, R., Thiene, M., & Hensher, D. A. (2010). Monitoring choice task attribute attendance in nonmarket valuation of multiple park management services: Does it matter? *Land Economics, 86*, 817–839.

Scarpa, R., Thiene, M., & Train, K. (2008). Utility in willingness to pay space: A tool to address confounding random scale effects in destination choice to the Alps. *American Journal of Agricultural Economics, 90*, 994–1010.

Scheufele, G., & Bennett, J. (2012). Response strategies and learning in discrete choice experiments. *Environmental & Resource Economics, 52*, 435–453.

Smith, V.K., & Pattanayak, S.K. (2002). Is meta-analysis a Noah's Ark for non-market valuation? *Environmental and Resource Economics, 22*, 271–296.

Smith, V. K., Pattanayak, S. K., & van Houtven, G. (2006). Structural benefit transfer: An example using VSL estimates. *Ecological Economics, 60*, 361–371.

Smith, V. K., van Houtven, G., & Pattanayak, S. K. (2002). Benefit transfer via preference calibration: 'Prudential algebra' for policy. *Land Economics, 78*, 132–152.

Swait, J., & Adamowicz, W. (2001). The influence of task complexity on consumer choice: A latent class model of decision strategy switching. *Journal of Consumer Research, 28*, 135–148.

Tait, P., Baskaran, R., Cullen, R., & Bicknell, K. (2012). Nonmarket valuation of water quality: Addressing spatially heterogeneous preferences using GIS and a random parameter logit model. *Ecological Economics, 75*, 15–21.

Tisdell, C., Wilson, C., & Swarna Nantha, H. (2006). Public choice of species for the 'Ark': Phylogenetic similarity and preferred wildlife species for survival. *Journal for Nature Conservation, 14*, 97–105.

van Bueren, M., & Bennett, J. (2004). Towards the development of a transferable set of value estimates for environmental attributes. *Australian Journal of Agricultural and Resource Economics, 48*, 1–32.

Von Haefen, R. H., Massey, D. M., & Adamowicz, W. L. (2005). Serial nonparticipation in repeated discrete choice models. *American Journal of Agricultural Economics, 87*, 1061–1076.

White, P. C. L., Gregory, K. W., Lindley, P. J., & Richards, G. (1997). Economic values of threatened mammals in Britain: A case study of the otter *Lutra lutra* and the water vole *Arvicola terrestris*. *Biological Conservation, 82*, 345–354.

Wielgus, J., Gerber, L. R., Sala, E., & Bennett, J. (2009). Including risk in stated-preference economic valuations: Experiments on choices for marine recreation. *Journal of Environmental Management, 90*, 3401–3409.

Chapter 11
Frontiers in Modeling Discrete Choice Experiments: A Benefit Transfer Perspective

Richard T. Carson, Jordan J. Louviere, John M. Rose and Joffre Swait

Abstract Given increasing survey costs, transferring model estimates obtained from one location and survey and applying them to another location is becoming increasingly appealing. The transfer of previously estimated model outputs to new application contexts has the potential to reduce the need for new large-scale data collection in the new application context as well as reduce the effort required to develop new models. As such, significant savings in cost and time can be achieved. Nevertheless, advantages in time and cost savings may be outweighed due to biases introduced if the transferred model does not adequately represent the behavior of individuals in the new application context. This chapter explores what benefits transfer means within the context of discrete choice experiments and outlines the challenges and possible improvements that could be made.

Keywords Benefits transfer · Discrete choice models · Discrete choice experiments · Challenges

11.1 Introduction

There is strong demand for benefit transfer (BT) in evaluating environmental policies because the number of benefit-cost assessments that need to be performed is large relative to the number of original benefit estimation studies. This is true

R.T. Carson
University of California, San Diego, CA, USA
e-mail: rcarson@ucsd.edu

J.J. Louviere (✉)
University of South Australia Business School, Sydney, Australia
e-mail: Jordan.Louviere@unisa.edu.au

J.M. Rose · J. Swait
Institute for Choice, University of South Australia, Sydney, Australia
e-mail: John.Rose@unisa.edu.au

J. Swait
e-mail: Joffre.Swait@unisa.edu.au

R.J. Johnston et al. (eds.), *Benefit Transfer of Environmental and Resource Values*,
The Economics of Non-Market Goods and Resources 14,
DOI 10.1007/978-94-017-9930-0_11

despite the existence of many studies estimating various types of environmental benefits. The Environmental Values Resource Inventory (EVRI) database, maintained by Environment Canada in conjunction with several other countries including Australia, France, the United Kingdom and United States, contains over 3000 benefit estimates, and many more exist in the literature.[1] The demand for benefit transfer is obviously driven by the cost and time required to perform high-quality original benefit assessments. A BT exercise typically tries to infer the value of environmental policies whose outputs differ from those of a policy that was the subject of a formal assessment, either with respect to (a) the population of interest, (b) the time period of interest, and/or (c) one or more attributes of the policy valued in the original study(s).[2]

A major source of valuing information used in BT exercises comes from discrete choice experiments (DCEs).[3] This paper addresses a narrow question related to DCEs and BT; namely, the increasing use of more advanced statistical approaches to model DCE data, which have focused on incorporating various aspects of consumer heterogeneity. Estimates from studies using DCEs are used in several distinct ways in a BT context, so it is useful to start with a brief overview of that literature.

The seminal formulation of the issues involved in undertaking a BT exercise were put forth in a 1992 symposium in *Water Resources Research* (Brookshire and Neill 1992). The recent Johnston and Rosenberger (2010) review and chapters in this handbook examine many key issues involved in benefit transfer.[4] Over time a key distinction has evolved in the literature dealing with "value transfer" from a study [e.g., mean willingness to pay (WTP) for a particular policy], or "function transfer" from a study (Loomis 1992).[5] The function transfer approach has bifurcated into two distinct branches: one that uses a utility/valuation function estimated from a single or small number of related DCE studies (often in conjunction with other information such as demographic characteristics of the new location) to

---

[1]https://www.evri.ca/.

[2]We use the word "policy" rather than "good" throughout to emphasize that most environmental goods are provided via some type of policy action. Differences in details of how a policy is implemented, including the perceived effectiveness of the government, can be important in carrying out a BT exercise.

[3]Louviere et al. (2000) provide the standard DCE treatment. Adamowicz et al. (1994) and Carson et al. (1990) provide initial DCE examples in the environmental literature. A recent review article focusing on environmental applications of DCE is Hoyos (2010). Carson and Louviere (2011) provide a common nomenclature for stated preference survey questions with an emphasis on environmental applications to try to clear up confusion over how various terms (including "DCE") are used.

[4]Other overviews include Boyle et al. (2010), Navrud and Ready (2007) and Rosenberger and Loomis (2003). Rolfe and Bennett (2006) look at using DCEs in a benefit-transfer context.

[5]In both cases, the issue arises as to whether one study or multiple studies should be used for benefit transfer. Adjustments are often made to account for factors that differ between the original estimate and the new situation, and they sometimes also are made with the valuation function when changes to the variable values that appear in the valuation function are not thought to adequately capture differences between original and new situations.

provide an estimate for the new policy; and a second that combines a sizeable number of estimates from disparate individual studies, often based on very different methodologies via a formal statistical meta-analysis approach.

In our view, transferring values for a policy directly or using those values in a meta-analysis look similar; that is, any modeling approach that produces higher quality estimates for the original policy should result in higher quality BT transfer estimates. It is far less obvious that estimation of a more advanced choice model capable of more flexibly estimating consumer heterogeneity will produce higher quality outputs that are more reliable when being inferred (i.e. transferred) to a different context. In part, this is due to the well-recognized statistical issue that highly parameterized models often fit in-sample data quite well, but produce lower quality out-of-sample estimates than simpler models. An example where consumer heterogeneity was modeled in an initial study and then transferred to a new situation that indicated the more complex model was more successful in transferring BT to the new situation is provided by Colombo et al. (2007). The generalizability of this result and how much it depends on the specific aspects of consumer heterogeneity modeled remain open research questions.

DCEs are being used more frequently in empirical environmental valuation applications; so they provide many of the more current benefit estimates and functions used in BT exercises. Further, DCEs that explicitly value policy attributes can substantially expand the range of policies that can be valued relative to contingent valuation (CV) surveys that typically value only one scenario. The latter property greatly expands the range of what attribute-based DCEs can be used to do in a BT, but not necessarily their accuracy. For example, Kaul et al. (2011) show in their meta-analysis of BT exercises that CV studies focused on valuing single policies produced lower average transfer errors than those derived from more complex attribute based DCEs.[6] Thus, it may be better to think of a DCE with a substantial number of attributes and levels as more of a direct competitor to a meta-analysis where all of the estimates come from a single valuation technique focused on one type of policy. Like most meta-analyses, the emphasis is not on valuing one policy scenario but rather a substantial range of policy scenarios.

Morrison et al. (2002) provided a pioneering empirical example of using a DCE for BT. They implemented DCEs in three surveys involving two Australian wetlands (Gwydir and Macquarie), and interviewed in two different locations (Moree, a rural area, and Sydney). The Sydney population was interviewed separately about both wetlands, so that it was possible to look at benefit transfers to different

[6]The policies being valued in a BT exercise are not the same, so one should not draw strong conclusions about the relative performance of different techniques as the "source" of differences in BT exercises. More direct comparisons clearly are needed to make a more informed judgment. In general, however, what likely occurs is a tradeoff: the analyst can construct a model with more parameters involving policy attributes that allows for predictions (without ad hoc adjustments) to a broad range of policy changes, but at some cost to the accuracy of those predictions. This may be due to fitting a larger number of parameters, greater reliance on functional form assumptions, and/or less comprehensive depiction of individual attributes and their levels, given the same survey length.

locations and different populations. Respondents were asked to answer five choice sets that each had three alternatives [the first was a status quo (SQ) alternative]. Morrison et al. (2002) fit a conditional logit model with alternative specific constants for the two hypothetical alternatives and included various interactions of respondent demographics with the choice of a non-status quo alternative. They compared implicit prices estimated from the three surveys as well a set of nine randomly chosen policies for which compensating surplus was calculated from a BT perspective, using the mean levels of the demographic variables at the target site. Tests suggested that many of the implicit prices did not differ statistically across the surveys, but there were some clear exceptions. They found that the average mean difference in model estimates was 32 % across the nine policy transfer scenarios, consistent with good quality transfers in the literature. Transfers for different wetlands using the same population involved less error than transfers for the same wetland using different populations.

Jiang et al. (2005) and Rolfe and Bennett (2006) provided other early BT tests using DCEs. These and other earlier benefit transfer applications were based on conditional and nested logit models and illustrate that there are a number of different comparisons that can be of policy interest, such as the ranking of policy options in addition to the usual implicit prices and WTP for particular attribute bundles. Colombo et al. (2007) provided the first BT comparison for a random parameters mixed logit model (Train 2009), which was used to capture stochastic heterogeneity in consumer preferences. They conducted two parallel surveys focusing on two similar policies involving soil erosion policies in two different regions of southern Spain, using them to predict each other's results. Like the Morrison et al. (2002) study, the Colombo et al. (2007) BT exercise should represent an ideal context for BT. They compared BT errors for 27 policy scenarios, randomly chosen from the full factorial, using a conditional logit specification and mixed logit specifications with and without correlated parameters.[7]

Colombo et al. (2007) found that the mixed logit specification without correlated errors produced smaller average transfer errors across the 27 transfer scenarios than the mixed logit model with correlated errors, which, in turn, dominated transfers from the conditional logit model specification.[8] On average the transfer error was 38 % lower using the uncorrelated mixed logit model than the conditional logit model, but there were cases where the simple conditional logit model dominated. Surprisingly, we could not find other benefit transfer tests in the literature using mixed logit or other ways to model consumer heterogeneity. Colombo et al. (2007) were careful to note that more tests would be required before one could rely on

[7]The study had six attributes, four of which were assumed normally distributed and two that were fixed, including cost. Similar to Morrison et al. (2002), demographic variables were interacted with an ASC and demographics from the target site were used in the transfer.

[8]Transfer errors in percentage terms were substantially larger on average in Colombo et al. (2007) than in Morrison et al. (2002), illustrating that a myriad of factors are likely at work beyond the particular valuation technique or modelling strategy used.

empirical regularities to determine the relative influence of different approaches to modelling DCE data on the performance of subsequent BT exercises.[9]

Many researchers have wanted to move beyond the workhorse conditional logit model (McFadden 1974) for one of two reasons.[10] The first reason is the independence of irrelevant alternatives (IIA) property that underlies the conditional logit model, making it computationally tractable even with modest computer resources. However, this comes at the cost of imposing strong restrictions on the pattern of substitution relationships between alternatives. The second is an interest in modeling different types of consumer heterogeneity. Initially researchers pursued models that directly relaxed the IIA property like nested logit, which is still quite popular, and multinomial probit, which for various reasons has never seen substantial applied use. For the purpose of modeling DCE data, nested logit models were first used in modeling environmental choice data (e.g., Carson et al. 1990) as a way to deal with the fact that the SQ alternative often behaves quite differently than other alternatives.

Although the use of nested logit models is still fairly common, recent work has put much more emphasis on relaxing the assumption that all consumers are identical except for a draw from the same error distribution. The motivation for this work is twofold: (1) relaxation of IIA, and (2) development of models that more realistically capture differences in consumers. There are three distinct ways to incorporate consumer heterogeneity. The first is to assume that at least some consumers have different preference parameters. Initially, this was achieved by interacting one or more attribute parameters with characteristics of consumers, effectively allowing different types of consumers to have different parameters, which accounts for systematic sources of heterogeneity. Although this often provides useful insights into consumer preferences, it does not allow for continuous (or discrete) distributions of preference parameters, nor does it directly attack the IIA issue.

More recently the random parameters logit [popularly known as a mixed logit (Train 2009)] has become the most popular approach for modeling consumer preference heterogeneity. This approach allows for a range of consumer preference parameters that are assumed to follow some distribution, typically normal or log-normal. This model effectively assumes that individual consumers follow a conditional logit model; hence each consumer is assumed to adhere to the IIA property. The mixed logit model is one way of dealing with IIA violations at the aggregate level if a violation is driven by the assumption that all consumers have the same preference parameters. Additionally, it can also deal with IIA at the individual level via an error-in-variables approach that defines shared stochastic effects through categorization dummies (e.g., private vs. public modes of transport). These dummies reflect IIA violations arising from shared unobserved attributes at the alternative level (see McFadden and Train 2000).

---

[9]There have been subsequent BT exercises using mixed logit (e.g., Baskaran et al. 2010), but these do not seem to have systematically tested the performance of variants of mixed logit versus conditional logit models.

[10]Hensher et al. (2005) provide a general overview of the properties of different choice models.

The second approach is to recognize that consumers may be characterized by draws from different error distributions, with each distribution having different scale parameters, reflecting a form of heteroscedasticity. This heteroscedasticity may be due to several factors including differential ability to choose between alternatives or differences in the importance of unobserved variables on choice behavior. The latter leads to scale heterogeneity models. The third approach is to assume that there are pure types of consumers and that sample members can be represented by these pure types (or mixtures of them), leading to what are popularly known as latent class models. It is possible to combine these different types of consumer heterogeneity, such that one can have models with both preference and scale heterogeneity as well as latent class models with preference and/or scale heterogeneity. However, as we discuss later, such models may not be well-identified statistically and can experience computational difficulties. In the extreme, one can have "individual"-level models whereby one estimates a model specific to each respondent in a DCE (Frischknecht et al. 2014).

Initial moves away from conditional logit models, such as nested logit, treated observations on choices as independent in the sense of observing only one choice per individual or, if more than one choice was observed, not linking those choices together. Models currently at the choice modeling research frontier exploit and often require multiple choice observations from the same individual under different conditions. This is typically accomplished using DCEs with multiple choice sets, but in principle revealed preference (RP) data of this type can also be collected (e.g., Swait et al. 2004). An example of this would be to have individuals record their fishing trips in a diary format. If one matches this with other data sources on temporal differences in fishing quality (crowding and other factors), one would have repeated choice occasions from the same individual under different conditions. If one has multiple choice occasions per individual, one can examine whether choice behavior changes across choice sets. The standard framework used assumes no systematic change across choice sets, which greatly simplifies statistical identification of key parameters. However, a number of empirical tests suggest that this assumption does not generally hold and several competing hypotheses have been put forward that predict specific types of changes.

Much of the work using more advanced choice models is associated with DCEs because the ability to control choice stimuli can greatly facilitate estimating more complex models. Consequently, we also discuss the role of experimental design for estimating advanced choice models. Another recent advance is the ability to collect extra preference information in each DCE choice set beyond that of the most preferred alternative. Thus, we also discuss a variant of best-worst preference elicitation that elicits most and least preferred choices within a set of alternatives (see for example Louviere et al. 2008; Marley et al. 2008). In addition to obtaining repeated stated preference (SP) observations, one can combine SP and RP data to obtain multiple choice occasions for an individual. This naturally raises questions about differences in "scale" in the two choice contexts (Swait and Louviere 1993), and ways to capture the differences and/or take them into account.

In the remainder of the chapter, we provide an overview of different ways to model consumer heterogeneity with an eye toward using these models in BT exercises. The starting point for our discussion is to lay out the behavioral and statistical assumptions underlying the conditional logit model that serves as a reference model for much applied work. Most of the issues we address later can be shown to follow from applying the same model to all choices and all individuals. Then we turn to a discussion of several advanced models currently in use, focusing on how each relaxes one or more key assumptions of the conditional logit model, and how they relate to each other. A critical issue in thinking about these models from a BT perspective is the role that covariates play (usually demographics that can be observed both at the donor site where the original study was performed and the target or transfer site). Next, we discuss experimental designs used to collect choice data and the impact they can have on the types of models that can be estimated and the precision of the associated parameter estimates. After that discussion, we consider combining different types of data: e.g., combining data from multiple DCEs in appropriate ways that can substantially extend the BT beyond that of the original models.

11.2 Behavioral and Statistical Framework Underlying the Conditional Logit Model

The foundational framework for econometric analyses of choosing one among J objects or alternatives in a choice set S is the utility-maximizing consumer. For each discrete object $j \in S$, say, a recreation site, the consumer n forms a judgment/evaluation/utility measure:

$$\mathrm{U_{nsj}} = \mathrm{U}(X_{nsj}, \mathrm{Z}_n | \beta_{nsj}, \theta_{nj}). \tag{11.1}$$

where X_{nsj} is a vector of quality attributes for the alternative including the price to "consume" it, Z_n is a vector of socio-demographic characteristics (e.g., income, age, gender) and other consumer specific information (e.g., attitudes) describing the individual, β_{nsj} is a deep vector of preference parameters associated with X_{nsj} and θ_{nj} is a vector of parameters associated with Z_n. Parameter vectors β_{nsj} and θ_{nj} may be constrained in various ways, such as being made generic across alternatives and fixed across respondents. Utility function (1) arises from a Lancasterian framework (Lancaster 1966), whereby consumers define preferences on the basis of benefits (termed "characteristics" by Lancaster) generated by the attributes of the good in question.[11] Our representation of the utility function above directly connects

[11]The Lancasterian model collapses downward to the standard framework used in most microeconomic theory if all goods are represented by only an alternative specific constant (ASC) and price.

attributes and prices to the overall judgment, and is therefore a simplified representation. In the final act of the decision process, the decision maker is assumed to select alternative $i^* \in S$ such that $U_{nsi}{}^* \geq U_{nsj}$ for all $j \neq i^*$. Note that this description is deterministic in nature.

The process description we made above is based on some critical assumptions, some of which we mention below.

1. The decision maker is aware of and uses all relevant information in making her judgment about the attractiveness of an alternative.
2. Further, relevant information is available for all alternatives.
3. The decision maker is exhaustive in her evaluations, not only in terms of information use, but also in terms of "looking at" all alternatives in *S*. This is done irrespective of the number of alternatives in S and of the cost of performing evaluations.
4. Selection of the preferred alternative is done by ranking all alternatives according to their utility, and choosing the highest valued alternative.

Although making these assumptions explicit may be unfamiliar and seem unnecessary, we do this to remind the reader that the underlying behavioral decision process adopted in extant models of choice, particularly all those to be examined in this paper, depicts decision makers as fully rational, all-knowing, inexhaustible utility maximizers. No matter how sophisticated the econometric approaches employed in the models we discuss, these (and other, very important) assumptions are [always] present.

Bringing the analyst into the mix requires us to introduce the possibility that the analyst does not have the full knowledge set available to the decision maker, and so we must allow for a stochasticcomponent to utility. We rewrite Eq. 11.1 below to reflect this:

$$U_{nsj} = V(X_{nsj}, Z_n|\beta_{nsj}, \theta_{nj}) + \varepsilon_{nsj}(\mu_n) \tag{11.2}$$

where $V(\bullet)$ is the deterministic component of utility (i.e., that part of total utility known to the decision maker that an analyst can specify), whereas $\varepsilon_{nsj}(\mu_n)$ is the stochastic utility that accounts for the difference between the decision maker's total utility and the deterministic component known to the analyst and μ_n is a deep parameter of the stochastic utility. It is perhaps more enlightening to rewrite Eq. 11.2 as follows to emphasize what actually gives rise to stochastic utility in random utility theory:

$$\varepsilon_{nsj}(\mu_n) = U_{nsj} - V(X_{nsj}, Z_n|\beta_{nsj}, \theta_{nj}). \tag{11.3}$$

It is in this sense that we can say that "analyst ignorance" about the underlying total utility gives rise to stochastic utility, and hence, to probabilistic (as opposed to deterministic) choice. The analyst's key relationship between the observed choice of alternative i∈M and the total utility construct Eq. 11.2 is this expression:

$$P_{nsi} = P\{U_{nsi} \geq U_{nsj}\} \quad \text{for all } i \neq j, i, j \in S, \tag{11.4}$$

where P_i is the probability that $i \in S$ is chosen, all other quantities as previously defined.

The crucial step in operationalizing expression Eq. 11.4 into different models of choice involves the stochastic specification of the vector $\varepsilon = (\varepsilon_{ns1}(\mu_n), \ldots, \varepsilon_{nsj}(\mu_n))$. As is well known, the workhorse Multinomial Logit (MNL) model (Ben-Akiva and Lerman 1985)

$$P_i = \exp(\mu_n V(X_{nsi}, Z_n | \beta_{nsi}, \theta_{ni})) \Big/ \sum\nolimits_{j \in S} \exp(\mu_n V(X_{nsj}, Z_n | \beta_{nsj}, \theta_{nj})) \tag{11.5}$$

results if we assume that the elements of ε are independent and identically distributed (IID) Type I Extreme Value random variables, that is,

$$\varepsilon_{nsj}(\mu_n) \sim F(w) = \exp(-\exp(-\mu_n w)), \quad -\infty < w < \infty, \mu \geq 0, \forall j \in S. \tag{11.6}$$

Ben-Akiva and Lerman (1985) and Hensher et al. (2005) also provide detailed discussions associated with the derivation of Eq. 11.5. It is important to consider two critical properties of model Eq. 11.5:

1. A confound exists between the deterministic (*V*) and stochastic utility components (parameter μ_n) because of the inseparable and multiplicative nature between the two parts: these always show up in the form $\mu_n V$.
2. As a result of the IID assumption, the MNL model has the Independence of Irrelevant Alternatives (IIA) property. This implies that the odds of choosing one alternative over another are influenced only by their own utilities, but not influenced by the utilities of other alternatives. This property can lead to counterintuitive implications about behavior in empirical contexts. To make the IIA assumption and its implications explicit, consider taking the ratio of the probabilities for two competing alternatives (often referred to as the odds ratio):

$$\begin{aligned} \frac{P_{nsi}}{P_{nsh}} &= \frac{\left(\exp(\mu_n V(X_{nsi}, Z_n | \beta_{nsi}, \theta_{ni}) \Big/ \sum_{j \in S} \exp(\mu_n V(X_{nsj}, Z_n | \beta_{nsj}, \theta_{nj})\right)}{\exp(\mu_n V(X_{nsh}, Z_n | \beta_{nsh}, \theta_{nh}) \Big/ \sum_{j \in S} \exp(\mu_n V(X_{nsj}, Z_n | \beta_{nsj}, \theta_{nj})}. \\ &= \frac{\exp(\mu_n V(X_{nsi}, Z_n | \beta_{nsi}, \theta_{ni})}{\exp(\mu_n V(X_{nsh}, Z_n | \beta_{nsh}, \theta_{nh})} \end{aligned} \tag{11.7}$$

As can be seen from Eq. 11.7, only the utility expressions of the two competing alternatives being considered actually matter; the utility functions of all other alternatives present in S drop out. Models such as the MNL force this property for all pairs of alternatives; however, issues arise when decision makers are more or less likely to substitute alternative i or h with another alternative $j \in S$, and hence the ratio of the probabilities are not independent of the presence or absence of other alternatives as implied by Eq. 11.7. In the following sections, we examine models that relax these characteristics, but in this section we discuss them from the perspective of BT.

To take these in order, let us first consider the consequences of the confound between the scalar $\mu_n \geq 0$ and the deterministic utility V. As first pointed out by Ben-Akiva and Lerman (1985), for a given set of observations and context, they are inseparable. Between contexts c_1 and c_2 (say, original measurement and transfer application contexts) two sources of differences can occur: scalars μ_{c1} and μ_{c2} (we drop subscript n here as it is common practice to assume that differences between μ are due to context effects rather than differences between consumers) differ, and/or V_{jc1} and V_{jc2} differ. That is, in a benefits transfer exercise one may not "transfer" successfully because (a) stochasticutility distributions differ between contexts, (b) systematic utility functions differ, or (c) both stochastic and systematic utilities differ. Swait and Louviere (1993) proposed a basic statistical test to determine if $\mu_{c1} = \mu_{c2}$ conditional on the assumption that $V_{jc1} = V_{jc2}$ for all j; if systematic utility functions differ between contexts, benefits transfer is a moot question. If both components differ between contexts, the confound between stochastic and systematic utilities implies that there is no way to determine why preferences do not transfer. This discussion makes it clear that at a basic level the deck is stacked against BT.

With respect to the IIA property of the MNL model, this is often thought to be too strong a theoretical restriction to hold in general. On the one hand, it should be noted that whether or not IIA is a reasonable assumption to impose on observed choices is an empirical issue. Additionally, IIA is an individual-level property that may or may not hold in aggregate choice data, whether or not it is reasonable to hold for any particular individual. A certain dataset may contain choices that display IIA-like behavior. On the other hand, choice model forms exist (e.g., nested logit, multinomial probit; see Swait 2006) that allow for flexibility in representing interdependence in substitution patterns. To make our discussion with respect to benefits transfer more concrete, consider this nested logit formulation:

$$P_{nsi} = \left(\exp(\mu_1 V_{nsi}) \Big/ \textstyle\sum_{j \in C_{(i)}} \exp(\mu_1 V_{nsj})\right) \left(\exp(\mu I_{C_{(i)}}) \Big/ \textstyle\sum_{h=1}^{H} \exp(\mu I_h)\right), \quad \forall i \in S, \tag{11.8a}$$

$$I_h = \frac{1}{\mu_1} \ln\left(\sum_{j \in C_h} \exp(\mu_1 V_{nsj})\right) \tag{11.8b}$$

where the set S is subdivided into H nests/clusters C_h ($C_{(i)}$ is the cluster to which alternative $i \in S$ belongs), which are collectively exhaustive and have no elements in common, $\mu_1 \geq 0$ is a scalar to be explained, I_h is a nest-specific inclusive value measure that is known to be the expectation of the maximum utility of the alternatives in C_h, and all other quantities are previously defined. This model arises if one assumes that the ε's are correlated within clusters C_h and uncorrelated between clusters instead of assuming that the stochastic utilities are IID Type I Extreme Value distributed. Specifically, the joint cumulative distribution function of the ε is given by

$$F(\varepsilon) = \exp(-G(\exp(-\varepsilon_{ns1}), \ldots, \exp(-\varepsilon_{nsJ})), \quad -\infty < \varepsilon < \infty, \tag{11.9a}$$

$$G(w_{ns1}, \ldots, w_{nsJ}) = \sum_{h=1}^{H} \left(\sum_{j \in C_h} w_{nsj}^{\mu_1} \right)^{\mu/\mu_1}, w_{nsj} \geq 0, j = 1, \ldots, J. \tag{11.9b}$$

This stochastic assumption allows us to specify the correlation between stochastic components as follows (Swait 2006):

$$\rho_{ih} = \begin{cases} 1 - \left({}^{\mu}/_{\mu_1} \right)^2 & if\ C_{(h)} = C_{(i)} \\ 0 & if\ C_{(h)} \neq C_{(i)} \end{cases} \quad i, h \in S. \tag{11.10}$$

Conceptually, this stochastic definition of the behavior of the ε's results in a covariance matrix for these random variables that is homoscedastic (equal diagonal terms, since μ applies to all alternatives), with non-zero covariances for all alternative pairs sharing cluster membership and zero off-diagonal terms for all pairs not sharing clusters. A model like multivariate probit has a more general covariance matrix,[12] but it will conceptually arrive at the same point of making choice probabilities exhibit a type of non-IIA responses. It is obvious that the nested logit model can capture non-IIA behavior if we form the odds ratio for alternatives h and i using Eq. 11.8a, which we find to depend upon whether or not the pair of alternatives share a cluster:

$$\frac{P_i}{P_h} = \frac{\exp(\mu_1 V_i)}{\exp(\mu_1 V_h)} \cdot \left\{ \begin{matrix} 1 & if\ C_{(h)} = C_{(i)} \\ \frac{\exp(\mu I_{C_{(i)}})}{\exp(\mu I_{C_{(h)}})} & if\ C_{(h)} \neq C_{(i)} \end{matrix} \right\} \quad i, h \in S \tag{11.11}$$

Thus, this straightforward extension to the MNL model allows one to avoid the IIA property empirically, but imposes preference homogeneity within clusters.

Now, let us consider the implication of this more general model with respect to BT. Clearly, if the structure of an IIA violation, i.e. the nests (or stochastic utility correlation structure) in measurement and transfer application contexts differ

[12] Bunch (1991) provides identification restrictions.

significantly between contexts, it may not be possible to successfully transfer between them. This effect is over and above that arising from the basic confound between stochastic and systematic utilities within each context. Even if the systematic utilities are identical across contexts, if the clustering structure is too dissimilar (e.g., it is place- or time-specific) or the quantity (μ/μ_1) is small (i.e. correlation is high, imposing pairwise requirements on stochastic utility for transfer to occur), BT may induce large errors, which may well be a key underlying difficulty with BT exercises that not seem to be clearly appreciated.[13]

11.3 Advanced Choice Models Available for Use with DCE Data

In this section, we move from a consideration of the nature of the underlying utility function to econometric issues that must be resolved for BT by examining how the estimates to be transferred arise. Of course, these are linked, but here we emphasize issues related to modeling data collected from samples of the population of interest where various sources of heterogeneity are considered. We focus on the general case, the likelihood function of discrete choice models, rather than examine all possible models that can be estimated.

Typically, the parameters β associated with each utility function V_{nsi} are unknown and must be estimated from data. Let y_{nsi} equal one if i is the chosen alternative in choice situation s shown to respondent n, and zero otherwise. In other words, y represents the outcomes of a discrete choice experiment. The parameters can be estimated by maximizing the likelihood function L,

$$L = \prod_{n=1}^{N} \prod_{s \in S_n} \prod_{j \in J_{ns}} (P_{nsi})^{y_{nsi}}, \tag{11.12}$$

where N denotes the total number of respondents and S_n is the set of choice situations faced by respondent n and P_{nsi}, is a choice probability. This choice probability is expressed as equation Eq. 11.5 for the conditional logit or MNL model and equation Eq. 11.8a for the nested logit model.

The log-likelihood functions of more advanced models can also be derived with a simple substitution of the appropriate choice probability for the model being estimated in Eq. 11.12. For example, in the mixed logit model, the parameters to be estimated are structural parameters representing the population moments of some underlying (multivariate) distribution (e.g., the mean(s) and standard deviation(s) of

[13]Note that transferring values via a meta-analysis does cannot solve this problem. The same good can be worth different amounts in different contexts where the difference between the contexts is not observed. The best that a meta-analysis can do in this instance is to average over a small number of values for the same good derived in different contexts.

a (multivariate) normal distribution). To estimate the parameters of the mixed logit model using simulated maximum likelihood involves taking draws from the random parameter distribution(s), calculating the expected or average logit choice probability over the simulation draws, and substituting this into equation Eq. 11.12. However, the underlying likelihood function is the same.

Rather than maximize the likelihood function, it is more common to maximize the log of the likelihood function because the product of a series of probabilities typically produces extremely small values that most computing software cannot adequately handle. Taking the logs of the probabilities produces large negative values, which when multiplied, produce even larger negative values. Consequently, the log-likelihood function of the model, shown below, is typically preferred:

$$LL = \ln\left[\prod_{n=1}^{N}\prod_{s\in S_n}\prod_{j\in J_{ns}}(P_{nsi})^{y_{nsi}}\right]. \tag{11.13}$$

If one assumes that the choice observations are independent over both respondents and choice situations and one uses the mathematical properties $\ln(n_1 n_2) = \ln(n_1) + \ln(n_2)$ and $\ln(n_1)^{y_{nsi}} = y_{nsi}\ln(n_1)$, and applies the same mathematical rules to choice tasks, s, and alternatives J, one can rewrite equation Eq. 11.13 in the more commonly recognized form:

$$LL = \sum_{n=1}^{N}\sum_{s\in S_n}\sum_{j\in J_{ns}} y_{nsi}\ln(P_{nsi}). \tag{11.14}$$

It is possible for some advanced models, such as the mixed logit model, to relax the assumption that responses are independent within a respondent. In this case, the log-likelihood function of the model becomes:

$$LL = \sum_{n=1}^{N}\ln\left(E\left(\prod_{s\in S_n}\prod_{j\in J_{ns}}(P_{nsi})^{y_{nsi}}\right)\right), \tag{11.15}$$

where $E\left(\prod_{s\in S_n}\prod_{j\in J_{ns}}(P_{nsi})^{y_{nsi}}\right) = E(P^*) = \int_\beta P_n^*(\beta)f(\beta|\theta)d\beta$, P^* is the probability of the chosen alternative, and $f(\beta|\theta)$ is the multivariate probability density function of β, given the structural parameters θ.

This chapter deliberately examines discrete choice models from the perspective of the log-likelihood function for two primary reasons. First, it is important to understand that the basics of the estimation process are fundamentally the same for all discrete choice models, and are independent of any particular assumed model specification. The reason for this is that if one views the log-likelihood function as described above, the logic can be extended to any discrete choice model by substituting the specific probability for the model of interest. In turn, this allows for a more general discussion of the relevant issues without getting bogged down in discussions of the specifics of all model forms. Second, a focus on the log-likelihood makes it clear that, regardless of model type, the estimation process involves nothing more than trying to find the parameter estimates (whether random or fixed) that maximize the probabilities over choice tasks for the alternatives observed to have been chosen in the data. The reason for the latter statement is that only the chosen alternative matters in the log likelihood function (i.e., if $y_{nsi} = 0$, the function is zero) and mathematically as $P_{nsi} \rightarrow 1$, the log of $P_{nsi} \rightarrow 0$, suggesting that as the log likelihood function approaches zero, the estimated parameters best predict the chosen alternatives.[14]

The latter point is particularly important, and so deserves more attention. As suggested by Eq. 11.4, the choice probabilities associated with any choice model are a function of the utilities specified by the analyst, which in turn are potentially a function of the design attributes, X_{nsi}, including the price to "consume" the alternative, socio-demographic characteristics of respondents, potential attitudes respondents have about the object(s) of interest, the decision context associated with the choice made, and the parameters to be estimated, β. The vast majority of DCE studies typically consider the design attributes, but generally do *not* consider socio-demographic characteristics. Even fewer consider attitudes and/or decision context(s).

Even if one tries to consider all possible decision influences, there are many possible ways that one can enter them into a final model specification. For example, one could enter socio-demographic variables linearly into the alternative specific constants (ASCs) of one or more utility functions (e.g., with the SQ alternative) and/or one can interact them with one or more of the design attributes. For more advanced models like latent class models, socio-demographic characteristics may enter the model as part of a class assignment model (Swait 1994), whereas in the case of nested logit one can extend Eq. (11.8b) to specify the inclusive value measure as a function of exogenous variables outside of any contained in V_{nsi}. Similar specification possibilities arise if one has attitudinal and/or decision context measures.

Thus, how one specifies the utility function of the model matters greatly from an estimation perspective, whether one is interested in BT or not. Unfortunately, the degree to which this matters is an empirical issue, which may vary from context to context. At a simple level, if a salient variable is omitted from the analysis, covariances between parameter estimates can adversely affect the parameter outcomes for those variables that entered the utility function. Unfortunately, one can learn about this only by specifically and systematically testing this effect by including all available variables into the model in all possible ways. Further, if interactions exist and are not incorporated into the model specification, then one runs a risk of introducing endogeneity bias into the estimation process.

[14]It should be noted though that the probabilities depend on all of the alternatives.

To elaborate, discrete choice models of all types assume that V_{nsi} and $\varepsilon_{nsi}(\mu)$ are orthogonal to one another. If one omits a salient interaction term involving one or more variables in the model specification, the observed component and the error term of the model become correlated. For example, if one estimates the parameters of a model where price *should* enter as both a main effect and an interaction in the model, but price *only* enters as a main effect, the price variable is associated with both the observed component and error term. In such cases, the parameters of the affected variables captured in the modeled component of utility will likely be biased, and in the immediately preceding price example, the price parameter will be biased. Naturally in such cases, all implicit prices also will be biased, and the amount of endogeneity bias can differ for different parameters, depending on where the problem resides in the utility function. Such problems are not limited to DCEs, and they are not limited by data collected in surveys. For example, if different respondents make different assumptions about a decision context and/or treat ambiguous attributes differently, or respondent attitudes are related to the true decision process but not observed, endogeneity bias may arise and require sophisticated methods to detect and deal with it.

The preceding discussion raises a very important, fundamental question for BT. That is, *before* one considers transferring outcomes from one location (or data source) to another, one must know if what is being transferred is correct (i.e., unbiased), regardless of the specific model type estimated. More specifically, one cannot ask what model type will provide the best outcomes for BT unless the estimates for *all* model types compared are based on the best representations (most correct approximations) of the underlying decision processes. To do otherwise is to merely compare the robustness of various model forms to violations similar to those discussed above. In all likelihood, the latter exercise merely poses an empirical question, the answer to which will depend on the type and degree of violation experienced.

Additionally, from a BT perspective the preceding discussion illustrates other broader issues that one faces when trying to transfer estimates obtained from one location (or data source) to another. For example, if data are available on all relevant decision variables in two different sites (data sources)—whether the data represent DCE design attributes, socio-demographic characteristics, attitudes or information about decision contexts—the first question one should ask is whether the same set of decision variables should enter the utility specifications of models that would be independently estimated from the two data sets. If the answer to the first question is "yes," one should then ask if they enter the utility functions of both models in *exactly the same way*. The latter differs slightly from the issue raised in the preceding paragraph because one is asking not only if the utility specification for the estimation site (data source) is correct, but also if it best represents the specification at the transfer site (data source). If the two utility specifications differ, then one needs to consider the particular effects or calculations being transferred, and whether such differences are likely to matter.

For example, if implicit prices are being transferred, problems can arise if there are interactions between design attributes and other exogenous variables and/or if different data transformations should be applied to variables entering the model. To illustrate this problem, consider the implicit price for the following utility specification (x_c the price, x_k a quality attribute):

$$V = \cdots + \beta_1 x_k + \beta_2 x_k x_c + \beta_3 x_c^2 + \cdots \quad (11.16)$$

The marginal implicit price for a one unit increase in x_k becomes

$$WTP_k = -\frac{\Delta x_c}{\Delta x_k} = -\frac{\frac{d}{dx_k}\left(\beta_1 x_k + \beta_2 x_k x_c + \beta_3 x_c^2\right)}{\frac{d}{dx_c}\left(\beta_1 x_k + \beta_2 x_k x_c + \beta_3 x_c^2\right)} = -\frac{\beta_1 + \beta_2 x_c}{\beta_2 x_k + 2\beta_3 x_c}. \quad (11.17)$$

Note that the value of the exogenous variable x_k does not drop out of the calculation; so, for the case of BT, one needs to know the value of the exogenous variable at the transfer site (data source) in addition to the parameters being transferred. An additional issue can arise in cases where one requires utility specifications that are highly non-linear in the attributes. In this case, one or more attributes in the estimation model may be irrelevant to the transfer site (data source); hence, one must carefully attend to what precisely is being transferred. Moreover, leaving aside the issue of models that allow for consumer heterogeneity in the parameter estimates, one also may need to consider the issue of heterogeneity (or range) in the *X*s associated with BT. Given the seriousness of these issues and the uncertainty likely to be associated with any particular BT application, one may be tempted to rely on simple models, such as simple linear in the parameter and linear in the attribute utility specifications. Unfortunately, however, our earlier discussion on potential biases that can occur when one misspecifies the true decision process highlights the fact that there clearly are potential risks in doing this.

Putting all the above aside, let us now consider a model specification in which both preferences, represented as vector β, and scale, represented by a scalar u_n, are assumed to vary over the sampled population, such that $V = u_n\beta$. It is possible to rewrite $V = \alpha_n = u_n\beta$, where the elements in α_n must be correlated as each term in β is multiplied by a common scalar u_n. When viewed in this light, the confound between scale and preference suggests that what is being modeled is also a form of correlation. If one or more parameters in α_n are fixed, or if all parameters are assumed to be randomly distributed but uncorrelated, then an analyst is making the [implicit assumption] that scale is homogenous across the sample. If one treats all parameters as random and correlated (via, say, a Cholesky decomposition of the parameter covariance matrix) in the mixed logit model, that model also allows for random scale. The question then becomes what is being transferred in BT exercises when one uses these advanced models.

Ideally, the model from the estimation site should capture both scale and preference heterogeneity, which implies allowing for correlated random parameters. Yet, one now must transfer not only mean estimates but also the entire covariance structure of random parameter terms, and assume that all these terms are similar to those at the transfer site, an important assumption that may not hold empirically.

To further complicate the BT issue, up to this point we have been deliberately vague about the specifics of how to appropriately model attitudinal data for discrete choice models. Typically, attitudinal data is collected using multi-attribute likert scales or some other similar type of approach. One often observes researchers entering such consumer-reported value(s) as an independent variable in the utility specification, but such an approach can be problematic. As is well known, consumers use rating scales differently, such that a rating of (say) two means different things to different consumers (see for example Lee et al. 2007). Thus, this is a type of measurement error in which the true underlying attitudes are measured by some latent unobserved value. More recently, Rungie et al. (2011) provide comprehensive statistical theory to integrate structural equation models and choice models, providing a theoretically appropriate way to incorporate latent variables in the latter. Despite these advances in incorporating latent constructs, such as attitudes in models, new issues arise for BT, as analysts now must transfer parameter estimates for the design attributes and parameters for the latent attitudinal variables. The latent variables now must be imputed for the transfer site somehow. In turn, this suggests that our earlier discussion on endogeneity also applies.

We end this particular discussion with an interesting and likely case-specific question; namely, what if the best models for two separate sites (data sources) are *not* the same. That is, what would be the likely outcome of a BT exercise if the model that best represented the decision processes at the estimation site has one particular model form (e.g., a mixed logit model), while the model that best fits the true decision processes at the transfer site (data source) is another model form (e.g., conditional or nested logit)? This would seem to be an important future research question that begs several other questions that, in turn, suggest that researchers need to consider other possible models, decision processes and even the prospect that different people use different decision processes.[15]

11.4 Experimental Design and Collection of SP Data for Estimating Choice Models

Thus far our discussion of models for choice data has been quite limited in the sense that data satisfying the conditions necessary for estimating such models have been assumed to be available. Data from DCEs typically come from stated preference surveys due to greater flexibility in controlling the stimuli seen by individuals, which in turn minimize certain statistical issues in estimating the desired model and

[15]It is important to note that none of these issues are avoided by using estimated WTP from DCEs for particular goods as inputs to the meta-analyses rather than making the transfer based directly on the utility function estimated using DCE data. That is because all the specification issues discussed as well as issues involving scale, can have substantial influence on WTP estimates from both the original study and estimates derived for BT exercises.

provide rigorous tests of particular hypotheses.[16] Typically, SP data result from some formal statistical design process, but there are exceptions. For example, some researchers still use random draws from the total possible sets of choice sets to construct SP surveys. To the extent that the number of draws is sufficiently large this will approximate the population of choice sets, but as is the case with any random sampling procedure, there can be serious departures from representativeness in small samples and the statistical identification of particular parameters may be tenuous (Carson et al. 2009).

SP data collection processes for DCEs can be specifically designed to maximize the power of tests of particular hypotheses and/or minimize estimation errors associated with specific parameters of particular models. Of course, there is no free lunch because SP studies face issues of external validity, due in no small part to incentive compatibility issues, hypothetical situations and a wide array of design-related and/or induced effects on outcomes.

11.4.1 Single Binary Discrete Choice CV Experiments

Contingent valuation (CV) methods have been used for over five decades, with a recent book citing thousands of studies (Carson 2012). This chapter views the standard single binary discrete choice (SBC) question recommended by Arrow et al. (1993) and used in many high profile CV studies (e.g., Carson et al. 2003) as the simplest special case of a DCE, whereby the choice options are limited to an SQ alternative and a substantive policy alternative. One attribute, typically the payment cost, is randomly varied. Payment costs are chosen to help identify the shape of the underlying distribution of WTP and minimize the confidence interval around key statistics like mean or median WTP. Generally speaking, there is no need for the values of the payment vehicle to be chosen randomly from a range. The number of payment amounts (often referred to as bid values) and their values depend critically on three factors: (a) the distributional assumption made about WTP, (b) prior information assumed about the value of the parameters of that distribution, and (c) the statistic(s) of interest. In the extreme case where a two-parameter distribution for WTP has been assumed and the values of the distribution's parameters are known with certainty, the typical result is that only two bid values should be used. Assumptions of more flexible WTP distributions or allowing for considerable uncertainty over parameter values generally results in more bid values being optimal, but optimal designs typically result in the use of four to eight bid values.

The SBC format allows one to trace out the WTP distribution in the population of interest if a parametric assumption about the WTP distribution is made (Cameron

[16]RP data also can be collected with DCEs in a laboratory setting, field tests or other contexts, such as internet websites selling products where it is possible to control and randomly vary stimuli seen by individuals. In this section, we assume that a DCE is used to collect SP data as that is by far the most common application.

1988; Cameron and James 1987) or one uses a nonparametric step-function to approximate it (Carson and Steinberg 1990; Haab and McConnell 2002). The SBC format with appropriate auxiliary conditions has desirable incentive properties (Carson and Groves 2007), but these properties arise mainly because the SBC format collects so little information from each respondent. For all practical purposes, preference parameters of individual respondents are unidentified beyond the interval in which their WTP lies, although some parameters related to observable individual characteristics (e.g., being an environmentalist) can be estimated at the average of members of the group. Technically, it is possible to estimate the difference in the population's WTP for two policies that differ by two levels of a single attribute by asking two statistically equivalent subsamples SBC questions that differ only on that dimension. This is known as an external scope test (Arrow et al. 1993), but it quickly gets prohibitively expensive when extended to multiple samples. The desire to value change in multiple attributes that individually have multiple levels pushes one in the direction of asking respondents to make choices in more than one choice set. Moving to DCEs with multiple choice sets also greatly facilitates more detailed modelling of consumer heterogeneity.

11.4.2 DCE with Multiple Choice Sets

DCEs were pioneered by Louviere and Hensher (1983) and Louviere and Woodworth (1983). DCEs evolved by drawing on work from several sources, including conjoint measurement (Luce and Tukey 1964), information integration theory (e.g., Anderson 1970), discrete multivariate analysis of contingency (crosstab) tables (e.g., Bishop et al. 1975), probabilistic discrete choice models (e.g., McFadden 1974) and the design of statistical experiments for linear models (e.g., Box et al. 2005). Basically a DCE is a sparse, incomplete contingency (crosstab) table. The table is constructed: i.e., it is purposefully designed a priori so that certain statistical properties are achieved. Typically, these properties include identification of certain effects associated with a particular statistical specification of research interest, the precision with which these effects should be estimated, and/or other properties. Statistical design theory provides theory and construction methods for achieving these purposes.

There are two basic types of DCE designs, namely generic and alternative-specific. Generic designs (GDs) are used when the choice options are unlabeled (e.g., Option A, Option B, etc.), and alternative-specific designs (ASDs) are used when the choice options are labeled (e.g., Manly Beach or Bondi Beach). Most of the advances in optimum design theory work for DCEs in the past decade has focused on GDs. Little work is available for ASDs outside of the transportation literature (although see Rose and Bliemer 2009). As noted below, the vast majority of DCE applications have used GDs; however, in many (but not all) cases, the researchers should have used ASDs.

A GD can be seen as a way to construct choice sets offering options described by a generic set of attributes and associated levels; such designs can be used to construct pairs, triples, etc., of options. Typically, environmental applications of DCEs present a status quo option and two or more competing options. So, the design problem is to select the "best" way to configure the choice sets to satisfy certain statistical criteria of interest. Typically, the criteria are identification (ensuring that all parameters of an assumed statistical model of interest can be estimated) and precision (the estimated parameters are minimum variance estimates). There can be other criteria, such as trying to maximize the fit of the model in- and/or out-of-sample. The onus is on the designer to insure that ALL relevant attributes are included (to avoid omitted variable bias), and that the levels are appropriate (ranges not too large or too small, qualitative levels correct and relevant). Louviere and Woodworth (1983) and Louviere et al. (2000) discuss these designs for cases where a constant option is present in each choice set (e.g., status quo).

Alternative-specific designs are not frequently seen in environmental applications, and may pose issues for BT. That is, to the extent that the modeling results from ASDs are specific to particular labeled options, the model parameter estimates and/or calculations based on them may not transfer to sites not included in the estimation set. The references cited in the preceding paragraph also discuss ASDs. Basically, such designs can be constructed by using designs for linear models as long as there is a constant option in each choice set, such as not choosing any of the options. The advantages of the design construction methods discussed in the cited references are that they allow one to estimate more general models and test violations of IIA.

Since 2000 there has been a rapidly growing literature in the design of DCEs. This literature has gone in three different directions. The first was an effort to examine the statistical properties of designs used in DCEs. This included efforts to improve the efficiency of different types of logit models, including the conditional logit (e.g., Bunch et al. 1996), the nested logit (Bliemer et al. 2009) and the mixed logit model (Bliemer and Rose 2010; Sándor and Wedel 2002), under various assumptions about the unknown parameters of the utility function, and also to study the implications of more complex models from a design perspective. The second was how to collect more information about preference in each choice set faced by respondents. The third was to ask whether particular designs used in DCEs influenced the choice behavior observed (see for example Bliemer and Rose 2011; Louviere et al. 2008).

Although orthogonal designs have been the traditional mainstay for DCEs and will continue to be so for many years to come, a number of researchers have begun to query the appropriateness of orthogonal designs for use in DCEs. Generally, the argument against the use of orthogonal designs is that the property of orthogonality may run counter to many of the desirable properties of the econometric models typically used to analyze DCE data (i.e., logit and probit models). Instead of merely looking at the correlation between the attribute levels, as with orthogonal designs, the latest theories related to the construction of experimental designs for DCEs seek to find designs that are statistically as efficient as possible in terms of predicted standard errors of the parameter estimates. Essentially, these designs try to maximize the information available from each choice situation and minimize the standard errors of the estimated parameters. Unlike linear models, the asymptotic variance-covariance (AVC) matrix of discrete choice models is derived by taking the negative inverse of the expected second derivatives of the log-likelihood function of the model (see for example McFadden 1974). Given that the log-likelihood function is itself a function of the choice probabilities which are in turn a function of the parameter estimates, it is necessary for the analyst to provide priors to determine the statistical efficiency of a design before the design used in the field.

Different researchers have made use of different prior parameters over the years, with the two main types being the use of locally optimal priors (fixed parameters; e.g., Huber and Zwerina 1996) or Bayesian priors (distributions of potential prior parameter estimates; e.g., Sándor and Wedel 2001). Within the literature dealing with locally optimal priors, there also exist several different research streams, including those that assume non-zero parameter values (e.g., Carlsson and Martinsson 2002) and those that assume that the parameters values are all zero (e.g., Street and Burgess 2007). It is interesting to note that for this latter design class, the choice probabilities will be $1/J$. In this case, the logit model will approximate a linear model and an orthogonal design will be optimal. The inverse case is also true; an orthogonal design is akin to assuming that the parameters will be zero.

Marley et al. (2008) suggested that previous work in Best-worst Scaling (e.g., Marley and Louviere 2005) could be extended to DCEs as a way to collect extra choice information about the choice options in each choice set. Specifically, they proposed using the order information from best and worst (most preferred and least preferred) choices in each choice set to expand the available data to additional sets of implied choices. Expanding (or exploding) the choices is based on Luce and Suppes (1965), and was applied to Random Utility Theory (RUT)-based choice models by Beggs et al. (1981) and Chapman and Staelin (1982) and others. Louviere et al. (2008) show how to use the extra order information to estimate models for single individuals. More recently, Louviere (2014) discusses the underlying ideas in much more detail, explicitly linking them to earlier work in DCEs, and Frischkneicht et al. (2014) discuss new ways to estimate models for single individuals that ensure accurate parameter recovery and convergence. Thus, we now have theory and methods for modeling single individuals using DCEs.

About five years ago, researchers in the Centre for the Study of Choice (CenSoC) in Australia noticed what seemed to be an unusually high number of experimental participants who exhibited deterministic choices, typically always choosing one level of one particular attribute (e.g., "always choose lowest price") in Street and Burgess (2007) designs. Further investigation led to the discovery that this was a widespread phenomenon. This led them to design 64 different DCEs to rigorously test differences in implied preferences between types of designs for DCEs, numbers of attributes, numbers of choice sets, and numbers of attribute levels. Results are emerging as we write this, but we can say that there is evidence for relatively high proportions of deterministic choices (30–40 %) in Street and Burgess (2007) and Statistical Analysis System (SAS) designs (Kuhfeld 2010), but virtually no evidence of this in the Balanced Incomplete Block Design (BIBD) approach proposed by Louviere et al. (2008) or in randomly constructed sets of choice sets. There also are systematic differences in attribute effects for five and eight attributes, with these effects frequently associated with quantitative attributes such as payment vehicles. Unfortunately, there is a growing trend for researchers not to test the results of DCE models against RP data, which is why this phenomenon may have gone undetected for some time. A clear and compelling research agenda emerging from these results is that not only models but types of designs need to be tested against RP data so that we can begin to understand which design strategies will produce the most accurate and reliable results for which purposes and contexts.

Closely related to the above comments is the fact that (as noted earlier) there seems to be far too much reliance on generic designs (Louviere et al. 2000, Chap. 4) in empirical research with DCEs. Indeed, in many cases, the appropriate design should be an alternative-specific design, but researchers seem to copy one another, and many researchers, particularly in applied economics (but also in marketing) routinely use generic designs. Generic designs are appropriate when one wants to generalize one's results to a generic class of options. However, researchers often focus on a specific class of problems, such as visits to particular competing recreational sites and/or particular competing treatments for a special health condition. Thus, it is fair to say that many researchers should be designing and implementing alternative-specific DCEs instead of generic DCEs.

Moreover, it is much harder to validate generic DCEs against RP data because RP data typically are associated with choices of particular competing options, such as products on store shelves in marketing applications, or transport modes in transport. Alternative-specific problems necessarily are narrower than generic problems, with the focus being on a particular set of choice options of interest. This set can include "any other options" and/or an outside good, but one rarely sees these included in DCEs. Alternative-specific designs have the decided advantage that they can simulate the features of real markets to any desired degree of accuracy required. That is, one can use such designs (literally) to create a variety of choice contexts such as store shelves or competing recreational opportunities.

Choice models for alternative-specific problems, not surprisingly, may or may not require alternative-specific utility functions, whereby J-1 of the competing options has its own utility function with potentially different parameters. The Jth option must be set equal to some constant for identification purposes. Louviere and Woodworth (1983) proposed that the latter option be the choice of none of the competing options, or another constant option like a status quo good. Such models can be specified as latent class models or random coefficient models, but one rarely sees this in practice (however, see Swait 1994).

From the standpoint of BT problems, one can clearly see the strong attraction of generic DCEs. Moreover, applied economics generally has tended to favor models without ASCs where possible, which contrasts strongly with marketing applications in which ASCs often are brand labels. Thus, a key question from a BT perspective is whether a donor site is something that can be characterized as a bundle of generic attributes or whether it is unique in ways that influence BT.[17] Clearly, this is an issue that also influences other approaches to BT, like meta-analysis. The DCE approach illustrates the specific technical nature of assumptions made in using generic designs and how it can bias transferred valuation estimates.

11.5 Issues in Combining Data from Multiple Sources

It is possible to combine data from multiple sources to improve the performance of individual models and extend the ability of a particular dataset to make predictions to additional situations. There has been a long standing interest in combining SP and RP data (Adamowicz et al. 1994; Ben-Akiva and Morikawa 1990; Cameron 1992). Moving from RP data to SP data was seen as a way to overcome the limitations that RP data have of dealing only with a narrow range of attribute levels and potentially having very highly correlated attributes, but it was immediately obvious that there was useful information in both types of data, so it was natural to try to combine them. Originally, RP data were seen as the "gold" standard representing the ideal measure of the behavior one wanted to predict, but this perspective has become more nuanced over time.

RP data are often collected by government agencies, non-profit organizations and commercial firms. Examples include trips taken for recreation purposes and purchases of different types of goods. A key issue to note is that most RP data is collected in surveys. RP data are subject to the general reporting error issues involved in collecting information in surveys, plus some special considerations like memory recall effects. Another source of potential divergence in behavior predicted by RP and SP data that one must be aware of arises from the nature of samples taken relative to the population of interest, and one frequently observes quite different selection bias between surveys collecting the two types of data.[18]

What is by now well-known is that there may be considerable differences in scale between RP and SP and that there are ways for taking this into account (Swait and Louviere 1993). The RP and SP choice environments may be characterized by very noisy levels, even though all or most of the parameter estimates are proportional between the two types of data (Louviere et al. 2000, Chap. 13; Swait et al. 1994).[19]

Calibration of SP estimates to predict RP choices seems clearly merited if two conditions hold. The first is that the information available in the RP environment

[17]Burgess et al. (2012) provide an initial investigation of this question, asking whether different types of landscape configurations can be adequately represented as a bundle of attributes.

[18]One way to avoid this issue is to collect both types of data in the same survey, but this may exacerbate recall issues, particularly if respondents are asked about behavior in the more distant past.

[19]Divergence from proportionality, when it occurs, tends to be concentrated in the ASC and cost parameters.

and the noise that characterizes it represent the context in which one wishes to predict choice behavior. This may not always be the case, particularly for many environmental decisions or new products in marketing, where one wants to predict the behavior of informed consumers.[20] The second condition occurs if the incentive structure of the SP question is inconsistent with truthful preference revelation, which is most likely to be the case for private and quasi-public goods.[21] Although RP data for pure public goods in the desirable incentive context of a coercive payment mechanism do not generally exist, RP data sometimes exist with voluntary payment mechanisms. This presents a difficult situation in which the incentive structure of the RP context should lead to well-known free-riding behavior. For SP data, Carson and Groves (2007) show that it is optimal to over-pledge in the survey to encourage the undertaking of the fundraising effort and then to free-ride on the actual effort. This discussion suggests that while there can be considerable benefit to combining RP and SP data, such data-pooling should be done with careful consideration of the processes generating both types of data.

It is also possible to combine SP data sets. While there is little experience with doing this, the principles are largely the same as combining RP and SP datasets. There need to be two or more common pairs of choice alternatives across the two datasets. Differences in scale between the two datasets will have to be controlled for in a manner similar to RP and SP data. The ability to combine multiple SP data sets opens the possibility of systematically collecting and merging SP datasets with an eye toward being able to value an increasing array of environmental goods in a consistent manner.[22]

11.6 Conclusions

This chapter represents an initial foray into many questions related to how different modeling strategies influence BT error rates. We noted that particular attention should be paid to the role of how observable differences in donor and target sites are handled in the transfer exercise. This includes determination of relevant population subgroups that occur across the two sites. In turn, this suggests considerable potential could be realized by reweighting the data at the donor site to match the

[20]Participants in DCEs that collect SP data often are given considerably more information than they are likely to have had in an RP context, where purchase decisions are infrequent or made only once.

[21]Carson and Groves (2007) provide a neo-classical framework for examining the incentive structure of various SP elicitation formats in different contexts. For new products, incentives in a SP context tend to encourage over-estimation of the propensity to purchase, whereas for existing products incentives may exist to understate the propensity to purchase.

[22]For instance, a government agency could adopt a plan to undertake a new SP study valuing various environmental goods every year where a small number of choice alternatives were common across years.

target site on key characteristics and then re-estimating the original model with reweighted data.[23]

A general difficulty with most empirical BT tests is that they are designed to eliminate many messy details involved in transfers actually performed for policy purposes. From a DCE perspective, a key difficulty is that attributes and levels may not match well between donor and target sites. It may also be that some demographic or attitudinal information collected in the survey is unavailable at the target or that a considerable length of time has passed between the original study and the BT exercise. No doubt some or all of these factors can impact the quality of the BT. A key research question is whether quality suffers differentially with different modeling strategies, which in turn points to the need to develop a more comprehensive theoretical and econometric framework for transferring the information contained in DCEs to help inform policy decisions. In the latter case the use of DCEs for direct BT is much less developed than the meta-analysis approach, but clearly has the potential to provide a coherent framework for BT. There is much conceptual work to be done, and empirical comparisons across donor and target sites using a range of modelling approaches and outside data are clearly necessary.

References

Adamowicz, W., Louviere, J. J., & Williams, M. (1994). Combining revealed and stated preference methods for valuing environmental amenities. *Journal of Environmental Economics and Management, 26*, 271–292.

Adamowicz, W., & Swait, J. (2012). Are food choices really habitual? Integrating habits, variety seeking and compensatory choice in a utility maximizing framework. *American Journal of Agricultural Economics, 95*, 17–41.

Anderson, N. H. (1970). Functional measurement and psychophysical judgment. *Psychological Review, 77*, 153–170.

Arrow, K., Solow, R., Portney, P., Learner, E., Radner, R., & Schuman, H. (1993). Report of the NOAA panel on contingent valuation. *Federal Register, 58*, 4601–4614.

Baskaran, R., Cullen, R., & Colombo, S. (2010). Testing different types of benefit transfer in valuation of ecosystem services: New Zealand winegrowing case studies. *Ecological Economics, 69*, 1010–1022.

Beggs, S., Cardell, S., & Hausman, J. A. (1981). Assessing the potential demand for electric cars. *Journal of Econometrics, 17*, 1–19.

Ben-Akiva, M., & Lerman, S. (1985). *Discrete choice analysis: Theory and application to predict travel demand*. Cambridge, MA: MIT Press.

[23]Demographic variables observable at both sites are the source of information for reweighting information. Other interesting alternatives are possible; for example, if it were known that information availability and decision makers (Hensher and Rose 2009; Swait et al. 2012) differed across the two sites and/or one knew the mix of decision rules (Adamowicz and Swait 2012) used by individuals, it would be possible to reweight on this source of heterogeneity across people. More generally, it might be possible to estimate individual level models at the donor site and then use some type of propensity score matching procedure to infer individual level utility functions at the transfer site.

Ben-Akiva, M., & Morikawa, T. (1990). Estimation of switching models from revealed preferences and stated intentions. *Transportation Research Part A, 24*, 485–495.

Bishop, S. E., Feinberg, S., & Holland, P. (1975). *Discrete multivariate analysis*. Cambridge, MA: MIT Press.

Bliemer, M. C. J., & Rose, J. M. (2010). Construction of experimental designs for mixed logit models allowing for correlation across choice observations. *Transportation Research Part B, 46*, 720–734.

Bliemer, M. C. J., & Rose, J. M. (2011). Experimental design influences on stated choice outputs: An empirical study in air travel choice. *Transportation Research Part A, 45*, 63–79.

Bliemer, M. C. J., Rose, J. M., & Hensher, D. A. (2009). Constructing efficient stated choice experiments allowing for differences in error variances across subsets of alternatives. *Transportation Research Part B, 43*, 19–35.

Box, G. G., Hunter, W. G., & Hunter, J. S. (2005). *Statistics for experimenters: Design, innovation, and discovery* (2nd ed.). New York: Wiley.

Boyle, K. J., Kuminoff, N. V., Parmeter, C., & Pope, J. C. (2010). The benefit-transfer challenges. *Annual Review of Resource Economics, 2*, 161–182.

Brookshire, D. S., & Neill, H. (1992). Benefit transfers: Conceptual and empirical issues. *Water Resources Research, 28*, 651–655.

Bunch, D. S. (1991). Estimability in the multinomial probit model. *Transportation Research, 25B*(1), 1–12.

Bunch, D. S., Louviere, J. J. & Anderson, D. A. (1996). A comparison of experimental design strategies for Multinomial Logit Models: The case of generic attributes. Working paper, Graduate School of Management, University of California at Davis.

Burgess, D., Finney, G. & Matthews, D. (2012, September). Landscape valuation: Choice experiments or contingent valuation. Paper presented at the 14th Annual International BIOECON Conference on Resource Economics, Biodiversity Conservation and Development, Cambridge, UK.

Cameron, T. A. (1988). A new paradigm for valuing non-market goods using referendum data: Maximum likelihood estimation by censored logistic regression. *Journal of Environmental Economics and Management, 15*, 355–379.

Cameron, T. A. (1992). Combining contingent valuation and travel cost data for the valuation of nonmarket goods. *Land Economics, 68*, 302–317.

Cameron, T. A., & James, M. D. (1987). Efficient estimation methods for closed-ended contingent valuation surveys. *Review of Economics and Statistics, 69*, 269–276.

Carlsson, F., & Martinsson, P. (2002). Design techniques for stated preference methods in health economics. *Health Economics, 12*, 281–294.

Carson, R. T. (2012). *Contingent valuation: A comprehensive bibliography and history*. Northampton, MA: Edward Elgar.

Carson, R. T., & Groves, T. (2007). Incentive and information properties of preference questions. *Environmental and Resource Economics, 37*, 181–210.

Carson, R. T., Hanemann, W. M., & Steinberg, D. (1990). A discrete choice contingent valuation estimate of the value of Kenai King salmon. *Journal of Behavioral Economics, 19*, 53–68.

Carson, R. T., & Louviere, J. J. (2011). The state of the art of environmental valuation with discrete choice experiments. *Environmental and Resource Economics, 49*, 539–559.

Carson, R. T., Louviere, J. J., & Wasi, N. (2009). A cautionary note on designing discrete choice experiments: A comment on Lusk and Norwood's "effect of experimental design on choice-based conjoint valuation estimates". *American Journal of Agricultural Economics, 91*, 1056–1063.

Carson, R. T. & Steinberg, D. (1990). Experimental design for discrete choice voter preference surveys. In *1989 Proceedings of the Survey Methodology Section of the American Statistical Association*. Washington, DC: ASA.

Carson, R. T., Mitchell, R. C., Hanemann, W. M., Kopp, R. J., Presser, S. & Ruud, P. A. (2003). Contingent valuation and lost passive use: Damages from the Exxon Valdez oil spill. *Environmental and Resource Economics, 25*, 257–286.

Chapman, R. G., & Staelin, R. (1982). Exploiting rank ordered choice set data within the stochastic utility model. *Journal of Marketing Research, 19*, 288–301.

Colombo, S., Calatrava-Requena, J., & Hanley, N. (2007). Testing choice experiment for benefit transfer with preference heterogeneity. *American Journal of Agricultural Economics, 89*, 135–151.

Frischknecht, B., Eckert, C., Louviere, J. J. & Geweke, J. (2014). Incorporating prior information to avoid complete separation problems in estimating discrete choice models. In S. Hess & A. Daly (Eds.), *Handbook of choice modelling*. London: Elsevier.

Haab, T. C., & McConnell, K. E. (2002). *Valuing environmental and natural resources: The econometrics of nonmarket valuation*. Northampton, MA: Edward Elgar.

Hensher, D. A., & Rose, J. M. (2009). Simplifying choice through attribute preservation or non-attendance: Implications for willingness to pay. *Transportation Research Part E, 45*, 583–590.

Hensher, D. A., Rose, J. M., & Greene, W. H. (2005). *Applied choice analysis: A primer*. New York: Cambridge University Press.

Hoyos, D. (2010). The state of the art of environmental valuation with discrete choice experiments. *Ecological Economics, 69*, 1595–1603.

Huber, J., & Zwerina, K. (1996). The importance of utility balance and efficient choice designs. *Journal of Marketing Research, 33*, 307–317.

Jiang, Y., Swallow, S. K., & McGonagle, M. P. (2005). Context-sensitive benefit transfer using stated choice models: Specification and contingent validity for policy analysis. *Environmental and Resource Economics, 31*, 477–499.

Johnston, R. J., & Rosenberger, R. S. (2010). Methods, trends and controversies in contemporary benefit transfer. *Journal of Economic Surveys, 24*, 479–510.

Kaul, S., Boyle, K. J., Pope, K. J., Parmeter, C. F. & Kuminoff, N. V. (2011, June). *Robust meta-analysis using median-quantile and nonparametric regression procedures: Investigating the validity of benefit transfers*. Paper presented at the Annual Summer Meeting of the Association of Environmental and Resource Economists, Seattle.

Kuhfeld, W. F. (2010). *Marketing research methods in SAS experimental design, choice, conjoint, and graphical techniques*. Cary, NC: SAS Institute.

Lancaster, K. (1966). A new approach to consumer theory. *Journal of Political Economy, 74*, 132–157.

Lee, J. A., Soutar, G., & Louviere, J. J. (2007). Measuring values using best-worst scaling: The LOV example. *Psychology and Marketing, 24*, 1043–1058.

Loomis, J. B. (1992). The evolution of a more rigorous approach to benefit transfer: Benefit function transfer. *Water Resources Research, 38*, 701–705.

Louviere, J. J. (2014). Modeling single individuals: The journey from Psych lab to the app store. In S. Hess & A. Daly (Eds.), *The state-of the art and the state-of the-practice*. London: Elsevier.

Louviere, J. J., & Hensher, D. A. (1983). Using discrete choice models with experimental design data to forecast consumer demand for a unique cultural event. *Journal of Consumer Research, 10*, 348–361.

Louviere, J. J., Hensher, D. A., & Swait, J. (2000). *Stated choice analysis: Methods and analysis*. New York: Cambridge University Press.

Louviere, J. J., Islam, T., Wasi, N., Street, D., & Burgess, L. (2008). Designing discrete choice experiments: Do optimal designs come at a price? *Journal of Consumer Research, 10*, 348–361.

Louviere, J. J., & Woodworth, G. (1983). Design and analysis of simulated consumer choice or allocation experiments: An approach based on aggregate data. *Journal of Marketing Research, 20*, 350–367.

Luce, R. D., & Suppes, P. (1965). Preference, utility and subjective probability. In R. D. Luce, R. Bush, & E. Galanter (Eds.), *Handbook of mathematical psychology* (pp. 249–410). New York: Wiley.

Luce, R. D., & Tukey, J. W. (1964). Simultaneous conjoint measurement: A new type of fundamental measurement. *Journal of Mathematical Psychology, 1*, 1–27.

Marley, A. A. J., Flynn, T. N., & Louviere, J. J. (2008). Probabilistic models of set-dependent and attribute-level best–worst choice. *Journal of Mathematical Psychology, 52*, 281–296.

Marley, A. A. J., & Louviere, J. J. (2005). Some probabilistic models of best, worst, and best–worst choices. *Journal of Mathematical Psychology, 49*, 464–480.

McFadden, D. (1974). Conditional logit analysis of qualitative choice behavior. In P. Zarembka (Ed.), *In frontiers in econometrics* (pp. 105–142). New York: Academic Press.

McFadden, D., & Train, K. (2000). Mixed MNL models for discrete response. *Journal of Applied Econometrics, 15*, 447–490.

Morrison, M., Bennett, J., Blamey, R., & Louviere, J. J. (2002). Choice modelling and tests of benefit transfer. *American Journal of Agricultural Economics, 84*, 161–170.

Navrud, S., & Ready, R. C. (2007). *Environmental value transfer: Issues and methods*. Dordrecht: Springer.

Rolfe, J., & Bennett, J. W. (Eds.). (2006). *Choice modelling and the transfer of environmental values*. Cheltenham, UK: Edward Elgar.

Rose, J. M., & Bliemer, M. C. J. (2009). Constructing efficient stated choice experimental designs. *Transport Reviews, 29*, 587–617.

Rosenberger, R. S., & Loomis, J. B. (2003). Benefit-transfer. In P. A. Champ, K. J. Boyle, & T. C. Brown (Eds.), *A primer on nonmarket valuation* (pp. 445–482). Northhampton, MA: Edward Elgar.

Rungie, C. M., Coote, L. V., & Louviere, J. J. (2011). Structural choice modelling: Theory and applications to combining choice experiments. *Journal of Choice Modelling, 4*, 1–29.

Sándor, Z., & Wedel, M. (2001). Designing conjoint choice experiments using managers' prior beliefs. *Journal of Marketing Research, 38*, 430–444.

Sándor, Z., & Wedel, M. (2002). Profile construction in experimental choice designs for mixed logit models. *Marketing Science, 21*, 455–475.

Street, D., & Burgess, L. (2007). *The construction of optimal stated choice experiments: Theory and methods*. New York: Wiley.

Swait, J. (1994). A structural equation model of latent segmentation and product choice for cross-sectional, revealed preference choice data. *Journal of Retailing and Consumer Services, 1*, 77–89.

Swait, J. (2006). Advanced choice models. In B. Kanninen (Ed.), *Valuing environmental amenities using stated choice studies: A common sense approach to theory and practice* (pp. 229–294). Dordrecht: Springer.

Swait, J., Adamowicz, V., & van Buren, M. (2004). Choice and temporal welfare impacts: Incorporating history into discrete choice models. *Journal of Environmental Economics and Management, 47*, 94–116.

Swait, J., & Louviere, J. J. (1993). The role of the scale parameter in the estimation and comparison of multinomial logit models. *Journal of Marketing Research, 30*, 305–314.

Swait, J., Louviere, J. J., & Williams, M. (1994). A sequential approach to exploiting the combined strengths of SP and RP data: Application to freight shipper choice. *Transportation, 21*, 135–152.

Swait, J., Popa, M. & Wang, L. (2012, July). A cost-benefit model of consumer information management and choice. Working Paper, Centre for the Study of Choice, University of Technology, Sydney.

Train, K. E. (2009). *Discrete choice models with simulation* (2nd ed.). New York: Cambridge Economic Press.

Chapter 12
Benefit Transfer for Ecosystem Service Valuation: An Introduction to Theory and Methods

Robert J. Johnston and Lisa A. Wainger

Abstract This chapter introduces the concepts and methods of benefit transfer applied to ecosystem service valuation. It integrates guidance provided in the ecosystem service valuation and benefit transfer literatures to provide introductory insights and guidelines. Building on the general benefit transfer introduction in prior chapters, the chapter provides a basic understanding of how these methods may be adapted to address the unique challenges of ecosystem service valuation. An illustrative application to fish habitat restoration illustrates some of the empirical methods that may be applied. The chapter also discusses some of the more common misuses of benefit transfer in this area and provides guidelines to promote validity and accuracy.

Keywords Ecosystem service · Unit value transfer · Benefit function transfer · Valuation · Ecological system

12.1 Introduction

Benefit transfer is often used to quantify values associated with ecosystem goods and services (henceforth, "ecosystem services"). As noted in prior chapters, benefit transfer is defined as the use of research results from preexisting primary studies at one or more sites (often called study sites) to predict welfare estimates, such as willingness to pay (WTP), for other, typically unstudied sites (often called policy sites). It is most often used when time, funding, data availability or other constraints

R.J. Johnston (✉)
Department of Economics and George Perkins Marsh Institute,
Clark University, Worcester, MA, USA
e-mail: rjohnston@clarku.edu

L.A. Wainger
Center for Environmental Science, University of Maryland,
Solomons, MD, USA
e-mail: wainger@umces.edu

R.J. Johnston et al. (eds.), *Benefit Transfer of Environmental and Resource Values*,
The Economics of Non-Market Goods and Resources 14,
DOI 10.1007/978-94-017-9930-0_12

preclude high-quality primary research, so that preexisting estimates must be used instead. The increasing focus among government agencies and others on the quantification of ecosystem service values (e.g., President's Council of Advisors on Science and Technology 2011), combined with a lack of time and resources required for high-quality primary research, has led to the increasing use of benefit transfer to quantify these values (Bateman et al. 2011b; Wainger and Mazzotta 2011).

The economic theory and methods that support ecosystem service valuation and benefit transfer are the same as those applicable to all market and non-market goods (Champ et al. 2003; Freeman et al. 2014; Hanley and Barbier 2009; Holland et al. 2010; Just et al. 2004). As noted by Hanley and Barbier (2009, p. 206): "ecosystems are assets that produce a flow of beneficial goods and services over time. In this regard, they are no different from any other asset in an economy, and in principle, ecosystem services should be valued in a similar manner." Despite these foundations, there are a variety of characteristics of ecosystem services that can complicate valuation and benefit transfer.

This chapter introduces the concepts and methods of benefit transfer applied to ecosystem service valuation. Building on the methodological introduction in Chap. 2, it provides a basic understanding of how benefit transfer methods may be adapted to address the unique challenges of ecosystem services. An illustrative application to fish habitat restoration illustrates empirical methods that may be applied. The chapter also highlights some of the primary challenges facing ecosystem service benefit transfer, recognizing that inaccurate transfers can lead to estimates that misrepresent public values.[1]

12.2 Ecosystem Service Valuation and Benefit Transfer

Ecosystem services may be defined as the outputs of natural systems that contribute to human welfare (cf., Brown et al. 2007; Daily 1997; Fisher et al. 2008, 2009; Millennium Ecosystem Assessment 2005). In the same way that humans combine capital, labor and technology to produce goods and services valued by people, ecosystems combine natural capital and processes to produce ecosystem services valued by people. These services can benefit people in different ways, either directly or in combination with other inputs such as human labor.

Economic valuation of ecosystem services can serve many different purposes. For example, valuation is a central component of cost-benefit analysis (CBA), which is

[1]Transfer *accuracy* is the extent to which a benefit transfer provides low-error estimates of value (i.e., the estimate is similar to the true underlying value); this is also called transfer reliability. Transfer accuracy is usually tested using convergent validity assessments in circumstances where a primary policy site study has been conducted. Benefit transfer estimates are then compared to the estimate provided by original research at the policy site. Transfer *validity* may be viewed from two different perspectives. Statistical validity implies that a transferred estimate is (or would be) statistically identical to a primary study estimate at the same site. Theoretical validity implies that the transferred estimate is grounded in appropriate economic theory.

used to judge whether a policy or action generates positive net economic benefits. If values for ecosystem services are not quantified within a CBA, these values are often presumed to be zero (Holland et al. 2010). Even in the absence of a full-scale CBA, ecosystem service values can be used to quantify the benefits of individual ecosystem services to different groups. Formal environmental welfare accounting "green GDP" systems also require estimates of these values (Boyd and Banzhaf 2007). Other uses include natural resource damage assessment, the support of advocacy for environmental protection or restoration, and broader sustainability evaluations. The need for precision within each of these uses varies (Kline and Mazzotta 2012; Navrud and Pruckner 1997). In these and other cases, appropriately quantified economic values can help ensure that decisions account for the economic benefits provided by ecosystems.

12.2.1 Sources of Error in Ecosystem Service Benefit Transfer

Given the increasing demand for information on ecosystem service values and the shortage of resources necessary to conduct primary research, benefit transfers are a common component of ecosystem service valuation. At the same time, benefit transfers are subject to error, which can diminish the accuracy of the resulting estimates. Two primary categories of error can occur in benefit transfer. The first is *measurement error* caused by underlying errors in the primary studies used for transfer (Rosenberger and Stanley 2006). That is, any errors in the primary studies used as a basis for benefit transfer will carry over to the transferred estimates. Because ecosystem services can present challenges when applying valuation tools (Bateman et al. 2011b; Hanley and Barbier 2009; Wainger and Mazzotta 2011), these errors can be significant.

The second category of error is *generalization error* related to a lack of similarity between study and policy contexts, and the ability of transfer methods to adjust for these differences. Accurate benefit transfer (or a lack of *generalization error*) requires correspondence or similarity between site characteristics, valuation context and populations at the study site(s) and those at the policy site(s). This includes similarity between the welfare-influencing quantities or qualities of ecosystem services at affected sites, both in primary studies from which values are estimated and in policy sites for which estimates are desired. Generalization errors can be large even if the primary studies used for benefit transfer are high quality, because these errors are due to differences between valuation contexts at the study and policy sites.

As an example of generalization error, consider that anglers targeting freshwater bass may value a water clarity change differently from those seeking trout in the same water bodies.[2] Therefore, transferring an estimated value for a water quality

[2]Unlike trout, bass can thrive in areas of reduced clarity and reductions in clarity can increase fishing value for bass anglers.

improvement from a study of trout anglers to a site used by bass anglers would introduce generalization error. Establishing the similarity of ecosystem service benefits across sites requires a clear understanding of the underlying service to be valued and how a change in that service contributes to human welfare. Valid transfers also require similarity in affected populations (or beneficiaries), as well as the scopes and scales at which values are quantified. Following standard nomenclature, here we define *scope* as the quantity or quality of an ecosystem service under consideration; *scale* is defined as the geographic area over which an analysis is conducted.

Requirements for similarity are most stringent for unit value transfers,[3] which provide little opportunity to adjust benefit estimates for differences between the study and policy context. Some differences between study and policy sites may be accommodated using benefit function or meta-analytic transfers.[4] Meta-analytic transfers adjust for differences between study and policy sites using an estimated benefit function that links coefficients to variables measured at diverse study sites (see Chap. 15). However, this approach is limited by the variables that can be included in the transfer function, which are in turn limited by the range of primary studies available to estimate the function. All of these factors influence generalization errors.

The choice to use benefit transfer instead of a primary study to estimate ecosystem service values will therefore affect the quality of results and applicability for decision making, because primary studies are expected to have lower error (Allen and Loomis 2008). However, benefit transfer can be appropriate when primary studies are not feasible, when monetary values are needed to inform decisions, and when highly precise estimates are not required.

To reduce error, the transfer of an ecosystem service value requires a utility-theoretic[5] understanding of the ways in which each ecosystem service influences human welfare (or provides benefit) across different sites, and the factors likely to cause variation in these benefits (or values) across sites. An understanding of these relationships enables the analyst to evaluate whether values at each site are framed and realized in a sufficiently similar manner to enable benefit transfer, and what adjustments are necessary to reduce generalization errors. These relationships also help to determine whether values are measured properly by underlying primary studies, in order to evaluate the possibility of primary study measurement error.

---

[3]Unit value transfers involve the transfer of a single number or set of numbers from preexisting primary studies.

[4]Benefit function transfers use a benefit function derived from a primary study or set of studies to calculate a welfare estimate calibrated to selected characteristics of a policy site (Loomis 1992; Rosenberger and Loomis 2003). There are two primary requirements for a benefit function transfer. The first requirement is a parameterized function that enables one to calculate the empirical outcome of interest, as a function of variables that include conditions observable at the policy site. Second, information on at least a subset of these variables is required for the policy site, in order to adjust the transferred function from the study site context to the policy site context.

[5]Utility-theoretic implies that the model is grounded in a model of human welfare consistent with economic welfare theory (Just et al. 2004).

Hence, an important component of any benefit transfer of ecosystem service values is a conceptual understanding of relationships between ecosystem processes and human benefits, together with an evaluation of the site and population characteristics that are expected to influence these relationships.

12.3 Conceptual Framework for a Transfer of Ecosystem Service Values

Within economic theory, all values are grounded—implicitly or explicitly—in an underlying utility or benefit function. The relationships in this function determine the factors that must be considered when conducting a benefit transfer or when evaluating transfer validity. To illustrate key concepts surrounding the transfer of ecosystem service values, we begin with an illustrative, direct utility function of the form $U_j(z_1, z_2, y)$. This simple example is used to illustrate the concepts underlying a transfer of ecosystem service values; actual examples are usually more complex.

Within economic theory, a utility function quantifies the total utility (or benefit) realized by an individual or group. Here, this function represents the benefit received by a representative individual at site j, as a function of a set of goods (z_1 and z_2) and an ecosystem service y. Following standard approaches in economics, we assume that the individual maximizes this utility function subject to her available income, m, and a vector of additional exogenous variables $\boldsymbol{x}_j$ that influence demand. These can include prices, characteristics of the site and individual, and other exogenous factors. The result of this maximization is an indirect (or maximized) utility function, in which maximum possible utility or benefit U_j^* is expressed directly as a factor of exogenous factors, ecosystem services, and income:

$$U_j^*\left(z_1\left(\boldsymbol{x}_j, m\right), z_2\left(\boldsymbol{x}_j, m, y\right), y\right) = U_j^*\left(\boldsymbol{x}_j, m, y\right). \qquad (12.1)$$

Within this illustrative framework, we assume that good z_1 is a composite commodity (e.g., all other goods combined), whereas z_2 is a non-market good (e.g., recreational fishing trips) whose value is influenced by ecosystem service y (e.g., fish abundance). For simplicity and illustrative purposes, we assume that only the demand for good z_2 is influenced by ecosystem service y.

This general specification allows ecosystem service y to influence utility both directly and indirectly. A *direct* effect on utility (or benefits) occurs when the individual would be willing to pay for improvement in y, even if there were zero consumption of all other goods and services (Johnston and Russell 2011). For example, some individuals may hold nonuse or existence values for improvements in fish populations that are unrelated to the consumption of any other goods (Johnston et al. 2012). Mathematically, a direct and positive impact on utility implies that

$$\frac{\partial U_j}{\partial y} > 0|_{z_1=0,z_2=0}. \tag{12.2}$$

However, ecosystem services may also have *indirect* impacts on utility, or impacts related to the production and/or consumption of related market goods. This is perhaps the most common way that ecosystem services benefit individuals. Such indirect impacts occur, for example, when ecosystem services are combined with other goods and services to produce valued commodities. This production can be generated by households (i.e., household production) or by firms. For simplicity, this example emphasizes household production rather than production by firms.[6] Indirect impacts also occur when an ecosystem service enhances the quality of a related market or non-market good. For example, fish abundance generally enhances the quality of fishing trips, making trips more highly valued by individuals.

In either case, *in the absence of direct effects on utility*, and assuming a positive (or complementary) relationship between the ecosystem service y and good z_2, indirect impacts imply formally that

$$\frac{\partial U_j}{\partial y} = 0|_{z_1=0,z_2=0}, \tag{12.3}$$

and

$$\frac{\partial U_j}{\partial y} > 0|_{z_1=0,z_2>0}. \tag{12.4}$$

Equation 12.3 simply states that there are no direct effects on utility—the effect of y on utility is zero when consumption of goods z_1 and z_2 are zero. Equation 12.4 states that the effect of y on utility is positive when consumption of good z_2 is also positive.[7]

Willingness to pay (WTP), or other related monetary measures of welfare, reflect money metric transformations of the underlying utility function. Intuitively, an individual's WTP for a change in y reflects that maximum amount of money or other resources that the individual would be voluntarily willing to give up in exchange for the total (direct plus indirect) utility improvement $\partial U_j/\partial y$. For example, returning to Eq. 12.1, WTP for a marginal improvement in y from y_{1_0} to y_{1_1} is defined theoretically by the utility relationship

[6]Bockstael and McConnell (2010) provide a good discussion of household production in the context of non-market valuation. Bateman et al. (2011b) discuss the use of ecosystem services by firms.

[7]Economists will recognize this as being similar to the weak complementary condition often used in revealed preference valuation, although similar relationships can apply within a household production framework (Bockstael and McConnell 2010).

$$U_j^*(\boldsymbol{x}_j, m, y_{1\_0}) = U_j^*(\boldsymbol{x}_j, m - WTP, y_{1\_1}). \tag{12.5}$$

WTP is the amount of money that, if paid by the individual, would exactly offset the utility gain from the improvement in y. Utility is exactly the same with the original level of the ecosystem service (y_{1_0}) as it is with the improved level (y_{1_1}) and income reduced by WTP. This is the most that a rational person would be willing to pay for the improvement. Although more sophisticated examples can be used to extend these concepts, the underlying theory is similar.[8]

Now assume that the analyst requires an estimate of WTP for a change in y at policy Site B ($j = B$), but that no primary study has been conducted at that site. Also assume that primary research at study Site A ($j = A$) has been conducted that estimates a benefit function predicting WTP for changes in y, for a representative individual at that site. This is often accomplished via estimation of a benefit function of the general form

$$\widehat{wtp}_A = g\left(\boldsymbol{x}_A, \Delta y, y_{1\_0}, \hat{\boldsymbol{\beta}}\right) \tag{12.6}$$

where $\widehat{wtp}_A$ is a predicted welfare (or WTP) estimate for the representative individual at Site A, $\boldsymbol{x}_A$ is a vector of exogenous conditions at Site A, $\hat{\boldsymbol{\beta}}$ is a vector of estimated parameters, and Δy is the change in ecosystem service y starting at baseline level y_{1_0}. For example, a simple linear benefit function might be

$$\widehat{wtp}_A = \hat{\beta}_0 + \sum_{k=1}^{K} \hat{\beta}_k x_{kA} + \hat{\beta}_{(K+1)}\Delta y + \hat{\beta}_{(K+2)} y_{1\_0} + \hat{\varepsilon}_A \tag{12.7}$$

where $K + 2$ is the number of nonintercept variables in the model and $\hat{\varepsilon}_A$ is a residual or error assumed to have a normal distribution with zero mean. Such functions are typically estimated using regression models, or mathematically derived from the results of these models. This simple, linear function is shown for illustrative purposes only. There are many assumptions implied by such linear functions, and many benefit functions are more sophisticated. Often, a benefit function such as (12.7) is used, in the original research or policy publication, to estimate aggregate or central tendency measure (e.g., mean or median) of welfare for the representative individual in the study sample, for a particular change in the ecosystem service. We denote such a point estimate $\overline{wtp}_A$.

Chapter 2 illustrates the variety of ways that benefit transfers may be conducted, based on such information, for the broader class of non-market goods and services. These may be simplified into two general approaches for ecosystem service benefit transfers. The first uses the unit value $\overline{wtp}_A$ estimated at Site A to approximate ecosystem service value at Site B, without using information in the benefit function

[8]Welfare measures for firms may be derived based on similar types of models and tradeoffs within production, where the goal is profits or producer surplus (Just et al. 2004).

(12.7) to adjust this estimate for differences between the two sites. This is referred to as *unit value transfer*. The second approach combines the benefit function (12.7) estimated at Site A (the study site) with available information from Site B (the policy site) to generate an updated or adjusted WTP estimate for Site B. An example would be updating the benefit function with information on exogenous conditions at Site B (x_{kB}) or a different baseline for the ecosystem service, y_{1_0}. The former would be accomplished by substituting x_{kB} for x_{kA} in (12.7), and then calculating WTP based on this updated equation. A parallel process would be used to update the baseline for the ecosystem service change. Approaches such as this are referred to as *benefit function transfer*.

Depending on the type of benefit function that has been estimated, benefit function transfers can enable a wide range of possible adjustments that are not possible using unit value transfer. For example, assume that a quadratic benefit function such as the following has been estimated at Site A:

$$\widehat{wtp}_A = \hat{\beta}_0 + \sum_{k=1}^{K} \hat{\beta}_k x_{kA} + \sum_{k=1}^{K} \hat{\mu}_k x_{kA} \Delta y + \hat{\theta}_1 \Delta y + \hat{\theta}_2 (\Delta y)^2 + \hat{\theta}_3 (y_{1\_0})(\Delta y) + \hat{\varepsilon}_A \tag{12.8}$$

In addition to the terms already introduced above, the $\hat{\theta}_h$ for $h = 1, 2, 3$ are parameters to be estimated, associated with the variables Δy, $(\Delta y)^2$ and the interaction $(y_{1\_0})(\Delta y)$. As above, Δy represents the change in the ecosystem service and y_{1_0} represents the baseline level. The variables in x_{kA} might include aspects that affect WTP including descriptors of the surveyed population (e.g., income), qualities of the ecosystem being affected (e.g., size), and elements of study design that have been demonstrated to alter responses (e.g., on- versus off-site surveys). The $\hat{\mu}_k$ are parameters to be estimated associated with the interaction $x_{kA}\Delta y$ (interactions between these exogenous factors and the change in the ecosystem service).[9] The form of the equation also allows for a nonlinear effect of ecosystem service quantity (Δy) on WTP, and for WTP per unit of y to depend on the baseline from which changes occur (y_{1_0}).

A benefit function such as (12.8) would enable mitigation of at least some of the scaling errors that might occur with a fixed unit value transfer, assuming that the quadratic form is appropriate.[10] Benefit function transfer would be conducted by plugging information on Δy, y_{1_0}, and x_{kB} (substituting for x_{kA}) from policy Site B into (12.8). The result would be an updated (typically per individual or per

[9]For example, individuals with higher income may be willing to pay more for a change in an ecosystem service, holding all else constant.

[10]For example, the quadratic functional form can be used to model relationships in which continued increases in an ecosystem service cause WTP to increase, but at a decreasing rate. Relationships such as this have been found for many ecosystem services (e.g., salmon abundance; Loomis and Richardson 2008).

household) WTP value for policy Site B. In some cases, information on relevant variables (e.g., in x_k) may not be available for the policy site. In this case, one generally retains the information on these variables from the study site (Site A).

When adjusting WTP for scale using a benefit function such as (12.8), analysts must be careful to scale only within or close to the range of the data used by the original study. Scaling beyond this range risks large generalization errors, because WTP can change dramatically with large changes in the abundance of a service. For example, Johnston et al. (2005) show that WTP for additional water quality improvements declines as baseline water quality improves, so that WTP for the first unit of quality improvement is greater than that for the second, third and so on. Additional discussion of scaling is provided later in the chapter.

Other benefit function transfer approaches are available; these methods can reduce transfer errors in some cases. These include methods that estimate benefit functions using information from multiple studies (e.g., meta-analysis or structural benefit transfers). The general conceptual basis for these approaches is similar to that presented above, except that the benefit function is estimated by the benefit transfer analyst using information from multiple primary studies. Simple descriptions of this process for meta-analytic benefit function transfer are provided by Johnston and Besedin (2009) and Rosenberger and Loomis (2003). An illustration of meta-analytic benefit transfer for an ecosystem service value is provided later in this chapter. Summaries of these methods are also provided in other chapters in this book, as well as in Boyle et al. (2010), Johnston and Rosenberger (2010), and Rosenberger and Loomis (2003).

12.4 Steps in a Benefit Transfer of Ecosystem Service Values

Chapter 2 details the steps involved in general benefit transfers. The following section supplements this material with additional steps and considerations necessary to address the unique challenges of ecosystem service valuation.

12.4.1 Define the Context for Ecosystem Service Valuation

The first step in any benefit transfer is to define the valuation and policy context under which benefit transfer will potentially occur, and to determine the type of economic information required, such as what values are relevant, what is the purpose of the analysis, and who will use the results? Answers to these questions will help determine the scope/scale of the analysis and the level of accuracy required (Bauer and Johnston 2013; Kline and Mazzotta 2012; Navrud and Pruckner 1997). For example, information that is considered appropriate to inform a nonprofit organization's decision of where to target restoration spending may not be appropriate for use within a formal legal proceeding, due to different standards of proof and precision.

The ecosystem service values chosen for analysis are likely to depend on still other questions. What types of policy changes, commodities and populations will be affected, and what types of information are likely to be available on these effects? Are there regulatory or other constraints on the analysis? For example, regulation or policy can limit the types of services that fall within a federal agency's mission and therefore restrict the ecosystem service values that can be used to justify agency actions. In addition, some types of evaluation may be legally mandated. Chapter 2 discusses this step in greater detail, as applied to general benefit transfers (cf., Desvousges et al. 1998; Rosenberger and Loomis 2003). Discussions with decision-makers and reviews of background documents typically take place to address these questions.

12.4.2 Establish the Need for and Feasibility of Benefit Transfer

Answers to the above questions will help determine whether benefit transfer is an appropriate means of valuation. Primary studies are generally preferred to benefit transfer, since values will reflect the site and policy specifics and thus are expected to have lower error (Allen and Loomis 2008). However, primary studies are not always possible or practical, given time and resource constraints. Factors influencing the choice of primary research versus benefit transfer include: (a) the time and resources available for analysis relative to those required for a primary study; (b) the availability of information necessary for a primary study; (c) the approvals or policy process constraints which restrict the collection of primary data or use of primary analysis;(d) the accuracy and other needs of the policy context and users of the information; (e) the size of policy impacts relative to the cost of a primary study; and (f) the availability of high-quality primary studies suitable for benefit transfer.

Although the *need* for benefit transfer is largely dictated by considerations (a) through (e), the *feasibility* of transfer is determined by (f)—the availability of primary studies of sufficient quality and similarity. As part of assessing the available primary study literature, the analyst should evaluate the consistency among the type, scope and scale of changes at potential study and policy sites, as discussed above. A common challenge is that available studies may represent services at superlative sites (e.g., remote iconic national parks), but value estimates from these sites be inappropriate for capturing values of changes in common ecosystem services at noniconic sites (Hoehn 2006; Rosenberger and Johnston 2009). Similarly, the scale of change for which values are desired may fall outside the range of changes evaluated by existing primary studies. For example, it is difficult to capture the value of an increase in the width of a narrow beach (used only by immediate neighbors) using studies that estimate values of changes in wide beaches (that serve as major recreation destinations). In some decision contexts, these studies might not serve as a defensible basis for benefit transfer. In summary, it is necessary to assess not only the *quality* of the study or studies available for transfer, but also the *relevance* of these studies to contexts (e.g., ecosystem services, sites, populations) for which estimates are required. Additional discussion of study quality and relevance is provided in Sect. 12.4.6 below.

12.4.3 Develop the Conceptual Basis for Valuation

Ecosystem service values can be realized through a number of direct and indirect channels, including the production and consumption of diverse market and non-market goods and services. Hence, an important step in any benefit transfer of these values is development of a conceptual model of relationships between ecosystem processes and human benefits, including the biophysical pathways through which benefits are realized and their connections to different beneficiary groups. Such a model should clarify linkages between the considered policy or actions, changes in ecosystem services, related market and non-market goods and services, and the direct and indirect influence of these changes on utility. This conceptual model is necessary to reduce the potential for large transfer errors, including those due to either omitted or double-counted values. It can also ameliorate problems related to inconsistencies between the ways that an ecosystem service is used and valued across sites.

For example, social benefits from a change in water quality will depend on ways in which the water is used (e.g., for drinking, fishing, boating). Hence, using values derived from a study of a drinking water reservoir to approximate values in an otherwise similar water body used primarily for fishing would likely generate large errors. The conceptual evaluation should also identify the formal theoretical properties of the welfare measures to be sought. For example, does the analysis require ecosystem service values realized by firms or individuals? If the latter, what type of welfare measures are required or appropriate (e.g., willingness to pay versus willingness to accept)? This conceptual model provides the basis for all subsequent steps in the benefit transfer. Note that this step is important for any type of ecosystem service valuation, whether using a primary study or benefit transfer. Bateman et al. (2011b), Boyd and Krupnick (2013) and Wainger and Mazzotta (2011) discuss the development of conceptual models such as these. Empirical illustrations are provided by many works in the ecosystem services literature (e.g., Zhao et al. 2013 for a case study of migratory fish restoration).

12.4.4 Define Ecosystem Services, Goods and Populations

Using the initial conceptual model, the next step is to identify the specific ecosystem service changes to be valued and the formal definitions of these services.

This includes a definition of the relevant population for benefit assessment and the scale of the analysis. Valuation is often targeted at ecosystem service changes that are expected to have the largest effects on human welfare; this is a function of both the number of affected people and the size of the anticipated per capita welfare change. Therefore, defining the human populations (or beneficiary groups) over which ecosystem service values will be estimated and aggregated is a particularly important component of the analysis. Analysts must consider a variety of questions when determining these populations. The first question is whether policy, institutional or legal constraints dictate the population to be considered. This defines the political jurisdiction for the analysis. For example, CBA for state government programs is often limited to state residents, regardless of whether residents in other states value the affected ecosystem services. The second question is the extent of the market, or the human populations for which values of an ecosystem service are expected to be nonzero. Unlike the political jurisdiction, determining the extent of the economic market in a benefit transfer study is an empirical question (Loomis 2000). Identifying the extent of the market within benefit transfer is often difficult, because this information is seldom provided by primary valuation studies, and the extent of the market for ecosystem services can vary between policy and study sites (Desvousges et al. 1998; Loomis and Rosenberger 2006).

Transfers of ecosystem service values face the additional challenge of defining and distinguishing the ecosystem services to be measured, building on the conceptual model. Discussions in the ecosystem services literature often try to emphasize ways in which ecosystems provide everything from basic life support to financial, cultural and social well-being. Associated typologies (e.g., de Groot et al. 2002; Millennium Ecosystem Assessment 2005) present extensive lists of the different services provided by ecosystems, with no formal means to account for overlapping services or causality.[11] For example, as described by Fisher et al. (2008, p. 2051), "in the Millennium Ecosystem Assessment, nutrient cycling is a supporting service, water flow regulation is a regulating service, and recreation is a cultural service. However, we see the first two as providing the same service, usable water, and the third (e.g., recreation on a clean, navigable river) turning the usable water into a human benefit (i.e., the endpoint that has a direct impact on human welfare). If all three Millennium Ecosystem Assessment services were to be individually valued and added to a cost–benefit analysis, we would commit the error of double counting, as the intermediate services are by default included in the value of the final service."

To avoid double counting, consistent estimates of ecosystem service values require careful definition of individual services in terms of their contribution to human welfare. Demonstrating relevance to human welfare requires differentiating intermediate ecosystem functions (e.g., fish habitat) from final ecosystem services (e.g., recreational fish abundance) so that values can be quantified and attributed to the appropriate beneficiaries. This differentiation also ensures that the benefit of

[11]This often occurs because these typologies fail to recognize distinctions between intermediate and final services, or inputs and outputs in production (Johnston and Russell 2011).

each distinct ecosystem condition or process, to each human beneficiary, is counted once and only once (Boyd and Krupnick 2013; Fisher et al. 2008; Johnston and Russell 2011; Johnston et al. 2013b).

For example, unless anglers hold separate and distinct values for fish habitat alone (or for other nonfishing services provided by fish habitat), all of the benefits of fish habitat to anglers should be captured in benefits that these anglers receive from improvements in fish abundance. The summation of values for both intermediate and final services for the same user population (e.g., value for fish habitat plus value for fish) double counts the contribution of the intermediate services to welfare. Careful definition and separation of the ecosystem services to be valued, in the context an underlying conceptual model, can help prevent or at least minimize double counting.

Given the many direct and indirect ways that ecosystem changes can influence people, it can be difficult to fully identify and disentangle all intermediate and final services so that all primary values are measured and no values are double counted. For example, habitat improvements may benefit recreational anglers due both to increases in fish abundance and to associated improvements in wildlife that eat fish (e.g., osprey or bear), to the extent that anglers also value wildlife viewing. In such cases, benefit transfers often focus on a subset of unique and nonoverlapping ecosystem service values expected to be largest for the relevant study populations. For example, when evaluating the benefits of water quality improvements to recreational anglers, a benefit transfer might focus solely on benefits realized though attendant improvements in recreational fishing, overlooking benefits that might be realized through other channels. Although primary studies can be designed specifically to fully distinguish and disentangle benefits from causal and overlapping ecosystem services (e.g., Johnston et al. 2013a), the modeling steps necessary for such analyses are rarely possible within benefit transfer (unless transferred benefit functions have been explicitly designed for such purposes).

12.4.5 Quantify Effects on Ecosystem Services

The next step in benefit transfer is to specify the exact changes in (or effects on) ecosystem services and related goods for which values will be estimated, based on the ecosystem service definitions formalized previously. This requires quantifying both the baselines and marginal changes, and may include changes in quantities, qualities or both. In some cases, the information required to predict ecosystem service changes is primarily biophysical. However, in most cases, effects on ecosystem services will depend on behavioral responses of people to policy or other changes. Failing to account for these behavioral reactions will bias any subsequent quantification of value.

Some valuation contexts will require welfare estimates only for a single policy option and set of changes in services. Others will require evaluation of multiple policies, changes and services. A related aspect of benefit transfer is whether there is uncertainty in the policy outcomes that must be addressed within benefit transfers; for example using expected values or sensitivity analysis. In such cases, benefit transfer may require information on both the possible policy outcomes and the probability of these outcomes. This is particularly relevant for ecosystem service valuation, given the challenges in predicting change in ecological systems.

12.4.6 Gather and Evaluate Valuation Data and Evidence

This step involves a review of available data and evidence on the outcome to be evaluated. It usually includes a literature review to identify prior empirical studies that address the general type of ecosystem services, policy effects and goods under study. The resulting set of studies is then screened for quality, relevance and correspondence to the specific policies and changes to be predicted by the transfer. As described above, correspondence (or consistency between a primary study and the valuation context) should be evaluated in terms of numerous factors, including the general policy context; pathways (e.g., specific goods and services) through which values are realized; relevant demand and household production functions; site, population and other exogenous characteristics; geospatial relationships and economic jurisdictions; and the scale and scope over which values are quantified.

Issues of scope and scale warrant particular attention (see Chap. 2). Ecosystem service analyses have sometimes attempted to value large scopes of services over planetary or other very large geographical scales, using data from primary studies conducted at much smaller scales (e.g., Costanza et al. 1997; Liu et al. 2010). The associated scaling of transferred benefit estimates beyond those assessed by the original primary studies sacrifices both accuracy and validity.[12] The loss of accuracy stems in part from a failure to take into account local differences in ecological or economic conditions. The loss of validity stems from a failure to incorporate established economic principles such as diminishing marginal returns and the fact that economic values are well-defined only for marginal changes from a known baseline.[13] To avoid such problems, the scope and scale of studies considered for benefit transfer should be generally consistent with those of the intended policy applications.

It is also important to evaluate the methodological quality of primary studies used as a basis for benefit transfer. Chapter 2 lists a set of general criteria that can be used to help evaluate quality. Quality evaluation is particularly critical for the transfer of ecosystem service values, given the methodological ambiguity and lack of scientific rigor that pervades much of the ecosystem service valuation literature.

---

[12]See discussions in, e.g., Bateman et al. (2011b), Bockstael et al. (2000), Fisher et al. (2008), Holland et al. (2010), Rolfe et al. (2011) and Toman (1998).

[13]In addition, many of these large-scale studies have used estimates of value such as replacement costs that are not supported by economic theory for most applications.

As noted by Bauer and Johnston (2013, p. xi), "the published [ecosystem services valuation] literature is plagued by a lack of clarity and consistency, particularly with regard to underlying theory and implications for the ways that well-defined ecosystem services are linked (and not linked) to human welfare." Any errors in the primary studies will carry over during transfer as measurement errors.

The analyst must also identify the type of ecosystem service values or other quantities estimated by each study. As noted above, total economic values (or WTP) for any type of outcome may be comprised of multiple components (e.g., market versus non-market values; use versus nonuse values; different types of non-market use values). Different methodologies may be used to evaluate different types of value (Champ et al. 2003; Freeman et al. 2014; Hanley and Barbier 2009; Holland et al. 2010). Analysts must exercise caution when comparing or aggregating values generated by different valuation methods, because these values may not be theoretically equivalent, and may sometimes overlap (Johnston et al. 2002).

12.4.7 Determine Benefit Transfer Method(s)

Based on the information provided by the prior research stages, the analyst must determine the benefit transfer methods that are most appropriate to policy needs and available data. As described in Chap. 2, the choice among different types of benefit transfer is dictated by a number of different factors, including the type of information and number of studies that are available, the type of value that is required, the general correspondence between the study and policy contexts, the level of analyst expertise, the time and resources available to develop transfer methods, and the precision necessary for different types of policy decisions (Bergstrom and DeCivita 1999; Navrud and Pruckner 1997). Among these issues, the most critical include the availability of studies, time and resources. Unit value transfers generally require fewer resources than benefit function transfer, but are usually less accurate. Because meta-analysis can be used to estimate a multidimensional value surface that combines information from many prior studies, it can often lead to improved transfer accuracy compared to unit value transfer or benefit function transfer that relies on information from a single study (Rosenberger and Phipps 2007). Yet meta-analyses are possible only if there is a sufficient number of existing primary studies (of the same or similar resources) to develop the needed metadata.

Unit value transfer is best suited to cases where a closely matching primary study can be found. However, the probability of finding a good fit between a single study site and a policy site is usually low (Boyle and Bergstrom 1992; Spash and Vatn 2006). Further, unadjusted unit value transfers are among the least accurate forms of benefit transfer (Johnston and Rosenberger 2010), and are appropriate only when the study and policy contexts are very similar (Bateman et al. 2011a; Kaul et al. 2013; Rosenberger and Stanley 2006), or for specialized cases such as the value of a statistical life (Brouwer and Bateman 2005; Mrozek and Taylor 2002; Viscusi and

Aldy 2003). For most ecosystem service applications, unit value transfers should be considered only when more flexible forms of transfer are infeasible.

12.4.8 Design and Implement Transfer(s)

Methods to design and implement the transfer will depend almost entirely on the specific transfer method(s) applied. General methods for unit value, benefit function, meta-analysis and structural benefit transfer are described in Chap. 2. Also see Desvousges et al. (1998), Johnston and Rosenberger (2010), Navrud and Ready (2007) and Rosenberger and Loomis (2003).

Regardless of the method applied, most ecosystem service benefit transfers must address variations in the scope and scale of ecosystem service changes between study and policy sites. A good discussion of the role of scope and scale in valuation is provided by Rolfe and Wang (2011). Formally, scope equivalence implies that the size of the change in the ecosystem service, Δy, is identical at Site A and Site B, or that estimated benefit functions provide a means to adjust for any differences. It also assumes that the baseline (or starting) level of y (y_{1_0}, where $\Delta y = y_{1\_1} - y_{1\_0}$) is identical. Scale equivalence implies that benefits are evaluated for changes that occur over geographical areas of similar sizes. For example, individuals will typically value an acre of future land preservation differently if that acre will be located in their home town, compared to a similar acre located somewhere in their home state. The reason is that policy scale differs between the two policies (Johnston and Duke 2009). Moreover, WTP is often inversely related the expected distance between a beneficiary and an ecosystem service change, and larger scales generally imply larger distances (Bateman et al. 2006).

Returning to the issue of scope consistency, Fig. 12.1 illustrates why both y_{1_0} and Δy must be consistent across sites, unless adjustments are possible based on the benefit function estimated at the original site. This figure illustrates possible demand (or marginal value) functions for an ecosystem service, as a function of the quantity of the ecosystem service available. For any given quantity of the good, these curves show a representative individual's marginal benefit (or MB) for the last, or marginal unit consumed.

For illustration, we portray three possible shapes for the marginal benefit function, given by MB_1, MB_2, and MB_3. Curve MB_1 reflects a standard, downward-sloping marginal benefit curve; this is the most common pattern that one might expect for ecosystem service values, reflecting diminishing but still positive marginal WTP as the quantity of the service increases. Curve MB_2 reflects the type of marginal benefits that might be expected for local, small-scale changes in an ecosystem service valued at a global level (such as carbon sequestration) or whose outputs are sold on large-scale global markets (such as agricultural commodities). In such cases, marginal WTP for additional units of the service will decline very slowly, leading to a flat marginal benefit curve (cf., Bateman et al. 2011b;

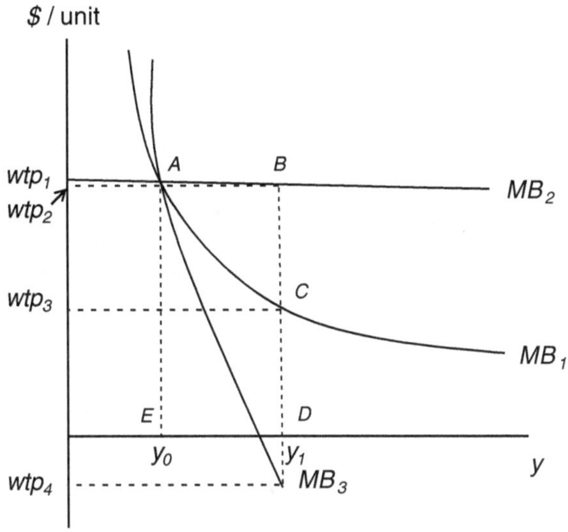

Fig. 12.1 Illustrative marginal benefit functions for an ecosystem service

Polasky et al. 2011). Finally, curve MB_3 reflects quickly diminishing WTP for an ecosystem service, such that WTP becomes negative after a certain threshold quantity. An example would be water quantity in a river. Up to a certain threshold, the marginal benefit of additional water is often positive. Once the river reaches flood stage, however, additional water can impose costs due to flooding.

As an example of how baseline quantity affect WTP (and implications for the type of benefit transfer that might be preferred), assume that identical marginal benefit curves exist at Site A (the study site) and Site B (the policy site)—a best case scenario for benefit transfer. Further assume that the primary study at Site A estimates WTP for a marginal (i.e., very small) change in y, beginning at a current level of y_0. At this point, marginal WTP is identical for all three curves (wtp_1). Now assume, however, that the policy analyst wishes to apply this marginal WTP estimate to another site, at which the current level is y_1. Assuming a marginal benefit given by MB_2, generalization errors will be small, because $wtp_1 \approx wtp_2$. However, if marginal benefit takes on a more common form such as MB_1, large generalization errors will occur from a simple unit value transfer, equal to $wtp_1 - wtp_3$. Even larger errors will occur with a benefit function such as MB_3. In this case actual WTP is negative at the policy baseline level of y_1, so that generalization error is equal to $wtp_1 - wtp_4$. In such cases, a more accurate approach to benefit transfer would be use a benefit function transfer that enables the analyst to account for the difference between marginal WTP at y_0 and y_1.

Similar biases will occur if one seeks to scale up a unit value marginal benefit estimate (e.g., wtp_1) to larger quantities. To illustrate this, we redefine the points in Fig. 12.1 to illustrate a *nonmarginal change* in ecosystem service y. In this new example, y_0 represents the baseline or starting point of the ecosystem service, and y_1 represents the projected final level. Hence, the nonmarginal change in y is equal to

$y_1 - y_0$. Assume that the marginal benefit curve is given by MB_1. The true total benefit of this nonmarginal change is the entire area under MB_1 between y_0 and y_1, or area ACDE. However, were one to simply scale up the marginal value per unit value estimated above (wtp_1) and apply it to the total nonmarginal change, the result would be $wtp_1 \times (y_1 - y_0)$, or approximately area ABDE. This estimate would overstate true value. Examples such as this demonstrate that failing to consider the shape of the marginal benefit curve and the current level of a service can lead to a large over- or under-estimate of value. Again, the use of benefit function transfer, including meta-analytic benefit function transfer, can help approximate the true shape of the underlying value function (in this example, MB_1) and hence improve the accuracy of value estimation—that is, to approximate the true benefit area ACDE rather than the incorrect scaled-up area $wtp_1 \times (y_1 - y_0)$.

12.4.9 Aggregate Values

Once per unit (or per individual) values are estimated, they must often be aggregated over relevant populations, areas and time periods. Although aggregation can be straightforward in some cases, it is also an area in which large errors can be introduced. Misspecification of the beneficiaries over which benefits are aggregated can have a much larger effect on the estimated value than errors in the per user values (Bateman et al. 2006). These aggregation challenges can be particularly significant for ecosystem service valuation, given the potential heterogeneity in both the provision and value of ecosystem services. In the simplest possible case in which marginal values are homogeneous (i.e., approximately identical across the population) or in which the benefit transfer provides an accurate estimate of mean value across the population, aggregation across populations can be as simple as multiplying a representative mean value per person by the size of the population. However, when populations are heterogenous, such simple aggregation can lead to transfer errors.

Heterogeneity can be related to either ecological factors (e.g., if ecosystem service delivery is heterogeneous over affected areas or population subgroups) or economic factors (e.g., if preferences are heterogeneous over population or beneficiary subgoups). For example, the value of a wetland's ability to trap sediment may be large when located upstream of a source-water reservoir used by many people, but may be inconsequential in an area remote from human population centers. Before attributing benefits to a given population, researchers must confirm that an ecosystem benefit is being delivered (and the level at which it is being delivered), and invest effort in appropriately estimating the number of beneficiaries. Some recent applications have developed somewhat more advanced benefit transfer approaches by applying GIS and other spatial tools, in an attempt to improve adjustments for differences in ecosystems, populations and policy contexts when aggregating values for large scale analyses (Bateman et al. 2011b; Polasky et al. 2011).

Heterogeneity in values may also be due to purely spatial phenomena such as the distance to an affected service or related threshold effects (e.g., whether an individual is in or out of an affected area). A common example is distance decay in WTP, in which values decline as one moves further from an ecosystem service, *ceteris paribus* (Bateman et al. 2006; Jørgensen et al. 2013; Schaafsma et al. 2012). In such cases, accurate benefit transfers require one to account for the distance of individuals from policy effects or the effect of other geospatial factors (e.g., national borders) that might influence benefits (Bateman et al. 2006; Johnston and Ramachandran 2014; Loomis and Rosenberger 2006; see also Chap. 18). Patterns such as these have been empirically estimated in many cases (Bateman et al. 2006; Hanley et al. 2003; Johnston and Duke 2009; Jørgensen et al. 2013; Loomis 2000; Martin-Ortega et al. 2012; Schaafsma et al. 2012).

Aggregation should also account, where feasible, for behavioral changes caused by changes in ecosystem services. For example, an increase in beach width may lead to an identifiable change in recreational value to existing beach visitors (see, e.g., Chap. 9). However, an increase in width may also cause new individuals to visit the beach, who did not visit previously. Multiplying a fixed per visitor WTP estimate by the original number of beach visitors will overlook values gained by the new beach visitors. Given the reliance of benefit transfers on existing data and models, estimating changes in beneficiaries or affected populations is often difficult, or can require strong assumptions. However, in some cases, existing studies provide information that can be used to approximate these changes, and hence improve transfer accuracy.

Finally, aggregation or comparison of benefits over time requires *discounting*, which accounts for the fact that the real value of any given outcome tends to decline over time when viewed from the present, *ceteris paribus*. For example, a payment of $100 is worth more if received today than it would be if received in 20 years (see Chap. 2). Additional discussion of methods and complications associated with the aggregation of benefits over time is provided by Arrow et al. (2012), Boardman et al. (2006) and Portney and Weyant (1999).

12.4.10 Conduct Sensitivity Analysis and Test Reliability

Sensitivity analysis quantifies the robustness of results to changes in the modeling approach and uncertainty about key parameters or data, including different potentially influential assumptions and model specifications (Boardman et al. 2006; Desvousges et al. 1998; Holland et al. 2010). In general, sensitivity analysis for a benefit transfer of ecosystem service values is similar to sensitivity analysis in other types of economic or ecological modeling. The most common approach is to recalculate ecosystem service values repeatedly under different assumptions, model specifications, or possible future scenarios. A potential difference is that the transfer of a policy's effect on ecosystem services may be subject to greater uncertainty than other types of biophysical policy effects, largely due to the uncertainties involved in predicting complex ecosystem changes. For example, the effect of new technology on pollution emissions by a manufacturing plant can often be quantified with fairly high certainty. In contrast, the effect of wetland restoration on nearby fish abundance is more difficult to predict.

Because ecosystems are complex, many locational and other uncertain factors can influence the ability of an action to generate an ecosystem service benefit. Sensitivity analysis can help characterize the impact of such uncertainty on benefit transfer results. Many factors may be relevant and uncertain; therefore, corresponding sensitivity analyses may need to capture variation associated with multiple dimensions. These include sensitivities related to both biophysical and economic factors.

Where possible, it is also useful to provide information characterizing the potential reliability of benefit transfer results (or their accuracy). Because the true value is unknown, a variety of indirect methods must be used. As described in Chap. 2, convergent validity tests may be used to evaluate the performance of similar types of transfer in cases for which a primary study has been conducted, and hence transfer errors can be calculated. For additional discussion of this topic, see Chap. 14.

12.4.11 Report Results

The final step in a benefit transfer is the reporting of results. Given that the accuracy of benefit transfer depends on the procedures and data that are applied, transparent description of these factors is crucial. Minimum features that should be reported include a description of: (a) steps of the transfer; (b) the conceptual model establishing linkages among ecosystem services, related goods and human values; (c) the affected policy site, populations and goods/services; (d) reasons for assumed correspondence among the site, populations and goods/services within the study and policy contexts; (e) quantities or qualities of goods/services for which values are estimated, including the specific units in which these are measured; (f) data sources used; (g) the specific type of value that is transferred, e.g., WTP, consumer surplus, etc.; (h) methods used to collect and screen data; (i) transfer methods; (j) statistical methods and assumptions; (k) any scaling that is conducted and implied assumptions; (l) final transferred unit and aggregated estimates of value or other outcomes, including assumptions involved in this aggregation; and (m) results of any sensitivity analyses, robustness tests and accuracy evaluations. Additional reporting requirements may apply for particular types of analyses (for example meta-analysis, as described by Stanley et al. 2013).

12.5 Using Decision Support Tools to Transfer Ecosystem Service Values

Some analysts who require benefit estimates have turned to off-the-shelf decision support tools marketed for ecosystem service valuation and transfer. These and other available tools vary widely across a wide range of evaluative factors, including the expertise and data required of users. Most have been subject to little or no systematic review (Bagstad et al. 2013). These ready-made tools are generally grounded in large-scale, complex, data-intensive models built around spatial modeling of ecosystem functions, often presented on GIS platforms.

Although these decision-support tools can be useful for visualizing services and understanding some tradeoffs, they have shortcomings when used for benefit transfer. Among them is a frequent reliance on simple unit value transfers that are unlikely to provide accurate approximations of value. When pre-coded benefit functions are used, these functions are often unable to account for issues discussed above, such as: (a) the different direct and indirect ways that ecosystem service changes influence welfare across different sites; (b) behavioral responses of individuals to changes in ecosystem services; (c) differences in these impacts across different beneficiary groups; (d) variations in values related to differences in scope and scale; and (e) other potential welfare-relevant inconsistencies between sites and populations.

These errors can persist even if these functions allow adjustments for biophysical conditions, or accord to readily available socioeconomic data such as population density, median/mean income and so on. The reason is that the one-size-fits-all approach of most decision support tools precludes the flexibility to use benefit functions chosen specifically for the characteristics of each policy site. Despite these limitations, there some instances in which off-the-shelf decision-support tools might provide approximations of value that are similar to those generated by alternative benefit transfer methods. An example is the social value of carbon sequestration, which is realized based on global consequences irrespective of the location where the carbon is sequestered. In such cases, value estimates provided by decision-support tools might provide useful approximations (presuming that the underlying social cost of carbon estimates is accurate). However, even here, these estimates would not likely account for co-benefits or losses that might arise, due to the methods of carbon sequestration applied in different areas (e.g., the amenity values of forests or farmland, which vary across regions).

12.6 An Illustrative Case Study: Transfer of Ecosystem Service Values Due to Improvements in Fish Habitat

As a simple example, we illustrate a benefit transfer of ecosystem service value resulting from improvements to brook trout habitat in the Merriland, Branch Brook and Little River (MBLR) Watershed in south coastal Maine, USA. We assume that the policy under consideration involves the restoration of riparian vegetation to enhance spawning habitat. We also assume that time or budget constraints prohibit the use of a primary study to estimate the resulting values. For conciseness, we give primary attention to aspects of ecosystem service valuation that are unique to benefit transfer, along with the assumptions that are required.

To assess the feasibility of benefit transfer, we first established the types of benefits and beneficiaries for which estimates are required, and then investigated the availability of existing economic studies to support benefit transfer. The transfer was then developed using a conceptual model that linked the proposed policies or actions to the ecosystem services that would be expected to change and to distinct groups of beneficiaries whose welfare would potentially be affected by those changes. Here, conceptual model development was supported by focus groups and interviews with residents and key informants—along with ecological data housed at the Wells National Estuarine Research Reserve (NERR)—that established linkages between riparian land restoration in the MBLR Watershed, the ecological changes that would be expected, and implications for changes in valued ecosystem services (Holland and Johnston 2014; Johnston et al. 2014; Wilson 2014). The primary ecosystem service changes that were expected to change measurably as a result of riparian land restoration and generate potentially significant value included (a) changes in aesthetic services provided by naturally vegetated riparian land, (b) changes in water quality and the ecological condition of area rivers, (c) changes in the safety of local waters for swimming, and (d) changes in the abundance of recreational fish, primarily brook trout (Holland and Johnston 2014). These and other potential changes could affect a variety of potential user or non-user groups such as homeowners, anglers, swimmers, and non-users who value improvements in stream health and trout populations.[14]

As noted above, a common strategy to make the most of limited resources is to focus the analysis on benefits that have the largest expected magnitude, that are most easily quantified, and/or that reflect tradeoffs viewed as most relevant to policy decisions. For the sake of illustration, here we focus only on a single ecosystem service realized by one beneficiary group—the benefit of increased brook trout abundance to recreational anglers. We choose this pathway of benefits because the affected fishery, trout fishing, is a high-value fishery to its users and it attracts many users in this region. Further, the economic literature includes hundreds of empirical studies of recreational fishing values, providing the data necessary to conduct a benefit transfer that is adjusted to reflect site conditions.

To obtain aggregate estimates of a change in ecosystem service values, a benefit transfer analysis must typically estimate both the change in value per beneficiary (i.e., affected person) and the total number of beneficiaries. In this example, we could estimate the number of beneficiaries by identifying the number of anglers in the region who would likely benefit from improvements in fish abundance. For a simple

[14]Some groups could also be harmed, for example if restored riparian vegetation impeded the water views of riparian homeowners.

analysis we might estimate this population based on the number of unique anglers observed to fish in the affected area during recent seasons (e.g., from creel surveys), and assume that this number would remain unchanged. To be more sophisticated and accurate, we would use methods that account for the possibility that the number of anglers fishing in a given area would likely increase as fish abundance or average catch rates improved (e.g., Anderson 1983; Englin and Lambert 1995).

Here, the number of affected anglers is unknown since we have no usage survey data. However, we are able to partially side-step this challenge because our ecological data (summarized below) enable a forecast of a specific number of fish added to the population. Our economic model, in turn, estimates value per fish caught by anglers. Given these two pieces of information, we are able to estimate an upper bound in additional ecosystem service value by assuming that all new fish added to the population (as a result of riparian restoration) are eventually harvested by recreational anglers.[15] This special case enables an estimate of ecosystem service value without knowing the exact number of anglers affected. Additional assumptions and limitations of this approach are detailed below.

Benefit transfer also requires the change in the ecosystem service to be quantified. Analysis of recreational fishing values, for example, would typically require a forecast of the change in fish abundance, catch rates, or other measure of fishing quantity or quality directly relevant to anglers. Predicted changes to habitat alone (e.g., proportion of riparian buffer restored) would not enable values for anglers to be estimated, because these changes alone would be insufficient to determine improvements in fishing quality.

In the present example, we use results reported by Johnston et al. (2014) and Wilson (2014) to quantify relationships between riparian land cover and brook trout abundance. These results, drawn from recent ecological research in the MBLR Watershed, suggest that each 1 % increase in riparian land tree canopy cover, on average, is associated with a 2.47 % increase in brook trout abundance in neighboring waterways. Over the entire watershed, a 1 % increase in tree canopy would imply an additional 47 acres of reforested riparian land. Given average (status quo) sampled brook trout abundance of 19 fish per 1000 ft.^{2} of river in the Watershed, each additional 47 acres of riparian reforestation is therefore forecast to provide an additional 0.469 brook trout per 1000 ft.^{2} of river (a 2.47 % increase over the baseline abundance of 19 fish). This quantifies the change in the ecosystem service of interest. Given our benefit transfer model below, the eventual change in estimated ecosystem service value will depend on the percentage of these added brook trout harvested by anglers.

We illustrate a function-based benefit transfer of ecosystem service values using the meta-analysis of recreational fishing values described by Johnston et al. (2006).

[15]Biophysical models are often available to quantify changes in fish abundance, but not expected catch rates. Because economic models often forecast values as a function of catch rates rather than underlying fish abundance, an additional step or set of assumptions is required to link changes in fish abundance to changes in catch rates, such as modeling (or assuming) the relationship between catch and abundance.

The meta-regression model (MRM) predicts WTP per fish based on species, region, baseline catch rate, and other relevant factors. The MRM was built using 391 observations from 48 individual studies conducted in the U.S. and Canada between 1977 and 2001. Observations in the metadata represent results of prior primary studies evaluating WTP per fish among recreational anglers, normalized to 2003 dollars.[16] These observations include per fish WTP associated with various freshwater, salt-water, and anadromous species. The majority of these estimates are under $20 per fish, but range from $0.05 to $612.79 with a mean of $16.82. For additional description of the data and MRM see Johnston et al. (2006) and Stapler and Johnston (2009).

The MRM leads to a benefit function of the following general form:

$$\ln(\widehat{wtp}) = \hat{\beta}_0 + \sum_{k=1}^{K} \hat{\beta}_k x_k + \sum_{g=1}^{G} \hat{\mu}_g f_g + \sum_{j=1}^{J} \hat{\theta}_j y_j, \tag{12.9}$$

where $\ln(\widehat{wtp})$ is the natural log of WTP per additional fish caught by recreational anglers, x_k is a set of methodological variables characterizing the valuation methods used by the primary study, f_g is a set of variables characterizing the anglers, species and regions where the fish were caught, and y_j is a set of variables characterizing baseline catch rates (further described in Johnston et al. 2006).

A random effects MRM with robust standard errors was used to generate estimates for the parameters $(\hat{\beta}_0, \hat{\beta}_k, \hat{\mu}_g, \hat{\theta}_j)$ in Eq. 12.9, associated with each model variable. The results are shown in Table 12.1. The analyst then assigns values for each of these model variables, in order to generate transferable WTP estimates. Values for variables characterizing the resource and policy context (f_g and y_j) are usually determined by the characteristics of the natural resource, site and policy for which values are desired. Values for methodological variables (x_k) (e.g., whether the original study used particular types of revealed or stated preference methods) are typically set at mean values from the metadata, because these variables cannot be observed at the policy site (Stapler and Johnston 2009).

The average baseline catch rate for brook trout in Maine is known (*spec_cr* = 1) and equal to 0.85 fish per angler trip (*cr_nonyear* = 0.85), based on the most recent brook trout catch rate data (from 2001) reported by the Maine Department of Inland Fisheries and Wildlife (2009). We assume that this average also applies to the MBLR Watershed. We also know that the region is not in the Great Lakes (*trout_GL* = 0). We do not have data on the average demographic characteristics of anglers in the MBLR Watershed, so we use average values for these characteristics from the metadata. An alternative source for some of these data would be surveys of brook trout anglers in the state of Maine (Edwards 2009). Trout fishing in the affected area takes place primarily from shore (*shore* = 1). This information,

[16]Given that the MRM estimates values in 2003 dollars, additional adjustments are required to transform these currently values into updated dollar equivalents (see Footnote 17).

Table 12.1 Meta-analysis variables, descriptive statistics and random effects regression results: willingness to pay per fish by recreational anglers

Variable	Description	Mean	Variable type in Eq. 12.9	Parameter estimate (Std. Dev.)
log_WTP	Natural log of the marginal value per fish, in constant 2003 dollars	1.8419	$\ln(\widehat{wtp})$	N/A
Intercept	Constant term in regression model	1.0000	$\hat{\beta}_0$	−1.4568 (1.0284)
SP_conjoint	Binary (dummy) variable indicating that the study used conjoint or choice experiment stated preference methodology	0.0435	x_k	−1.1672*** (0.3973)
SP_dichot	Binary (dummy) variable indicating that the study used stated preference methodology with a dichotomous choice elicitation format	0.1739	x_k	−0.9958*** (0.2455)
TC_individual	Binary (dummy) variable indicating that the study used a travel cost model based on the number of trips taken by individual respondents to recreational sites	0.1074	x_k	1.1091* (0.5960)
TC_zonal	Binary (dummy) variable indicating that the study used a zonal travel cost model based on the aggregate number of trips taken to recreational sites by visitors who live within specified distance ranges	0.0409	x_k	2.0480*** (0.6444)
RUM_nest	Binary (dummy) variable indicating that the study used a nested random utility model	0.2353	x_k	1.3324** (0.6377)
RUM_nonnest	Binary (dummy) variable indicating that the study used a non-nested random utility model	0.3043	x_k	1.7892*** (0.6131)
SP_year	If the study used stated preference methodology, this variable represents the year in which the study was conducted, converted to an index by subtracting 1976; otherwise, this variable is set to zero	4.6036	x_k	0.0875*** (0.0259)

(continued)

Table 12.1 (continued)

Variable	Description	Mean	Variable type in Eq. 12.9	Parameter estimate (Std. Dev.)
TC_year	If the study used travel cost methodology, this variable represents the year in which the study was conducted, converted to an index by subtracting 1976; otherwise, this variable is set to zero	0.7315	x_k	−0.0397 (0.0319)
RUM_year	If the study used RUM methodology, this variable represents the year in which the study was conducted, converted to an index by subtracting 1976; otherwise, this variable is set to zero	9.3734	x_k	−0.00291 (0.0195)
SP_mail	Binary (dummy) variable indicating that the study was a stated preference study administered by mail	0.0512	x_k	0.5440 (0.4608)
SP_phone	Binary (dummy) variable indicating that the study was a stated preference study administered by phone	0.1304	x_k	1.0859*** (0.4098)
high_resp_rate	Binary (dummy) variable indicating that the sample response rate was greater than 50 %	0.3581	x_k	−0.6539** (0.2779)
inc_thou	Household income of survey respondents in thousands of dollars. If the study does not list income values, *inc_thou* was imputed from Census data	46.7008	f_g	0.00387 (0.0140)
age42_down	Binary (dummy) variable indicating that the mean age of sample respondents was less than 43. If the mean sample age was greater than or equal to 43, or was not reported, this variable was set equal to zero	0.0972	f_g	0.9206*** (0.2612)
age43_up	Binary (dummy) variable indicating that the mean age of sample respondents was 43 or greater. If the mean sample age was less than 43, or was not reported, this variable was set equal to zero	0.2711	f_g	1.2221*** (0.2369)

(continued)

Table 12.1 (continued)

Variable	Description	Mean	Variable type in Eq. 12.9	Parameter estimate (Std. Dev.)
trips19_down	Binary (dummy) variable indicating that the mean number of fishing trips taken each year by sample respondents was less than 20. If the mean number of trips was not reported, this variable was set equal to zero	0.1100	f_g	0.8392*** (0.2230)
trips20_up	Binary (dummy) variable indicating that the mean number of fishing trips taken each year by sample respondents was 20 or greater. If the mean number of trips was not reported, this variable was set equal to zero	0.3350	f_g	−1.0112** (0.4381)
nonlocal	Binary (dummy) variable indicating that no respondents in the sample were local residents	0.0051	f_g	3.2355*** (0.4666)
big_game_pac	Binary (dummy) variable indicating that the target species was big game in the California or Pacific Northwest regions	0.0077	f_g	2.2530*** (0.4048)
big_game_natl	Binary (dummy) variable indicating that the target species was big game in the North Atlantic or Mid-Atlantic regions	0.0486	f_g	1.5323*** (0.4544)
big_game_satl	Binary (dummy) variable indicating that the target species was big game in the South Atlantic or Gulf of Mexico regions	0.0205	f_g	2.3821*** (0.5356)
small_game_pac	Binary (dummy) variable indicating that the target species was small game in the California or Pacific Northwest regions	0.0281	f_g	1.6227*** (0.3488)
small_game_atl	Binary (dummy) variable indicating that the target species was small game in the North Atlantic, Mid-Atlantic, South Atlantic, or Gulf of Mexico regions	0.1611	f_g	1.4099** (0.7094)

(continued)

Table 12.1 (continued)

Variable	Description	Mean	Variable type in Eq. 12.9	Parameter estimate (Std. Dev.)
flatfish_pac	Binary (dummy) variable indicating that the target species was flatfish in the California or Pacific Northwest regions	0.0179	f_g	1.8909*** (0.4826)
flatfish_atl	Binary (dummy) variable indicating that the target species was flatfish in the North Atlantic, Mid-Atlantic, South Atlantic, or Gulf of Mexico regions	0.0997	f_g	1.3797*** (0.3373)
other_sw	Binary (dummy) variable indicating that the target species was bottom fish or other saltwater species	0.2276	f_g	0.7339* (0.3902)
musky	Binary (dummy) variable indicating that the target species was muskellunge	0.0026	f_g	3.8671*** (0.3507)
pike_walleye	Binary (dummy) variable indicating that the target species was northern pike or walleye	0.0307	f_g	1.0412*** (0.3469)
bass_fw	Binary (dummy) variable indicating that the target species was largemouth bass or smallmouth bass	0.0358	f_g	1.7780*** (0.4301)
trout_GL	Binary (dummy) variable indicating that the target species was trout in the Great Lakes region	0.0128	f_g	1.8723*** (0.2620)
trout_nonGL	Binary (dummy) variable indicating that the target species was trout in states outside the Great Lakes region	0.1253	f_g	0.8632*** (0.3034)
salmon_pacific	Binary (dummy) variable indicating that the target species was salmon on the Pacific coast	0.0844	f_g	2.3570*** (0.4205)
salmon_atl	Binary (dummy) variable indicating that the target species was salmon on the Atlantic coast	0.0051	f_g	5.2689*** (0.4100)

(continued)

Table 12.1 (continued)

Variable	Description	Mean	Variable type in Eq. 12.9	Parameter estimate (Std. Dev.)
salmon_GL	Binary (dummy) variable indicating that the target species was salmon in the Great Lakes	0.0230	f_g	2.2135*** (0.2722)
steelhead_pac	Binary (dummy) variable indicating that the target species was steelhead on the Pacific coast	0.0358	f_g	2.1904*** (0.5635)
steelhead_GL	Binary (dummy) variable indicating that the target species was steelhead in the Great Lakes	0.0051	f_g	2.3393*** (0.2198)
cr_nonyear	For studies that present catch rate on a per-hour, per-day, or per-trip basis, this variable represents the baseline catch rate for the target species, expressed in fish per day or fish per trip; otherwise, this variable is set to zero. See text for calculation details	2.1038[a]	y_j	−0.0814 (0.0681)
cr_year	For studies that present catch rate on a per year basis, this variable represents the baseline catch rate for the target species, expressed in fish per year; otherwise, this variable is set to zero	41.2277[a]	y_j	−0.0521*** (0.0145)
catch_year	Binary (dummy) variable indicating that the study expressed catch rates on a per-year basis	0.0716	y_j	1.2693*** (0.4888)
spec_cr	Binary (dummy) variable indicating that the study presents information on the baseline catch rate	0.8440	y_j	0.6862*** (0.2323)
shore	Binary (dummy) variable indicating that all respondents in the sample fished from shore	0.1458	f_g	−0.1129 (0.1299)

(continued)

Table 12.1 (continued)

Variable	Description	Mean	Variable type in Eq. 12.9	Parameter estimate (Std. Dev.)
$-2\ LnL\ \chi^2\ (df)$				231.8*** (41)
σ_u^2				1.25×10^{-19}
σ_e^2				0.6581
N				391

Results from Johnston et al. (2006)
***Denotes significance at $p < 0.01$, ** denotes significance at $p < 0.05$, * denotes significance at $p < 0.10$
[a]These values represent mean values and standard deviations *only* for those observations in which the variable value was specified (i.e., zero values are suppressed for the purposes of calculating the mean and standard deviation only)

combined with mean values for methodological model variables (e.g., averages of from the metadata), is plugged into Eq. 12.9 as illustrated by Table 12.2.

Once variable values are selected by the analyst, all that is required to forecast WTP per fish for expected catch improvements are simple spreadsheet calculations. Coefficient estimates for each variable, taken from MRM results in Table 12.1, are entered into column A of Table 12.2. Variable levels chosen above are entered into column B. Column C shows the arithmetic product of columns A and B for each model variable. The sum of these products for the illustrated policy example is 1.4012. This value is equivalent to the quantity $[\hat{\beta}_0 + \sum_{K=1}^{K} \hat{\beta}_k x_k + \sum_{g=1}^{G} \hat{\mu}_g f_g + \sum_{j=1}^{J} \hat{\theta}_j y_j]$ in Eq. 12.9, and is given the label D in Table 12.2. This value represents the predicted natural log of WTP for the illustrated ecosystem service change, per fish caught.

The final step uses a standard formula to transform this predicted natural log into the desired WTP estimate,

$$WTP = e^{\left(D+\sigma_\varepsilon^2/2\right)}, \tag{12.10}$$

where e is the exponential operator, D is defined above and σ_ε^2 is the residual variance (0.6581) from the regression model; addition of the term $(\sigma_\varepsilon^2/2)$ corrects for log transformation bias. Applying this formula generates WTP = \$5.64, which represents projected per angler WTP for a one fish increase in catch, tailored to the specific policy context characterized above.[17] This represents an estimate that could be transferred to approximate ecosystem service value for the illustrated policy change, in the absence of original study results.

[17]As dollar values in all source studies were adjusted to June 2003 dollars prior to model estimation, the MRM provides benefit estimates in June 2003 dollars. Results may be adjusted to other base years by using an appropriate consumer price index (CPI).

Table 12.2 Using a meta-analysis benefit function to estimate willingness to pay for a marginal ecosystem service change

Variable	(A) Parameter estimates (from Table 12.1)	(B) Selected variable values	(C) = (A) × (B)
Intercept	−1.4568	1.0000	−1.4568
SP_conjoint	−1.1672	0.0435	−0.0508
SP_dichot	−0.9958	0.1739	−0.1732
TC_individual	1.1091	0.1074	0.1191
TC_zonal	2.0480	0.0409	0.0838
RUM_nest	1.3324	0.2353	0.3135
RUM_nonnest	1.7892	0.3043	0.5445
SP_year	0.0875	4.6036	0.4030
TC_year	−0.0397	0.7315	−0.0290
RUM_year	−0.00291	9.3734	−0.0273
SP_mail	0.5440	0.0512	0.0279
SP_phone	1.0859	0.1304	0.1416
high_resp_rate	−0.6539	0.3581	−0.2342
inc_thou	0.00387	46.7008	0.1808
age42_down	0.9206	0.0972	0.0895
age43_up	1.2221	0.2711	0.3313
trips19_down	0.8392	0.1100	0.0923
trips20_up	−1.0112	0.3350	−0.3388
Nonlocal	3.2355	0.0051	0.0165
big_game_pac	2.2530		
big_game_natl	1.5323		
big_game_satl	2.3821		
small_game_pac	1.6227		
small_game_atl	1.4099		
flatfish_pac	1.8909		
flatfish_atl	1.3797		
other_sw	0.7339		
musky	3.8671		
pike_walleye	1.0412		
bass_fw	1.7780		
trout_GL	1.8723	0.0000	0.0000
trout_nonGL	0.8632	1.0000	0.8632
salmon_pacific	2.3570		
salmon_atl	5.2689		
salmon_GL	2.2135		
steelhead_pac	2.1904		
steelhead_GL	2.3393		
cr_nonyear	−0.0814	0.8500	−0.0692

(continued)

Table 12.2 (continued)

Variable	(A) Parameter estimates (from Table 12.1)	(B) Selected variable values	(C) = (A) × (B)
cr_year	−0.0521		
catch_year	1.2693		
spec_cr	0.6862	1.000	0.6862
shore	−0.1129	1.000	−0.1129
D = Sum of Column (C)			1.4012
$E = \sigma_e^2$ (from Table 12.1)			0.6581
WTP per fish = $e^{(D+E/2)}$			$5.64

The total value of the increase in fish catch resulting from riparian canopy restoration in the MBLR Watershed depends on the proportion of the change in fish population caught by anglers (e.g., on an annual basis). An upper bound for this value is obtained if we assume (a) an increase in annual recreational catch equal to 100 % of the modeled increase in trout abundance, and (b) that this increase in annual catch is distributed evenly across anglers, so that no one angler catches more than one additional fish per year. In this case, each additional 47 acres of riparian canopy restored would lead to an assumed increase in catch of 0.469 fish per year, per 1000 ft.^{2} of river. Given a transferred value of $5.64 per angler/fish, on the margin, this would lead to a value increase of $2.65 per year, per 1000 ft.^{2} of river in the Watershed. Adjusting this value to 2014 dollars using the September 2014 CPI Detailed Report on the Historical Consumer Price Index for All Urban Consumers (CPI-U): U.S. city average, all items (available from http://www.bls.gov/cpi/) leads to an updated value of $3.42 per 1000 ft.^{2} of river. This would be an upper bound estimate of the change in ecosystem service value related to the increase in brook trout abundance, resulting from the restoration of riparian canopy. If only a portion of the forecast increase in fish abundance is caught per year, the estimated ecosystem service value declines proportionally.

As is often the case with benefit transfers of ecosystem service values, these estimates imply strong assumptions. For example, given a lack of information linking expected angler catch rates to fish abundance, the estimated upper bound values described above are based on assumptions regarding this relationship (i.e., that 100 % of the increase is caught). Other assumptions include that: (1) all anglers enjoy a similar change in catch, (2) anglers' behavior or catch of other species will not change as a result of the projected changes in fishing quality for brook trout. The validity of these and other assumptions will influence the accuracy of estimated values. This estimate also captures only the portion of ecosystem service value related to the modeled change in brook trout harvest. Other benefits of riparian reforestation (e.g., effects on water quality or clarity) remain unquantified. In cases such as this, it is important to clarify that only some aspects of value have been captured (in this case, likely only a small proportion).

As highlighted by the above example, the use of benefit transfer can reduce the time and cost required to estimate values, particularly when one has access to a high-quality benefit function (e.g., from a high-quality meta-analysis). However, the availability of such a benefit function does not free the analyst from the challenges and questions of benefit aggregation and bioeconomic modeling required to estimate the underlying change in ecosystem services. The accuracy of the resulting economic value estimates depends on each step in the process linking proposed policy changes to effects on ecosystem services to changes in economic value. Because of these caveats, it is important to be transparent when describing the methods and limitations of any benefit transfer of ecosystem service values.

12.7 Conclusion

This chapter has introduced the concepts and methods of benefit transfer applied to ecosystem service valuation, with the goal of promoting better-informed practice. Benefit transfer is the most common valuation method applied to ecosystem services worldwide. Without these methods, the value of many services would remain unquantified, and would likely be omitted from formal, quantitative evaluations such as CBA. Given the lack of time and resources necessary to conduct primary valuation studies for most ecosystem services, benefit transfer remains a primary tool for valuation, allowing ecosystem service values to be estimated when primary studies are infeasible.

At the same time, enthusiasm for the ecosystem services concept has led to many empirical applications that sacrifice scientific rigor in ways that provide inaccurate information on economic values (Bauer and Johnston 2013). Many of these have involved benefit transfer. Some justify these applications under the guise that some number is better than no number, or that raising awareness of the value of ecosystems is more important than the accuracy or validity of empirical results. However, errors that dramatically inflate values will likely be rejected as irrelevant by decision makers, while simultaneously eroding confidence in the validity of ecosystem service valuation. Errors that deflate values by failing to consider location-specific or unique conditions (e.g., high land values, irreplaceable ecosystems) can lead to decisions that fail to reflect the full value provided by ecosystems. In either case, inaccurate results can promote policies or actions which reduce human welfare.

Benefit transfer is an often necessary tool for analysts seeking to quantify the value of ecosystem services, but is valuable only if it provides information that enables decision-makers to better understand the impacts of their decisions on human welfare and make more informed choices. Information from inaccurate transfers can lead to decisions that fail to meet ecological and social goals because they fail to distinguish the best opportunities to improve welfare. The many challenges of both ecosystem service valuation and benefit transfer, together with the likelihood that the transfer of ecosystem service values will continue to be an important part of agency cost benefit analyses, underscore the need for future work and awareness in this area.

References

Allen, B. P., & Loomis, J. B. (2008). The decision to use benefit transfer or conduct original valuation research for benefit-cost and policy analysis. *Contemporary Economic Policy, 26*, 1–12.

Anderson, L. G. (1983). The demand curve for recreational fishing with an application to stock enhancement activities. *Land Economics, 59*, 279–286.

Arrow, K. J., Cropper, M. L., Gollier, C., Groom, B., Heal, G. M., Newell, R. G., et al. (2012). How should benefits and costs be discounted in an intergenerational context? The Views of an expert panel. RFF Discussion Paper RFF DP 12–53, Washington, DC.

Bagstad, K. J., Semmens, D. J., Waage, S., & Winthrop, R. (2013). A comparative assessment of decision-support tools for ecosystem services quantification and valuation. *Ecosystem Services, 5*, e27–e39.

Bateman, I. J., Brouwer, R., Ferrini, S., Schaafsma, M., Barton, D. N., Dubgaard, A., et al. (2011a). Making benefit transfers work: Deriving and testing principles for value transfers for similar and dissimilar sites using a case study of the non-market benefits of water quality improvements across Europe. *Environmental and Resource Economics, 50*, 365–387.

Bateman, I. J., Day, B. H., Georgiou, S., & Lake, I. (2006). The aggregation of environmental benefit values: Welfare measures, distance decay and total WTP. *Ecological Economics, 60*, 450–460.

Bateman, I. J., Mace, G. M., Fezzi, C., Atkinson, G., & Turner, K. (2011b). Economic analysis for ecosystem service assessments. *Environmental and Resource Economics, 48*, 177–218.

Bauer, D. M., & Johnston, R. J. (2013). The economics of rural and agricultural ecosystem services: Purism versus practicality. *Agricultural and Resource Economics Review, 42*, iii–xv.

Bergstrom, J. C., & De Civita, P. (1999). Status of benefit transfer in the United States and Canada: A review. *Canadian Journal of Agricultural Economics, 47*, 79–87.

Boardman, A. E., Greenberg, D. H., Vining, A. R., & Weimer, D. L. (2006). *Cost-benefit analysis: Concepts and practice* (3rd ed.). Upper Saddle River, NJ: Prentice Hall.

Bockstael, N. E., Freeman, A. M., Kopp, R. J., Portney, P. R., & Smith, V. K. (2000). On measuring economic values for nature. *Environmental Science and Technology, 34*, 1384–1389.

Bockstael, N. E., & McConnell, K. E. (2010). *Environmental and resource valuation with revealed preferences. A theoretical guide to empirical models*. Dordrecht: Springer.

Boyd, J., & Banzhaf, S. (2007). What are ecosystem services? The need for standardized environmental accounting units. *Ecological Economics, 63*, 616–626.

Boyd, J. W., & Krupnick, A. (2013). Using ecological production theory to define and select environmental commodities for nonmarket valuation. *Agricultural and Resource Economics Review, 42*, 1–32.

Boyle, K. J., & Bergstrom, J. C. (1992). Benefit transfer studies: Myths, pragmatism and idealism. *Water Resources Research, 28*, 657–663.

Boyle, K. J., Kuminoff, N. V., Parmeter, C. F., & Pope, J. C. (2010). The benefit transfer challenges. *Annual Review of Resource Economics, 2*, 161–182.

Brouwer, R., & Bateman, I. J. (2005). Benefits transfer of willingness to pay estimates and functions for health-risk reductions: A cross-country study. *Journal of Health Economics, 24*, 591–611.

Brown, T. C., Bergstrom, J. C., & Loomis, J. B. (2007). Defining, valuing and providing ecosystem goods and services. *Natural Resources Journal, 47*, 329–376.

Champ, P. A., Boyle, K. J., & Brown, T. C. (Eds.). (2003). *A primer on nonmarket valuation*. Dordrecht: Kluwer Academic Publishers.

Costanza, R., d'Arge, R., de Groot, R., Farber, S., Grasso, M., Hannon, B., et al. (1997). The value of the world's ecosystem services and natural capital. *Nature, 387*, 253–260.

Daily, G. C. (1997). *Nature's services*. Covelo, CA: Island Press.

de Groot, R. S., Wilson, M. A., & Boumans, R. M. J. (2002). A typology for the classification, description and valuation of ecosystem functions, goods and services. *Ecological Economics, 41*, 393–408.

Desvousges, W. H., Johnson, F. R., & Banzhaf, H. S. (1998). *Environmental policy analysis with limited information: Principles and applications of the transfer method*. Cheltenham, UK: Edward Elgar.

Edwards, M. (2009). *Brook trout angling in Maine: 2009 Survey results*. University of Maine cooperative extension: Franklin County Office.

Englin, J., & Lambert, D. (1995). Measuring angling quality in count data models of recreational fishing. *Environmental and Resource Economics, 6*, 389–399.

Fisher, B., Turner, R. K., & Morling, P. (2009). Defining and classifying ecosystem services for decision making. *Ecological Economics, 68*, 643–653.

Fisher, B., Turner, K., Zylstra, M., Brouwer, R., de Groot, R., Farber, S., et al. (2008). Ecosystem services and economic theory: Integration for policy relevant research. *Ecological Applications, 18*, 2050–2067.

Freeman, A. M. III., Herriges, J. A., & Kling, C. L. (2014). *The measurement of environmental and resource values: Theory and methods*. Washington, DC: Resources for the Future.

Hanley, N., & Barbier, E. B. (2009). *Pricing nature: Cost benefit analysis and environmental policy*. Cheltenham, UK: Edward Elgar.

Hanley, N., Schlapfer, F., & Spurgeon, J. (2003). Aggregating the benefits of environmental improvements: Distance-decay functions for use and nonuse values. *Journal of Environmental Management, 68*, 297–304.

Hoehn, J. P. (2006). Methods to address selection effects in the meta regression and transfer of ecosystem values. *Ecological Economics, 60*, 389–398.

Holland, B., & Johnston, R. J. (2014). Spatially-referenced choice experiments: Tests of individualized geocoding in stated preference questionnaires. American Applied Economics Association (AAEA) Annual Meeting. Minneapolis, MN, July 27–29.

Holland, D. S., Sanchirico, J. N., Johnston, R. J., & Joglekar, D. (2010). *Economic analysis for ecosystem based management: Applications to marine and coastal environments*. Washington, DC: Resources for the Future.

Johnston, R. J., & Besedin, E. Y. (2009). Estimating willingness to pay for aquatic resource improvements using benefits transfer. In H. W. Thurston, M. T. Heberling, & A. Schrecongost (Eds.), *Environmental economics for watershed restoration* (pp. 95–122). New York: CRC Press.

Johnston, R. J., Besedin, E. Y., Helm, E. & Stapler, R. (2013a). *Overlaps and causality in aquatic ecosystem impacts of energy facilities: Evaluating benefits of cooling water intake regulations under section 316b of the clean water act*. Presented at 3rd Annual Association of Environmental and Resource Economists (AERE) Summer Conference. Banff, Alberta, Canada, June 6–8.

Johnston, R. J., Besedin, E. Y., Iovanna, R., Miller, C., Wardwell, R., & Ranson, M. (2005). Systematic variation in willingness to pay for aquatic resource improvements and implications for benefit transfer: A meta-analysis. *Canadian Journal of Agricultural Economics, 53*, 221–248.

Johnston, R. J., & Duke, J. M. (2009). Willingness to pay for land preservation across states and jurisdictional scale: Implications for benefit transfer. *Land Economics, 85*, 217–237.

Johnston, R. J., Grigalunas, T. A., Opaluch, J. J., Diamantedes, J., & Mazzotta, M. (2002). Valuing estuarine resource services using economic and ecological models: The Peconic Estuary System study. *Coastal Management, 30*, 47–66.

Johnston, R. J., Holland, B., Feurt, C., Wiley, P., Wilson, K., Smith, T., et al. (2014). Integrating ecological and economic data for ecosystem service valuation: The Wells riparian land project. Wells National Estuarine Research Reserve Ecosystem Service Value Project Results Meeting, NOAA, Silver Spring, MD. August 8.

Johnston, R. J., & Ramachandran, M. (2014). Modeling spatial patchiness and hot spots in stated preference willingness to pay. *Environmental and Resource Economics, 59*, 363–387.

Johnston, R. J., Ranson, M. H., Besedin, E. Y., & Helm, E. C. (2006). What determines willingness to pay per fish? A meta-analysis of recreational fishing values. *Marine Resource Economics, 21*, 1–32.

Johnston, R. J., & Rosenberger, R. S. (2010). Methods, trends and controversies in contemporary benefit transfer. *Journal of Economic Surveys, 24*, 479–510.

Johnston, R. J., & Russell, M. (2011). An operational structure for clarity in ecosystem service values. *Ecological Economics, 70*, 2243–2249.

Johnston, R. J., Schultz, E. T., Segerson, K., Besedin, E. Y., & Ramachandran, M. (2012). Enhancing the content validity of stated preference valuation: The structure and function of ecological indicators. *Land Economics, 88*, 102–120.

Johnston, R. J., Schultz, E. T., Segerson, K., Besedin, E. Y., & Ramachandran, M. (2013b). Stated preferences for intermediate versus final ecosystem services: Disentangling willingness to pay for omitted ecological outcomes. *Agricultural and Resource Economics Review, 42*, 98–118.

Johnston, R. J., Swallow, S. K., Bauer, D. M., & Uchida, E. (2014b). Connecting ecosystem services to land use: Implications for valuation and policy. In J. M. Duke & J. Wu (Eds.), *The Oxford handbook of land economics* (pp. 196–225). New York: Oxford University Press.

Jørgensen, S. L., Olsen, S. B., Ladenburg, J., Martinsen, L., Svenningsen, S. R., & Hasler, B. (2013). Spatially induced disparities in users' and non-users' WTP for water quality improvements—Testing the effect of multiple substitutes and distance decay. *Ecological Economics, 92*, 58–66.

Just, R. E., Hueth, D. L., & Schmitz, A. (2004). *The welfare economics of public policy: A practical approach to project and policy evaluation*. Cheltenham, UK: Edward Elgar.

Kaul, S., Boyle, K. J., Kuminoff, N. V., Parmeter, C. F., & Pope, J. C. (2013). What can we learn from benefit transfer errors? Evidence from 20 years of research on convergent validity. *Journal of Environmental Economics and Management, 66*, 90–104.

Kline, J. D., & Mazzotta, M. J. (2012). Evaluating tradeoffs among ecosystem services in the management of public lands. Gen. Tech. Rep. PNW-GTR-865. Portland, OR: U.S. Department of Agriculture, Forest Service, Pacific Northwest Research Station. 48 p. http://www.fs.fed.us/pnw/pubs/pnw_gtr865.pdf.

Liu, S., Costanza, R., Troy, A., D'Aagostino, J., & Mates, W. (2010). Valuing New Jersey's ecosystem services and natural capital: A spatially explicit benefit transfer approach. *Environmental Management, 45*, 1271–1285.

Loomis, J. B. (1992). The evolution of a more rigorous approach to benefit transfer: Benefit function transfer. *Water Resources Research, 28*, 701–705.

Loomis, J. B. (2000). Vertically summing public good demand curves: An empirical comparison of economic versus political jurisdictions. *Land Economics, 76*, 312–321.

Loomis, J., & Richardson, L. (2008). Technical documentation of benefit transfer and visitor use estimating models of wildlife recreation, species and habitats. Fort Collins, CO: Colorado State University Dept. of Agricultural and Resource Economics. Retrieved from http://dare.agsci.colostate.edu/outreach/tools/#BTT.

Loomis, J. B., & Rosenberger, R. S. (2006). Reducing barriers in future benefit transfers: Needed improvements in primary study design and reporting. *Ecological Economics, 60*, 343–350.

Martin-Ortega, J., Brouwer, R., Ojea, E., & Berbel, J. (2012). Benefit transfer and spatial heterogeneity of preferences for water quality improvements. *Journal of Environmental Management, 106*, 22–29.

Millennium Ecosystem Assessment. (2005). *Ecosystems and human well-being: Synthesis*. Washington, DC: Island Press.

Mrozek, J. R., & Taylor, L. O. (2002). What determines the value of life? A meta-analysis. *Journal of Policy Analysis and Management, 21*, 253–270.

Navrud, S., & Pruckner, G. J. (1997). Environmental valuation—To use or not to use? A comparative study of the United States and Europe. *Environmental and Resource Economics, 10*, 1–26.

Navrud, S., & Ready, R. (Eds.). (2007). *Environmental value transfer: Issues and methods*. Dordrecht: Springer.

Polasky, S., Nelson, E., Pennington, D., & Johnson, K. A. (2011). The impact of land-use change on ecosystem services, biodiversity and returns to landowners: A case study in the state of Minnesota. *Environmental and Resource Economics, 48*, 219–242.

Portney, P., & Weyant, J. (Eds.). (1999). *Discounting and intergenerational equity*. Washington, DC: Resources for the Future.

President's Council of Advisors on Science and Technology. (2011). *Sustaining environmental capital: Protecting society and the economy*. Washington, DC: Executive Office of the President.

Rolfe, J., Bennett, J., Johnston, R. J., & Kerr, G. (2011). *'Yes we can…': Using benefit transfer to inform environmental policymaking*. Canberra, Australia: EERH Policy Brief, Environmental Economics Research Hub, Australian National University.

Rolfe, J., & Wang, X. (2011). Dealing with scale and scope issues in stated preference experiments. In J. Bennett (Ed.), *The international handbook on non-market environmental valuation* (pp. 254–272). Cheltenham, UK: Edward Elgar.

Rosenberger, R. S., & Johnston, R. J. (2009). Selection effects in meta-analysis and benefit transfer: Avoiding unintended consequences. *Land Economics, 85*, 410–428.

Rosenberger, R. S., & Loomis, J. B. (2003). Benefit transfer. In P. A. Champ, K. J. Boyle, & T. C. Brown (Eds.), *A primer on non-market valuation* (pp. 445–482). Dordrecht: Kluwer.

Rosenberger, R. S., & Phipps, T. T. (2007). Correspondence and convergence in benefit transfer accuracy: Meta-analytic review of the literature. In S. Navrud & R. Ready (Eds.), *Environmental value transfer: Issues and methods* (pp. 23–43). Dordrecht: Springer.

Rosenberger, R. S., & Stanley, T. D. (2006). Measurement, generalization and publication: Sources of error in benefit transfers and their management. *Ecological Economics, 60*, 372–378.

Schaafsma, M., Brouwer, R., & Rose, J. (2012). Directional heterogeneity in WTP models for environmental valuation. *Ecological Economics, 79*, 21–31.

Spash, C. L., & Vatn, A. (2006). Transferring environmental value estimates: Issues and alternatives. *Ecological Economics, 60*, 379–388.

Stanley, T. D., Doucouliagos, H., Giles, M., Heckemeyer, J. H., Johnston, R. J., Laroche, P., et al. (2013). Meta-analysis of economics reporting guidelines. *Journal of Economic Surveys, 27*, 390–394.

Stapler, R. W., & Johnston, R. J. (2009). Meta-analysis, benefit transfer, and methodological covariates: Implications for transfer error. *Environmental and Resource Economics, 42*, 227–246.

Toman, M. (1998). Why not to calculate the value of the world's ecosystem services and natural capital. *Ecological Economics, 25*, 57–60.

Viscusi, W. K., & Aldy, J. E. (2003). The value of a statistical life: A critical review of market estimates throughout the world. *Journal of Risk and Uncertainty, 27*, 5–76.

Wainger, L., & Mazzotta, M. (2011). Realizing the potential of ecosystem services: A framework for relating ecological changes to economic benefits. *Environmental Management, 48*, 710–733.

Wilson, K. R. (2014). Choices for our land and water: What is the value of ecological buffers in the Merriland River and Branch Brook? The 5th Annual Solstice Summit (municipal official and stakeholder meeting), Wells National Estuarine Research Reserve, Wells, ME, June 19, oral presentation.

Zhao, M., Johnston, R. J., & Schultz, E. T. (2013). What to value and how? Ecological indicator choices in stated preference valuation. *Environmental and Resource Economics, 56*, 3–25.

Chapter 13
Ecosystem Services Assessment and Benefit Transfer

Silvia Ferrini, Marije Schaafsma and Ian J. Bateman

Abstract Ecosystem service assessments aim to integrate the natural environment into decision-making by developing linked biophysical and economic models that demonstrate how changes in the environment affect human welfare. When these analyses inform national level, strategic choices, large-scale analyses are required. Such assessments, embracing multiple ecosystem services, will often rely on the transfer of either economic or biophysical models, or both. This chapter discusses the main concepts of ecosystem service assessments and illustrates the conceptual framework with examples from the UK National Ecosystem Assessment. An analysis of the recreational and carbon values arising from land use changes shows how differences in ecological, socioeconomic or climatic factors result in high spatial heterogeneity in ecosystem services and how this variation can be incorporated within transfer values.

Keywords Ecosystem services · Benefit transfer · Value mapping

S. Ferrini (✉) · I.J. Bateman
Centre for Social and Economic Research on the Global Environment,
School of Environmental Sciences, University of East Anglia, Norwich, UK
e-mail: s.ferrini@uea.ac.uk

I.J. Bateman
e-mail: i.bateman@uea.ac.uk

S. Ferrini
University of Siena, Siena, Italy

M. Schaafsma
Geography and Environment & Centre for Biological Sciences,
University of Southampton, Southampton, UK
e-mail: m.schaafsma@soton.ac.uk

R.J. Johnston et al. (eds.), *Benefit Transfer of Environmental and Resource Values*,
The Economics of Non-Market Goods and Resources 14,
DOI 10.1007/978-94-017-9930-0_13

13.1 Introduction

Despite a nearly universal recognition that human society relies on nature for basic needs such as food and fresh air, numerous assessments have shown that management of the natural environment has not been sufficiently integrated to foster sustainable development (e.g., Millennium Ecosystem Assessment (MEA) 2005). A more integrated approach is necessary to ensure that national and local planning agencies maintain the environment such that it can continue to provide benefits to society. The so called "ecosystem services approach" (MEA 2005) seeks to address this need. This approach requires that agencies consider nature and its services at all stages in the decision-making process. At the core of ecosystem service assessments is the objective of incorporating a holistic consideration of ecosystem services and their value into decision-making (e.g., Department for Environment, Food and Rural Affairs (DEFRA) 2007).

The incorporation of sustainable development goals at national levels has propelled interest in large scale assessments of ecosystem services. The increased demand for quantification and valuation of the benefits that nature provides to society has driven environmental economists and social scientists to seek greater cooperation with natural scientists, and vice versa. Integration can also be sought in valuation studies, including benefit transfer, where biophysical values can be considered explicitly. An inclusive, multidisciplinary approach is imperative when multiple ecosystems and their services are considered. Linking biophysical analyses with socioeconomic valuation is vital for assessing situations where tradeoffs and synergies between ecosystem services may occur in the face of changes in ecosystems and biodiversity (The Economics of Ecosystems and Biodiversity (TEEB) 2010).

Inevitably, large scale assessments cannot rely on primary data collection alone. The transfer of models across space is one of the fundamental but challenging building blocks of the methodology of ecosystem assessment. Benefit transfer methods are hence likely to play an important role in ecosystem assessments, and these transfers may involve both biophysical and economic models.

The UK National Ecosystem Assessment (UK NEA) study of 2011 shows the wide scope that such large scale assessments can cover (and the resources they require; in the UK NEA, over five hundred natural scientists worked together intensively with more than fifty social scientists). In this chapter, we use the UK NEA as a case study to demonstrate how large-scale ecosystem assessments can use benefit transfer to provide policy-relevant information for sustainable development decision making. Scenario development and spatial analysis form the basis of ecosystem services assessment, recognizing that ecosystem services are highly context-specific and change over space and time.

Benefit transfer techniques play a central role in the UK NEA. Rather than a small area or local site, the primary "study site" is here represented by countries of the UK. The method applied takes data from different countries and relates them to local characteristics so as to build models which can then be applied to every location within these countries, as well as to adjacent countries in the UK for which no primary data are available. So in this case the "policy site" is not an independent site but a wider geographical area for which ecosystem services and local variables are likely to be similar. Moreover, the UK NEA acknowledges that ecosystem models are the drivers of values. Therefore, spatially explicit models are developed for both biophysical ecosystem services and their economic value, and transferred across space and time.

In this chapter, the use of benefit transfer approaches to value two particular ecosystem services, carbon sequestration and open-access recreation, will be discussed. These combine biophysical and economic models and make use of different spatial transfer methods. The carbon example shows how biophysical models can be transferred across space to predict CO_2 emission levels in multiple locations, to which economic values can be assigned. The results show spatial variation in the final benefit maps, even though carbon has a fixed price per quantity unit which is unlikely to exhibit diminishing marginal values across the range of provisional levels considered (Bateman et al. 2011). The recreation example demonstrates that both economic and biophysical outputs can vary across space: both visitation numbers and values per trip vary with different habitats. Furthermore, substitution effects across different recreation sites (of both the same and different ecosystem types) need to be incorporated to allow for the likelihood of diminishing marginal values (Bateman et al. 2011). We start this chapter with an introduction of the concepts of ecosystem service assessment and the role of mapping and scenarios. This summary of ecosystem services assessment sheds light on the role of economic valuation of non-market goods. Subsequently, the section on large scale assessments points at the complexity of using primary economic methods (e.g., contingent valuation, travel cost) and introduces the approach of developing spatially explicit, transferable functions for assessing both ecosystem services and their monetary values. The function transfer approach is used to understand and maintain the biophysical link between spatially explicit characteristics of the natural environment and human systems as these jointly determine ecosystem service values. Two examples from the large-scale ecosystem assessment of the UK NEA describe the transfer of ecosystem service values across space. This source of analyses is retained for a final scenario mapping section which provides a formal illustration of transfers across time under alternative policy scenarios.

13.2 Ecosystem Service Assessment

13.2.1 Framework

Working with the framework of ecosystem service assessment requires a stronger focus on natural sciences than is common among environmental economists. One of the key messages of The Economics of Ecosystems and Biodiversity (TEEB) study is that "any ecosystem assessment should first aim to determine the service delivery in biophysical terms, to provide solid ecological underpinning to the economic valuation or measurement with alternative metrics" (2010, p. 3). This integration allows better accounting for ecosystem functioning and interrelations between ecosystem services in economic analysis, and provides vital information for evaluating the sustainability of systems (Bateman et al. 2011).

Various frameworks, definitions and terminology have been put forward to describe how ecosystems can produce services and goods that are of human benefit, through ecosystem processes and functions as additional capital inputs (e.g. Millennium Ecosystem Assessment 2005; The Economics of Ecosystems and Biodiversity 2010; Bateman et al. 2011). In the UK National Ecosystem Assessment study (2011, p. 12), an ecosystem is defined as "a complex where interactions among the biotic (living) and abiotic (non-living) components of that unit determine its properties and set limits to the types of processes that take place there." Thus an ecosystem can be regarded as a "stock" of ecosystem assets, which generates a "flow" of ecosystem services (Mäler et al. 2008; Barbier 2009). Conceptualizing ecosystem services using stock and flow notions highlights the importance of the sustainable use of renewable and non-renewable resources. For the former, an optimal harvesting of their services is the key point, whereas for the latter the attention is on optimal depletion and reinvestment (Barbier 2011; Bateman et al. 2011).

The different biological, physical and chemical components of an ecosystem and their interactions determine the functioning of the ecosystem processes from which ecosystem services result (see Fig. 13.1, which expands upon the ecosystem services framework of UK NEA 2011). Fisher and Turner (2008) define ecosystem services as the aspects of ecosystems utilized (actively or passively) to produce human well-being. These services can be subdivided into final services, which directly contribute to the goods that are valued by people, and intermediate services, which underpin the final services. In many cases, these final ecosystem services have to be combined with other resource inputs, such as manufactured or human capital, to generate valuable goods.

In the framework of Fig. 13.1, "goods" can be tangible or non-tangible, and marketed or non-marketed; their main characteristic is that they are at least partly produced by an ecosystem. Most of these goods can be given a [monetary] value to reflect the well-being they provide, using economic valuation methods.[1] The current status of ecosystems and their associated human well-being effects depends on factors related to the demographic, economic and environmental situation as well as the management regime in place. The future development of ecosystem service delivery depends on changes in these drivers. In the UK NEA, alternative policy

[1]Although many ecosystem service assessment frameworks highlight intrinsic and community or shared social values, this chapter will focus on the benefits that can be assessed at the level of the individual and expressed in monetary terms. Bateman et al. (2011) discusses cases where reliable monetary values might not be available.

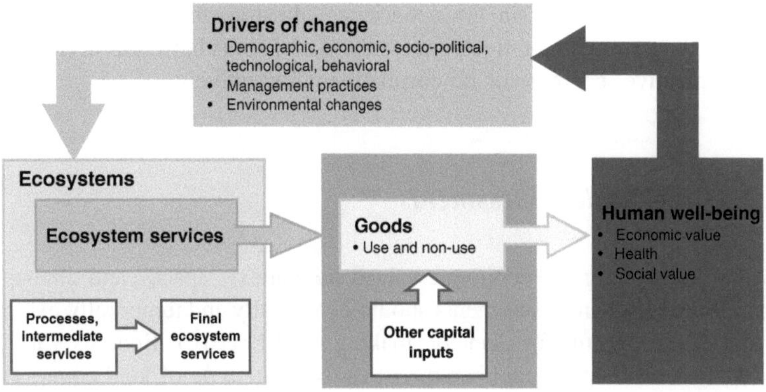

Fig. 13.1 Ecosystem assessment framework

scenarios were developed to demonstrate how human well-being would be affected by different political regimes.

Ecosystem valuation is meaningful only when "marginal changes" are considered. The concept of "marginal changes" is valid if the ecosystem is operating above some safe minimum standard (SMS) which guarantees its functional integrity (Fisher et al. 2008; Turner et al. 2010). Unfortunately, this SMS point is often unknown and the complexity and non-linearity of the interrelated ecosystem processes is often poorly understood. One aspect of ecosystem complexity is that ecosystems may respond to environmental changes in an unpredictable or irreversible way, shifting from one state to another when the SMS threshold is passed. The economics of thresholds are complex and are beyond the purview of this chapter (Johnston and Sutinen 1996; Arrow et al. 2003; Polasky et al. 2011). In addition, the final estimates of the economic benefits of ecosystem services will have confidence intervals that affected not only by uncertainties in economic models, but also variation and model uncertainty in the biophysical assessment of the provision levels of ecosystem services. As a result, because of limited scientific knowledge, wide confidence intervals must be placed on estimates of the change in ecosystem services and physical quantities arising from a change.

The interdependencies between different ecosystem services require careful attention in ecosystem assessments to avoid double counting. Double counting can occur when (a) a service is valued as an intermediate or supporting service as well as a final service, and both values are included in the cost-benefit analysis; or (b) two competing services are valued separately and included in a cost-benefit analysis (Turner et al. 2010). The risk of case (a) can be reduced by using clear definitions of ecosystem services, and focusing on final ecosystem services for valuation (De Groot 2006; Hein et al. 2006). Sufficient understanding of the different processes, functions and services of the ecosystem and their interactions is paramount. The latter case (b) refers to situations where these services cannot be delivered in one "bundle" (TEEB 2010) and have to be traded off. To avoid double counting and include only final services in the economic valuation, the UK NEA developed matrices of services to qualitatively assess the correlations between services, with +, − or 0 for positive, negative or no correlations, respectively.

13.2.2 Large Scale Assessments

Ecological functioning and economic values are context-, space-, and time-specific, and therefore ecosystem assessments should be spatially and temporally explicit at a scale that is meaningful for decision making (TEEB 2010). Benefits vary across space, along with biophysical characteristics (e.g., the type of land cover, climate, altitude) and socioeconomic factors (e.g., population density and distribution, road network, income, land ownership, land use). The same good can generate very different benefits depending on its context and timing of delivery. Mapping and quantifying the linkages between primary processes, intermediate and final ecosystem services through to beneficial goods is therefore a core component of an ecosystem assessment (Bateman et al. 2011; Fisher et al. 2011).

Fisher et al. (2011) present a spatially explicit ecosystem services approach that is based on creating various model-based maps of stocks, production, flow, beneficiaries, benefits and costs (see Table 13.1).

The first step of the assessment (Fig. 13.2) is an inventory of the ecosystems and their contexts, including those factors that drive environmental change (cf. Fisher et al. 2011). The layer of service production shows what the ecosystem provides and maps the service at the location of production in biophysical units. The related service flow map demonstrates where these services are flowing and can be enjoyed, reflecting that not all services are "consumed" where they are "produced." For instance, in the case of water quality, one area may collect and purify water while another consumes it. Since services generate value only when there are people to enjoy their benefits, a separate layer highlights the relevant stakeholders and their socio-demographic characteristics from the population. Finally, the biophysical and socioeconomic components are brought together in an economic

Table 13.1 Overview of changes in population, income and land cover under the UK NEA WM and GPL scenarios

	Baseline	GPL	WM
Change in population (%)	0.0	2.0	21.0
Change in real income (%)	0.0	2.0	2.0
Urban (%)	6.7	6.7	14.3
Heathlands (%)	13.8	14.6	11.7
Grasslands (%)	15.9	25.3	13.7
Conifer (%)	5.3	3.8	6.2
Broadleaf (%)	6.3	11.1	5.3
Farmland (%)	43.5	29.3	39.3
Other (%)	8.3	9.1	9.5

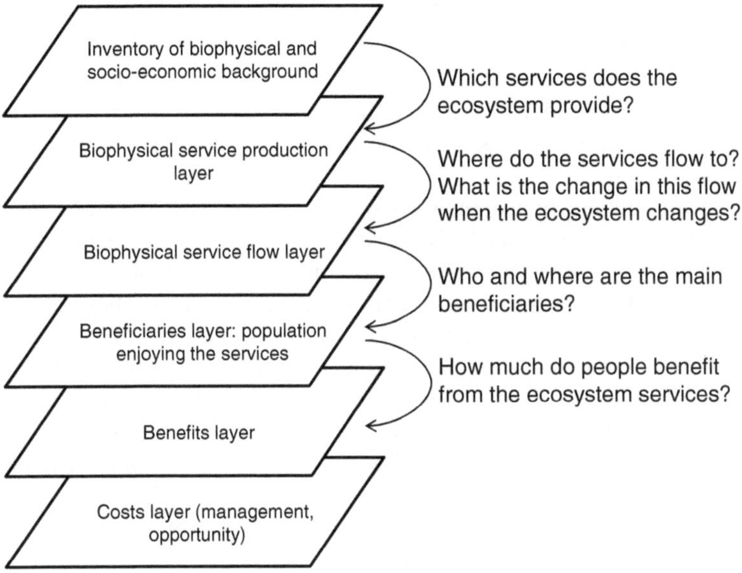

Fig. 13.2 Spatially explicit ecosystem services assessment

valuation exercise to produce maps of benefits as well as costs (e.g. management and opportunity costs).

Mapping the full set of ecosystem services in large-scale, nation-wide assessments requires spatially explicit, transferable models (Bateman et al. 2011). The larger the scale of the assessment, the less likely it is that primary data are available for each area and services of interest. Spatially explicit, transferable models recognize that ecosystem services are context-specific and can be used to transfer analyses to the scale of interest (Brander et al. 2011).

Working from the various information types seen in Fig. 13.2, the modeling framework of an ecosystem assessment is split into:

(a) Biophysical modeling, in which ecosystem services expressed in biophysical units are linked to explanatory factors reflecting the spatial and temporal context; and
(b) Economic modeling, in which monetary values per unit of biophysical output are scaled using both spatial and temporal context variables.

These models use input data of both biophysical and socioeconomic processes and characteristics, typically based on geographical information systems (GIS) layers. The two sets of models are combined to assess and map the overall welfare impacts across wider areas and different scales, as required by the level of decision-making. This integration of the economic and ecological models forms an essential part of ecosystem assessment and a departure from environmental economic analyses in which biophysical service provision is taken as a given and ecological heterogeneity largely ignored.

Understanding the spatial distribution of the costs and benefits of changes in land management allows policymakers to spatially target those sites and actions which yield positive net social welfare changes. Overlaying cost and benefit maps also allows for the identification of individual winners and losers, which, combined with socioeconomic information, is important for policymaking as it informs distributional considerations.

One main strength of mapping ecosystem services is its capacity to support scenario analysis. Scenario analysis is a key component of ecosystem assessments. Scenarios reflect hypothetical but internally consistent and (biophysically) plausible story lines with feasible outcomes in terms of land use changes. They are, however, different from forecasts based on time-series analysis. Scenario analysis recognizes that costs and benefits are best measured as a function of changes between counterfactual scenarios (marginal values). Differences between alternative scenarios, resulting from different policy decisions, are often more informative for policymaking than total value estimates (Swetnam et al. 2011). By examining the tradeoffs of alternative future states of the world, the option that offers the highest net benefits to society can be selected and its distributional impacts evaluated. Maps enable the valuation and comparison of benefits and costs related to changes in land use and ecosystem management under alternative options or scenarios in a spatially explicit manner.

The scenario story lines are based on possible changes in the drivers of environmental change, including knowledge and technology, legislation at national and international levels, policies, institutions, governance, societal behavior, markets and incentives, and industrial practice (see Fig. 13.1). Within the spatial analysis framework, these story lines are translated into changes in biophysical outcomes that can be mapped, including future land use maps and population predictions. The models of current behavior are then combined with the future input data layers to predict future ecosystem service benefits, assuming that the functional relationships and parameter estimates remain constant over time.

13.3 Examples from the UK National Ecosystem Assessment

13.3.1 The UK National Ecosystem Assessment

The UK NEA was initiated by the UK government after the publication of the Millennium Ecosystem Assessment study in 2005. It provides a unique synthesis of current knowledge regarding UK ecosystems and explores the inter-linkages among habitats, ecosystem services and biodiversity. This peer-reviewed showcase of the state and value of the UK's natural environmental assets supports decision makers in developing policies that correspond with an ecosystem services approach. The UK NEA makes no claim to be a comprehensive assessment of all services; in part it highlights knowledge gaps regarding habitats, ecosystems and valuation. The analyses reflect a joint collaboration between scientists from the natural and social sciences, while the wider UK NEA process involved not only academics, but also government, nongovernmental organizations (NGOs) and private sector institutions.

The UK NEA aims to assess major ecosystem services in a spatially explicit manner. The results and outcomes of various scenarios are presented as maps, showing how Scotland, North Ireland, Wales and England might fare in the future under various policy directions and climate-change scenarios. The maps are created by transferring spatially explicit models, estimated by using data from representative areas of the UK over the entire country.

Two examples drawn from the UK NEA are presented. The goal is to illustrate ways in which benefit transfers are combined with large-scale ecosystem assessments to predict future outcomes for human welfare under alternative biophysical and policy scenarios. The first example refers to services provided by agricultural land use to greenhouse gases (GHGs) emissions and the second to open-access recreation sites. These two services were chosen because they represent contrasting examples: whereas GHG values vary only across space because of biophysical differences across land uses, recreational benefits are spatially heterogeneous because both ecological and economic factors affect their economic value.

13.3.2 Greenhouse Gas Emissions

The UK government has set out its GHG emission strategy in the Climate Change Act 2008. The Act aims to reduce carbon emissions by at least 34 % by 2020 and at least 80 % by 2050. The implementation of this Act requires a broad understanding of the terrestrial carbon cycle and its determinants.

The carbon cycle from ecosystems is determined by carbon flow (fluxes) and changes in stocks. Carbon fluxes are determined by carbon emission/sequestration due to changes in carbon stocks by direct emissions from human activities and the natural environment. The carbon stock is the quantity of carbon stored in live biomass, above and below soil, and in the soil as organic carbon, which is primarily composed of various bacteria and fungi. The ability of soils to store carbon depends on many factors such as type of soil, land use, topography, hydrology and climatic factors.

Agricultural management is one of the human activities with a considerable impact on greenhouse gases. Agricultural land uses affect carbon storage, whereas livestock numbers and agricultural activities (e.g., tillage, harvesting) influence carbon fluxes through terrestrial GHG emissions, including methane and nitrous oxide. Agriculture accounts for approximately 77 % of land and roughly 9 % of the UK's net GHG emissions (Thomas et al. 2011). Therefore, sustainable land management and reducing on-farm emissions is part of the implementation plan supporting the UK Climate Change Act.

Forest, woodlands and other (semi-) natural habitats are net sinks for GHG regulation. Over the last 50 years, there has been a slight increase in carbon storage in woodlands in the UK, due to peat land restoration and extensive tree planting projects (Dyson et al. 2009). One of the main objectives of decisions related to future woodland management may be to set land aside for long-term carbon sinks. However, there is no incentive mechanism in place to internalize agricultural GHG emissions into farmers' land use choices. Farmers currently base their land use decisions mainly on agricultural profit maximization objectives. Since farmers influence GHG emissions through agricultural land management and conversion, the inclusion of these land use choices and land-management activities is an integral part of carbon assessment. To capture the spatial variation in the contribution of agricultural activities to climate change through agricultural activities, and therefore GHG changes, a spatially explicit model of farmers' land use decisions is required that reflects the effect of differences in climate and soil conditions across Great Britain (GB). Further, a benefit transfer approach is necessary to assess the impact of farmers' decisions on GHG regulation for the entire UK (including North Ireland) and for valuing future climate and political scenarios. Therefore, the UK NEA GHG case study presents a benefit transfer example of GHG regulation services across space and time. In the UK NEA, the biophysical analysis of GHG emissions consists of a spatially explicit agricultural land use model (Fezzi and Bateman 2011) combined with an assessment of carbon stocks and flows across various habitats and land use types (Abson et al. 2010). Figure 13.3 presents a schematic overview of the model used for transferring the biophysical and economic GHG values across space and time.

The agricultural land use model reflects how climate and land use types influence farmers' profits and therefore the way they use their land. It disaggregates the broad category of "agricultural land" in the GB land cover map into various types of land uses related to different crops and livestock. The land use model considers farmers' outputs produced in GB over the last 40 years, prices of those outputs,

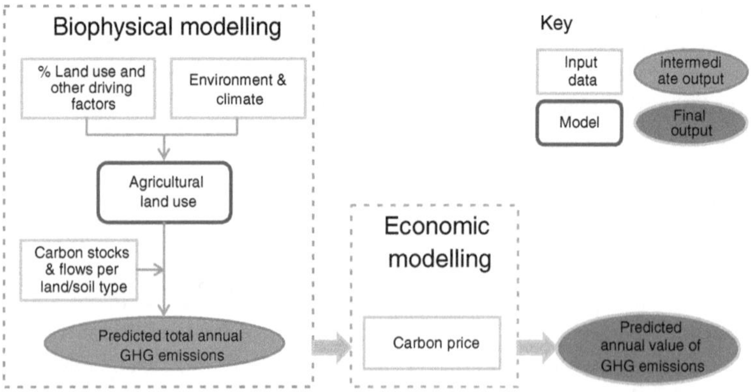

Fig. 13.3 Economic assessment of GHG emissions

costs of inputs, policy and market drivers of farmers' choices and a set of environmental and climate variables. All of these input variables in the land use model are collected at a detailed spatial resolution (2 km grid square) for GB and only partial information are available for North Ireland. The model describes how changes in these factors result in farms allocating different shares of their available land to different activities. The main land categories analyzed are cereals, oilseed rape, root crops, temporary grassland rough grazing and a bundle of other land uses, such as on-farm woodland. The numbers of livestock per grid square are also included in the model to account for direct GHG emissions from agricultural activities. The different land uses as predicted by the land use model are, in turn, associated with different levels of carbon stocks and fluxes. The total area considered in this analysis (farmland forest and woodland) accounts for approximately 88 % of GB terrestrial area, representing the majority of GB land (Abson et al. 2010). Three major categories of GHG emissions were considered and converted to CO_2 equivalents:

1. Direct and indirect emissions from land use and management;
2. Annual flows of carbon from soils due to land use changes;
3. Emissions and accumulations of carbon in terrestrial vegetative biomass.

The carbon fluxes are determined by estimating the emission levels from typical farming practices for different agricultural crops and the manure and enteric fermentation (animal digestive process responsible for methane emissions) due to livestock density. For each crop, a typical farming practice is assumed and the relative CO_2 emissions are calculated for different land shares. Further, changes in land uses are associated with annual GHG fluxes. For example, a conversion of arable land to permanent grass was estimated to produce an average accumulation of soil organic carbon (SOC), whereas a change of rough grazing to permanent grass will imply a loss of SOC. Accumulation of SOC continues until a new equilibrium state is reached; this equilibrium state varies by land use type. The time period over which the change occurs is taken into account in calculating the change in biomass stocks and the relative GHG annual fluxes. Essentially, combining literature findings and case specific assumptions a mean benefit transfer is conducted for determining the GHG quantity in each soil type. More details are available in Abson et al. (2010); the following briefly explains the approach.

The stock of carbon is determined as a function of land uses and woodland density, with estimates of these GHG categories drawn from the literature (e.g., Milne et al. 2001; Bradley et al. 2005; Worrall et al. 2009) and combined with assumptions detailed in Abson et al. (2010). For SOC, a distinction is made between organic and non-organic soils. Abson et al. (2010) assume that peat soils under rough grazing (organic soil) have an average soil carbon density of 1200 tC/ha and that non-peat soils have a density of 224 tC/ha. Furthermore, SOC varies across regions in the UK with, for example, the average SOC (up to 1 m) value in England being 132.6 tC/ha and 212.2 tC/ha in Scotland (Bradley et al. 2005). Other estimates are provided for Wales and North Ireland. Average SOC levels per land use are adjusted for different regions by assuming that the SOC per land use is proportional to the regional average. For example, crops lands are assumed to have 84 % of the non-peat SOC of the same soils under temporary and grassland (Cruickshank et al. 1998). This implies that given that in England the average SOC estimates for temporary and grassland for non-peat soils is 133 tC/ha, the resulting average SOC for crops is 111 tC/ha. Similar estimates have been produced for the other regions.

The biomass stocks in different land use types are based on estimates from the literature (see Abson et al. 2010 for details), whereas for terrestrial carbon storage in woodlands, estimates given by Thomson et al. (2007) are used. The sum of the SOC and vegetative biomass per each 2 km grid cells represents the UK distribution of carbon stock in terrestrial ecosystems.

The annual GHG emissions from agricultural land in each grid are the sum of the annual soil organic carbon and biomass carbon (crops and woodland) fluxes and the estimated emissions from agricultural activities, where the spatial variation is dictated by the predicted land use shares. To check the validity of the biophysical model results, different out-of-sample tests have been conducted for the land use model (Fezzi and Bateman 2011) and a comprehensive literature review of estimates of carbon stocks and fluxes was carried out. Although satisfactory, the comparison of GHG estimates is less robust than the out-of-sample tests for the land use model, and the mean benefit transfer for GHG quantity could introduce biases into the biophysical model which cannot be tested easily. The results show that the estimated annual GHG emissions from terrestrial ecosystems are roughly 26 million tons of carbon dioxide equivalent in the year 2000. These emission levels are highly heterogeneous across GB, which demonstrates the sensitivity of ecosystem services quantification to spatial and contextual characteristics. Figure 13.4 shows that areas with high impact agricultural practices are mainly in the western coastline of the country.

Based on these underlying biophysical data, a benefit transfer approach is used to predict GHG emissions and resulting economic costs for North Ireland.[2] As in a standard transfer exercise, England, Wales and Scotland represent the "study sites" and North Ireland the "policy site" for which the annual value of GHG emissions must be predicted.

The biophysical relationship between land uses and carbon stocks and flows is captured in the biophysical model. The annual quantity of GHG emissions in North Ireland is predicted by (a) estimating the land use shares in North Ireland using the land use model combined with secondary data of agricultural drivers and environmental and climatic variables for the policy area, and (b) applying stocks and flows carbon estimates to these values. For step (a), the functional transfer approach is likely to produce low error given that policy and study site present similar farm

---

[2]It is worth observing that this exercise aims at transferring biophysical values and not benefits. Therefore under the UK NEA, the correct term for the methodology used would be "value transfer" and not "benefit transfer."

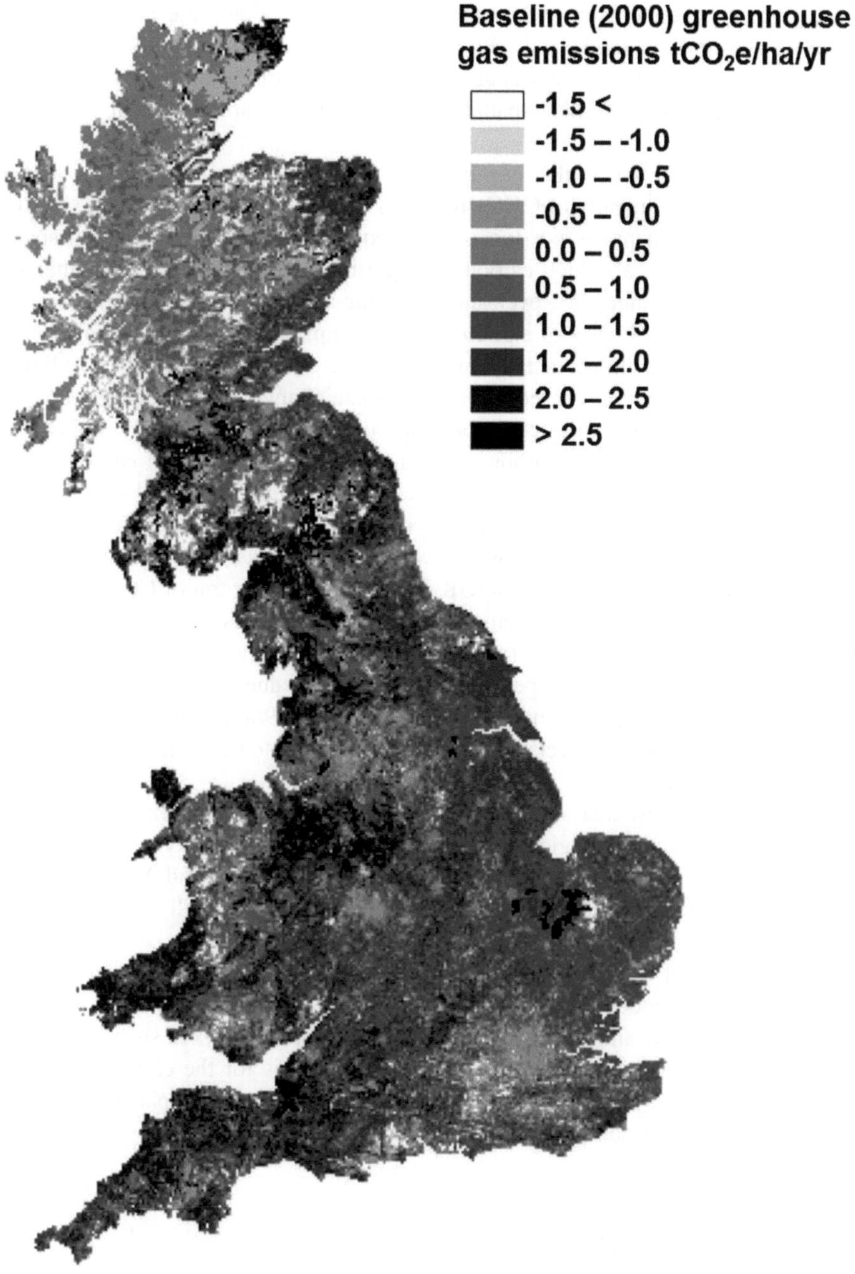

Fig. 13.4 Predicted annual quantity of GHG emissions

management characteristics and technological standards, and the policy site data entering the land use model fall within the range of data of the study site. For the step (b), greater uncertainty exists about the reliability of the mean benefit transfer estimates which do not reflect spatial variability in soil carbon content within the same soil type.

Finally, the economic value of current agricultural GHG emissions is obtained by multiplying the quantity of GHG emissions in each grid cell by the price of CO_2 equivalents. Carbon prices per ton of CO_2 equivalent do not vary across space, because the location at which carbon is sequestered or emitted does not alter the effect on climate change. However, the economic assessment of GHG emissions in terms of the marginal costs of carbon is not a simple task; the welfare impacts of climate changes are influenced by many factors, such as uncertainty in climate change effects, economic consequences of climate change, ecosystems response to climate change, etc. The most commonly applied approaches for estimating carbon prices are the social cost of carbon and marginal abatement cost of carbon (Stern 2007; Department of Energy and Climate Change (DECC) 2009). From the perspective of an economic cost-benefit analysis it is the former value which is greater relevance for welfare evaluations. However, the official non-traded marginal abatement cost of carbon set by the UK DECC (£41.28 per ton of CO_2-equivalent in 2010 prices) falls within the range of published estimates of the social costs of carbon reported by Tol (2010) (whose meta-analysis yields an average value of around £33/tCO_2 with an upper 95 % percentile value of around £123/tCO_2, although the modal value is much lower at just over £9/tCO_2 suggesting that our chosen values, while policy relevant, may be considered to be on the high side from a welfare perspective).

The findings show that average costs from agricultural GHG emissions in GB are £94 per hectare, but regional analysis shows great variability across country, with higher values along the western coastline of England and Wales that are mainly dominated by intensive agricultural practices, principally beef livestock (Fig. 13.5).

The example of GHG emissions shows the importance of linking social sciences and biophysical modeling in ecosystem services assessments and the role that transfer of economic and biophysical models across space plays in such large scale analyses. When the economic value is constant, as it is for the carbon price, the overall carbon values still vary across space following the spatial pattern predicted by the biophysical model.

13.3.3 Recreational Benefits

Recreational opportunities are one of the clearest examples of non-consumptive benefits that the natural environment and ecosystem services provide to human beings. Open-access recreation is valued in excess of £20 billion annually in England alone (Sen et al. 2011). These values are highly variable across space. For

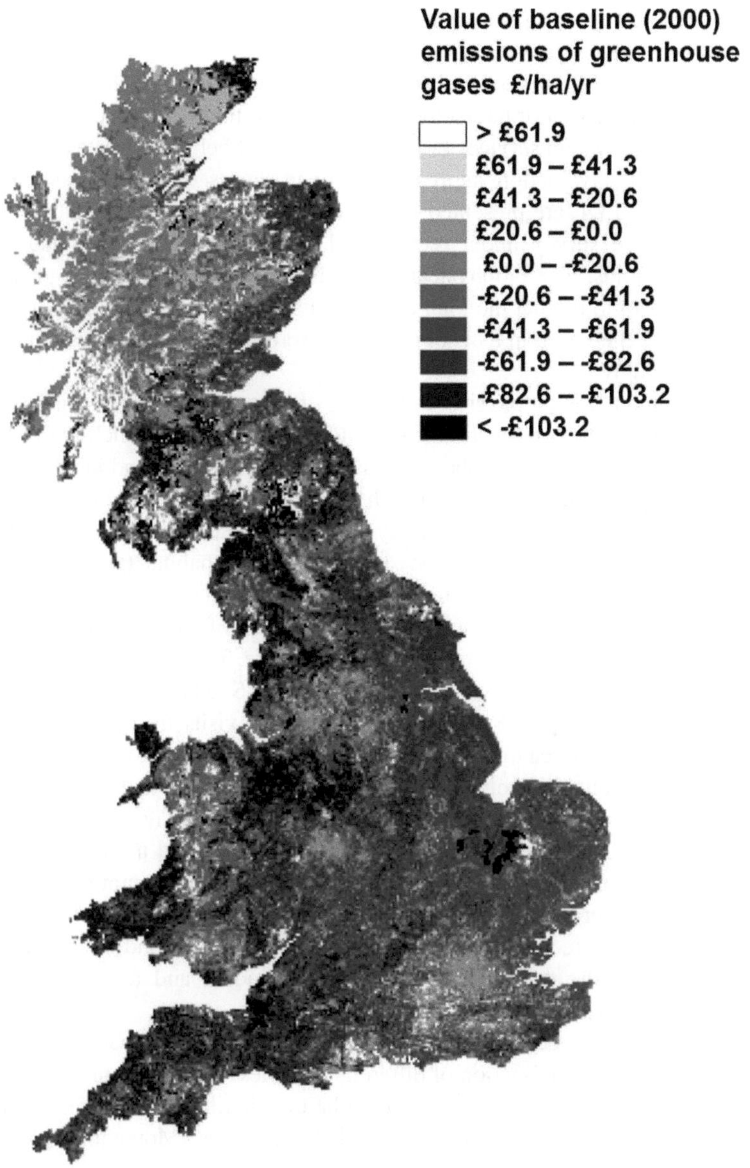

Fig. 13.5 Predicted annual values of GHG emissions

example, the recreation services provided by a river can yield a much higher value when located nearby a highly populated area than for a biophysically similar river located in a remote area. As a consequence, the number of visits to a recreation area is highly non-random and driven by local characteristics. Therefore, different aspects of recreational benefits such as distance to urban area, habitat characteristics and the availability of substitute recreational sites should be taken into account when valuing open-access resources.

In the UK NEA, the first step of the recreation analysis consisted of an inventory of sites and examination of the factors that determine their existence. Subsequently, the impact of ecosystem services flows to recreational behavior was estimated. The full details of the valuation approach for open-access recreation, schematically depicted in Fig. 13.4, are presented in Sen et al. (2010). The biophysical model is based on a large survey about recreational behavior among more than 45,000 English households. The biophysical output is combined with a meta-analysis on the value per trip to predict the total annual value across different types of habitats. The models are then used in a benefit transfer exercise to predict recreational values in Scotland and Wales.

In this example, the biophysical modelling of ecosystem services consists of two elements: a site prediction function (SPF) and a trip generation function (TGF). The current recreation sites and number of related visits are known and identified by the survey data (Monitor of Engagement with the Natural Environment (MENE) 2010). However, given that sites in non-surveyed areas and under different scenarios are not known, the relationship between site location and habitat type is statistically analyzed. The SPF predicts the number of recreational sites in an area as a function of type of natural resources at the site, the distribution of the population around the site and the travel time from that population to the site. This model is used to predict recreational sites for the policy site areas (Wales and Scotland) or in different states of the world. Next, the TGF models the number of visits from each UK Census Lower Super Output Area to any given recreational site as a function of: population characteristics, availability of potential substitutes, distance and type of habitats. Data for all of these analyses are obtained at different spatial resolutions using GIS (see Sen et al. 2010). The output of the TGF is the predicted number of visits per site which has been multiplied by the number of sites per cell (output of the SPF) to produce the number of visits per week to all 1 km grid square cells across the current estimates of recreational sites in England under the land cover map 2000. These values are the output of the biophysical model and are calibrated with observed visits to sites in England.

The results of the model, reported in Sen et al. (2010), show that the variation in the number of visits is a function of different variables such as location and its main habitat characteristics, road network, population distribution and characteristics, substitutes and complements of different habitats types. Mountains, coasts and freshwater sites and woodlands have a significant positive effect on the number of visits. Retired and richer people have higher levels of participation in recreation activities.

Unlike the fixed prices per unit of carbon, recreational values per trip are likely to be context-specific. Therefore, a meta-analytical regression model was developed based on revealed and stated preference estimates of willingness to pay per person per trip from nearly 250 previous studies on open-access recreational sites worldwide.[3] The trip value is modeled as a function of ecosystem types, controlling for the sample size, valuation method, valuation unit (e.g. household or person) and country. This generates a model of the site-specific willingness to pay value per person per visit, which varies according to the habitat type characteristics of the visited site. Further, details about this meta-analysis study can be found in Sen et al. (2010).

Multiplying the value per trip by the predicted number of visits to a site in that cell produces the annual recreational (or access) value. Although the value of, say, mountain visits is high, the number of visits is low and therefore the annual total value reflects these two components. Further, since cells can contain various types of habitat, the overall habitat value of each cell is obtained by multiplying the coverage of the different habitats by their money measure. For example, given that the value per person per trip to woodland is estimated at £6.10 and that to wetlands £6.88, if in a 1 km cell the coverage is 50 % woodlands and 50 % wetlands, the per trip value of that cell is given by adding £3.44 + £3.05.

The predicted average annual number of visits per each 5 km grid cell is 394,000. This corresponds to over 2.9 billion annual visits, representing more than £8.9 billion in economic benefits for England. These recreational values change according to the natural environment of the area, the availability of substitutes, the infrastructure and the distribution of the population around that area. The models are therefore highly transferable and results can be aggregated across any desired spatial unit (e.g. county, region, and catchment) and scenario. In order to test the robustness of the biophysical results, out-of sample tests have been conducted and an improved version of the biophysical modeling approach is published in Sen et al. (2014).

In the UK NEA, the model has been used for a benefit transfer exercise to predict the annual value of visits to (semi-) natural habitats for the UK, where England is used as the study site and values are predicted for Scotland and Wales. Spatial information on habitat types, travel times and land uses were collected for Scotland and Wales, and coupled with the parameters of the TGF, to predict the annual number of visits to these policy sites.

Figure 13.7 presents the resulting distribution of the TGF model, showing that variation in number of visits is predicted to vary with distance to populated area, habitat and land use types. For example, number of visits to Highland Scotland is relatively low compared to those of England, because the distance to populated areas is high.

In the final step of the analysis, the annual total value of recreational visits is obtained by combining the distribution of recreational visits with the results of the meta-regression model (Fig. 13.6). Since the economic values of outdoor recreation are spatially sensitive, the distribution in Fig. 13.8 differs from that in Fig. 13.7. Figure 13.8 shows that some remote areas, such as the Scottish Highlands, for

[3]The use of meta-analysis for valuing recreational trips is a well developed area of research and interested readers are referred to Rosenberger and Loomis (2000).

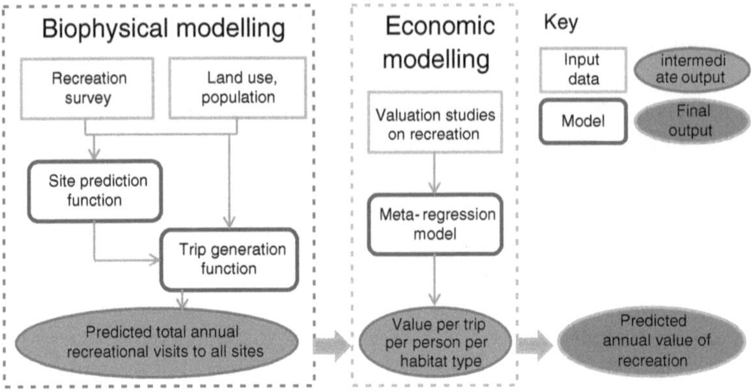

Fig. 13.6 Economic assessment of recreational sites

which the number of visits is relatively low (in the range of 10,000–100,000), are nevertheless associated with relatively higher annual recreational values (greater than £100,000). This is because the habitats in these areas are highly appreciated and therefore the value per visit is higher than for other types of habitat.

13.3.4 Scenario Mapping

13.3.4.1 The Scenarios

The scenario analyses consist of a comparison of ecosystem services in the 2000 baseline (prices in 2010) with various future states in 2060 generated by the UK NEA scenarios team. The baseline is set to a reference year and prices are adjusted to 2010 levels. The scenario analysis uses benefit transfer for valuing ecosystem services under different states of the world by transferring the estimated functions describing both ecosystem services and their values. The scenario analysis proceeds by applying these functions to the same geographical area, but with the physical attributes of that area altered in line with expectations formed through the scenario generation process as described by Haines-Young et al. (2011). The latter study generated a number of scenarios as likely to arise under differing policies formulations. These are further perturbed by climate drivers described by the UK Climate Impacts Programme (UK-CIP) reported in Murphy et al. (2009). For simplicity we focus upon just two scenarios, both of which assume a 'high emissions' trajectory.

The first scenario, "Green and Pleasant Land" (GPL), envisions that economic growth is driven mainly by secondary and tertiary sectors. Pressures on rural areas are assumed to be declining as a result of increased concern for the conservation of biodiversity and landscape. Here, as biodiversity preservation is a key objective for policy makers, sometimes habitats will be preserved and conserved primarily to improve the aesthetic appeal of landscape and countryside. Arable lands decline and

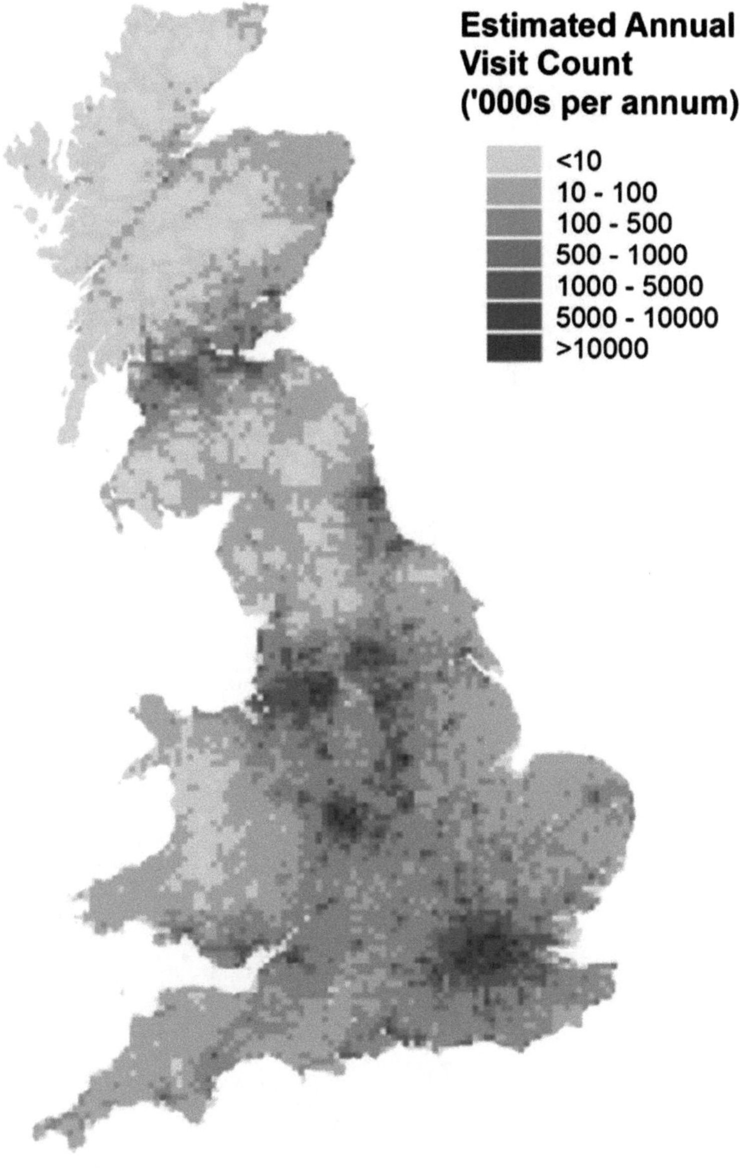

Fig. 13.7 Predicted annual number of visits

the biodiversity and aesthetic values of landscapes are enhanced by increases in improved grassland (temporary or permanent grassland with reduced fertilizer), semi-natural grassland and conifer woodland. This implies a decrease in food production which is compensated for by increased imports to offset the demands of a larger population.

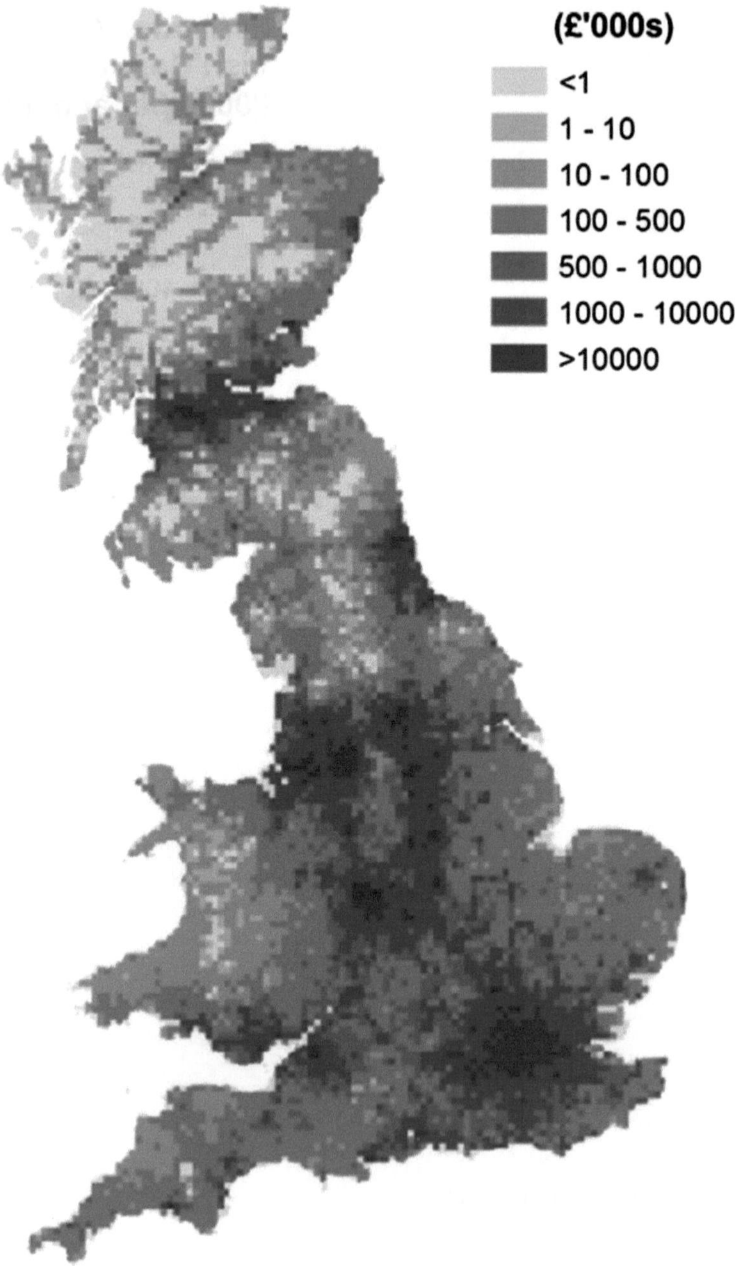

Fig. 13.8 Predicted annual values of recreation

In the second scenario, called "World Market" (WM), a 30 % population increase is envisioned and concomitant changes in land uses are substantial. A complete liberalization of trade is assumed, which implies an end to agricultural subsidies, increases in international trade of resources/goods and reduced rural and urban planning regulation. Consequently, the proportion of arable lands increases and improved grassland and semi-natural areas decrease to accommodate urban growth. Biodiversity declines and technological development is pursued mainly by private companies.

Table 13.1 gives an overview of the land cover and population changes under the WM and GPL scenarios, drawing from results presented in UK NEA (2011). Within the spatial analysis framework, these scenario storylines are translated into future land use maps and population predictions. The models of current behavior are combined with these future input data layers to predict future ecosystem service benefits, assuming that the functional relationships and parameter estimates remain constant over time.

13.3.4.2 GHG Emissions

In this example, the benefit transfer is applied to value changes under the two scenarios reported in Table 13.1. Using the standard nomenclature of benefit transfer, the "study site" is here represented by the current state of the world and the "policy site" is not a different area, but the same area under future foreseen changes. Therefore, changes in Table 13.1 represent the hypothesised values for the scenarios/scenario input valuables available for the "policy site." A function transfer approach is applied to determine the predicted annual quantity of GHG emissions. The biophysical model predicts changes in agricultural GHG emissions due to land use changes assumed under the GPL and WM scenarios. For woodland planted between 2000 and 2060, average annual flows were assumed following Haines-Young et al. (2011). Carbon stocks and fluxes are calculated using the new land use shares and assumptions presented in Sect. 13.3.2.

Figure 13.9 describes changes in terrestrial ecosystem emissions (tons of CO_2e/ha/year) between the baseline and 2060 under the two scenarios (cf. UK NEA 2011). Darker colors in Fig. 13.7 show where the changes in GHG emissions are going to be most substantial. Scotland and the north of England are predicted to show the highest increase in emissions of agricultural GHG emissions, due to the conversion of rough grazing to more intensive agricultural land uses.

As expected, in the GPL scenario there is a relatively uniform decrease in GHG emission equivalent of roughly 8 million tons of CO_2e/year. This reduction stems mainly from lowland areas, where arable land and improved grasslands are converted to semi-natural and rough grazing. This in turn results in lower density of beef and sheep livestock and therefore lower emissions from fertilizer than in the baseline. In the upland areas, there is a moderate increase in GHG emissions, mainly driven by increased livestock numbers and decreased carbon accumulation

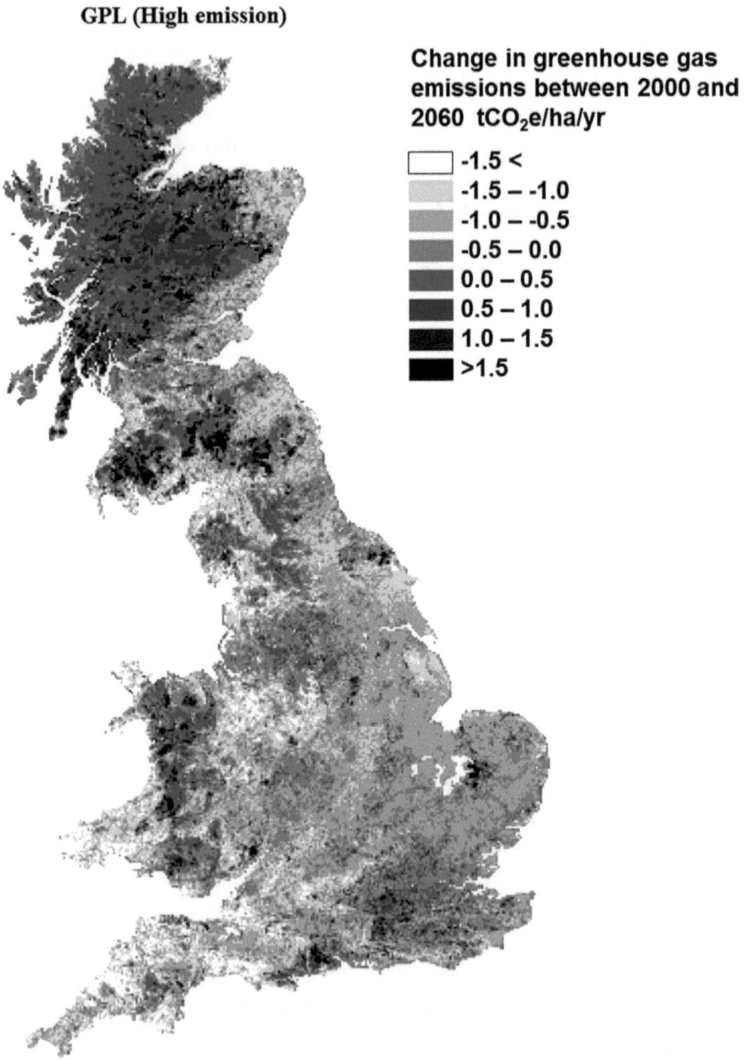

Fig. 13.9 GHG emissions: Scenario analysis and changes between 2060 and baseline. *Left* GPL (high emission). *Right* WM (high emission)

in forests. The latter is assumed to happen because the rate of carbon uptake will decline when numerous conifer plantations reach maturity. Overall, the GPL scenario presents a positive impact in terms of GHG emission reductions.

The WM scenario presents a contrasting result. Here emission levels increase by roughly 6 million tons of CO_2e/year compared to the baseline. The main drivers of this change are reductions in the extent of woodlands, due to the envisioned high pressure of urban expansion, and moderate expansion of arable and dairy production, largely at the expense of semi-natural grasslands.

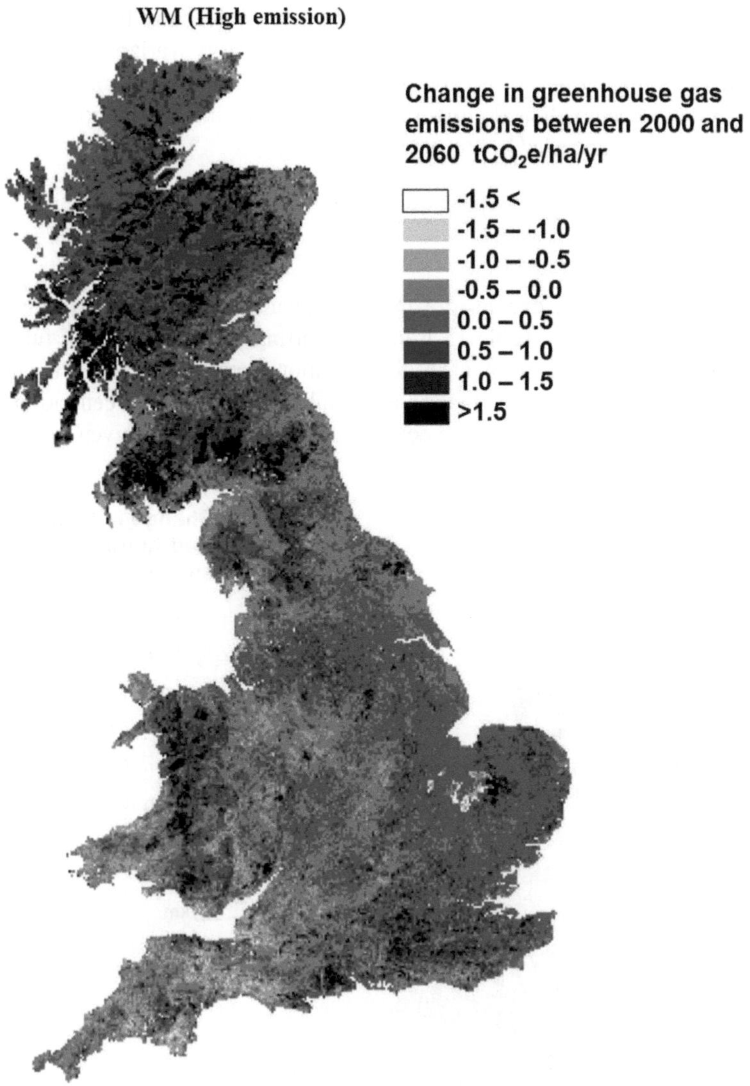

Fig. 13.9 (continued)

The total value of annual GHG emissions is obtained by applying the carbon price to the emission levels under the two scenarios. The results suggest that GB would save more than £2 billion annually in terms of GHG emissions costs under the GPL scenario, while increased emissions under the WM scenario would imply a loss of societal welfare of more than £2 billion annually.

The scenario analysis shows how benefit transfer methods can be used to express significant differences between the GPL and the WM scenario in a spatially explicit manner, and visualize which areas are going to contribute most to the net change in

GHG emissions. Benefit transfer was used to predict values for the future scenarios assuming that the relationships between the explanatory variables and outcome variable as estimated in the biophysical and economic models remain robust over time as does the unexplained variability. Violation of this assumption may invalidate the results of the GHG comparison.

13.3.4.3 Recreation

Again a benefit transfer exercise for the same area across time is performed using the changes described in Table 13.1 and transferring the SPF and TGF function to predict the annual number of visits to recreation sites under the GPL and WM scenarios. The changes in predicted visit numbers for the GPL scenario are visualized in Fig. 13.10, which can be compared with the baseline given in Fig. 13.7.

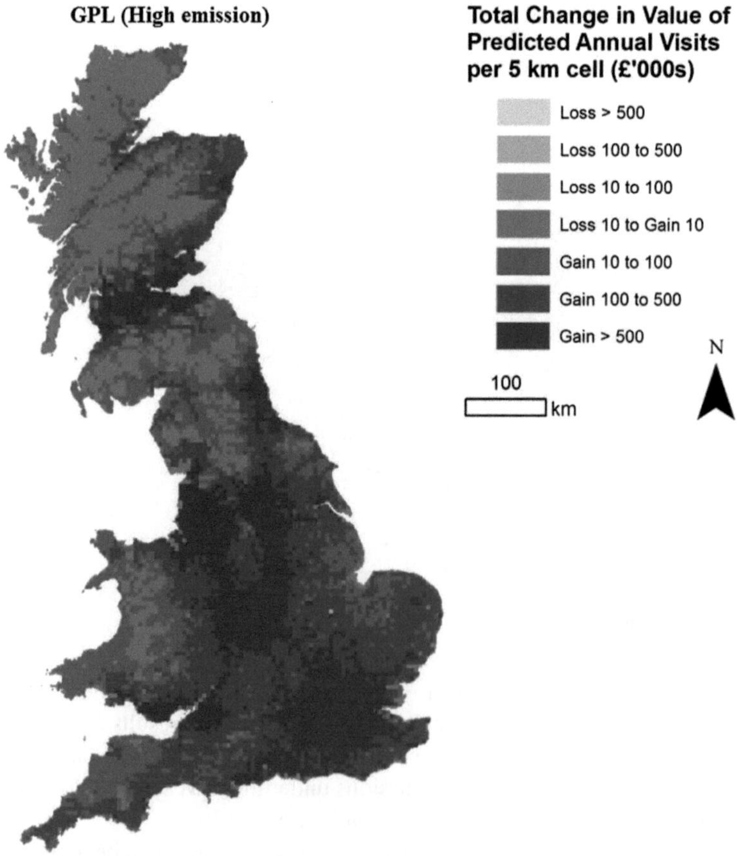

Fig. 13.10 Recreation: Scenario analysis in 2060, changes between 2060 and baseline. *Left* GPL (high emission). *Right* WM (high emission)

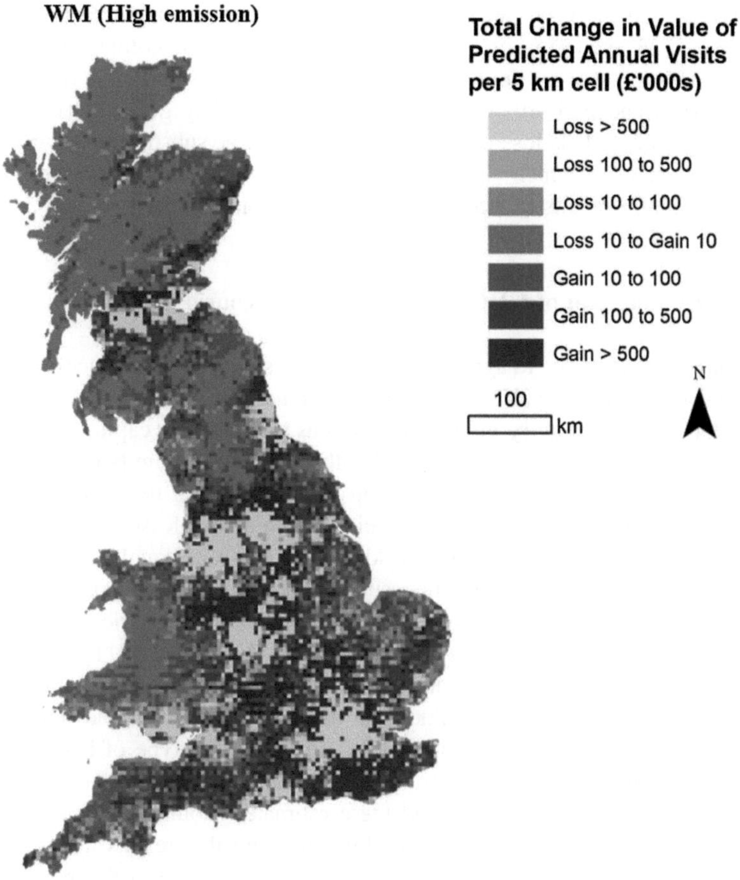

Fig. 13.10 (continued)

The changes in the number of visits occur for a variety of reasons, including changes in the availability of substitutes; variation in habitat characteristics (e.g., change of arable land to improved grassland); and changes in medium household income and population characteristics (e.g., higher proportion of retired people). Under the GPL scenario, the number of visits per year is predicted to increase substantially, mainly around urban areas. Indeed, only remote areas fail to experience increased recreational visit numbers. The values of these changes depicted in Fig. 13.10 are obtained by applying the meta-analytical values per trip to these future visitor numbers, as illustrated (UK NEA 2011). Overall, the GPL scenario delivers a substantial increase in recreational values over the baseline. This effect is driven mainly by a decrease in primary sector production and an increase in aesthetical landscape conservation and protection.

In contrast to the GPL scenario, the distribution of annual recreational values under the WM scenario shows a significant decrease in comparison with the baseline. Major losses are found near to large urban areas and arise due to reductions in urban and peri-urban recreational areas (including green belt) envisioned under the scenario. This loss of resource results in some substitution towards more rural areas, many of which show increased recreational values. However, overall the WM scenario results in major losses of recreational value.

13.3.4.4 Comparison of GHG and Recreation Scenario Analyses

The changes in value due to agricultural GHG emissions and open-access recreation visits under the GPL and WM scenario are summarized in Table 13.2. Values are aggregated for each country in GB and given as total and per capita sums. Results show that the GPL scenario is associated with higher ecosystem benefits per capita (£61 million p.a. for recreation values and £37 million p.a. per GHG emissions), whereas the WM scenario is always associated with annual losses, even though the real income in the two states of the world is assumed to be equal (see Table 13.2).

The numbers in Table 13.2 are the results of the benefit transfer exercises and are based on a range of assumptions, not only those of the scenarios, but also that models can be transferred across time and space (UK NEA 2011). The transfer across time applies the current functional models for GHG and recreation ecosystem services and values, and assumes that the independent variables change over time under two alternative scenarios: GPL and WM. This type of benefit transfer exercise is surrounded with considerable uncertainty and ideally, some information about the confidence intervals around these estimates should be given. However, producing such confidence intervals is difficult given that results are based on a combination of various models, especially when non-linearity in ecosystem

Table 13.2 Total and per capita value of changes in annual ecosystem service values from baseline year for annual recreation visits and GHG emissions under two 2060 UK NEA scenarios

Region	Recreation (million £ p.a.)		GHG (million £ p.a.)	
	GPL	WM	GPL	WM
England	4451	−678	2193	−381
Scotland	556	−61	−52	−1192
Wales	149	−84	268	−101
Great Britain	5156	−823	2410	−1675
GB population (millions)	66	72	66	72
GB per capita values (£ p.a.)	61	−57	37	−23

Note Negative values represent an annual loss. The GHG and recreational value changes are obtained following slightly different approaches: GHG values are based on the difference in GHG quantity and multiplied by 2010 prices, whereas the recreational values reflect the difference in the number of visits for 2000 and 2060 multiplied by the WTP per trip in 2000 prices

functioning may play a role, when there is uncertainty related to the future values of the input variables of the models and uncertainty regarding the stability of preferences over generations of people. The results may be considered as general trends in value change arising from scenarios.

13.4 Conclusions

The two case studies from the UK-NEA presented in this chapter reflect the complexity of large-scale ecosystem services assessment and the central role that benefit transfer methods play in these exercises. Firstly, both ecosystem services and their values are explicitly modeled and the resulting spatially explicit statistical functions are based on highly detailed primary data. Secondly, the transfer takes place at national level using primary models at nation level and transferring that to adjacent areas. Thirdly, the models are used for scenario analysis and transferred across time.

The ecosystem service approach and the UK assessment demonstrate a means to unite natural sciences with economic assessments to estimate the value of changes under different states or development pathways, thereby informing decision-making regarding strategies for improving societal welfare. Ecosystem assessment requires the combination of sound biophysical modeling of ecosystems, services and their processes and interactions, along with detailed socioeconomic analyses of the final ecosystem services and the value that they provide to humans. It is this combination which forms a necessary evolution for environmental valuation from earlier valuation work, where ecosystem functioning was often simplified to very basic levels and interrelations with other ecosystems were often ignored.

The key to ecosystem assessments is that services are not considered in isolation, but in combination, showing where tradeoffs have to be made or synergies can be achieved in ecosystem management. The integration of disciplines in scenario analysis allows for the evaluation of current levels of ecosystem use, and can help elucidate tradeoffs among alternative policy options which can ultimately lead to more sustainable futures with higher ecosystem service benefits. For instance, the analysis of GHG emissions from agricultural land and the quantification of associated costs may be a first step in understanding where emissions reductions can be achieved most efficiently and developing a mechanism to internalize these costs within land-management decisions.

The benefit transfer exercise presented in the chapter relies on spatially sensitive transferable functions for biophysical and economic aspects of ecosystem valuation and ensures that analyses account for the locational context of ecosystem values. Furthermore, in order to minimize errors in modelling and subsequently in transferring ecosystem service values, data from across a large area, in this case GB, at a very fine level of resolution are analysed for different ecosystem types. This suggests that with greater data availability, benefits transfer exercises may be based on spatially explicit models which can better capture variability in ecosystem services.

In order to recognize the importance of spatial context, the UK NEA ecosystem assessments rely heavily on GIS-based maps, visualizing the results of spatially explicit biophysical and economic models. Thus, benefit transfer methods support the incorporation of ecosystem values in policy making, and can provide information about costs and benefits of ecosystem services at a high spatial resolution, even at the national scale of countries the size of the GB.

It should be noted that spatially explicit large scale assessments are complicated by conceptual and practical issues. First, the spatial boundaries of ecosystems and their services are not clear-cut; ecosystems vary widely in spatial scale and their key processes operate across a range of rates that are overlapping in time and space (TEEB 2010). In addition, ecological, economic, social and political boundaries may not match. Second, data availability at the relevant scale or precision may be limited and data collection can be resource-intensive, thereby limiting the accuracy of the analysis or the variables that can be included in spatially explicit models. A high level of GIS information as well as modeling capacity is required. The associated investments in the start-up phase may be considerable, but the results can be used by various stakeholders at any given scale of assessment. At the same time, large scale analyses based on benefit transfers might raise non-trivial questions about the reliability of the predicted values and the related errors. Particularly, where local scale models are applied to larger scales, e.g., to national levels, without (the possibility to do) reliability checks, or vice versa, the assumptions of stability of preferences across space may be challenged. Further, the combination of biophysical and economic models requires that are both well specified and spatially explicit, because where both estimates are associated with large errors, the multiplication of the estimates may introduce considerable transfer bias in the ultimate welfare estimates. Therefore, the results of large-scale ecosystem services assessment based on transferred values may be suitable for initial stages of decision-making, whereas later stages nearing implementation of projects or policies, where higher reliability of value estimates is required, may require more reliability checks or primary valuation studies.

One of the remaining issues in the UK NEA is the assessment of sustainability. Sustainability assessments require the comparison of actual use to regeneration levels, i.e., the impact of service flow changes on the levels of stocks of relevant ecosystem services. In the case of timber use, projections of carbon emissions over time also have to take into account the lifecycle of products made of timber. Both carbon storage and sequestration in woodlands and carbon storage in timber products are excluded in this chapter. Current scientific knowledge is not sufficient for a robust assessment of the sustainability of the current resource use, but this issue is on the list of future research themes.

Of academic and political importance is the need to develop more rigorous testing of the reliability of these large-scale transfers, which may require new primary data collection or temporal stability tests of transferred data. Furthermore, the results presented in this chapter do not consider the confidence intervals of the biophysical and economic models and their effect on transferred values. The use of the benefit transfer for ecosystem services valuation involves a trade-off between scale of analysis and accuracy. The reliability of benefit transfer across time and space builds upon a range of assumptions. Most notably are the assumed similarity of sites, whereas sites across a nation are likely to vary considerably; the stability of preferences across space, whereas there are likely to be economic and socio-cultural differences between people within a country; and stability of preferences over time, whereas there are likely to be changes in preferences and economic demand over longer timeframes. More localised, short-term decision-making may require more accurate results for which the costs of additional primary data may be justified. The type of large scale assessments, as presented in this chapter, is mainly suitable to inform long-term, strategic policy making at higher levels, using contrasting scenarios to show the direction in which various policies may result in different policy outcomes and associated economic welfare estimates.

Acknowledgments We are grateful to David Abson, Amii Harwood and Antara Sen for their help and comments and for providing the GIS maps. All errors are the sole responsibility of the authors.

References

Abson, D., Termansen, M., Pascual, U., Aslam, U., Fezzi, C. & Bateman, I. J. (2010). *Valuing regulating services (climate regulation) from UK terrestrial ecosystems*. Report to the Economics Team of the UK National Ecosystem Assessment, School of Earth and Environment, University of Leeds, Department of Land Economy, University of Cambridge and Centre for Social and Economic Research of the Global Environment (CSERGE), University of East Anglia.

Arrow, K. J., Dasgupta, P., & Mäler, K.-G. (2003). Evaluating projects and assessing sustainable development in imperfect economies. *Environment and Resource Economics, 26*, 647–685.

Barbier, E. B. (2009). Ecosystems as natural assets. *Foundations and Trends in Microeconomics, 4*, 611–681.

Barbier, E. B. (2011). *Capitalizing on nature: Ecosystems as natural assets*. New York: Cambridge University Press.

Bateman, I. J., Mace, G. M., Fezzi, C., Atkinson, G., & Turner, R. K. (2011). Economic analysis for ecosystem service assessments. *Environment and Resource Economics, 48*, 177–218.

Bradley, R. I., Milne, R., Bell, J., Lilly, A., Jordan, C., & Higgins, A. (2005). A soil carbon and land use database for the United Kingdom. *Soil Use and Management, 21*, 363–369.

Brander, L. M., Bräuer, I., Gerdes, H., Ghermandi, A., Kuik, O., Markandya, A., et al. (2011). Using meta-analysis and GIS for value transfer and scaling up: Valuing climate change induced losses of European wetlands. *Environmental and Resource Economics* (online first). doi:10.1007/s10640-011-9535-1.

Cruickshank, M. M., Tomlinson, R. W., Devine, P. M., & Milne, R. (1998). Carbon in the vegetation and soils of Northern Ireland. *Biology and Environment: Proceedings of the Royal Irish Academy, 98B*, 9–21.

Department of Energy and Climate Change (DECC). (2009). *Carbon appraisal in UK policy appraisal: A revised approach*. UK: Climate Change Economics, Department of Energy and Climate Change.

Department for Environment, Food and Rural Affairs (DEFRA). (2007). *An introductory guide to valuing ecosystem services*. UK: Department for Environment, Food and Rural Affairs Publications.

De Groot, R. S. (2006). Function-analysis and valuation as a tool to assess land use conflicts in planning for sustainable, multifunctional landscapes. *Landscape and Urban Planning, 75*, 175–186.

Dyson, K. E., Thomson, A. M., Mobbs, D. C., Milne, R., Skiba, U., Clark, A., et al. (2009). Inventory and projections of UK emissions by sources and removals by sinks due to land use, land use change and forestry. Annual report July 2009. Edinburgh: Centre for Ecology and Hydrology, 279 pp. (CEH Project No. C03116, DEFRA Contract GA01088).

Fezzi, C., & Bateman, I. J. (2011). Structural agricultural land use modeling for spatial agro-environmental policy analysis. *American Journal of Agricultural Economics, 93*, 1168–1188.

Fisher, B., & Turner, R. K. (2008). Ecosystem services: Classification for valuation. *Biological Conservation, 141*, 1167–1169.

Fisher, B., Turner, R. K., Zylstra, M., Brouwer, R., De Groot, R., Farber, S., et al. (2008). Ecosystem services and economic theory: Integration for policy-relevant research. *Ecological Applications, 18*, 2050–2067.

Fisher, B., Turner, R. K., Burgess, N. D., Swetnam, R. D., Green, J., Green, R. E., et al. (2011) Measuring, modeling and mapping ecosystem services in the Eastern Arc Mountains of Tanzania. *Progress in Physical Geography*, *35*(5). 595–611. ISSN 0309-1333.

Haines-Young, R., Paterson, J., Potschin, M., Wilson, A. & Kass, G. (2011). Scenarios: Development of storylines and analysis of outcomes. In *The UK National Ecosystem Assessment Technical Report*. Cambridge: UNEP-WCMC. Also available from http://UKNEA.unep-wcmc.org.

Hein, L., van Koppen, K., de Groot, R. S., & van Ierland, E. C. (2006). Spatial scales, stakeholders and the valuation of ecosystem services. *Ecological Economics, 57*, 209–228.

Johnston, R. J., and Sutinen, J. G. (1996). Uncertain biomass shift and collapse: implications for harvest policy in the fishery. *Land Economics*, *72*(4), 500–518.

Mäler, K. G., Aniyar, S., & Jansson, Å. (2008). Accounting for ecosystem services as a way to understand the requirements for sustainable development. *Proceedings of the National Academy of Sciences of the United States of America (PNAS), 105*, 9501–9506.

Millennium Ecosystem Assessment (MEA). (2005). *Ecosystems and human well-being: A framework for assessment*. Washington, D.C.: Island Press.

Milne, R., Tomlinson, R. W. & Gauld, J. (2001). The land use change and forestry sector in the 1999 UK greenhouse gas inventory. In R. Milne (Ed.), *UK emissions by sources and removals by sinks due to land use and land use change and forestry activities* (pp. 11–59). Annual report for DETR Contract EPG1/1/160.11. http://www.nbu.ac.uk/ukcarbon.

Monitor of Engagement with the Natural Environment. (2010). *The national survey on people and the natural environment—Annual Report from the 2009–10 survey*, NECR049. Sheffield, UK: Natural England.

Murphy, J. M., Sexton, D. M. H., Jenkins, G. J., Boorman, P. M., Booth, B. B. B., Brown, C. C., et al. (2009). *UK climate projections science report: Climate change projections*. Exeter: Met Office Hadley Centre.

Polasky, S., de Zeeuw, A., & Wagener, F. (2011). Optimal management with potential regime shifts. *Journal of Environmental Economics and Management, 62*, 229–240.

Rosenberger, R. S., and Loomis. J. B. (2000). Using meta-analysis for benefit transfer: in-sample convergent validity tests of an outdoor recreation database. *Water Resources Research*, *36*(4): 1097–1107.

Sen, A., Darnell, A., Crowe, A., Bateman, I. J., Munday, P. & Foden, J. (2010). *Economic assessment of the recreational value of ecosystems in Great Britain*. Report to the Economics Team of the UK National Ecosystem Assessment, School of Earth and Environment, University of Leeds, Department of Land Economy, University of Cambridge and Centre for Social and Economic Research of the Global Environment (CSERGE), University of East Anglia.

Sen, A., Harwood, A. R., Bateman, I. J., Munday, P., Crowe, A., Brander, L., et al. (2014). Economic assessment of the recreational value of ecosystems: Methodological development and national and local application. *Environmental and Resource Economics, 57*, 233–249.

Stern, N. (2007). *The economics of climate change: The Stern review*. Cambridge: Cambridge University Press.

Swetnam, R. D., Fisher, B., Mbilinyi, B. P., Munishi, P. K., Willcock, S., Ricketts, T., et al. (2011). Mapping socio-economic scenarios of land cover change: A GIS method to enable ecosystem service modelling. *Journal of Environmental Management, 92*, 563–574.

The Economics of Ecosystems and Biodiversity (TEEB). (2010). Integrating the ecological and economic dimensions in biodiversity and ecosystem service valuation. In *The economics of ecosystems and biodiversity: The ecological and economic foundations* (Chapter 1). http://www.teebweb.org.

Thomas, J., Thistlethwaite, G. MacCarthy, J., Pearson, B., Murrells, T., Pang, Y., et al. (2011). *Greenhouse gas inventories for England, Scotland, Wales and Northern Ireland: 1990–2009.* Didcot: Department of Energy and Climate Change (DECC), The Scottish Government, The Welsh Government, The Northern Ireland Department of Environment.

Thomson, A. M., Mobbs, D. C., & Milne, R. (2007). Annual inventory estimates for the UK. In A. M. Thomson & M. Van Oijen (Eds.), *Inventory and projections of UK emissions by sources and removals by sinks due to land use, land use change and forestry*. London: Centre for Ecology and Hydrology/DEFRA.

Tol, R. S. J. (2010). The economic impact of climate change. *Perspektiven der Wirtschaftspolitik, 11*, 13–37.

Turner, R. K., Morse-Jones, S., & Fisher, B. (2010). Ecosystem valuation: A sequential decision support system and quality assessment issues. *Annals of the New York Academy of Sciences, 1185*, 79–101.

UK National Ecosystem Assessment. (2011). *The UK national ecosystem assessment: Synthesis of the key findings*. Cambridge: UNEP-WCMC.

Worrall, F., Bell, M. J., & Bhogal, A. (2009). Assessing the probability of carbon and greenhouse gas benefit from the management of peat soils. *Science of the Total Environment, 408*, 2657–2666.

Chapter 14
Benefit Transfer Validity and Reliability

Randall S. Rosenberger

Abstract Benefit transfer is subject to a variety of potential errors that affect confidence in its use. Convergent validity testing is the typical focus of validity and reliability assessments of transfer applications. This chapter summarizes 38 studies that assess statistical validity and reliability (i.e., the magnitude of transfer error) in various contexts. Evidence suggests that function transfers (median of 36 % transfer error) out-perform value transfers (median of 45 % transfer error) in minimizing transfer error. However, validity tests generally reject value, model, and parameter equality. Included is a discussion of errors identified in the literature, as well as the relationship between transfer errors and the possibility of unrecognized selection biases in the valuation literature. Implications for benefit transfer practice and future research needs are also identified.

Keywords Benefit transfer · Convergent validity · Reliability · Selection effects · Transfer error

14.1 Introduction

Previous chapters have defined benefit transfer methodologies and applications. This chapter reviews tests and empirical evidence on the validity and reliability associated with benefit transfer. Benefit transfers are subject to a variety of potential errors, many of which are directly or indirectly related to issues discussed throughout this book. See also Johnston and Rosenberger (2010) for a comprehensive overview of

R.S. Rosenberger (✉)
Department of Forest Ecosystems and Society, College of Forestry,
Oregon State University, Corvallis, OR, USA
e-mail: R.Rosenberger@oregonstate.edu

R.J. Johnston et al. (eds.), *Benefit Transfer of Environmental and Resource Values*,
The Economics of Non-Market Goods and Resources 14,
DOI 10.1007/978-94-017-9930-0_14

issues associated with benefit transfer.[1] For this chapter, benefit transfer may be succinctly defined as the use of pre-existing research outcomes to predict estimates needed in other policy applications. Benefit transfer validity and reliability assess the precision and accuracy of transfers using statistical tests (i.e., validity) and transfer error (i.e., reliability), both of which fall under the general heading of convergent validity testing.

Several studies have assessed the statistical validity and magnitude of transfer error in various contexts (Bergstrom and DeCivita 1999; Brouwer and Spaninks 1999; Rosenberger and Loomis 2003; Morrison and Bergland 2006; Rosenberger and Stanley 2006; Johnston and Rosenberger 2010), as summarized below. Others have emphasized that acceptable transfer error (i.e., reliability) is context dependent, whereas transfer validity is purely a statistical issue in matching data and estimates from study sites (the source of information) with needed information at policy sites. Statistical validity is established in several ways (Hanley et al. 2006; Johnston and Duke 2008; Kristofersson and Navrud 2005; Muthke and Holm-Mueller 2004), whereas reliability depends on the acceptable level of error for a given policy context (Ben-Akiva 1981; Desvousges et al. 1998; Bergstrom and DeCivita 1999; Colombo and Hanley 2008; Kristofersson and Navrud 2005; Stapler and Johnston 2009). For example, higher degrees of precision and consequently lower transfer errors are needed as one moves from broad cost-benefit analyses for information gathering or screening of projects and policies to calculation of compensatory amounts in negotiated settlements and litigation (Navrud and Pruckner 1997).

Two general categories of benefit transfer error have been identified (Rosenberger and Stanley 2006). These categories include measurement error and generalization error. Measurement errors are associated with primary study methods and assumptions, which may lead to divergences between a true underlying value and an estimate of that value in a primary study. Measurement errors rest in the original study from which transfer estimates are derived (Bockstael and Strand 1987; McConnell 1992; Rosenberger and Stanley 2006). Any given empirical study or body of evidence has associated errors—some sources of error are systematic and related to researcher decisions while others are random. As Rosenberger and Stanley (2006, p. 374) note, "even if the process of benefit transfer were without error, the transferred value would be expected to differ from the actual value by the square root of the sum of the estimation variances of [the] two sites." Thus, in order to fully embrace measurement error in benefit transfer, analysts should evaluate information

[1]The general issues discussed in Johnston and Rosenberger (2010) include commodity consistency (Bergstrom and DeCivita 1999; Desvousges et al. 1998; Rosenberger and Stanley 2006; Smith et al. 2002), primary study methodologies (Moeltner et al. 2007; Rosenberger and Loomis 2000; Smith and Osborne 1996; Stapler and Johnston 2009), spatial patterns (Bateman et al. 2006; Desvousges et al. 1998; Hanley et al. 2003; Loomis 1996, 2000; Smith 1993), temporal trends (Brouwer 2006; Eiswerth and Shaw 1997; Zandersen et al. 2007), international transfer issues (Brouwer and Bateman 2005; Lindhjem and Navrud 2008; Ready et al. 2004; Rozan 2004), and site similarity (Colombo and Hanley 2008; Johnston 2007; Leon-Gonzalez and Scarpa 2008; Rosenberger and Phipps 2007).

sources according to research design, statistical sampling and theory, and modeling of the data generating process. For example, see Boyle et al.'s (2009) discussion on separability and specification conditions for valid transfers.

Generalization or transfer errors are errors related to the transfer process itself (Brookshire and Neill 1992; Rosenberger and Stanley 2006). Generalization errors are related to such factors as the correspondence between sites and populations, the commensurability of non-market goods and policy contexts, and the benefit transfer methods applied (Rosenberger and Phipps 2007). Thus, minimizing transfer error is in part a function of matching the contexts of study sites and a policy site (Boyle et al. 2009, 2010). Strict adherence to protocol as necessary conditions for valid transfers is not sufficient to ensure low transfer error, but may provide a greater degree of confidence in the validity and reliability of the transfer. The remainder of this chapter summarizes the benefit transfer literature on reliability and validity testing, and discusses issues associated with selection effects in a literature that may confound measurement and generalization errors in benefit transfers.

14.2 Evidence from Reliability Testing

Percent transfer error is the typical focus of reliability testing (Rosenberger and Stanley 2006). In these tests, transfer estimates are compared to a primary study estimate for the site in question. A smaller difference between the calibrated transfer estimate and a primary estimate specific to the transfer site suggests increased transfer reliability or accuracy. While transfer error is a form of convergent validity testing, it is commonly denoted reliability testing (Johnston and Rosenberger 2010). Reliability testing is an assessment of the accuracy or potential reliability of benefit transfer in repeated applications. Results of reliability testing may provide some level of confidence when conducting benefit transfers in real situations within which policy site estimates—and hence generalization errors—are not known. However, evidence from these tests is mixed, leading some authors to argue that conventional benefit transfer practices are unreliable (Smith et al. 2002).

Percentage transfer error (PTE) is calculated as:

$$\begin{aligned} \mathrm{PTE} &= [(\mathrm{V_T} - \mathrm{V_P})/\mathrm{V_P}] \times 100, \text{ or} \\ \mathrm{PTE} &= [(\mathrm{V_T}/\mathrm{V_P}) - 1] \times 100, \end{aligned} \tag{14.1}$$

where V_T is the transfer estimate and V_P is the known, or actual, estimate for the policy site. PTE then measures the degree of difference between the transferred estimate and the actual estimate at the policy site. As such, PTE requires both estimates be available, typically within the context of a primary study that has derived them.

Table 14.1 summarizes the benefit transfer literature that reported PTE, or provided enough information to calculate PTE. Table 14.1 also reports summary statistics for the absolute value of PTE (|PTE|), including the median, mean, and

Table 14.1 Convergent validity summary—absolute percentage transfer error (|PTE|)

Reference	Resource	Method[a]	Transfer type[b]	Median |PTE|	Mean |PTE|	Range	N
Loomis (1992)	Recreation	TCM	Value	20	20	4–39	10
			Function	5	6	1–18	10
Parsons and Kealy (1994)	Recreation	RUM	Value	50	48	16–75	4
			Function	3	5	1–15	7
Bergland et al. (2002)	Water quality	CVM-DB	Value	32	34	25–45	4
			Function	27	28	18–41	4
Loomis et al. (1995)	Recreation	TCM	Function	55	85	1–475	104
Downing and Ozuna (1996)	Recreation	CVM-DC	Value	38	54	0–577	552
Bowker et al. (1997)	Recreation	TCM	Value	59	84	25–341	20
			Function	38	57	0–302	20
Kirchhoff et al. (1997)	Recreation	CVM-PC	Value	38	42	24–69	12
			Function	31	63	2–228	12
Brouwer and Spaninks (1999)	Farm land	CVM-PC, OE	Value	30	31	27–36	3
			Function	28	30	22–40	3
Morrison and Bennett (2000), Morrison et al. (2002)	Wetland ecosystems	CE	Value	34	45	4–191	18
			Value-IP	42	56	13–146	8
Rosenberger and Loomis (2000)	Recreation	MA	Function	38	48	0–319	118
Piper and Martin (2001)	Water supply	CVM-OE	Function	18	39	3–149	8
Shrestha and Loomis (2001)	Recreation	MA	Function	20	22	1–51	6
VandenBerg et al. (2001)[c]	Water quality	CVM-PC	Value-sites	na	42	1–239	132
			Function-sites	na	44	0–298	132
			Value-pooled	na	31	0–105	12
			Function-pooled	na	18	1–56	12
Barton (2002)	Water quality	CVM-DB	Value	21	20	10–30	8
			Function	22	21	2–29	8
Chattopadhyay (2003)	Air quality	HPM	Value	38	159	8–1491	78
			Function	36	127	9–929	78
Shrestha and Loomis (2003)	Recreation	MA	Function	58	84	12–411	34
Jeong and Haab (2004)	Recreation	RUM	Function	36	39	11–66	20
Muthke and Holm-Mueller (2004)	Water quality	CVM-DB	Value	59	220	13–946	32
			Function	146	269	1–858	8

(continued)

Table 14.1 (continued)

Reference	Resource	Method[a]	Transfer type[b]	Median |PTE|	Mean |PTE|	Range	N
Ready et al. (2004), Ready and Navrud (2007)	Human health	CVM-IB, PC	Value	33	37	20–81	14
			Function	31	37	20–83	7
Rozan (2004)	Air quality	CVM-IB, OE	Function	28	25	16–30	4
Brouwer and Bateman (2005)	Human health	CVM-OE	Value	31	46	0–123	12
			Function	31	34	3–73	12
Groothius (2005)	Recreation	TCM	Value	32	29	1–69	30
			Function	18	22	1–64	30
		CVM-DC	Value	29	34	2–136	30
			Function	26	34	0–135	30
Jiang et al. (2005)	Coastal land	CE	Value-IP	55	57	19–101	25
			Function	64	68	53–85	5
Hanley et al. (2006)	Aquatic ecosystem	CE	Value-IP sites	109	172	58–419	12
			Value-IP pooled	54	71	23–212	24
Kerr and Sharp (2006)	Aquatic ecosystem	CE	Value-IP	59	96	2–704	22
			Function-IP	61	108	2–704	22
Morrison and Bennett (2004, 2006)	Aquatic ecosystem	CE	Value	33	120	3–1366	40
			Value-IP	18	35	1–171	28
Colombo et al. (2007)	Soil	CE	Value	72	207	8–4575	108
			Value-IP	18	29	0–257	30
Eshet et al. (2007)	Waste stations	HPM	Value	20	18	1–46	16
Johnston (2007)	Develop	CE	Value-IP	32	37	7–101	24
Kristofersson and Navrud (2007)	Aquatic ecosystem	CVM-PC	Value	68	96	9–319	18
			Function	48	75	8–210	18
Zandersen et al. (2007)	Recreation	RUM	Function	37	53	1–229	52
Birr-Pedersen (2008)	Forest	HPM	Value	58	75	11–247	60
Colombo and Hanley (2008)	Farm land	CE	Value	72	446	2–7496	288
Lindhjem and Navrud (2008)	Forest ecosystem	MA	Value-average	93	192	4–1157	51
			Value-best point	12	71	1–482	25
			Function-all	70	126	10–596	26
			Function-optimized	37	47	2–266	26

(continued)

Table 14.1 (continued)

Reference	Resource	Method[a]	Transfer type[b]	Median |PTE|	Mean |PTE|	Range	N
Matthews et al. (2009)	Recreation	CVM-DB	Value	22	34	0–160	42
			Function	16	27	0–125	84
Stapler and Johnston (2009)[a]	Recreation	MA	Function-mean	69	100	0–736	372
			Function-best	47	80	0–1023	372
Baskaran et al. (2010)	Ecosystem services	CE	Value	39	41	1–95	48
			Value-IP	43	71	0–868	96
Londoño and Johnston (2012)	Coral reefs	MA	Function-all	57	101	na	na
			Function-CVM	45	165	na	na
Total			Value	45	140 (10.6)[d]	0–7496	1792
			Function	36	65 (4.0)[d]	0–929	756

[a]Method abbreviations include TCM (travel cost model), RUM (random utility model), HPM (hedonic property method), CVM (contingent valuation method), CVM-DB (CVM double-bounded), CVM-DC (CVM dichotomous choice), CVM-PC (CVM payment card), CVM-IB (CVM iterative bidding), CVM-OE (CVM open-ended), CE (choice experiment), and MA (meta-analysis)
[b]Transfer types include value-based transfers and function-based transfers. Value transfers include across individual sites, across grouped sites (i.e., "pooled"), and using averages, point estimates, or implicit prices (Value-IP). Function transfers include across sites, across grouped sites (i.e., "pooled"), and using mean functions, best functions, and various model specifications (i.e., full vs. optimized)
[c]Not included in totals
[d]Standard error of the mean

range, as well as the number of PTE estimates provided (N). Each study is further classified by resource type, primary valuation method, and benefit transfer category (i.e., value or function transfer). There are 38 studies listed in Table 14.1, with nearly half (n = 18) of them evaluating value and function transfers within the same context (recreation). These 38 evaluations provide 1792 PTE for value transfers (inclusive of point estimate, average value, and implicit prices) and 756 PTE for function transfers (inclusive of benefit, demand, and meta-analysis functions).

Value transfer |PTE| ranged from very low (<1 %) to very high (>7000 %), with a mean |PTE| of 140 % and standard error of 10.6. Function transfer |PTE| ranged from very low (<1 %) to high (>900 %), with a mean |PTE| of 65 % and standard error of 4.0. Based solely on these statistics, the null hypothesis that value transfer and function transfer |PTE| are equal is rejected using a t-test (p-value < 0.01); this evidence supports the popular conclusion that function transfers perform better (i.e., generally have lower PTE) than value transfers (Johnston and Rosenberger 2010). The median |PTE|'s for value transfers (45 %) and function transfers (36 %) are

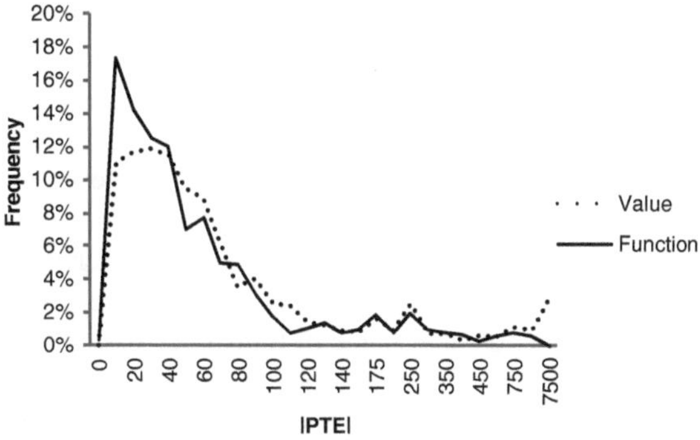

Fig. 14.1 Absolute value of percentage transfer error (|PTE|) distributions for value transfers and function transfers across studies in Table 14.1

similar in magnitude, although function transfer median |PTE| is still statistically lower than value transfer median |PTE| using the Mann-Whitney U-test (p-value < 0.01). The differences between the mean and median values for the distributions of |PTE| for value and function transfers is illustrated in Fig. 14.1. This figure shows that function transfer |PTE| has greater mass in the lower range, whereas value transfer |PTE| has greater mass in the upper range, as noted previously.

It is generally assumed that the use of benefit transfer best practices will reduce transfer error. The definition of these practices is a common theme in benefit transfer reviews (Boyle et al. 2009, 2010; Johnston and Rosenberger 2010). However, a quantitative assessment of factors associated with varying transfer errors is not available. Nonetheless, a few patterns are suggested from the benefit transfer literature. First, errors are generally found to be smaller in cases where sites and populations are more similar (Rosenberger and Phipps 2007). Studies that illustrate the importance of site correspondence include Barton (2002), Colombo and Hanley (2008), Johnston (2007), Loomis (1992), Morrison and Bennett (2004), (2006), Morrison et al. (2002), Piper and Martin (2001), Rosenberger and Loomis (2000), and VandenBerg et al. (2001).

Second, function transfers perform generally better than value transfers. Intra-study comparisons of value transfer versus function transfer show that function transfers result in lower mean and median |PTE| than value transfers the majority of the time (e.g., Bergland et al. 2002; Bowker et al. 1997; Boyle et al. 2010; Groothius 2005; Kirchhoff et al. 1997; Kristofersson and Navrud 2007; Loomis 1992; Matthews et al. 2009; Parsons and Kealy 1994; VandenBerg et al. 2001). There are a few studies, however, (e.g., Barton 2002; Muthke and Holm-Mueller 2004) that have found value transfer to systematically have lower PTE than function transfers. This difference may be at least partially explained by recent work of Bateman et al. (2011, p. 383), who argue that "the choice of [value vs. function transfer] depends crucially upon the degree of similarity of the sites under consideration." They find that value transfer can outperform function transfer for highly similar sites and commodities, but that as sites become less similar the relative performance of value transfer declines. Nonetheless, when comparing |PTE| for value versus function transfers across the entire literature, function transfers seem to outperform value transfers, as noted previously.

Third, the summary |PTE| reported in Table 14.1 demonstrates the possible range in transfer errors. However, these studies are experiments where the underlying policy site value is also estimated. In these experiments some estimates are compared because they *can* be compared, not because they should—these naïve transfers illustrate the potential range of transfer errors, including sometimes extremely large errors. Some of these represent errors that would result from transfers that would not be used, or would rarely be used, in policy settings. For example, Lindhjem and Navrud (2008) compare several transfers, including average value from the literature to selecting the best representative sample from the literature to meta-analysis function transfers. Their results show that PTE magnitude and range are reduced when screening for best fit or using a meta-analysis function to predict policy site values. And finally, interpretations of PTE as validity indicators in transfers is weak because study site values are themselves estimated with error, leaving real transfer errors largely unknown (Boyle et al. 2010).

14.3 Evidence from Validity Testing

Another set of convergent validity tests applied in the literature are referred to as transfer validity tests (Johnston and Rosenberger 2010). Transfer validity tests evaluate the statistical equality of transferable components, including model parameters, implicit prices, and welfare estimates. In transfer validity testing, the assessed components are presumed equal unless statistical tests reject a null hypothesis of equality. Results of the tests often differ depending on the measure considered (Johnston 2007). In contrast to transfer error, which provides a general indication of transfer performance, validity testing imposes strict conditions on assessing transferability. Although, as noted below, there is little correlation between passing validity tests and minimizing transfer error due to a multitude of confounding factors.

There are two general types of convergent validity tests—the difference in value test and the difference in model coefficients test (Bergstrom and DeCivita 1999). Simply stated, the difference in value test statistically compares the transfer value V_T from a study site with the policy site value V_P, testing for statistical equality using t-test or F-test, overlapping confidence intervals (Kirchhoff et al. 1997) or convolutions (Colombo et al. 2007; Poe et al. 1994, 2005; Rozan 2004).

$$H_0 : V_T = V_P. \tag{14.2}$$

In this case, the V_T estimate may be an unadjusted point estimate; point estimate adjusted for income, purchasing power, or exchange rate differences; or an average point estimate (Brouwer and Bateman 2005; Brouwer and Spaninks 1999; Muthke and Holm-Mueller 2004; Ready et al. 2004). For choice experiments, equality of value estimates are statistically tested by comparing implicit prices (H_0: $IP_T = IP_P$) or value estimates for policy scenarios, as in Eq. 14.2 (Colombo et al. 2007; Hanley et al. 2006; Jiang et al. 2005; Johnston and Duke 2009; Morrison et al. 2002). Value estimates predicted from function transfers, whether benefit functions (Muthke and Holm-Mueller 2004) or meta-analysis functions (Shrestha and Loomis 2001, 2003), have similarly been tested:

$$H_0 : V_T(\beta_S, X_P) = V_P, \tag{14.3}$$

where the function fit to the study site data with parameters β_S is adapted through measures of policy site variables X_P to predict the policy site value.

Complementary to the difference in value test is the difference in model coefficients test. There are essentially two forms of this test: equality of parameters and equality of models (Bergland et al. 2002; Brouwer and Bateman 2005; Brouwer and Spaninks 1999; Muthke and Holm-Mueller 2004). For the equality of parameters test, the parameters for the function derived for a study site are assumed equal to the parameters of the policy site:

$$H_0 : \beta_S = \beta_P, \tag{14.4}$$

and may be tested using Wald, Chow, Lagrange Multiplier or Likelihood Ratio tests (Brouwer and Spaninks 1999). Alternatively, the models estimated for the study site and policy site may be tested whether they belong to a common function:

$$H_0 : \beta = \beta_S = \beta_P, \tag{14.5}$$

where β is the pooled model and may be tested using Wald, Chow, Lagrange Multiplier or Likelihood Ratio tests (Brouwer and Spaninks 1999).

Table 14.2 summarizes a sample of convergent validity hypothesis tests from the benefit transfer literature. The table reports the number of hypothesis tests and the percent of those tests that were rejected at the 95 % confidence level. The average of median absolute percentage transfer error (|PTE|) for a study from Table 14.1 is reported as well. The types of difference in means tests (Eqs. 14.2 and 14.3) and difference in model coefficients tests (Eqs. 14.4 and 14.5) are not differentiated, nor are other features such as the resource, valuation method, or transfer type.

The number of difference in value tests varied widely across the studies depending on the number of study sites, implicit price comparisons, and/or the combination of policy scenarios evaluated. The equality of transfer estimate and policy site estimate rejection rate ranged from no rejections (0 %) to full rejection (100 %). The average rejection rate across studies (i.e., each study has equal weight) is 63 %, whereas the average rejection rate across all tests (i.e., each test has equal weight) is 55 %. Thus, evidence from the benefit transfer literature suggests that convergent validity in value estimates is rejected most of the time.

Table 14.2 Convergent validity summary—hypothesis tests

Reference	H_0: $V_T = V_P$ H_0: $V_T(\beta_S, X_P) = V_P$ H_0: $IP_T = IP_P$		H_0: $\beta_S = \beta_P$ H_0: $\beta = \beta_S = \beta_P$		Average median |PTE|
	No. tests	% rejected	No. tests	% rejected	
Parsons and Kealy (1994)			12	67	26
Loomis et al. (1995)			7	100	55
Downing and Ozuna (1996)	61	55	256	35	38
Bowker et al. (1997)	20	40			48
Kirchhoff et al. (1997)	30	67	3	100	35
Brouwer and Spaninks (1999)	6	83	8	38	29
Shrestha and Loomis (2001)	6	33			20
VandenBerg et al. (2001)	78	44[a]	78	35[a]	34[b]
Barton (2002)	16	88	8	50	22
Bergland et al. (2002)	4	100	1	100	30
Morrison et al. (2002)	17	53	2	100	38
Shrestha and Loomis (2003)	2	100	4	25	58
Muthke and Holm-Mueller (2004)	40	100	4	50	102
Rozan (2004)	4	100	2	0	28
Brouwer and Bateman (2005)	6	50	12	100	31
Groothius (2005)—TCM			30	87	25
Groothius (2005)—CVM			30	20	27
Jiang et al. (2005)	5	100	5	80	60
Hanley et al. (2006)	6	100	2	100	82
Kerr and Sharp (2006)	11	0	1	100	60
Colombo et al. (2007)	69	83	3	100	45
Johnston (2007)	42	21	6	33	32
Kristofersson and Navrud (2007)	42	86			58
Zandersen et al. (2007)	104	12	2	100	37
Birr-Pedersen (2008)	60	43			58
Colombo and Hanley (2008)	144	72			72
Johnston and Duke (2009)	24	29			na
Baskaran et al. (2010)	72	22	6	100	41
Total	869	55	482	44	
Correlation with |PTE|		0.26		0.17	

[a]All tests at $p < 0.05$ except VandenBerg et al. (2001) where $p < 0.10$
[b]Average |PTE| based on mean |PTE| from Table 14.1

Not surprisingly, the rejection rate of estimate equality is positively correlated with the average median |PTE| across studies; however, this correlation is not very high (r = 0.26).

Similar patterns are found for the difference in model coefficients tests. The number of tests varied widely across studies depending on the number of sites, models, and parameter comparisons. The equality of parameters and models rejection rates ranged from no rejections (0 %) to full rejection (100 %). The average rejection rate across studies (i.e., each study has equal weight) is 69 %; however, the average rejection rate across tests (i.e., each test has equal weight) is 44 %. Thus, if evaluating the literature based on aggregate measures per study, one would conclude that model coefficient and model equality are rejected the majority of the time. But if evaluating based on individual tests of model coefficient and model equality, the few studies that make several comparisons (e.g., Downing and Ozuna 1996; Groothius 2005; VandenBerg et al. 2001) on average would fail to reject model coefficient and model equality. The difference in model coefficient test rejection rates are weakly, positively correlated with |PTE| (r = 0.17).

Contrary to general conclusions on benefit transfer reliability drawn from PTE calculations (Table 14.1), Table 14.2 seems to suggest that benefit transfer generally fails convergent validity hypothesis testing. However, validity testing often ignores the context of benefit transfers and acceptable level of accuracy, as well as the counterintuitive result that less efficient statistical estimates (i.e., larger standard errors) have a higher probability of failing to reject equality compared to more efficient estimates (i.e., smaller standard errors), implying greater transferability (Kristofersson and Navrud 2005). Nonetheless, standard hypothesis tests remain the norm in the benefit transfer literature (Lindhjem and Navrud 2008).

Alternatives have been proposed, but are not yet widely adopted (Ben-Akiva 1981; Desvousges et al. 1998; Lerman 1981; Lindhjem and Navrud 2008; Spash and Vatn 2006). An example is the literature that applies equivalence testing within benefit transfer (Hanley et al. 2006; Johnston 2007; Johnston and Duke 2008; Kristofersson and Navrud 2005; Muthke and Holm-Mueller 2004). Equivalence testing changes the burden of proof in traditional hypothesis testing by reversing the null and alternative hypotheses; i.e., estimates are assumed different unless hypothesis tests show the difference is smaller than a specified tolerance limit and probability value. In benefit transfer applications, the tolerance limit is specified as the maximum acceptable transfer error in which the transfer and policy estimates are considered equivalent. Recent advances include variants of the traditional equivalence test based on the difference between independent empirical distributions that permit valid inference for non-normal distributions (Johnston and Duke 2008). Kristofersson and Navrud (2007) and Johnston and Duke (2008) show limited support for transfers using tolerance limits of 40 %, and Baskaran et al. (2010) at 50 and 80 %. However, there are too few empirical applications of equivalence testing upon which to base any general conclusions or implications for benefit transfer.

14.4 Selection Effects

Benefit transfer depends upon information available in the literature, or the breadth, depth and quality of primary research (Loomis and Rosenberger 2006). Implicit within benefit transfers applications are key assumptions about the literature, including that it is a random, unbiased sample of the population of empirical estimates, and that empirical estimates are unbiased representations of the true, underlying values (Rosenberger and Phipps 2007). However, if a literature contains selection effects, then transfers based on it may be biased (Rosenberger and Johnston 2009). Selection effects reflect a type of measurement error, or the ability of the literature as a whole to provide an unbiased representation of true value distributions. Such concerns are most often noted for the case of meta-analysis (Rosenberger and Johnston 2009), but apply to all types of transfer.

Discussions of selection effects are sparse and scattered throughout the literature—Bergstrom and Taylor (2006), Florax (2002), Hoehn (2006), Nelson and Kennedy (2009), Rosenberger and Johnston (2009), Rosenberger and Stanley (2006), and Smith and Pattanayak (2002). Rosenberger and Johnston (2009) identify four potential sources of selection bias in any given body of literature, including research priority selection, methodology selection, publication selection, and metadata sample selection. By not accounting for selection biases in the literature, these biases may carry over into empirical benefit transfers.

Research priority selection may be driven by sociopolitical circumstances such as the societal awareness and perceived importance of a particular resource or valuation context that leads to a willingness for sponsors to fund, and scientists to study, them (Hoehn 2006; Rosenberger and Johnston 2009). The primary issue with research priority selection is that a collection of empirical estimates (i.e., the literature) may not exhibit a representative range of possible estimates. In other words, empirical studies do not randomly select the resources and contexts to study and therefore do not represent the full domain (or population) of potential estimates. For example, if all studies that estimate the value of whitewater rafting are based on samples of users for high-profile or high-value locations, then a transfer estimate derived from this literature would arguably overestimate the value of whitewater rafting at lower-profile locations. This overestimation, in fact, is what Hoehn (2006) measured for wetland research in the U.S.—a four-fold exaggeration of value if the wetland valuation literature was used in an assessment of a generic wetland resource. However, this does not imply that all literatures are biased due to research priority selection. Some literatures or databases of non-market values may have covered a range of resource contexts that more closely resembles a random sample of empirical estimates; e.g., recreation values (Rosenberger and Johnston 2009).

Methodology selection, in turn, affects estimation and further complicates benefit transfers when methodological characteristics are significant determinants of the variation in estimates. Researchers must make choices when designing valuation studies, including modeling approach, model estimation, survey design and implementation, and treatments of the data (Rosenberger and Johnston 2009).

These choices, in turn, affect research outcomes, sometimes in systematic ways as shown in meta-analyses (Bateman and Jones 2003; Brouwer et al. 1999; Johnston et al. 2003, 2005, 2006a, b; Moeltner et al. 2007; Poe et al. 2001; Rosenberger and Loomis 2000; Smith and Osborne 1996; Stapler and Johnston 2009). The primary issue with methodological selection in benefit transfers is how to address it within benefit transfers. The appropriate response, at least in part, depends on whether these effects are interpreted as pervasive biases or expected theoretical and empirical patterns (Johnston and Rosenberger 2010). Regardless of what one believes, systematic patterns in methodological effects influence transfer validity and reliability. Even when these patterns are measured in meta-analysis a question remains on how to address them in subsequent benefit transfers (Johnston et al. 2006a; Moeltner et al. 2007; Rosenberger and Johnston 2009; Stapler and Johnston 2009).

Publication selection may bias a literature, as found in many areas of inquiry (Florax 2002; Rosenberger and Johnston 2009; Rosenberger and Stanley 2006; Stanley 2005, 2008). Once again, publication selection bias leads to a nonrandom sample of empirical evidence. Card and Kreuger (1995, p. 239) identify three potential sources of publication selection bias in economics: (1) reviewers and editors may be predisposed to accept papers consistent with the conventional view; (2) researchers may use the presence of conventionally expected results as a model selection test or are unwilling to report estimates outside the range of previously reported values; and (3) everyone may possess a predisposition to treat "statistically significant" results more favorably. Smith and Pattanayak (2002, p. 273) add (4) most journals in environmental and resource economics are not interested in new estimates of values for their own sake, instead selecting manuscripts based primarily on methodological innovations and contributions. The primary issue of publication selection for benefit transfer is that empirical evidence may be skewed, leading to over- or underestimation of true, underlying values (Rosenberger and Johnston 2009; Stanley 2005, 2008).

Several methods have been devised for identifying, measuring, and correcting for publication selection bias (Florax 2002; Stanley 2005, 2008). Methods range from simply suggesting a comprehensive search of the literature, including the gray literature (e.g., theses, dissertations, working papers, and reports), to weighting schemes based on the standard error of value estimates or other transferable empirical estimates. The standard error of empirical estimates is gaining a central role in the issue of dealing with publication selection bias by giving greater weight to more precise (i.e., higher quality, larger primary study sample sizes) estimates in a literature (Stanley and Doucouliagos 2010; Stanley et al. 2010).

A logical strategy that benefit transfer analysts may take, given priority and publication selection, is to selectively choose studies that most closely correspond to the policy context and/or have greater precision in estimates—a strategy intended to reduce transfer error and increase validity. However, important information may be lost by systematically excluding empirical evidence. Just as sample selection is known to bias estimates of value if not corrected in primary data models (Bateman et al. 2002; Garrod and Willis 1999), it is also a relevant concern for benefit transfers (Bergstrom and Taylor 2006; Moeltner and Rosenberger 2008; Rosenberger and Johnston 2009). The primary importance of sample selection for benefit transfer is how narrow to define the policy context and other selection criteria. The tradeoff is identifying too few estimates to fully characterize the distribution of estimates (Moeltner et al. 2007), potentially exacerbating other selection effects, with additional information and variance in a broader dataset—the optimal scope problem of Moeltner and Rosenberger (2008).

Selection effects comprise a relatively new area of inquiry for benefit transfer, although it has a long history in meta-analysis for a variety of disciplines. Systematic evaluations of the various types of selection effects, including methods to detect and correct for their biases, and their relative pervasiveness within a given literature are needed. The literature also needs to expand so that broad representations of values and their contexts are available, as well as re-sampling and re-estimating of values over time. In the meantime, benefit transfer analysts should be cognizant of potential selection effects in any literature and use good judgment when deriving transferable information from it.

14.5 Conclusions

Benefit transfer is a widely used method for deriving information from existing research outcomes, but many analysts remain skeptical of its validity and reliability. This is in large part due to the variability in transfer errors (Table 14.1) and lack of statistical equivalence of estimates and models in benefit transfer experiments (Table 14.2). If analysts cannot demonstrate benefit transfer to be a valid and reliable method when the target (i.e., policy) estimate is known, then how confident will they be in contexts for which the target value is not known? There is a need for more applications and systematic testing of benefit transfer methods along with an analysis of relationships between attributes of benefit transfer (e.g., methods, contexts, and data) and transfer error (Boyle et al. 2009; Rosenberger and Phipps 2007). A few systematic patterns have emerged of which analysts should be aware (Johnston and Rosenberger 2010), but there is much yet to be discovered, including assessment strategies that go beyond existing convergent validity measurement and testing (Boyle et al. 2010; Spash and Vatn 2006). A logical outcome of broad assessments of benefit transfer applications would be best practice guidelines, as called for by Smith (1992) and reiterated over the past two decades (Bergstrom and DeCivita 1999; Johnston and Rosenberger 2010; Wilson and Hoehn 2006). The development of and adherence to best practice guidelines, however, is no panacea for valid and reliable benefit transfers (Boyle et al. 2009). But they do have the potential to increase confidence in applied transfers by mitigating factors known to increase transfer error. The importance of benefit transfer to policy assessments coupled with its controversies in use lends support to the need for future research targeting benefit transfer validity and reliability.

The availability and accessibility to data limits assessments and use of them in benefit transfers (Morrison 2001), including the potential effects of a variety of selection biases (Rosenberger and Johnston 2009) that may affect benefit transfer validity. However, the solution does not reside solely in improved primary study reporting and data accessibility, as suggested by Loomis and Rosenberger (2006), although this is important for the overall credibility of benefit transfers. Primary studies should be designed that not only adequately address relevant research questions, but also with an eye toward their potential future uses in benefit transfer, which may increase their transfer validity and reduce transfer error by rigorously defining the study context, reporting of details, and generalizing the scope of outcomes. Augmenting transfer functions with secondary data (e.g., demographic profiles, GIS information) also has been shown to be an important predictor in benefit transfer models (Brander et al. 2006; Eade and Moran 1996; Moeltner et al. 2009; Troy and Wilson 2006) and has the potential to increase the relevance of primary study outcomes and reduce transfer error.

And finally, more contextually-relevant research is needed for applied benefit transfer. Not all decisions need the same degree of accuracy or level of precision, but all transfers will be subject to some error. Under what conditions, and at what level, is uncertainty acceptable in a policy context? Several approaches to deal with uncertainty have been proposed, such as Boyle et al.'s (2010) bounding approach or Akter and Grafton's (2010) risk and simulation approach. Both approaches embrace transfer error and other sources of uncertainty in the decision model. The more we learn about applied research, its outcomes, and how they are distributed across space, time and other dimensions, the better off we will be when applying benefit transfers to real world issues.

References

Akter, S., & Grafton, R. Q. (2010). Confronting uncertainty and missing values in environmental value transfer as applied to species conservation. *Conservation Biology, 24*, 1407–1417.

Barton, D. N. (2002). The transferability of benefit transfer: Contingent valuation of water quality improvements in Costa Rica. *Ecological Economics, 42*, 147–164.

Baskaran, R., Cullen, R., & Colombo, S. (2010). Testing different types of benefit transfer in valuation of ecosystem services: New Zealand winegrowing case studies. *Ecological Economics, 69*, 1010–1022.

Bateman, I. J., Brouwer, R., Ferrini, S., Schaafsma, M., Barton, D. N., Dubgaard, A., et al. (2011). Making benefit transfers work: Deriving and testing principles for value transfers for similar and dissimilar sites using a case study of the non-market benefits of water quality improvements across Europe. *Environmental and Resource Economics, 50*, 365–387.

Bateman, I. J., Carson, R. T., Day, B., Hanemann, M., Hanley, N., Hett, T., et al. (2002). *Economic valuation with stated preference techniques: A manual*. Cheltenham, UK: Edward Elgar.

Bateman, I. J., Day, B. H., Georgiou, S., & Lake, I. (2006). The aggregation of environmental benefit values: Welfare measures, distance decay and total WTP. *Ecological Economics, 60*, 450–460.

Bateman, I. J., & Jones, A. P. (2003). Contrasting conventional with multi-level modeling approaches to meta-analysis: Expectation consistency in U.K. woodland recreation values. *Land Economics, 79*, 235–258.

Ben-Akiva, M. (1981). Issues in transferring and updating travel-behavior models. In P. R. Stopher, A. H. Meyburg, & W. Borg (Eds.), *New horizons in travel-behavior research* (pp. 665–686). Lexington, MA: Lexington Books.

Bergland, O., Magnussen, K., & Navrud, S. (2002). Benefit transfer: Testing for accuracy and reliability. In R. J. G. M. Florax, P. Nijkamp, & K. G. Willis (Eds.), *Comparative environmental economic assessment* (pp. 117–132). Cheltenham, UK: Edward Elgar.

Bergstrom, J. C., & DeCivita, P. (1999). Status of benefit transfer in the United States and Canada: A review. *Canadian Journal of Agricultural Economics, 47*, 79–87.

Bergstrom, J. C., & Taylor, L. O. (2006). Using meta-analysis for benefits transfer: Theory and practice. *Ecological Economics, 60*, 351–360.

Birr-Pedersen, K. (2008). *Measurement and benefit transfer of amenity values from afforestation projects—A spatial economic valuation approach using GIS technology*. Ph.D. Thesis. Denmark: National Environmental Research Institute, University of Aarhus, 197p.

Bockstael, N. E., & Strand, I. E. (1987). The effect of common sources of regression error on benefit estimates. *Land Economics, 63*, 11–20.

Bowker, J. M., English, D. B. K. & Bergstrom, J. C. (1997). *Benefit transfer and count data travel cost models: An application and test of a varying parameter approach with guided whitewater rafting*. Working paper FS 97-03. Athens, GA: University of Georgia.

Boyle, K. J., Kuminoff, N. V., Parmeter, C. F., & Pope, J. C. (2009). Necessary conditions for valid benefit transfers. *American Journal of Agricultural Economics, 91*, 1328–1334.

Boyle, K. J., Kuminoff, N. V., Parmeter, C. F., & Pope, J. C. (2010). The benefit-transfer challenges. *Annual Review of Resource Economics, 2*, 161–182.

Brander, L. M., Florax, R. J. G. M., & Vermaat, J. E. (2006). The empirics of wetland valuation: A comprehensive summary and a meta-analysis of the literature. *Environmental and Resource Economics, 33*, 223–250.

Brookshire, D. S., & Neill, H. R. (1992). Benefit transfers: Conceptual and ethical issues. *Water Resources Research, 28*, 651–655.

Brouwer, R. (2006). Do stated preference methods stand the test of time? A test of the stability of contingent values and models for health risks when facing an extreme event. *Ecological Economics, 60*, 399–406.

Brouwer, R., & Bateman, I. J. (2005). Benefits transfer of willingness to pay estimates and functions for health-risk reductions: A cross-country study. *Journal of Health Economics, 24*, 591–611.

Brouwer, R., Langford, I. H., Bateman, I. J., & Turner, R. K. (1999). A meta-analysis of wetland contingent valuation studies. *Regional Environmental Change, 1*, 47–57.

Brouwer, R., & Spaninks, F. A. (1999). The validity of environmental benefits transfer: Further empirical testing. *Environmental and Resource Economics, 14*, 95–117.

Card, D., & Kreuger, A. B. (1995). Time-series minimum-wage studies: A meta-analysis. *American Economic Review, 85*, 238–243.

Chattopadhyay, S. (2003). A repeated sampling technique in assessing the validity of benefit transfer in valuing non-market goods. *Land Economics, 79*, 576–596.

Colombo, S., Calatrava-Requena, J., & Hanley, N. (2007). Testing choice experiment for benefit transfer with preference heterogeneity. *American Journal of Agricultural Economics, 89*, 135–151.

Colombo, S., & Hanley, N. (2008). How can we reduce the errors from benefits transfer? An investigation using the choice experiment method. *Land Economics, 84*, 128–147.

Desvousges, W. H., Johnson, F. R., & Banzhaf, H. S. (1998). *Environmental policy analysis with limited information: Principles and applications of the transfer method*. Cheltenham, UK: Edward Elgar.

Downing, M., & Ozuna, T, Jr. (1996). Testing the reliability of the benefit function transfer approach. *Journal of Environmental Economics and Management, 30*, 316–322.

Eade, J. D. O., & Moran, D. (1996). Spatial economic valuation: Benefits transfer using geographical information systems. *Journal of Environmental Management, 48*, 97–110.

Eiswerth, M., & Shaw, W. (1997). Adjusting benefit transfer values for inflation. *Water Resources Research, 33*, 2381–2385.

Eshet, T., Baron, M. G., & Schechter, M. (2007). Exploring benefit transfer: Disamenities of waste transfer stations. *Environmental and Resource Economics, 37*, 521–547.

Florax, R. J. G. M. (2002). Methodological pitfalls in meta-analysis: Publication bias. In R. J. G. M. Florax, P. Nijkamp, & K. G. Willis (Eds.), *Comparative environmental economic assessment* (pp. 177–207). Cheltenham, UK: Edward Elgar.

Garrod, G., & Willis, K. G. (1999). *Economic valuation of the environment: Methods and case studies*. Cheltenham, UK: Edward Elgar.

Groothius, P. A. (2005). Benefit transfer: A comparison of approaches. *Growth and Change, 36*, 551–564.

Hanley, N., Schlapfer, F., & Spurgeon, J. (2003). Aggregating the benefits of environmental improvements: Distance-decay functions for use and nonuse values. *Journal of Environmental Management, 68*, 297–304.

Hanley, N., Wright, R. E., & Alvarez-Farizo, B. (2006). Estimating the economic value of improvements in river ecology using choice experiments: An application to the water framework directive. *Journal of Environmental Management, 78*, 183–193.

Hoehn, J. P. (2006). Methods to address selection effects in the meta regression and transfer of ecosystem values. *Ecological Economics, 60*, 389–398.

Jeong, H. & Haab, T. (2004). *The economic value of marine recreational fishing: Applying benefit transfer to marine recreational fisheries statistics survey (MRFSS)*. Working paper AEDE-WP-0039-04. Columbus, OH: Ohio State University.

Jiang, Y., Swallow, S. K., & McGonagle, M. P. (2005). Context-sensitive benefit transfer using stated choice models: Specification and convergent validity for policy analysis. *Environmental and Resource Economics, 31*, 477–499.

Johnston, R. J. (2007). Choice experiments, site similarity and benefits transfer. *Environmental and Resource Economics, 38*, 331–351.

Johnston, R. J., Besedin, E. Y., Iovanna, R., Miller, C., Wardwell, R., & Ranson, M. (2005). Systematic variation in willingness to pay for aquatic resource improvements and implications for benefit transfer: A meta-analysis. *Canadian Journal of Agricultural Economics, 53*, 221–248.

Johnston, R. J., Besedin, E. Y., & Ranson, M. H. (2006a). Characterizing the effects of valuation methodology in function-based benefits transfer. *Ecological Economics, 60*, 407–419.

Johnston, R. J., Besedin, E. Y., & Wardwell, R. F. (2003). Modeling relationships between use and nonuse values for surface water quality: A meta-analysis. *Water Resources Research, 39*, 1363–1372.

Johnston, R. J., & Duke, J. M. (2008). Benefit transfer equivalence tests with non-normal distributions. *Environmental and Resource Economics, 41*, 1–23.

Johnston, R. J., & Duke, J. M. (2009). Willingness to pay for land preservation across states and jurisdictional scale: Implications for benefit transfer. *Land Economics, 85*, 217–237.

Johnston, R. J., Ranson, M. H., Besedin, E. Y., & Helm, E. C. (2006b). What determines willingness to pay per fish? A meta-analysis of recreational fishing values. *Marine Resource Economics, 21*, 1–32.

Johnston, R. J., & Rosenberger, R. S. (2010). Methods, trends and controversies in contemporary benefit transfer. *Journal of Economic Surveys, 24*, 479–510.

Kerr, G. N., & Sharp, B. M. H. (2006). Transferring mitigation values for small streams. In J. Rolfe & J. Bennett (Eds.), *Choice modelling and the transfer of environmental values* (pp. 136–163). Cheltenham, UK: Edward Elgar.

Kirchhoff, S., Colby, B. G., & LaFrance, J. R. (1997). Evaluating the performance of benefit transfer: An empirical inquiry. *Journal of Environmental Economics and Management, 33*, 75–93.

Kristofersson, D., & Navrud, S. (2005). Validity tests of benefit transfer – are we performing the wrong tests? *Environmental and Resource Economics, 30*, 279–286.

Kristofersson, D., & Navrud, S. (2007). Can use and non-use values be transferred across countries? In S. Navrud & R. Ready (Eds.), *Environmental values transfer issues and methods* (pp. 207–225). Dordrecht, The Netherlands: Springer.

Leon-Gonzalez, R., & Scarpa, R. (2008). Improving multi-site benefit functions via Bayesian model averaging: A new approach to benefit transfer. *Journal of Environmental Economics and Management, 56*, 50–68.

Lerman, S. R. (1981). A comment on interspatial, intraspatial, and temporal transferability. In P. R. Stopher, A. H. Meyburg, & W. Borg (Eds.), *New horizons in travel-behavior research* (pp. 628–632). Lexington, MA: Lexington Books.

Lindhjem, H., & Navrud, S. (2008). How reliable are meta-analyses for international benefit transfers? *Ecological Economics, 66*, 425–435.

Londoño, L. M., & Johnston, R. J. (2012). Enhancing the reliability of benefit transfer over heterogeneous sites: A meta-analysis of international coral reef values. *Ecological Economics, 78*, 80–89.

Loomis, J. B. (1992). The evolution of a more rigorous approach to benefit transfer: Benefit function transfer. *Water Resources Research, 28*, 701–705.

Loomis, J. B. (1996). How large is the extent of the market for public goods: Evidence from a nationwide contingent valuation survey. *Applied Economics, 28*, 779–782.

Loomis, J. B. (2000). Vertically summing public good demand curves: An empirical comparison of economic versus political jurisdictions. *Land Economics, 76*, 312–321.

Loomis, J. B., Roach, B., Ward, F., & Ready, R. (1995). Testing transferability of recreation demand models across regions: A study of corps of engineer reservoirs. *Water Resources Research, 31*, 721–730.

Loomis, J. B., & Rosenberger, R. S. (2006). Reducing barriers in future benefit transfers: Needed improvements in primary study design and reporting. *Ecological Economics, 60*, 343–350.

Matthews, D. I., Hutchinson, W. G., & Scarpa, R. (2009). Testing the stability of the benefit transfer function for discrete choice contingent valuation data. *Journal of Forest Economics, 15*, 131–146.

McConnell, K. E. (1992). Model building and judgment: Implications for benefit transfers with travel cost models. *Water Resources Research, 28*, 695–700.

Moeltner, K., Boyle, K. J., & Paterson, R. W. (2007). Meta-analysis and benefit transfer for resource valuation–Addressing classical challenges with Bayesian modeling. *Journal of Environmental Economics and Management, 53*, 250–269.

Moeltner, K., Johnston, R. J., Rosenberger, R. S., & Duke, J. M. (2009). Benefit transfer from multiple contingent experiments: A flexible two-step model combining individual choice data with community characteristics. *American Journal of Agricultural Economics, 91*, 1335–1342.

Moeltner, K. & Rosenberger, R. S. (2008). Predicting resource policy outcomes via meta-regression: Data space, model space, and the quest for 'optimal scope'. *The B.E. Journal of Economic Analysis and Policy*. doi:10.2202/1935-1682.2028.

Morrison, M. (2001). Non-market valuation databases: How useful are they? *Economic Analysis and Policy, 31*, 33–55.

Morrison, M., & Bennett, J. (2000). Choice modelling, non-use values and benefit transfer. *Economic Analysis and Policy, 30*, 13–32.

Morrison, M., & Bennett, J. (2004). Valuing New South Wales rivers for use in benefit transfer. *Australian Journal of Agricultural and Resource Economics, 48*, 591–611.

Morrison, M., & Bennett, J. (2006). Valuing New South Wales rivers for use in benefit transfer. In J. Rolfe & J. Bennett (Eds.), *Choice modelling and the transfer of environmental values* (pp. 71–96). Cheltenham, UK: Edward Elgar.

Morrison, M., Bennett, J., Blamey, R., & Louviere, J. (2002). Choice modeling and tests of benefit transfer. *American Journal of Agricultural Economics, 84*, 161–170.

Morrison, M., & Bergland, O. (2006). Prospects for the use of choice modelling for benefit transfer. *Ecological Economics, 60*, 420–428.

Muthke, T., & Holm-Mueller, K. (2004). National and international benefit transfer with a rigorous test procedure. *Environmental and Resource Economics, 29*, 323–336.

Navrud, S., & Pruckner, G. J. (1997). Environmental valuation–to use or not to use? A comparative study of the United States and Europe. *Environmental and Resource Economics, 10*, 1–26.

Nelson, J. P., & Kennedy, P. E. (2009). The use (and abuse) of meta-analysis in environmental and resource economics: An assessment. *Environmental and Resource Economics, 42*, 345–377.

Parsons, G. R., & Kealy, M. J. (1994). Benefits transfer in a random utility model of recreation. *Water Resources Research, 30*, 2477–2484.

Piper, S., & Martin, W. E. (2001). Evaluating the accuracy of the benefit transfer method: A rural water supply application in the USA. *Journal of Environmental Management, 63*, 223–235.

Poe, G. L., Boyle, K. J., & Bergstrom, J. C. (2001). A preliminary meta-analysis of contingent values for ground water revisited. In J. C. Bergstrom, K. J. Boyle, & G. L. Poe (Eds.), *The economic value of water quality* (pp. 137–162). Northampton, MA: Edward Elgar.

Poe, G. L., Giraud, K. L., & Loomis, J. B. (2005). Computational methods for measuring the difference in empirical distributions. *American Journal of Agricultural Economics, 87*, 353–365.

Poe, G. L., Severance-Lossin, E. K., & Welsh, M. P. (1994). Measuring the difference (X−Y) of simulated distributions: A convolutions approach. *American Journal of Agricultural Economics, 76*, 904–915.

Ready, R., & Navrud, S. (2007). Morbidity value transfer. In S. Navrud & R. Ready (Eds.), *Environmental values transfer issues and methods* (pp. 77–88). Dordrecht, The Netherlands: Springer.

Ready, R., Navrud, S., Day, B., Dubourg, R., Machado, F., Mourato, S., et al. (2004). Benefit transfer in Europe: How reliable are transfers between countries? *Environmental and Resource Economics, 29*, 67–82.

Rosenberger, R. S., & Johnston, R. J. (2009). Selection effects in meta-analysis and benefit transfer: Avoiding unintended consequences. *Land Economics, 85*, 410–428.

Rosenberger, R. S., & Loomis, J. B. (2000). Using meta-analysis for benefit transfer: In-sample convergent validity tests of an outdoor recreation database. *Water Resources Research, 36*, 1097–1107.

Rosenberger, R. S., & Loomis, J. B. (2003). Benefit transfer. In P. A. Champ, K. J. Boyle, & T. C. Brown (Eds.), *A primer on non-market valuation* (pp. 445–482). Dordrecht, The Netherlands: Kluwer.

Rosenberger, R. S., & Phipps, T. T. (2007). Correspondence and convergence in benefit transfer accuracy: Meta-analytic review of the literature. In S. Navrud & R. Ready (Eds.), *Environmental value transfer: Issues and methods* (pp. 23–43). Dordrecht, The Netherlands: Springer.

Rosenberger, R. S., & Stanley, T. D. (2006). Measurement, generalization and publication: Sources of error in benefit transfers and their management. *Ecological Economics, 60*, 372–378.

Rozan, A. (2004). Benefit transfer: A comparison of WTP for air quality between France and Germany. *Environmental and Resource Economics, 29*, 295–306.

Shrestha, R. K., & Loomis, J. B. (2001). Testing a meta-analysis model for benefit transfer in international outdoor recreation. *Ecological Economics, 39*, 67–83.

Shrestha, R. K., & Loomis, J. B. (2003). Meta-analytic benefit transfer of outdoor recreation economic values: Testing out-of-sample convergent validity. *Environmental and Resource Economics, 25*, 79–100.

Smith, V. K. (1992). On separating defensible benefit transfers from smoke and mirrors. *Water Resources Research, 28*, 685–694.

Smith, V. K. (1993). Nonmarket valuation of environmental resources: An interpretive appraisal. *Land Economics, 69*, 1–26.

Smith, V. K., & Osborne, L. (1996). Do contingent valuation estimates pass the scope test? A meta analysis. *Journal of Environmental Economics and Management, 31*, 287–301.

Smith, V. K., & Pattanayak, S. K. (2002). Is meta-analysis a Noah's Ark for non-market valuation? *Environmental and Resource Economics, 22*, 271–296.

Smith, V. K., van Houtven, G., & Pattanayak, S. K. (2002). Benefit transfer via preference calibration: Prudential algebra for policy. *Land Economics, 78*, 132–152.

Spash, C. L., & Vatn, A. (2006). Transferring environmental value estimates: Issues and alternatives. *Ecological Economics, 60*, 379–388.
Stanley, T. D. (2005). Beyond publication bias. *Journal of Economic Surveys, 19*, 309–345.
Stanley, T. D. (2008). Meta-regression methods for detecting and estimating empirical effect in the presence of publication selection. *Oxford Bulletin of Economics and Statistics, 70*, 103–127.
Stanley, T. D., & Doucouliagos, H. (2010). Picture this: A simple graph that reveals much ado about research. *Journal of Economic Surveys, 24*, 170–191.
Stanley, T. D., Jarrell, S. B., & Doucouliagos, H. (2010). Could it be better to discard 90 % of the data? A statistical paradox. *The American Statistician, 64*, 70–77.
Stapler, R. W., & Johnston, R. J. (2009). Meta-analysis, benefit transfer, and methodological covariates: Implications for transfer error. *Environmental and Resource Economics, 42*, 227–246.
Troy, A., & Wilson, M. A. (2006). Mapping ecosystem services: Practical challenges and opportunities in linking GIS and value transfer. *Ecological Economics, 60*, 435–449.
VandenBerg, T. P., Poe, G. L., & Powell, J. R. (2001). Assessing the accuracy of benefits transfers: Evidence from a multi-site contingent valuation study of ground water quality. In J. C. Bergstrom, K. J. Boyle, & G. L. Poe (Eds.), *The economic value of water quality* (pp. 100–120). Cheltenham, UK: Edward Elgar.
Wilson, M. A., & Hoehn, J. P. (2006). Valuing environmental goods and services using benefit transfer: The state-of-the-art and science. *Ecological Economics, 60*, 335–342.
Zandersen, M., Termansen, M., & Jensen, F. S. (2007). Testing benefits transfer of forest recreation values over a twenty-year time horizon. *Land Economics, 83*, 412–440.

Part III
Meta-analysis

Chapter 15
Meta-analysis: Statistical Methods

Jon P. Nelson

Abstract Meta-analysis is the quantitative synthesis of multiple primary studies containing estimates of similar empirical magnitudes or effect sizes. Meta-analysis allows generalizations about the underlying population of effects and increases the power of statistical tests. Meta-regression analysis can control statistically for factual heterogeneity, methodological diversity, and possible biases among the primary studies. In the context of benefit transfers, meta-analysis can produce reduced-form functions that identify and test systematic influences of study, economic, and resource attributes on willingness to pay and other environmental valuations. This chapter provides an introduction to basic statistical methods employed in meta-analysis, including weighted-averages and meta-regressions. The chapter identifies and discusses solutions to several econometric problems commonly associated with metadata, including heterogeneity, heteroskedasticity, correlated effects, and publication bias. Basic statistical concepts and methods are illustrated using a sample of estimates for the value of a statistical life, including within-sample and out-of-sample forecasts. Benefit-transfer errors are assessed using several alternative statistical measures.

Keywords Meta-analysis · Meta-regression analysis · Benefit transfer · Environmental valuation functions · Benefit-transfer errors · Publication bias · Value of a statistical life

15.1 Introduction

Meta-analysis is the use of statistical methods to combine, analyze, and synthesize results from multiple, related empirical studies with the objective of drawing general conclusions. The term "meta-analysis" was introduced by Glass (1977, p. 352),

J.P. Nelson (✉)
Department of Economics, Pennsylvania State University, University Park, PA, USA
e-mail: jpn@psu.edu

R.J. Johnston et al. (eds.), *Benefit Transfer of Environmental and Resource Values*,
The Economics of Non-Market Goods and Resources 14,
DOI 10.1007/978-94-017-9930-0_15

who noted that "the accumulated findings of ... studies should be regarded as complex data points, no more comprehensible without the full use of statistical analysis than hundreds of data points in a single study." The early development of statistical methods for quantitative syntheses is credited to Pearson (1904) and Fisher (1932). Pearson developed a method for combining correlation coefficients and Fisher developed a method for combining *p*-values from statistically independent tests of the same hypothesis. In particular, Fisher argued that statistical tests often failed to reject the null hypothesis because they lacked statistical power. However, if the tests were combined, their cumulative power would be greater. These and other early methods were largely forgotten until the mid-1970s when Glass and others began working on the problem of combining results from studies using different scales of measurement. This work culminated with a book, *Statistical Methods for Meta-Analysis* (Hedges and Olkin 1985), which summarized the work-to-date and provided rigorous proofs of basic and advanced statistical concepts.

The use of meta-analysis in economics can be traced to 1989–90 when analyses appeared by Smith and Kaoru (1990), Stanley and Jarrell (1989), and Walsh et al. (1989). Using estimates of consumer surplus obtained from travel cost demand relationships, Smith and Kaoru summarized estimates of recreation benefits from over two hundred empirical studies. They recognized that synthesis of observational studies must account for a complex data-generating process. This includes statistical issues of factual heterogeneity (site or context), methodological diversity, heteroskedasticity, and non-independence of observations. Walsh et al. analyzed results from 120 recreation demand studies that used either travel cost or contingent valuation (CV) methodologies. Their work was the first to recognize the potential application of meta-analysis to benefit transfer, and they drew comparisons between estimates from their synthesis and recreation use values employed by the U.S. Forest Service.[1]

Since the two analyses of recreation demand, over 125 meta-analyses have summarized empirical results in environmental and resource economics (Nelson and Kennedy 2009), and about half incorporate either within-sample predictions or out-of-sample benefit transfers. The analyses cover many areas of interest in environmental economics, including air pollution, water pollution, hazardous waste, global warming, endangered species, wetlands valuation, environmental regulations, and so forth. It should be emphasized that meta-analysis is more than a set of statistical methods. It also includes systematic procedures for data collection, coding, quality assessment, graphing, analysis, application, and reporting of methods and outcomes (Cooper 2010; Stanley 2001). However, as discussed in Nelson and Kennedy (2009) and Smith and Pattanayak (2002), many existing meta-analyses in economics are deficient in either the application of statistical methods,

[1]Previous articles that discuss methodological issues concerning benefit transfer and meta-analysis include Bergstrom and Taylor (2006), Johnston and Rosenberger (2010), Johnston et al. (2006), Lindhjem and Navrud (2008), Rosenberger and Loomis (2000a), Rosenberger and Phipps (2007), Rosenberger and Stanley (2006), Shrestha and Loomis (2001, 2003), Shrestha et al. (2007), Smith and Pattanayak (2002), Stapler and Johnston (2009), and Van Houtven et al. (2007).

reporting of procedures for research synthesis, or the consistency of economic concepts being summarized. This limits their usefulness or accuracy for benefit transfer, project evaluation, and evidence-informed public policy.

This chapter serves as an introduction to the use of meta-analysis as applied to econometric studies and data, where the ultimate objective is a benefit transfer. In order for this to be possible, the research design in a meta-analysis must first meet minimum procedural and statistical standards that help guarantee its *internal validity*. Once internal validity is assured, it is possible to consider *external validity* and policy applications such as a benefit transfer. The chapter concentrates on research designs that satisfy tests of internal validity, but with an eye toward application. In order to illustrate basic concepts, the chapter provides a worked example using 28 estimates of the value of a statistical life (VSL), drawn from metadata in Bellavance et al. (2009) and U.S. EPA (2010). This is a small sample, but sample size is often an issue in meta-analysis. The VSL analysis is tentative and exploratory, and does not claim to be a definitive treatment of this important concept. A prime attraction of meta-analysis is that a benefit transfer could be based on a function estimate, rather than a point or unit value estimate (Rosenberger and Phipps 2007). However, under some circumstances, point estimates derived through meta-analysis may be valuable for policy purposes. In conjunction with the review of statistical methods, the chapter offers guidelines for application of meta-analysis to economic studies.[2]

15.2 Basic Concepts

Consider any common econometric problem, such as estimation of the price elasticity for gasoline demand; the willingness to pay (WTP) to avoid loss of endangered species; the hedonic price for the effect of aircraft noise on housing values; and the value of a statistical life. Empirical studies of these relationships produce related estimates, which can be expressed in common units of measurement. The empirical studies containing estimates are the "primary studies" and the common metric found in each study is the estimated "effect size." To avoid confusion, researchers conducting the primary studies are referred to as "investigators" and researchers conducting the meta-analysis are referred to as "analysts." Let N represent the population of primary studies of a particular relationship, $i = 1,\ldots,N$,

[2]Borenstein et al. (2009) is an excellent introduction to basic statistical models employed in meta-analysis; see also Cooper (2010) and Cooper et al. (2009). Specialized software available for meta-analysis includes CMA (Biostat 2005), SAS, and Stata (Sterne 2009). Many basic calculations can be implemented using Excel or other statistical software, although standard errors are not always correctly computed in non-specialized software packages (Konstantopoulos and Hedges 2009; Rhodes 2012). Monte Carlo comparisons of alternative models presented below are found in Rhodes (2012).

and let n represent a sample of studies containing estimated effect sizes, $i = 1, \ldots, n$. For the present, the n estimates are assumed to be independent of each other. The true effect size in the population is denoted by β_i and the estimate reported in the i-th primary study is denoted by Y_i. The true effect size may or may not vary across the population. Effect sizes in economics are typically coefficient estimates from a regression or can be derived from the regressions reported in the primary studies. Given the standard errors on the coefficients, metadata are obviously heteroskedastic. More precise estimates should be given greater weight in the analysis, and this can be accomplished through weighting of the estimates or by use of weighted-least squares in a meta-regression. The population or subgroup of interest is largely defined by the research question posed by the analyst. For example, the objective might be to estimate the VSL for North American workers.

Given a sample of studies and estimated effect sizes, there are several possible objectives of a meta-analysis: (1) estimate an average effect size in the sample, its confidence interval, and generalize about the distribution of effects in the population; (2) explain the variation in the sample of effects using a meta-regression model, without regard to possible selection bias in the sample; (3) estimate a meta-regression model that adjusts for sample selection bias; and (4) apply the model using an out-of-sample forecast such as a benefit-function transfer. Most meta-analyses in economics have focused on the second objective, which ignores selection bias in the sample of primary studies or what is referred to as "publication bias." Selection bias is one reason for emphasis on inclusiveness in sampling of primary studies, but as discussed below there is a tradeoff between inclusiveness and heterogeneity in the primary data.

The first step in a meta-analysis is the establishment of a protocol for determining the primary studies to be included, coding of effect sizes, and reporting on metadata (Cooper 2010; Doucouliagos and Stanley 2012; Stanley 2001).

Suggested Protocol: Begin by providing a clear, transparent protocol for procedures leading to the sample of primary studies. Identify the research question, population of interest, and report criteria for inclusion and exclusion of studies. Report keywords, search engines (EconLit, RePEc, Google-Scholar, etc.), and other bibliographic and archival materials used for locating relevant studies, including searches in the "grey literature" such as unpublished research papers, dissertations, government documents, and other technical reports. The search protocol should be reported in the analysis or made available on-line.[3] Report also the coding procedures for assembling the basic data and consider using a double-blind procedure for coding of large complex samples. Effect sizes and other variables should be given clear conceptual definitions. Provide a table containing the metadata or make it available on-line. Provide complete references for all primary studies.

[3]As noted by White (2009, p. 61), "one does hear of [and encounter] innocents who think that database or Web searches retrieve everything that exists on a topic."

15.3 Homogeneous and Heterogeneous Effects: Weighted-Means

15.3.1 Fixed-Effect and Random-Effects Weighted-Means

Suppose every primary study produces a single unbiased estimate of the same population value, and each study has been conducted in a similar fashion, so design and estimation features of the studies do not affect the expected value of the estimates. Suppose also the estimates are stochastically independent of each other. The fixed-effect size (FES) model postulates that the estimates share a common, or fixed, population effect size. The assumption of a homogeneous effect may seem restrictive, but can be defended based on the defined population or because the FES model has better small sample properties (Rhodes 2012). A basic model is given by the population mean plus the measurement or sampling error in each primary study

$$Y_i = \beta + e_i, \quad i = 1, \ldots, n \tag{15.1}$$

where $\beta_1 = \beta_2 = \cdots = \beta_n = \beta$ and where e_i is a sampling error, assumed to be normally distributed with mean zero and variance σ^2. If the within-study variances are "known" because the primary estimates, s_i^2, are available or can be approximated, a weighted-mean can be calculated with inverse variances as weights, denoted by $w_i = 1/s_i^2$. Estimates with smaller variances are given more weight because they contain more precise information. As summary statistics, the FES weighted-mean, $\bar{\beta}$ and its estimated variance, $\hat{\sigma}^2$, are given by (Borenstein et al. 2009)

$$\bar{\beta} = \sum w_i Y_i \Big/ \sum w_i, \quad \hat{\sigma}^2 = 1 \Big/ \sum w_i \tag{15.2}$$

where all summations are from $i = 1, \ldots, n$. The mean and variance depend on the size of the primary estimates, their precision, and the number of estimates.

Alternatively, there may be reason to believe there is heterogeneity in the true effects, either a priori or on the basis of a statistical test (described below). Suppose also that the heterogeneity is not measurable using regressors. This view is sometimes justified due to a large number of possible artifacts in primary studies, which are unobservable by the analyst. In this case, each effect size is modeled as a random draw from a distribution of effects with an unknown mean and variance, typically a normal distribution. Formally, the heterogeneity is modeled as $Y_i = \beta_0 + u_i$, where the mean outcome across studies is given by β_0 and u_i is an error term with mean zero and variance σ_u^2. The true effect varies from study to study, so β_0 is the mean of a super-distribution of true effects (Borenstein et al. 2009, p. 79). The error term captures unmeasured heterogeneity in the true effect sizes. Adding this to Eq. 15.1 yields the random-effects size (RES) model

$$Y_i = \beta_0 + u_i + e_i, \quad i = 1, \ldots, n \tag{15.3}$$

where e_i and u_i are assumed to be independent. An estimate of the between-study variance, $\hat{\sigma}_u^2$, can be obtained by comparing the dispersion of estimated effects to the dispersion that would be expected given knowledge of the s_i^2 values (Borenstein et al. 2009; Rhodes 2012). The RES variance of each primary observation is given by the composite error, $\upsilon_i^2 = s_i^2 + \hat{\sigma}_u^2$, so weighting by the inverse of υ_i^2 produces the RES weighted-mean and variance. Compared to the fixed-effect model, the RES model gives relatively greater weight to less precise estimates.

Suggested Protocol: The starting point for statistical analysis is the calculation of fixed- and random-effect size weighted-means. These summary averages require estimates for the effect sizes and variances (or standard errors, t-statistics, p-values). The variances can be difficult to obtain or estimate, so special effort may be required to complete this task as part of the data collection phase. In some cases, missing variances are derived from an indirect regression on the sample sizes (Bellavance et al. 2009) or by other means such as the delta method (Cavlovic et al. 2000) and bootstrapping. An alternative set of weights can be based on the sample size of each estimate (Borenstein et al. 2009; Hunter and Schmidt 2004; Van Houtven et al. 2007). It is crucial that the effect sizes are measured on a common scale and represent identical theoretical concepts. The analysis should include a clear reporting of the theoretical model and basic econometric model that underlies the primary estimates.

15.3.2 Heterogeneity Analysis: The Q-Statistic and Weighted-Means for VSL

The U.S. EPA (2010) and Viscusi and Aldy (2003) address the important issue of the value of mortality risk reductions. They review the econometric literature, including primary VSL estimates based on stated preference and hedonic wage methodologies. The metadata used in this chapter draw on hedonic wage estimates reported in Bellavance et al. (2009, p. 459) and the U.S. EPA (2010, p. 85). The VSL estimates in Bellavance et al. are in 2000 dollars. EPA's estimates are in 2009 dollars, but also include an adjustment based on real income growth and an assumed income elasticity of 0.5. The two datasets are combined using the metadata in Bellavance et al., but expressed in 2009 \$ (the inflationary adjustment is 1.2459 based on the CPI-U index). Following the U.S. EPA (2010, p. 42), estimates are dropped if standard errors are unavailable, the risk measures are based on the Society of Actuaries (SOA) data, or the sample size is less than 100. Two VSL values (\$0.67 and \$76 million) are outliers and there also are two studies with extreme risk estimates or poor precision. These four observations are dropped from the analysis. Table 15.1 shows the VSL values, standard errors, and selected sample characteristics for each of the 28 primary studies. The VSL estimates range from

Table 15.1 Metadata for analysis of VSL

Study no. (data year)	VSL (mil. 2009 $)	SE (mil. 2009 $)	Mean income (2009 $)	Mean risk ($\times 10^4$)	Smpl size	White-only (=1)	NA (=1)
1 (1967)	11.50	4.79	36,167	1.250	3183	1	1
3 (1969)	3.04	1.75	39,810	1.182	496	0	1
5 (1978)	15.42	6.20	41,749	0.951	5993	0	1
na (1976)	10.24	1.77	37,930	1.040	3977	0	1
6 (1975)	7.54	1.67	32,910	0.930	5509	0	0
8 (1978)	14.66	6.19	26,956	0.576	1697	0	1
10 (1979)	4.11	1.95	37,012	1.200	879	1	1
11 (1974)	12.54	5.31	44,471	1.420	1529	1	1
12 (1977)	5.22	2.89	33,421	1.400	514	0	1
13 (1981)	10.43	4.28	15,999	1.280	4225	0	0
14 (1981)	20.45	4.41	37,393	1.080	2863	0	1
15 (1982)	11.42	2.98	33,090	0.792	1349	1	1
16 (1981)	5.04	2.91	56,455	1.900	718	0	0
18 (1980)	9.49	1.66	52,722	0.970	22,837	0	1
23 (1974)	8.91	2.71	42,417	1.340	1502	0	1
24 (1979)	5.99	0.58	36,951	0.764	32,713	0	0
25 (1986)	3.92	1.18	36,038	2.500	4352	0	0
26 (1983)	17.67	8.41	19,470	0.332	1353	0	0
28 (1981)	13.84	2.60	36,819	1.102	1528	0	1
31 (1991)	22.40	1.71	33,188	0.680	18,850	0	0
32 (1998)	1.93	0.40	20,577	4.850	321	0	0
33 (1986)	2.93	0.76	28,336	1.800	1503	0	0

(continued)

Table 15.1 (continued)

Study no. (data year)	VSL (mil. 2009 $)	SE (mil. 2009 $)	Mean income (2009 $)	Mean risk ($\times 10^4$)	Smpl size	White-only (=1)	NA (=1)
34 (1980)	38.32	7.70	36,350	0.500	3608	0	0
35 (1995)	30.35	4.31	37,014	1.670	2014	0	0
37 (1996)	3.39	0.75	30,973	0.976	45,001	0	1
38 (1997)	20.11	1.90	37,936	0.362	83,625	1	1
39 (1997)	6.36	0.75	37,428	0.402	99,033	0	1
na (1997)	5.77	1.77	66,568	0.490	43,261	0	1
Mean (sd)	11.54 (8.68)	3.01 (2.19)	36,648 (10,374)	1.205 (0.87)	14,087 (25,283)	0.18	0.79

Notes The VSL, SE, and Income data are from Bellavance et al. (2009) with an inflationary adjustment to 2009 $. The study no. refers to their appendix table. Following U.S. EPA (2010), studies are screened to delete those with small samples, Society of Actuaries (SOA) risk data, and outliers in the VSL and risk data. *SE* standard error of VSL; *White only* white workers only sample; *NA* North America studies (Canada, U.S.); and *Data year* mid-point year of the data

\$1.93 to \$38.32 million, with an unweighted mean of \$11.54 million (sd = 8.68) and a median of \$9.87 million. The FES weighted-mean is \$5.00 million (0.23), with a 95 % confidence interval of \$4.55–\$5.46 million. The RES mean is \$9.77 million (1.03), with a 95 % confidence interval of \$7.75–\$11.79 million. The common between-study variance is \$21.4 million (12.1).[4]

Are the VSL estimates homogeneous? Clearly there is a wide range of estimates in the raw data. There are obvious differences that might affect the analysis: the estimates come from hedonic wage studies that employ different data sources, sample sizes, countries of origin, model specifications, and so forth. Borenstein et al. (2009, p. 106) restrict the definition of heterogeneity to variation in the true effect sizes, while differences in the estimates due to, say, econometric methods are described as "methodological dispersion." Their terminology is adopted here, although in practice a regression is required to separate these sources of variation.

An analysis of heterogeneity first asks the question: "Is the observed variance in effect sizes statistically different from that expected from sampling error alone?" The standard statistical test for homogeneity is based on Cochran's Q-statistic, which has a chi-square distribution with $n - 1$ degrees of freedom (Borenstein et al. 2009, p. 109)

$$Q = \sum w_i (Y_i - \bar{\beta})^2 = \sum ((Y_i - \bar{\beta})/s_i)^2 \tag{15.4}$$

which is the weighted sum-of-squares (WSS) on a standardized scale, i.e., the ratio of the observed variation to the within-study error. If Q is large, the null hypothesis is rejected and the fixed-effect model may be inappropriate. The expected WSS under the assumption of a common effect is $(n - 1) = df$, so $Q - df$ is a standardized measure of excess variation.[5] However, use of the Q-statistic is tempered by three considerations. First, the test is known to have low power since the number of primary studies is often small, the number of data points accumulated across studies is small, or the within-study variances are large. Second, the test can be affected by outliers in the data, which might be due to low-quality primary studies or other anomalies. Third, when n is small, the confidence interval for $\hat{\sigma}_u^2$ is usually large and a model that assumes homogeneity may have better small sample properties (Rhodes 2012). As an aid to understanding heterogeneity and statistical tests based on Q, it is common to include graphical displays as part of the meta-analysis such as forest plots and Galbraith diagrams; see Borenstein et al. (2009).

---

[4]For the RES model, the *prediction interval* describes the possible distribution of true effect sizes, given estimates of the between-study variance and the RES variance; see Borenstein et al. (2009). For these data, the 95 % prediction interval is $9.77 \pm 1.96\,(21.4 + 1.06)^{1/2}$ or \$0.48–19.06 million per life.

[5]An alternative measure of heterogeneity is $I^2 = ((Q - df)/Q) \times 100$, which is 92.5 % for the VSL data. This statistic describes the proportion of observed variance due to real differences in the estimates rather than chance, i.e., the excess dispersion divided by the total dispersion. Values above 75 % are considered "high."

The Q-statistic for the VSL data is 360.9 ($p < 0.001$; df = 27), so the test easily rejects the null. The heterogeneity analysis can be extended by dividing the data categorically. The metadata in Table 15.1 includes 17 estimates for the U.S.; five estimates for Canada, and six estimates for other countries (Australia, Austria, South Korea, U.K.). A fixed-effect model might apply to the North American (NA) subgroup of estimates. The FES and RES weighted-means for the NA subgroup are \$5.92 (0.29) and \$8.79 million (0.97). The weighted-means for the non-NA subgroup are \$3.44 (0.38) and \$14.98 million (4.72). The Q-statistic for the NA sample is 164.6 ($p < 0.001$; df = 21), so there is evidence of heterogeneity within the NA subgroup. For the non-NA subgroup, the Q-statistic is 169.5 ($p < 0.001$; df = 5). Summing these Q values and subtracting from the sample total is a test of the grouping factor (i.e., country) as a significant contributor to the variance in effect sizes. The difference is 26.8, which is distributed as chi-square with one degree of freedom. The critical value at the 95 % level is 3.84, so the null hypothesis of equal effect sizes in the two subgroups is rejected. Labor market and other socioeconomic conditions are reflected in the VSL estimates, leading to factual heterogeneity. Country of origin is therefore a potential explanatory variable in a meta-regression.

While adding the non-NA studies increases the sample size, it also adds variation that may be difficult to explain. Van Houtven (2008, p. 904) argues "if the scope is too broadly defined, between-study heterogeneity will make it difficult to meaningfully combine and jointly analyze the research results … if the scope is too narrow, the number of included results will be too small for meaningful statistical analysis." The modifier "too small" can apply to the number of studies or the number of positive observations for a particular study characteristic, such as the number of primary studies based on a "white-only" sample. When the number of studies is small, the between-studies variance will have poor precision and estimates based on a random-effects model will have wider confidence intervals. Hence, there is a possible tradeoff between sample size and heterogeneity, which the analyst must consider as a potential limiting factor for a benefit transfer (Moeltner and Rosenberger 2008).

It is of course possible to think of other groupings of these data, such as subgroups by union status, mean risk level, or year of data. Meta-analysts in other disciplines often examine categorical differences using an analysis of variance (ANOVA) model. However, it is natural to think of estimating a meta-regression model, where explanatory variables (moderators, covariates, or regressors) represent characteristics and factors that explain the measurable parts of heterogeneity and dispersion. The explanatory variables may be continuous (e.g., income) or categorical, such as a binary dummy variable for NA countries. Again, a risk of this approach is that the meta-analyst seeks a large, diverse sample of primary studies, which are less representative of the population of interest. For example, a meta-analysis for VSL might combine studies using the hedonic wage approach with studies using stated preference (SP) methods, which is strongly discouraged (Smith and Pattanayak 2002; U.S. EPA 2006, 2010).

Suggested Protocol: Benefit transfers using meta-analysis are facilitated by thinking carefully about the relevant population and effect size concept. Ideally, this should be done a priori in conjunction with the data collection phase.

15.3.3 Power Analysis for Meta-analysis

A *Type I error* occurs when the researcher concludes the null hypothesis ($\hat{\beta} = 0$) is false when it is in fact true (*false positive*); a *Type II error* occurs when the conclusion is the null hypothesis is true when it is false (*false negative*). Traditional testing methodology in econometrics calls for specifying the probability of a Type I error (the significance level, typically set at 5 %), and then attempting to maximize the power of the test (one minus the probability of a *Type II* error) by choice of a consistent estimation method with desirable asymptotic properties (Kennedy 2008, p. 67). The power of a test is a standardized measure of the likelihood of obtaining a significant result, where 80 % is traditionally regarded as good statistical power. In general, power depends on the statistical significance level (α); the magnitude of the effect in the population; the sample size used for the test; and the statistic chosen for the test. It is widely believed that failure to reject the null (e.g., insignificant regression coefficients) is often due to low power.

An advantage of meta-analysis is that power is usually high, provided the primary estimates are independent of each other and the number of studies is sufficient. In the fixed-effect model, power depends on the cumulative number of observations and number of studies (Borenstein et al. 2009). For example, the cumulative sample size in Table 15.1 exceeds 400,000 observations. In the random-effects model, power depends on the cumulative sample size, the number of studies, and the between-study variance. An analyst could use a power analysis to facilitate grouping of the metadata into homogeneous subgroups. Suppose the analyst is interested in testing whether a VSL weighted-mean is different from a default value of \$7.9 million, expressed in 2009 dollars (U.S. EPA 2010, p. 10). Suppose also there are only five comparable primary studies with at least one hundred observations: each has a standard error that is one-half of the effect size and the between-study variance equals the (uniform) within-study variance. Following Borenstein et al. (2009, p. 268) and using Excel, a comparison based on the FES mean yields a *Power* = 1 − NORMSDIST (1.96 − 4.473) + NORMSDIST (−1.96 − 4.473) = 0.994 or 99.4 %. The power of the random-effects model is less clear, given the size of the between-study variance. A similar calculation yields a *Power* = 70.5 % for the RES mean. Lastly, following Borenstein et al. (2009, p. 275), the *Q*-test reported above has a *Power* = CHIDIST (40.113/(1 + 1/1), 27) = 0.828 or 82.8 %. This value exceeds the 80 % goal, reflecting the number of studies and the cumulative sample size available for the VSL analysis.

For the meta-regressions reported below, the power analysis is less clear when applied to individual covariates. Except for the income elasticity, it is difficult to form prior expectations about coefficient sizes or variances. For instance, there are only five studies in Table 15.1 that are based on a white workers only sample. Borenstein et al. (2009, pp. 188 and 269) recommend at least ten studies for each study-level covariate, but this suggestion is based on an average sample size of only 25 observations per study.

Suggested Protocol: Use a fixed-effect average if the primary studies are "identical" or if the goal is simply to summarize the sample of studies and not to generalize to other populations. Consider separating the metadata into homogeneous subgroups for policy applications. A power analysis can help to decide if there are sufficient studies in subgroups. Homogeneity should be tested statistically prior to a regression analysis using the *Q*-statistic or related measures. Graphical plots of the data are helpful to provide context for more advanced analyses. In most applications, economic data are likely to be heterogeneous or reflect substantial methodological diversity due to publication policies of academic journals. If the between-study variance is statistically significant, a more complex analysis, such as a meta-regression model, should be considered.

15.4 Meta-regression Models: Explaining Heterogeneity and Dispersion

This section considers two extensions of the basic models and associated estimation procedures. The fixed-effect meta-regression model assumes that the true effect size is homogeneous *conditional* on a set of explanatory variables that capture observed or measured sources of heterogeneity and dispersion (Borenstein et al. 2009, p. 207; Rhodes 2012). For any set of values for the covariates, there is one population effect size. The explanatory variables represent factual (site or contextual) differences among the primary studies and important methodological differences. For economists, a fixed-effect regression has been the work horse model for meta-analyses. In contrast, the random-effects regression model assumes that unobserved or unmeasured factors also influence the estimated effect sizes, which affect the estimated slope coefficients and standard errors. This model is a more general statement of the data-generating process, although in practice there can be good reasons to rely on a fixed-effect regression. For example, with a small number of primary studies, the between-study variance is unlikely to be estimated precisely. This section develops both models and presents estimates for the VSL example. As tests of validity, both within- and out-of-sample forecasts are reported.

15.4.1 Fixed-Effect Meta-regression Model: Measured Heterogeneity

Suppose there are multiple factors, denoted by the vector X_i, that are measurable and which a priori are believed to influence the true effect size or its estimation in the primary studies. Suppose also there are no unmeasured sources of heterogeneity or methodological dispersion. The explanatory variables are measured at the study level, but differ generally from the covariates used in the primary study regressions. Some variables in the meta-regression describe the setting or context of the primary data, while others describe the empirical methodology and econometric methods employed by the primary investigators. For example, Randall et al. (2008) conduct a meta-analysis of agricultural conservation programs, with a benefit transfer objective. They estimate separate meta-regressions according to type of environmental services (wetlands, terrestrial habitat, surface water quality), where the effect size is the estimated WTP from a sample of 40 valuation studies. Sample sizes for the ordinary least-squares (OLS) meta-regressions are: wetlands, 72 observations; terrestrial, 23; and surface water, 98. They consider a wide variety of explanatory variables, but many are dropped from the final analysis. The three meta-regressions include socioeconomic variables (e.g., income), environmental type and size (wetlands type, open space, waterbody size), environmental services provided (water quality, viewing, fishing), and valuation methodology (type of survey, protest bids excluded, outliers excluded). This scope of heterogeneity is typical in a meta-regression.

Given the primary data and the analyst's conceptualization of the data-generating process, the FES meta-regression model is a simple extension of Eq. 15.1

$$Y_i = \beta + X_i\alpha_1 + e_i \tag{15.5}$$

where α_1 is a vector of regression coefficients and where the coefficient estimates are obtained by generalized least-squares to account for heteroskedastic disturbances.[6] As is well known, failure to correct for heteroskedasticity will result in inefficient, but unbiased, parameter estimates. In the FES regression model, an estimate or prediction of the average effect size in the population is obtained by substituting for $\overline{X}$, but the confidence interval will depend on the precision of the parameter estimates. More generally, potential sources of forecast error include specification errors, conditioning errors in the X_i values, sampling errors in the parameter estimates, and random errors (Kennedy 2008, p. 332). Viewing benefit

[6]Smith and Kaoru (1990, p. 425) express doubts about the use of variance weights for a meta-regression. They argue that the "weighting implicitly assumes that the estimates based on incorrect modeling assumptions remain unbiased but simply have less informational content." Alternatives include use of robust standard errors (e.g., Huber-White); weighting by the sample size; inclusion of the sample size as a regressor; and inclusion of regressors that describe the possible biases. See Nelson and Kennedy (2009) and Rhodes (2012) for additional discussion.

transfer as a forecasting problem also leads to several alternative criteria for judging forecast accuracy, but benefit transfers have traditionally relied on the mean-absolute-percent error as the gold standard for validity (Lindhjem and Navrud 2008; Rosenberger and Loomis 2000a; Rosenberger and Stanley 2006; Stapler and Johnston 2009).

15.4.2 Random-Effects Meta-regression Model: Unmeasured Heterogeneity

Suppose instead there is both measured and unmeasured heterogeneity in the true effect sizes. The measured heterogeneity is again captured by the X_i variables for each of the primary studies. Denote the unmeasured sources of variation by a study-level vector Z_i. Unless the omitted Z variables are orthogonal to the included regressors, the parameter estimates are biased (Rhodes 2012). The magnitude and direction of biases are unknown in a multivariate context. For example, due in part to statistical insignificance of regional dummy variables, Randall et al. (2008, p. 5) express concern about a lack of farm-level descriptors for their meta-analysis. Hence, their benefit transfer application lacks a potentially important spatial component. This might be justification for estimation of the random-effects model, which attempts to account for unobserved or unmeasured sources of heterogeneity and dispersion.[7]

Given the missing information, the RES meta-regression model (also known as a mixed-effect model) is specified as

$$Y_i = \beta_0 + X_i\alpha_1 + u_i + e_i \tag{15.6}$$

where u_i is the part of $Z_i\alpha_2$ that is orthogonal to X_i (Rhodes 2012). For any set of values for the covariates, there is a distribution of effect sizes, reflecting the between-study error term. Estimation of Eq. 15.6 requires an estimate of the *residual* between-study variance, $\hat{\sigma}_u^2$, which can be obtained simultaneously with the regression coefficients or iteratively. The simplest (and most common) procedure is to estimate a fixed-effect regression and use the parameter estimates to derive a value for the Q-statistic, which is the method-of-moments procedure. The Q-statistic is used to derive the between-studies variance, and the regression model is re-estimated with the new weights. Alternatively, estimation is possible using iterative maximum likelihood, restricted maximum likelihood, or the empirical

[7]As discussed by Kennedy (2008, p. 339), misspecification is not always a disaster. Although the estimated coefficients are biased, a parsimonious model can still provide better forecasts as the biased parameters incorporate some of the information in the unobserved or omitted variables. Many existing meta-analyses in environmental economics focus on "taking stock of the literature" through parameter estimation for a host of explanatory variables, but this does not guarantee that the models can generate good forecasts for a benefit-function transfer.

Bayes method. Standard errors for the regression coefficients can be calculated using either parametric (e.g., Stata *metareg* default) or nonparametric methods. Random-effects regressions will tend to produce wider confidence intervals for the coefficients, so less precise results should be expected generally.

Suggested Protocol: Meta-regression analysis should be used to explain sources of factual heterogeneity and methodological dispersion. The fixed-effect regression assumes homogeneity of population effects *conditional* on study-level covariates. Corrections for heteroskedasticity should always be made, either through the use of weighted least-squares or provision of robust standard errors. The random-effects regression also accounts for between-study variation that is not observed or measurable. The *residual between-study variance* can be obtained only after a regression has been estimated, but several estimation procedures are available. However, with small samples, precision of regression parameters is likely to be an issue, regardless of model and estimation method. For benefit transfer, there is also a potential tradeoff between complexity of the regression specification and forecasting ability. A complete meta-regression analysis should therefore include an assessment of within- and out-of-sample forecasts.

15.4.3 Meta-regressions: VSL Metadata

Four alternative meta-regressions were estimated using the VSL metadata: (1) OLS with robust standard errors; (2) fixed-effect regression using weighted-least squares (WLS) and inverse variance weights; (3) random-effects regression using method-of-moments (MM); and (4) random-effects regression using restricted maximum likelihood (REML). The MM and REML regressions are based on the Stata macro *metareg* (Sterne 2009). There are a number of possible explanatory variables in Bellavance et al. (2009), but only a few have sufficient observations. As a data-fitting exercise, sample size and publication date (or data year) are possible covariates, but these are ignored here. The explanatory variables are two factual variables (log of income, North American origin) and one methodological descriptor (white-workers only sample).[8] Except for the OLS regression, the effect size standard error is also included as a regressor. In a weighted-regression, this is equivalent to including the estimate's precision, $1/s_i$, as a regressor, which corrects for publication bias in the sample (Doucouliagos and Stanley 2009; Nelson 2011).

[8]Point estimates for the VSL are usually preferred for ethical reasons, although different life-saving benefits may be given different values (Kenkel 2003). Meta-regressions can be used to correct for methodological dispersion, obtain a summary value for the income elasticity, correct for publication bias, examine the influence of labor market imperfections, or examine situational differences in VSL. A range of estimates also is valuable for sensitivity analysis in benefit-cost studies and other project evaluations.

Table 15.2 Meta-regression Results for VSL

Variable	(1) OLS	(2) Fixed-effect	(3) Random-effects	(4) Random-effects
Intercept	−56.47 (0.64)	−84.29 (6.79)*	−51.61 (0.99)	−55.07 (1.03)
Log income	7.189 (0.80)	8.645 (6.92)*	5.693 (1.11)	6.040 (1.15)
White-only smpl	2.476 (0.72)	2.921 (2.77)*	1.828 (0.56)	1.346 (0.41)
NA sample	−9.800 (1.13)	−3.271 (4.58)*	−5.570 (1.51)	−5.670 (1.47)
Precision	–	2.422 (8.64)*	2.540 (3.51)*	2.531 (3.68)*
R-sq.	0.136	0.531	0.687	0.614
RMSE (unwt)	7.919	6.352	6.113	6.090
F-stat. (p-value)	0.520 (0.52)	6.510 (0.001)	4.160 (0.01)	4.400 (0.01)
J-B residual	5.069	4.677	5.181	5.319
Test (p-value)	(0.08)	(0.10)	(0.07)	(0.07)
Between-var. estimate	–	–	14.73	28.97
Est. method	OLS	WLS	MM	REML
Std error est.	Robust HC3	Hedges	Default	Default

Notes Absolute value of t-statistics in parentheses; asterisks indicate statistically significant at the 95 % confidence level. Standard errors for the fixed-effect model calculated by dividing the WLS standard errors by the WLS RMSE. Conventional R-sqs. for the OLS and fixed-effect models (weighted); R-sqs. for the random-effects models are based on the percent reduction in the true variance; see Borenstein et al. (2009, p. 202). All estimates calculated using Stata/IC 11.2 and the Stata *metareg* macro

Table 15.2 shows the four meta-regression estimates.[9] In addition to the coefficient estimates and t-statistics, the table contains the R-square, root-mean-squared error (RMSE), F-statistic and p-value, and between-study variance estimate. The residuals are analyzed using the Jarque-Bera (J-B) test, which always fails to reject the normality null at the 5 % significance level. In column (1), the OLS regression performs poorly, but this regression mimics what is often estimated in poorly designed meta-analyses. The fixed-effect regression fits the data better and all coefficients are significant. The two random-effect regressions provide similar estimates, which is a common outcome (Borenstein et al. 2009, p. 207). However, some of the estimates are different compared with the fixed-effect regression (e.g., Intercept, Income). The implied income elasticity at the sample mean for the fixed-effect regression is 8.645/11.536 = 0.75, while the random-effect estimates are 0.49 and 0.52.

[9] I also experimented with a dummy variable for four studies where the standard errors were obtained by an indirect regression on sample size. The dummy coefficient for these studies was insignificant. Although the precision variable corrects for publication bias, its interpretation is somewhat different in the random-effect regressions.

Viscusi and Aldy (2003, p. 67) report a consensus income elasticity in the range 0.5–0.6, but some estimates are about 0.76. The meta-regression estimates in Table 15.2 span this range.[10]

15.4.4 Forecasting Accuracy: Within- and Out-of-Sample VSL Forecasts

A test of internal validity is based on within-sample predictions of each model. A test of external validity is based on *ex ante* or out-of-sample forecasts. In the benefit transfer literature, out-of-sample performance is referred to as reliability or validity testing and the forecast errors are generalization or transfer errors (Johnston and Rosenberger 2010, p. 486). The meta-regression models should outperform naive models, so the questions are: (1) by how much; and (2) do the predictions satisfy criteria for potential benefit transfer applications? Table 15.3 reports three prediction statistics: root-mean-squared error; mean-absolute-percent error; and median-absolute-percent error. The range of errors also is reported. These statistics are reported for two unweighted averages; two weighted averages; and the four regression models. The RMSE is smallest for the random-effects regressions. The larger RMSE value for the fixed-effect average is due to two large VSL values of $38 and $30 million in Table 15.1. However, the fixed-effect average does better than the OLS meta-regression using the mean error, 54.1 % compared to 91.2 %. This result shows that simplicity can beat complexity when forecasting (Kennedy 2008, p. 335). The fixed-effect regression provides the smallest mean percent error, 49.3 %, and the smallest median percent error, 30.7 %. As a quality criterion, many benefit transfers using meta-analysis find transfer errors in the range 30–60 % (Lindhjem and Navrud 2008), so some results in Table 15.3 are at the upper limit of this range. For the fixed-effect regression, 14 of the 28 predictions have errors of 30 % or less compared to only 6 of 28 predictions for the fixed-effect mean. Incorporating some additional information would therefore appear to have a good payoff in this example.

In order to develop out-of-sample forecasts, I employ data from Costa and Kahn (2004). They provide hedonic wage estimates for VSL for the years 1940, 1950, 1960, 1970, and 1980. Their estimates illustrate a rising value of a statistical life during a period when fatal accident rates fell substantially and life expectancies rose dramatically. Expressed in 2009 dollars, their linear-model estimates increased from

[10]Using their metadata for 32 VSL estimates, I also estimated the MM model in Bellavance et al. (2009, p. 455). Using *metareg*, I could (approximately) reproduce their coefficient and between-study variance estimates, but most of my t-statistics were smaller. The REML model for these data failed to converge. Their reported income elasticity estimates ranged from 0.72–0.86 for a restricted sample and 0.84–1.08 for the full sample. A VSL income elasticity of 0.7–0.9 is reported in Lindhjem et al. (2011), which is reduced to 0.3–0.4 for restricted samples.

Table 15.3 Forecast errors for VSL analysis

Estimate	RMSE	Mean absolute % predict error (sd)	Median absolute % predict error	Range of % errors (%)
Unweighted mean	8.519	96.7 (114.6)	50.8	0.3–496.4
Unweighted median	8.681	81.2 (91.0)	53.4	3.7–409.9
Fixed-effect mean	10.735	54.1 (31.8)	56.3	0.6–158.6
Random-effects mean	8.700	80.4 (89.5)	52.9	2.9–404.8
OLS regression	7.919	91.2 (129.5)	51.1	1.4–671.7
Fixed-effect regression (WLS)	6.352	49.3 (50.7)	30.7	0.9–179.5
Random-effects regression (MM)	6.113	53.0 (49.9)	39.9	0.4–208.2
Random-effects regression (REML)	6.090	52.9 (50.0)	38.5	3.6–207.5

Notes RMSE for the fixed-effect mean is influenced by two VSL values

$1.63 million in 1940 to $8.78 million per life in 1980. Forecasts were prepared using the fixed-effect regression in Table 15.2, with the Precision variable set equal to the sample median in Table 15.1 for NA studies. The Income variable was calculated for each forecast year using the manufacturing weekly wage in the *Statistical Abstract*. In 2009 dollars, the primary estimates in Costa and Kahn are $1.63, $2.88, $3.42, $6.15, and $8.78 million per life. The forecast values are $3.19 (se = 3.35), $5.91 (2.32), $7.79 (1.60), $8.89 (1.39), and $9.01 (1.38) million per life. The percent forecast errors are 95.7, 105.2, 127.8, 26.7, and 2.6 %. The larger errors for the earlier years illustrate the general principle that transfers that are far outside the range of the data involve larger transfer errors.

Suggested Protocol: As a technique for benefit-function transfers, meta-regression analysis can be used to prepare out-of-sample forecasts that are sensitive to alternative circumstances and settings. The ability to provide accurate forecasts is partly conditional on the reporting methods in the primary studies, which may lack crucial information regarding data and methods. Insuring consistent definition of variables across primary studies is important, where standardization and conversion may be necessary. Accounting for missing data in primary studies may be required. Use of proxy variables also may be required, which will introduce measurement error. Estimation of meta-regression models for benefit transfer requires a sensitivity analysis that examines parameter quality and forecasting outcomes for alternative model specifications and estimation methods. Among other likely candidates for a sensitivity analysis are treatment of outliers and missing data, functional form, and methods used to handle non-independent metadata; see Lindhjem et al. (2011) for an excellent example.

15.5 Meta-regression Analysis: Complex Data Structures

As discussed above, meta-regression analysis represents a general statistical approach that can account for between-study differences in methodology, data, measurement, quality, participants, and other possible economic, social, and environmental differences. The purpose of this and the following section is to delve further into the use of meta-regression analysis as applied to the types of observational data that are commonly found in economic studies. Some of the issues and methods considered have not seen widespread discussion in the literature.

15.5.1 Conceptual and Observational Equivalence of Effect Sizes

A basic principle in meta-analysis is that the effect sizes represent the same conceptual concept and are measured on an identical scale. A well-performed analysis will take steps to ensure that this is at least approximately true. However, a common criticism of meta-analysis is that the analyst ends up combining different types of measures (*apples and oranges*), without sufficient recognition of important conceptual or observational differences.[11] As noted by Smith and Pattanayak (2002, p. 274), "synthesis requires an ability to define a common concept to be measured … [and] meta-analyses summarizing non-market valuation studies have often *not* met the goal of measuring 'identical' concepts." They argue that a higher standard of consistency must be applied for benefit transfers, and they provide a table summarizing effect sizes used in 15 meta-analyses of environmental resources, including outdoor recreation, air quality, water quality, wetlands, and endangered species. Smith and Pattanayak report inconsistencies in the effect size concepts in half of the analyses, such as combining Marshallian and Hicksian benefit measures without adjusting for income effects. Hence, it is crucial for benefit transfer that the analyst transforms the primary effect estimates to a common definition or provides adequate controls. In some instances this can be accomplished by inclusion of study-level covariates, but it also might require dropping some studies, grouping the studies, or adjusting the estimates using information from the primary studies or other sources. The adjustment process should be clearly reported and transparent.

A related problem is expressing the primary estimates using a consistent measurement scale, such as a price elasticity, willingness to pay, hedonic price, or value of a statistical life. Compared to other disciplines, this has not been a major obstacle in economics because the primary estimates are commonly measured as a pure number (elasticities) or in monetary units. Translation is sometimes hindered by

[11] Rhodes (2012) argues that the seemingly-unrelated regressions (SUR) model can be used when studies report multiple outcomes from the same data set.

reporting procedures in the primary studies, such as a failure to adequately document data sources. However, Stanley and Rosenberger (2009) report a problem in the stated preference literature, where nonmarket values (consumer surpluses) are transformed from estimated price parameters. Briefly, they demonstrate that the transformation mandates a negative price parameter and renders the consumer surplus value and its standard error dependent upon each other. This dependence invalidates significance tests and also worsens the problem of publication bias, i.e., calculated consumer surpluses and standard errors are *smaller* when the reported price coefficients are larger, which is the opposite of expectations. Instead of inverse variances as weights for meta-analysis, they advocate the use of weights based on the sample size. More generally, the paper by Stanley and Rosenberger alerts meta-analysts to potential problems if effect sizes and standard errors are derived from the primary estimates, rather than adopted directly. Only access to the raw data from the primary studies might fully resolve this "generated regressand" issue.

15.5.2 Non-independent (Correlated) Metadata

In most econometric studies, the primary investigator reports more than one set of regression estimates. Sometimes the estimates are for largely independent subgroups, such as separate regressions for visitor and non-visitor participants in a CV survey. In other cases, the estimates are based on the same data and are not independent–they are correlated. For example, the investigator might report multiple estimates using different model specifications or econometric methods, reflecting specification searches ("data mining") and sensitivity analyses; see Kennedy (2008, p. 365) for a discussion of the difference. When a meta-analysis is based on non-independent data and the estimates are positively correlated, the summary standard errors are biased downward, resulting in confidence intervals that are too narrow. If the estimates happen to be negatively correlated, the summary standard errors are biased upward (Borenstein et al. 2009, p. 227). The summary effect size also is likely to be biased because a study that reports 10 estimates will tend to be assigned more weight than a study that reports only one or two estimates. Furthermore, cross-study correlations can arise due to use of the same primary data, which will bias estimates of the between-study variance. Numerous meta-analyses in environmental economics have failed to recognize these problems (Nelson and Kennedy 2009). Fortunately, a statistical method is available—cluster robust standard errors—that resolves part of the problem. However, this solution does not address the problem of excess weight assigned to primary studies with multiple estimates or the problem of bias in the between-study variance. This section first describes in greater detail the problem of correlated estimates and then reviews several methods that have been used to deal with non-independent metadata.

15.5.2.1 Sources of Non-independent Metadata

An obvious first source of non-independent metadata is the tendency by primary investigators to report multiple estimates for the same data, a practice that is more common in economics than it is in other disciplines. Nelson and Kennedy (2009) report that the median meta-analysis in environmental economics used three observations per primary study (mean of seven). Second, ostensibly independent estimates within a primary study may share an unobservable characteristic, such as "anchoring" by CV survey respondents. Third, investigators often use common primary data such as aggregate time-series data or public surveys. Indeed, Cameron and Trivedi (2005, p. 830) argue that public survey data are usually correlated. Fourth, several primary studies may share an unobservable characteristic such as similar management of an environmental commodity at different locations. Fifth, several primary studies may share an observable characteristic, such as an identical functional form, omission of a key explanatory variable, or data drawn from the same study location or time period. When a study characteristic is observable, it can be dealt with by including study-level regressors such as a dummy variable for common data sources. In the other cases, subsets of estimates are correlated, either within- or between-studies, leading to data clusters.

Suggested Protocol: Analysts should place a high priority on detecting and dealing with non-independent effect sizes. One obvious solution is to base the meta-analysis on only one observation per primary study (e.g., Nelson 2004), which is sometimes referred to as "best evidence" synthesis. However, using only one observation discards potentially valuable information and may appear arbitrary, unless the analyst provides a good road map. Alternatively, the meta-sample could be a random sample of primary observations, possibly with repeated sampling. A procedure recommended in the general literature is to create a synthetic effect size for each study, defined as the mean effect size in the study, with a variance that accounts for the correlation among the multiple estimates; see Borenstein et al. (2009). Lastly, it is a common practice to weight multiple observations by the reciprocal of the number of estimates by study, which insures that each primary study receives equal weight; see Lindhjem et al. (2011) for an example.

15.5.2.2 Methods for Dealing with Non-independent Metadata

Clustered errors are a common occurrence in microeconometrics, so a number of estimation methods are available.[12] Several possible methods are briefly reviewed here, with technical discussions omitted; see also Cameron and Trivedi (2005) and Nelson and Kennedy (2009). The first method is the use of cluster-robust standard

[12]Florax (2002) proposes use of Moran's I and Moran's scatter-plot as methods for visualizing within-study and between-study dependence. Stata also contains a number of tests and procedures for clustered data.

errors, which is easily implemented using the *clustered robust error* option in Stata or other software options. This estimator converges to the true error as the number of clusters (J) approaches infinity, where a large number is $J > 50$. However, if the number of clusters is very small ($J < 10$) or unbalanced, it is possible for the cluster-robust errors to be biased downward. This method is now a common solution for correlated metadata (e.g., Lindhjem et al. 2011; Nelson 2011), but the analyst needs to be aware of potential problems if the number of clusters is small or unbalanced.

A second method is to use panel-data econometrics as demonstrated by Jeppesen et al. (2002) and Rosenberger and Loomis (2000b). By interpreting each study (or each "grouping" of observations) as providing a panel of observations, panel-data software can be used for estimation. It is not necessary that the panels are based solely on individual studies as several possible stratifications are usually possible. Because the number of observations obtained from each grouping is unlikely to be uniform, the primary data form an unbalanced panel. In the fixed-effects (FE) panel-data model, the intercepts for each panel are viewed as fixed parameters. This avoids the bias due to correlation, but can be costly in terms of degrees of freedom if the number of panels is large. The random-effects (RE) model for panel data treats the intercepts as random draws from a distribution. The RE model matches the multilevel model discussed below, so this is a familiar approach for economists. Moreover, panel-data modeling lays stress on testing for correlation between the heterogeneity and the regressors, via a Hausman specification test, based on differences between FE- and RE-models estimates (Wooldridge 2006, p. 497). Hence, analysts using panel-data software are unlikely to overlook testing for correlated metadata and unobserved heterogeneity.

A third method is the use of a hierarchical, or multilevel, regression model. This method has seen widespread use for meta-analyses, especially by non-economists; see Hox and de Leeuw (2003). Examples of multilevel analyses by economists include Johnston et al. (2003) and Nelson (2011). Hierarchical regression allows the coefficients to vary randomly across groups, creating composite errors that are quite complicated. The potential for allowing the slopes to vary randomly makes this modeling approach very flexible, but the more common application is one where the intercept is random, but the slopes are not. This creates the mixed-effects regression model described in Eq. 15.6, but in the context in which the effect size estimates are correlated within groups. A drawback of this approach is that it assumes that the heterogeneity, as represented by the random intercepts, is uncorrelated with the regressors. This assumption is rarely mentioned in the literature on multilevel models, but such correlation is a common occurrence in economics (Nelson and Kennedy 2009, p. 356).

Suggested Protocol: Meta-analysis with non-independent data is so common that the analyst needs to have good understanding of the problem, its potential sources in the primary data, and alternative methods for dealing with it. It is important to recognize that there can be both within-study correlation (e.g., multiple estimates from a given primary study) and between-study correlation produced by observable and unobservable features of the studies and data. In addition to correlated data, outliers are a common occurrence in metadata. This problem may be evident in the raw data, but also arises as "influential data points" in the meta-regressions. A sensitivity analysis for outliers should be a standard practice for meta-analysis.

15.6 Meta-regression Analysis: Publication Bias

Publication bias is defined as the publication or non-publication of empirical results depending on the direction, statistical significance, and magnitude of the results. If the published literature is a biased sample of all relevant studies or contains other systematic biases, summary statistics for a meta-analysis will reflect this bias. In the absence of selection and heterogeneity, observed effects should vary randomly about the true effect size, independent of the estimated standard errors. However, due to emphasis on statistical significance, published studies are likely to be skewed toward larger effects, especially when mainstream theory supports a specific effect or there is an overwhelming professional consensus. In order to achieve statistical significance (e.g., t-statistics greater than 2.0), studies with less precise effects are more likely to report larger effects, reflecting specification searches and other estimation artifacts. Meta-analysis summaries are biased if the primary investigator searched among possible results to find one that meets standard criteria for statistical significance or publishability. This bias will carry over to benefit transfers, so external validity is compromised. The purpose of this section is to illustrate possible publication bias in the VSL metadata. A detailed discussion is beyond the scope of this chapter, but including the standard error in the meta-regression is a start on correcting for publication bias. More detailed discussions are found in Doucouliagos and Stanley (2009, 2010), Doucouliagos et al. (2012), Hoehn (2006), Nelson (2011), Roberts and Stanley (2005), Rosenberger and Stanley (2006), and Stanley (2008).

A standard method for detecting publication bias is the funnel graph. Briefly, the effect magnitudes are plotted on the horizontal axis and standard errors (or inverses) on the vertical axis. More precise estimates appear toward the upper part of the graph and less precise estimates toward the bottom. In the *absence* of publication bias, the funnel graph should be symmetric about the summary mean, i.e., there is absence of a relationship between the magnitude of the effect and its standard error. In the *presence* of publication bias, the funnel graph will be asymmetric, with more studies missing toward the bottom-left if there is positive-bias asymmetry. This reflects the notion that less precise estimates are more likely to get published if they have *larger* than average effects. The graphical display can be extended by using the nonparametric trim-and-fill procedure, which imputes the missing observations necessary for symmetry to be achieved (Nelson 2011). Figure 15.1 shows the funnel graph for the 22 VSL estimates for North America, where the summary mean is the random-effects mean of $8.80 million (shown by the diamond on the horizontal axis). There is evidence of positive-bias asymmetry. Figure 15.2 shows the funnel graph with the addition of seven imputed observations (shaded dots). The recomputed random-effects mean is $6.05 million (shaded diamond), which is

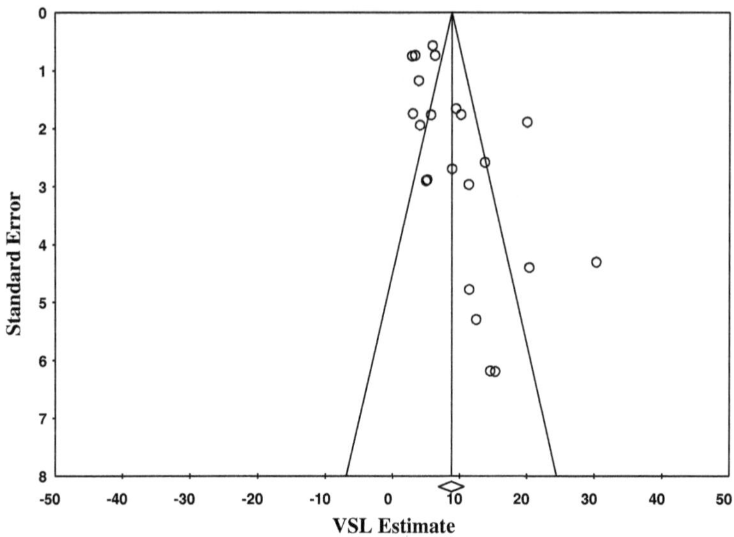

Fig. 15.1 Funnel plot of North American VSL data

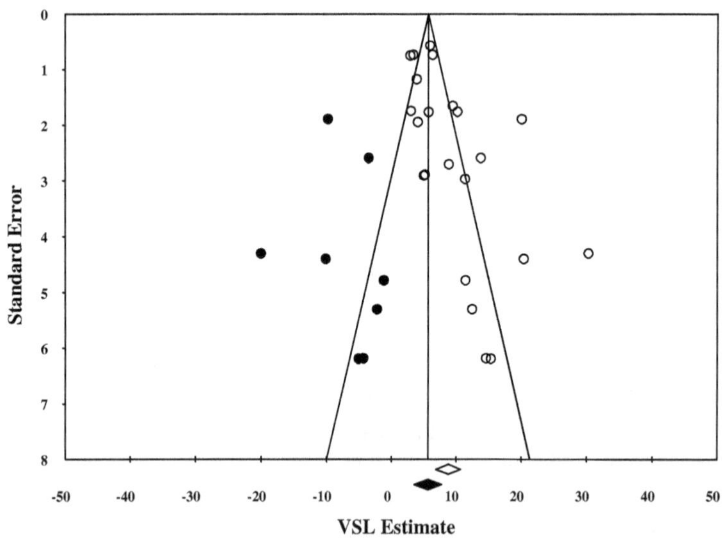

Fig. 15.2 Trim-and-fill funnel plot

a 31 % reduction in size. The recomputed fixed-effect mean (not shown) is $5.27 million, which is an 11 % reduction. However, the imputed values lie toward the middle and lower portions of the graph and all are negative, suggesting that other factors could be at work. Note that negative values are not required for a finding of no publication bias, only symmetric distribution of actual observations about a summary mean. It is important to recognize that asymmetry can arise for reasons other than publication bias, such as dispersion due to study methodology or genuine differences in population effect sizes. A more detailed analysis would be necessary to separate bias from other sources of heterogeneity and dispersion.[13]

Suggested Protocol: Publication bias has potentially important and damaging effects on attempts to use meta-analysis for benefit transfers. In order to ensure external validity, analysts need to employ statistical methods for detecting and correcting for this bias. A starting point is the construction of funnel graphs and inclusion of standard errors in meta-regressions. Bias also can occur in the form of investigators' interpretation or dissemination of results, which can carry over to external literature reviews and benefit transfers; see Nelson (2011) for examples. Lastly, one suggestion is to focus statistical analyses on data points in the symmetric upper part of the funnel graph (Stanley et al. 2010).

15.7 Conclusion

Earlier reviews by Smith and Pattanayak (2002) and Bergstrom and Taylor (2006) emphasized that benefit transfers based on meta-analysis must adhere to a strict set of rules, which require theoretically strong models, predictive tests of validity, and a disciplined approach to meta-analysis and benefit transfer. Smith and Pattanayak (2002, p. 282) conclude that a "research synthesis, which is conducted systematically, should help in isolating the 'best' of the current information available." However, they find that many existing meta-analyses fail to meet consistency requirements for internal and external validity. Nelson and Kennedy (2009) reach a similar conclusion for a much larger sample of meta-analyses in environmental economics. Bergstrom and Taylor (2006, p. 358) also stress the need for analysts to follow strict, systematic protocols, including exacting procedures for consistent measurement, coding, and analysis of primary data. This chapter has provided guidelines for the types of models, estimation procedures, and statistical problems likely to be encountered in a meta-analysis. The consistency required for benefit transfer shifts the focus of a meta-regression analysis from "taking stock of the literature" to predicting reliable values that can be transferred across time, space, or environmental commodity.

Acknowledgments This paper is dedicated to the memory of my colleague and long-time friend, Peter Kennedy, who passed away suddenly in 2010. He was a great source of wisdom about econometric methods and about life. I would also like to acknowledge comments and suggestions received from Ed Coulson, Mark Roberts, and Jim Tybout.

[13]Nelson (2011) illustrates the use of a truncated regression model for dealing with metadata containing selection bias; see Wooldridge (2006) for an introduction to this model.

References

Bellavance, F., Dionne, G., & Lebeau, M. (2009). The value of a statistical life: A meta-analysis with a mixed effects regression model. *Journal of Health Economics, 28*, 444–464.

Bergstrom, J. C., & Taylor, L. O. (2006). Using meta-analysis for benefits transfer: Theory and practice. *Ecological Economics, 60*, 351–360.

Biostat (2005). *Comprehensive meta-analysis* (Version 2.0). Englewood, NJ: Biostat, Inc.

Borenstein, M., Hedges, L. V., Higgins, J. P. T., & Rothstein, H. R. (2009). *Introduction to meta-analysis*. Chichester, UK: Wiley.

Cameron, A. C., & Trivedi, P. K. (2005). *Microeconometrics: Methods and applications*. Cambridge: Cambridge University Press.

Cavlovic, T. A., Baker, K. H., Berrens, R. P., & Gawande, L. (2000). A meta-analysis of environmental Kuznets curve studies. *Agricultural and Resource Economics Review, 29*, 32–42.

Cooper, H. (2010). *Research synthesis and meta-analysis: A step-by-step approach* (4th ed.). Los Angeles: Sage.

Cooper, H., Hedges, L. V., & Valentine, J. C. (2009). *The handbook of research synthesis and meta-analysis* (2nd ed.). New York: Russell Sage Foundation.

Costa, D. L., & Kahn, M. E. (2004). Changes in the value of life, 1940–1980. *Journal of Risk and Uncertainty, 29*, 159–180.

Doucouliagos, H., & Stanley, T. D. (2009). Publication selection bias in minimum-wage research? A meta-regression analysis. *British Journal of Industrial Relations, 47*, 406–428.

Doucouliagos, H., & Stanley, T. D. (2010). Picture this: A simple graph that reveals much ado about research. *Journal of Economic Surveys, 24*, 170–191.

Doucouliagos, H., & Stanley, T. D. (2012). *Meta-regression analysis in economics and business*. New York: Routledge.

Doucouliagos, H., Stanley, T. D., & Giles, M. (2012). Are estimates of the value of a statistical life exaggerated? *Journal of Health Economics, 31*, 197–208.

Fisher, R. A. (1932). *Statistical methods for research workers*. London: Oliver & Boyd.

Florax, R. J. G. M. (2002). *Accounting for dependence among study results in meta-analysis: Methodology applications to the valuation and use of natural resources*. Retrieved March 23, 2012, from Free University of Amsterdam. Web site: http://dare.ubvu.vu.nl/handle/1871/1643

Glass, G. V. (1977). Integrating findings: The meta-analysis of research. *Review of Research in Education, 5*, 351–379.

Hedges, L., & Olkin, I. (1985). *Statistical methods for meta-analysis*. Orlando: Academic Press.

Hoehn, J. P. (2006). Methods to address selection effects in the meta regression and transfer of ecosystem values. *Ecological Economics, 60*, 389–398.

Hox, J. J., & de Leeuw, E. D. (2003). Multilevel models for meta-analysis. In S. P. Reise & N. Duan (Eds.), *Multilevel modeling: Methodological advances, issues, and applications* (pp. 90–111). Mahwah, NJ: Lawrence Erlbaum.

Hunter, J. E., & Schmidt, F. L. (2004). *Methods of meta-analysis: Correcting error and bias in research findings* (2nd ed.). Thousand Oaks, CA: Sage.

Jeppesen, T., List, J. A., & Folmer, H. (2002). Environmental regulations and new plant decisions: Evidence from a meta-analysis. *Journal of Regional Science, 42*, 19–49.

Johnston, R. J., Besedin, E. Y., & Ranson, M. H. (2006). Characterizing the effects of valuation methodology in function-based benefits transfer. *Ecological Economics, 60*, 407–419.

Johnston, R. J., Besedin, E. Y., & Wardwell, R. F. (2003). Modeling relationships between use and nonuse value for surface water quality: A meta-analysis. *Water Resources Research, 39*, 1363–1371.

Johnston, R. J., & Rosenberger, R. S. (2010). Methods, trends and controversies in contemporary benefit transfer. *Journal of Economic Surveys, 24*, 479–510.

Kenkel, D. (2003). Using estimates of the value of a statistical life in evaluating consumer policy regulations. *Journal of Consumer Policy, 26*, 1–21.

Kennedy, P. (2008). *A guide to econometrics* (6th ed.). Malden, MA: Blackwell Publishing.
Konstantopoulos, S., & Hedges, L. V. (2009). Analyzing effect sizes: Fixed-effects models. In H. Cooper, L. V. Hedges, & J. C. Valentine (Eds.), *The handbook of research synthesis and meta-analysis* (2nd ed., pp. 279–293). New York: Russell Sage Foundation.
Lindhjem, H., & Navrud, S. (2008). How reliable are meta-analyses for international benefit transfer? *Ecological Economics, 66*, 425–435.
Lindhjem, H., Navrud, S., Braathen, N. A., & Biausque, V. (2011). Valuing mortality risk reductions from environmental transport and health policies: A global meta-analysis of stated preferences. *Risk Analysis, 31*, 1381–1407.
Moeltner, K., & Rosenberger, R. S. (2008). Predicting resource policy outcomes via meta-regression: Data space, model space, and the quest for 'optimal scope'. *The BE Journal of Economic Analysis & Policy.* DOI:10.2202/1935-1682.2028
Nelson, J. P. (2004). Meta-analysis of airport noise and hedonic property values: Problems and prospects. *Journal of Transport Economics and Policy, 38*, 1–28.
Nelson, J. P. (2011). Alcohol marketing, adolescent drinking and publication bias in longitudinal studies: A critical survey using meta-analysis. *Journal of Economic Surveys, 25*, 191–232.
Nelson, J. P., & Kennedy, P. E. (2009). The use (and abuse) of meta-analysis in environmental and resource economics: An assessment. *Environmental and Resource Economics, 42*, 345–377.
Pearson, K. (1904). Report on certain enteric fever inoculation statistics. *British Medical Journal, 3*, 1243–1246.
Randall, A., Kidder, A., & Chen, D.-R. (2008). *Meta-analysis for benefits transfer—Toward value estimates for some outputs of multifunctional agriculture*. Paper presented at the 12th Congress of the European Association of Agricultural Economics, Ghent. Web site: http://ageconsearch.umn.edu/bitstream/43648/2/013.pdf
Rhodes, W. (2012). Meta-analysis: An introduction using regression models. *Evaluation Review, 36*, 24–71.
Roberts, C. J., & Stanley, T. D. (2005). *Meta-regression analysis: Issues of publication bias in economics*. Oxford: Blackwell.
Rosenberger, R. S., & Loomis, J. B. (2000a). Using meta-analysis for benefit transfer: In-sample convergent validity tests of an outdoor recreation database. *Water Resources Research, 36*, 1097–1107.
Rosenberger, R. S., & Loomis, J. B. (2000b). Panel stratification in meta-analysis of economic studies: An investigation of its effects in the recreation valuation literature. *Journal of Agricultural and Applied Economics, 32*, 459–470.
Rosenberger, R. S., & Phipps, T. (2007). Correspondence and convergence in benefit transfer accuracy: Meta-analytic review of the literature. In S. Navrud & R. Ready (Eds.), *Environmental value transfer: Issues and methods* (pp. 23–43). Dordrecht: Springer.
Rosenberger, R. S., & Stanley, T. D. (2006). Measurement, generalization, and publication: Sources of error in benefit transfer and their management. *Ecological Economics, 60*, 372–378.
Shrestha, R. K., & Loomis, J. B. (2001). Testing a meta-analysis model for benefit transfer in international outdoor recreation. *Ecological Economics, 39*, 67–83.
Shrestha, R. K., & Loomis, J. B. (2003). Meta-analytic benefit transfer of outdoor recreation economic values: Testing out-of-sample convergent validity. *Environmental and Resource Economics, 25*, 79–100.
Shrestha, R., Rosenberger, R., & Loomis, J. (2007). Benefit transfer using meta-analysis in recreation economic valuation. In S. Navrud & R. Ready (Eds.), *Environmental value transfer: Issues and methods* (pp. 161–177). Dordrecht: Springer.
Smith, V. K., & Kaoru, Y. (1990). Signals or noise? Explaining the variation in recreation benefit estimates. *American Journal of Agricultural Economics, 72*, 419–433.
Smith, V. K., & Pattanayak, S. K. (2002). Is meta-analysis a Noah's arc for non-market valuation? *Environmental and Resource Economics, 22*, 271–296.
Stanley, T. D. (2001). Wheat from chaff: Meta-analysis as quantitative literature review. *Journal of Economic Perspectives, 15*, 131–150.

Stanley, T. D. (2008). Meta-regression methods for detecting and estimating empirical effects in the presence of publication selection. *Oxford Bulletin of Economics and Statistics, 70*, 103–127.

Stanley, T. D., & Jarrell, S. B. (1989). Meta-regression analysis: A quantitative method of literature surveys. *Journal of Economic Surveys, 3*, 161–170.

Stanley, T. D., Jarrell, S. B., & Doucouliagos, H. (2010). Could it be better to discard 90 % of the data? A statistical paradox. *American Statistician, 64*, 70–77.

Stanley, T. D. & Rosenberger, R. S. (2009). *Are nonmarket values systematically underestimated? Reducing publication bias for benefit transfer*. Presented at the 2009 Meta-Analysis of Economic Research Workshop. Retrieved April 28, 2012, from Hendrix College. Web site: http://www.hendrix.edu/uploadedFiles/Departments_and_Programs/Business_and_Economics/AMAES/RootnMRA.pdf

Stapler, R. W., & Johnston, R. J. (2009). Meta-analysis, benefit transfer, and methodological covariates: Implications for transfer error. *Environmental and Resource Economics, 42*, 227–246.

Sterne, J. A. C. (2009). *Meta-analysis in Stata: An updated collection from the Stata Journal*. College Station, TX: Stata Press.

U.S. Environmental Protection Agency (2006). *SAB advisory on EPA's issues in valuing mortality risk reductions*, EPA-SAB-08-001. Retrieved March 10, 2012, from http://yosemite.epa.gov/ee/epa/eerm.nsf/vwRepNumLookup/EE-0494?OpenDocument

U.S. Environmental Protection Agency. (2010). *Valuing mortality risk reductions for environmental policy: A white paper*. Washington, DC: EPA. Retrieved February 15, 2012, from http://www.epa.gov

Van Houtven, G. (2008). Methods for the meta-analysis of willingness-to-pay data. *PharmacoEconomics, 26*, 901–910.

Van Houtven, G., Powers, J., & Pattanayak, S. K. (2007). Valuing water quality improvements in the United States using meta-analysis: Is the glass half-full or half-empty for national policy analysis? *Resource and Energy Economics, 29*, 206–228.

Viscusi, W. K., & Aldy, J. E. (2003). The value of a statistical life: A critical review of market estimates throughout the world. *Journal of Risk and Uncertainty, 27*, 5–76.

Walsh, R. G., Johnson, D. M., & McKean, J. R. (1989). Issues in nonmarket valuation and policy applications: A retrospective glance. *Western Journal of Agricultural Economics, 14*, 178–188.

White, H. D. (2009). Scientific communications and literature retrieval. In H. Cooper, L. V. Hedges, & J. C. Valentine (Eds.), *The handbook of research synthesis and meta-analysis* (2nd ed., pp. 51–71). New York: Russell Sage Foundation.

Wooldridge, J. M. (2006). *Introductory econometrics: A modern approach* (3rd ed.). Mason, OH: Thompson South-Western.

Chapter 16
Meta-analysis: Rationale, Issues and Applications

John Rolfe, Roy Brouwer and Robert J. Johnston

Abstract This chapter reviews the key reasons for using meta-analysis for benefit transfer and provides an illustrative case study application. The case study involves a meta-analysis of values for improved river health in Australia from 2000 to 2009. To minimize potential problems of commensurability and methodology, we restrict the analysis to consider only values drawn from choice experiments. Different measures and scales of river health across studies were reconciled by transforming implicit prices into a comparable standard of willingness to pay (WTP) per kilometer of river in good health. Ordinary least squares and random effects meta-regression models were used to identify systematic relationships between the dependent variable (WTP/km) and explanatory variables characterizing sites, populations, affected resources, and primary study methodology. The case study illustrates both advantages and challenges involved in the application of meta-analysis to benefit transfer.

Keywords Benefit transfer · Choice experiments · River health · Willingness to pay · Meta-regression analysis

J. Rolfe (✉)
School of Business and Law, CQUniversity, Rockhampton, QLD, Australia
e-mail: j.rolfe@cqu.edu.au

R. Brouwer
Department of Environmental Economics, Institute for Environmental Studies, Vrije Universiteit, Amsterdam, The Netherlands
e-mail: roy.brouwer@vu.nl

R.J. Johnston
Department of Economics and George Perkins Marsh Institute, Clark University, Worcester, USA
e-mail: rjohnston@clarku.edu

R.J. Johnston et al. (eds.), *Benefit Transfer of Environmental and Resource Values*, The Economics of Non-Market Goods and Resources 14,
DOI 10.1007/978-94-017-9930-0_16

16.1 Introduction

Meta-analysis (MA), defined as the systematic quantitative summary of evidence across empirical studies (Glass et al. 1981), has been advanced as a way of generating more robust and reliable estimates of values for use in benefit transfer (BT) (Bateman and Jones 2003; Bergstrom and Taylor 2006; Nelson and Kennedy 2009; Johnston and Rosenberger 2010; Bateman et al. 2011; Kaul et al. 2013). Applications of MA in environmental economics have become increasingly common, with the technique used to pool estimates from nonmarket valuation studies to generate a predictive function. Nelson and Kennedy (2009) report 140 studies in environmental economics, of which more than half were conducted since 2003.

A meta-analysis for benefit transfer is, at its simplest, a regression analysis in which the left-hand side or dependent variable is the willingness to pay (WTP) for a change in a specified commodity and the right-hand side variables are the moderator variables that help to explain the variation in WTP across studies, including differences between sites and populations. The data for these MAs come from existing primary valuation studies; an MA can be conducted only where there are sufficient source studies valuing the same good or amenity to enable statistical analysis of these values. The purpose of an MA is to predict the WTP for a change in the amenity of interest while controlling for the various independent parameters that can vary across sites. The result of the MA regression provides the benefit function that can then be applied to target sites of interest.

Meta-analysis has been widely applied across a range of disciplines and policy issues (Nelson and Kennedy 2009). Within environmental economics, MA has been used to assess the values of environmental amenities such as wetlands (e.g., Brander et al. 2006; Brouwer et al. 1999; Woodward and Wui 2001), water (e.g., Johnston et al. 2003; Van Houtven et al. 2007), aquatic resources (e.g., Johnston et al. 2005) and forests (e.g., Barrio and Loureiro 2010; Zandersen and Tol 2009). It has been used to pool values estimated with a single nonmarket valuation method (e.g., Brouwer et al. 1999; Rolfe and Brouwer 2013; Smith and Osbourne 1996), or, more commonly, to pool values estimated with different nonmarket valuation techniques, including both revealed and stated preference techniques. It can be used to pool values from international case studies (e.g., Johnston and Thomassin 2010; Lindhjem and Navrud 2008), or to source studies from a limited geographic region (e.g., Rolfe and Brouwer 2013).

There are a number of reasons why a MA is used for benefit transfer. The most commonly identified advantages over unit value transfer and single site benefit function transfer are that information from multiple studies can be incorporated and synthesized, and outlier values—the effects of site and population heterogeneity and the effect of methodological differences—can be identified and controlled (Rosenberger and Loomis 2000). As a result, it is generally assumed that MA should generate more accurate values for benefit transfer (Rosenberger and Stanley 2006), as subsequent value functions are more easily adjusted to the characteristics of potential target sites. However, the evidence for this is mixed, with some studies identifying challenges in accurate BT with MA (Johnston and Rosenberger 2010; Smith and Pattanayak 2002). Nelson and Kennedy (2009) identify a number of methodological reasons why the quality of an MA can vary, and recommend best-practice guidelines for future analysis (also see Chap. 15).

Although benefit functions can be compiled from a single source study, particularly those generated from choice experiments (Rolfe 2006), MA provides an attractive alternative for generating a benefit function, allowing for the incorporation of both factual and methodological heterogeneity. The synthesis of values provides at least some reassurance that transfer values should be robust and free from major errors (Shrestha et al. 2007), in part because transferred values are less sensitive to the attributes and characteristics of individual studies (Moeltner et al. 2007). However, results can still be sensitive to included/excluded variables, study selection, and model specification choices; Chaps. 14, 15 and 17 discuss these issues in greater detail, including ways that model robustness can be evaluated.

Benefit functions can be developed from MA in two main ways (Moeltner et al. 2007). The first, and most common approach, is to identify all suitable studies that predict the value category of interest, and then develop a predictive function by regressing values against all suitable explanatory variables. The second is to build a more targeted BT function by choosing only very similar source studies and a small number of explanatory variables. Moeltner et al. (2007) demonstrate the use of Bayesian modelling to deal with the sparse data availability of the second approach. Both the general and targeted MA benefit functions have the potential to be more broadly applicable than the individual component studies (Moeltner et al. 2007; Stapler and Johnston 2009), as well as to generate smaller transfer errors (Shrestha et al. 2007).

In this chapter, the reasons for conducting MA are reviewed and approaches to improving accuracy and validity are illustrated with the help of a case study approach. Particular attention is given to choices and assumptions required to pool data from heterogeneous primary studies into metadata suitable for regression analysis. These practical issues are a central component of all valuation meta-analyses, but are often unreported in published documents. We also give attention to the choice of statistical methods for MA. These issues are given less emphasis than other modelling issues, however, as they are addressed elsewhere in this volume. The primary goal is to illustrate and discuss a practical application of meta-analysis to benefit transfer, including both the advantages of the approach and challenges that are commonly encountered.

The chapter is structured as follows. In the next two sections, the different reasons for and challenges in performing an MA are discussed. A case study application involving the synthesis of river protection values from multiple choice experiments (CE) is described in Sect. 16.4, with results and discussion presented in Sects. 16.5 and 16.6.

16.2 Reasons for Conducting a Meta-analysis

Meta-analysis is used in many discipline areas to summarize and analyze results of prior primary studies. Glass (1976, p. 3) identifies the purpose of conducting an MA as "integrating the findings" and "providing a rigorous alternative to the casual, narrative discussion of research studies." Nelson and Kennedy (2009) describe the rapid growth of MA in economics, with applications in environmental economics representing only one area of popular use. The following sections discuss some of the most important uses and advantages of MA for BT applications.

16.2.1 Synthesis of Values

Smith and Pattanayak (2002) note that a key role of meta-analysis is to synthesize results across the "flood of numbers" generated from microeconomic studies. A large number of value estimates have been developed in microeconomics for a variety of resource commodities and issues. Some of these are consistent (e.g., in terms of theoretical properties and empirical methods), while others diverge in important ways. Meta-analysis helps to summarize those data in a systematic manner, allowing conclusions to be drawn about a pool of studies in a particular topic area. An implicit assumption of this synthesis role is that the accuracy of benefit transfer can be improved by pooling data from a number of studies to allow such systematic analysis (Bateman and Jones 2003; Rosenberger and Loomis 2000).

16.2.2 Identifying Outliers

The variable of interest within valuation meta-analyses typically exhibits substantial variation across studies (Nelson and Kennedy 2009; Woodward and Wui 2001). For example, Stapler and Johnston (2009) report that values for recreational fishing from 390 observations ranged from $0.05 to $612.79 per fish, or 12,256 times; Zandersen and Tol (2009) report that values for forestry recreation from 290 observations ranged from €0.66 to €112 per trip, or 170 times; and Dekker et al. (2011) report that Value of a Statistical Life estimates from 290 observations ranged from $0.13 million to $33.58 million, or 258 times. Other studies (e.g., Brander et al. 2006, 2007) report high levels of skewness in value estimates, with mean values several times higher than median values.

Concerns that outlier values may be unrealistic or have a disproportionate effect on study results and/or that value distributions may be non-normal have led to different control strategies (Nelson and Kennedy 2009). Some researchers (e.g., Woodward and Wui 2001) have omitted outlier observations, some (e.g., Brander et al. 2006, 2007) convert the dependent variable to log form, some (e.g., Smith and Osbourne 1996; Stapler and Johnston 2009) apply multilevel models to account for potential correlations among studies, some (e.g., Johnston et al. 2006) use a weighted least squares approach to minimize the contribution of extreme values, while others (e.g., Kaul et al. 2013) use nonparametric models to address concerns about normality failures. Regardless of the strategy, a key benefit of MA is that outlier observations can be identified, and transfer errors from using these cases for point source transfers can be avoided.

16.2.3 Addressing Site and Population Heterogeneity

A key goal of an MA is to go beyond finding a simple mean for the variable of interest towards understanding and explaining heterogeneity in the observations (Woodward and Wui 2001; Colombo et al. 2007). Some heterogeneity can be explained by differences in site and population characteristics among studies. This is described by Christensen (2003) and Nelson and Kennedy (2009) as factual heterogeneity, where divergence among sites and populations is generally expected to lead to larger differences in the predictor values. A standard MA will include site and population characteristics as moderator variables in the regression; these variables allow adjustments in the subsequent benefit function to extrapolate values to a target site and population (Brouwer 2000; Johnston and Rosenberger 2010; Moeltner et al. 2007; Nelson and Kennedy 2009).

16.2.4 Controlling for Methodological Variations

Most meta-analyses combine the results of different studies for analysis, including studies that apply different types of study methods. For example, values included in an MA might have been estimated using different types of revealed preference and stated preference valuation techniques, each involving different methods for collecting and analyzing the data. Christensen (2003) and Nelson and Kennedy (2009) characterize differences in value estimates that arise from these factors as methodological heterogeneity. A number of MA studies have found that methodological study attributes help to explain variations in the value parameter of interest, confirming that it is important to include these factors in an MA (Moeltner et al. 2007; Nelson and Kennedy 2009; Stapler and Johnston 2009).

16.3 Challenges in Application

There are a number of challenges to performing a successful meta-analysis (Johnston and Rosenberger 2010). Five of the most important challenges are (1) identifying suitable source studies, (2) identifying a consistent definition of the dependent variable that is used in the analysis so that the primary value estimates are commeasurable, (3) identifying relevant study and methodological variables that may explain variations in the dependent variable, (4) ensuring that data for these independent variables in the MA are commeasurable, and (5) addressing potential statistical complications related to factors such as sample selection effects, primary data heterogeneity, heteroskedasticity and nonindependence of multiple observations from individual studies.

The combined suitability, quality and commensurability of source studies is widely identified in the literature as a critical factor in the successful conduct of benefit transfer. The statistical analysis for a successful MA requires substantial data; Nelson and Kennedy (2009) report in their quantitative summary of 140 meta-analyses that the mean number of primary studies and observations included was 42 and 191, respectively. However, the field of primary valuation studies is sparse in many areas, requiring tradeoffs between the quality/commensurability of studies and the number of studies available for analysis. Analysts who select more comprehensive samples (to increase the number of observations for statistical purposes) also tend to increase heterogeneity in the sample and the need for more regressors in the explanatory function. Moeltner et al. (2007) refer to this as the N versus K problem. Stanley et al. (2013) detail a number of criteria for selecting primary studies for MA, while Nelson and Kennedy (2009) recommend that protocols for searching and selecting primary studies be clearly identified, and that the treatment of the data should be carefully explained.

Another key challenge in MA is achieving a consistent definition of the dependent variable (values for the amenity changes of interest) (Boyle and Bergstrom 1992; Loomis and Rosenberger 2006; Nelson and Kennedy 2009). Johnston and Rosenberger (2010) note that analysts performing MAs are required to make assumptions about the required level of commensurability in the dependent variable of interest. These definitional issues occur across two key dimensions.

The first involves commodity consistency, recognizing that commodities may be defined in subtly different ways, particularly in environmental studies, making it difficult to ensure that the values from different primary studies refer to similar underlying commodities. For example, values for water quality may vary according to whether water is being used for drinking, recreation or environmental protection purposes, where water quality may be a proxy for a higher-order good. In some cases, differences in commodities can be reconciled by including appropriate explanatory variables as regressors in the analysis. However, when fundamentally different concepts are being measured (e.g., human health or environmental protection), then analysts need to address this in the source study selection protocol.

The second issue involves welfare consistency, reflecting similarity in theoretical welfare measures being estimated across source studies. Welfare *inconsistency* occurs when there are variations in the dependent variable related to variations in the theoretical properties of welfare measures. These are often caused by differences in valuation techniques or methods are being employed by source studies (Bergstrom and Taylor 2006; Nelson and Kennedy 2009; Smith and Pattanayak 2002). For example, the pooling of revealed and stated preference data often involves implicit comparison of Hicksian and Marshallian welfare measures. The appropriateness of pooling these measures within a single MA is the subject of some disagreement (Nelson and Kennedy 2009; Johnston and Moeltner 2014; Londoño and Johnston 2012; Smith and Pattanayak 2002).

Other sources of welfare inconsistency can occur when methodological variations in the applications, such as the payment vehicle and length of the payment period, underpin expected differences in the welfare estimates. Similar issues can arise when willingness to pay (WTP) and willingness to accept (WTA) estimates are pooled, where use, nonuse or total value estimates are pooled, or where publication effects underpin subtle methodological variations among experiments (Smith and Pattanayak 2002). Nelson and Kennedy (2009) identify a pragmatic approach to concerns about commensurability, where best-practice guidelines include recommendations that the same economic concepts are being measured in the selected primary studies and that potential sources of heterogeneity in observations are incorporated into a meta-analysis.

There are also challenges in identifying, describing and reconciling the moderator variables to include in an MA. There are a number of factors that can explain variation in the dependent variable of interest across studies and observation, and these must be captured by the analyst so that these differences can be reconciled and the subsequent MA function can be used for prediction. Johnston and Rosenberger (2010) classify these moderators as variables that identify the resource, policy, context and population attributes of the primary studies, whereas Nelson and Kennedy (2009) characterize them as characteristics of the environmental issue or site of interest, characteristics of the primary study methods and the analyses used, and context variables such as income, location and time period. However, the need to capture the different influences that are important can intersect poorly with the limited availability and diversity of source studies, as well as inadequate reporting of data and methods within primary study publications (Loomis and Rosenberger 2006).

The choice of statistical methods for MA also involves tradeoffs and challenges. A common approach to analyzing economic data is ordinary least squares regression (OLS). However, simple OLS is inadequate for MA for several reasons (Nelson and Kennedy 2009). First, heteroskadasticity may be generated by variations in sample sizes and other effects that cause differences in variances among samples. A typical correction involves weighting observations according to their estimated reliability, with variables such as standard errors, variances, primary study sample size, or the number of observations in the primary study used as the weighting variable. Here, greater weight is given to studies that estimate values with greater precision (e.g., lower standard errors). Nelson and Kennedy (2009), however, note that information on variances and other data is often missing from source studies, and that weightings across different valuation studies is often inconsistent because of variations in specifications across and within source studies. Hence, defensible weighting is not often possible in MA of estimated values.

A second concern is that many data sets involve multiple observations (e.g., split samples) from a single study, so that observations are correlated within broader studies. Corrections for the panel nature of these datasets, such as the use of fixed effects panel data or random effects panel data models, are needed. Third, there are likely to be correlations between the heterogeneity in the dependent variable and the independent variables. This can be addressed by using a robust standard errors estimator, such as a White or Huber-White estimator (Nelson and Kennedy 2009).

16.4 Case Study Application: Valuing River Protection in Australia

The advantages, performance and challenges of a meta-analysis can be most effectively illustrated with a case study. Such illustrations can help to clarify the steps involved in the analysis, as well as the procedures required to deal with some of the main application challenges. The case study we use for illustration uses MA to predict the WTP per kilometer of rivers in good health in Australia. This is an important issue for policy makers in the country, with measures to protect and restore river health widely debated in recent years. That interest has driven a number of primary case studies to estimate economic values for river health, providing the data necessary for MA.

There were three major justifications for developing the MA reported in this case study. First, it provided a benefit transfer function that could be rapidly applied to approximate values for river health across a wide range of potential circumstances. This enabled the economic values of potential improvements to river protection to be included in a large number of potential policy applications. Second, it allowed reconciliation of some large differences in per kilometer values of river health that had been found in prior valuation studies. This systematic analysis helped to demonstrate to policy makers that variations in value estimates could be explained and provided more confidence in the use of value estimates. Third, it provided a basis for conducting comparative quality control for unit value transfers. The large variations in value estimates described above led to the potential for unit value transfers for river health improvements to vary greatly depending on the source studies used for transfer. The illustrated MA provided a basis for comparison, allowing the identification of unit value transfers that provided comparatively high or low estimates.

The MA was performed by collating WTP estimates from primary valuation studies for improved river health in Australia over an 11-year period, from 2000 to 2010,[1] and then analyzing them with different regression models to generate a predictive function for the WTP value. The meta-analysis involved 154 individual value estimates of WTP for river health from 17 separate valuation studies across six states (Table 16.1).

[1]The data sourced for the MA have been described in more detail in Rolfe and Brouwer (2013).

Table 16.1 Overview of studies included in the meta-analysis

	Authors	Study year	River catchment	State	Split-samples	Implicit price (WTP)
1	Van Bueren and Bennett (2004)	2000	All waterways (not specified)	National QLD WA	6	$/hh/year per 10 km restored waterway for fishing or swimming
2	Morrison and Bennett (2004, 2006)	2000	Bega, Clarence, Georges, Gwydir, Murrumbidgee	NSW	36	$/hh/year and one-time-off per % of river covered with healthy native vegetation/per fish species/for fishable/ swimmable water quality whole river/per waterbird/per other fauna species
3	Rolfe et al. (2002)	2000	Fitzroy, Dawson, Comet-Nogoa-Mackenzie	QLD	7	$/hh/year per km of waterways in the catchment remaining in good health
4	Rolfe and Windle (2003)	2001	Fitzroy	QLD	3	$/hh/year per km of waterways in the catchment remaining in good health
5	Windle and Rolfe (2004)	2003	Fitzroy	QLD	3	$/hh/year and one-time-off per km of waterways remaining in good health
6	Windle and Rolfe (2006)	2005	SE Queensland Fitzroy Murray-Darling Mackay Whitsunday Great Barrier Reef	QLD	18	$/hh/year per % of waterways in good health
7	Kragt et al. (2007)	2006	Goulburn	NSW	16	$/hh one-time-off per % native fish species and population level/for % of river length with healthy native vegetation/per native waterbird and animal species
8	Bennett et al. (2008a)	2006	Murray	NSW VIC	5	$/hh/year per % of pre-European fish numbers/ % of healthy flooded vegetation (river red gums)
9	Bennett et al. (2008b)	2006	Moorabool Gellibrand Goulburn	NSW VIC	12	$/hh one-time-off per % native fish species and population level/for % of river length with healthy native vegetation/per native waterbird and animal species/
10	Rolfe and Bennett (2009)	2002	Fitzroy	QLD	1	$/hh/year per km of waterways in the catchment remaining in good health

(continued)

Table 16.1 (continued)

	Authors	Study year	River catchment	State	Split-samples	Implicit price (WTP)
11	Kragt and Bennett (2009a)	2008	George	TAS	3	$/hh/year per km of river length with healthy native vegetation
12	Kragt and Bennett (2009b)	2008	George	TAS	2	$/hh/year per km of river length with healthy native vegetation
13	Kragt and Bennett (2010)	2008	George	TAS	4	$/hh/year per km of river length with healthy native vegetation
14	Mazur and Bennett (2009)	2008	Lachlan, Namoi, Hawkesbury-Nepean	NSW	7	$/hh/year per km of healthy waterways
15	Mazur (2011)	2008	Hawkesbury-Nepean	NSW	5	$/hh/year per km of healthy waterways
16	Mazur and Bennett (2010)	2008	Hawkesbury-Nepean	NSW	7	$/hh/year per km of healthy waterways
17	Hatton MacDonald and Morrison (2010)	2010	Murray	NSW VIC SA	18	$/hh/year per % of healthy vegetation/per % of original population of native fish

NSW New South Wales, *QLD* Queensland, *SA* South Australia, *TAS* Tasmania, *VIC* Victoria, *WA* Western Australia

16.4.1 Establishing the Dependent Variable

Issues concerning the commensurability and quality of source studies were addressed in part by selecting only value estimates from choice experiment (CE) studies. Restricting source data in this way helped generate a consistent base for the data-pooling and minimize methodological variations. Issues of welfare consistency were further addressed by using marginal tradeoffs in terms of implicit prices for river health (also known as part-worths) as the dependent variable. This avoided scale parameter concerns when comparing results across studies (Swait and Louviere 1993), and allowed only values for the relevant attribute to be reported. The alternative of using compensating surplus estimates as the dependent variable was not practical because of the difficulty in establishing future protection scenarios that were consistent across case studies. (In cases where only unit changes in single attributes were involved, however, estimates of compensating surplus collapsed to implicit prices.) Other issues of welfare commensurability did not arise; all values were expressed in WTP (rather than WTA), terms, and all captured both use and nonuse values.

Issues of commodity consistency were much more challenging. The aspects of river health that were measured varied across case studies, so that the value estimates were for slightly different amenities, each relating to improved river health or a related ecological function. This is a common challenge in valuation MA. Out of 150 separate studies, 51 assessed values for river health directly, 23 assessed values for healthy waterways for recreation, 40 for healthy vegetation in waterways, 30 for healthy birdlife, and 10 for healthy fish stocks (Fig. 16.1). In the MA, these value

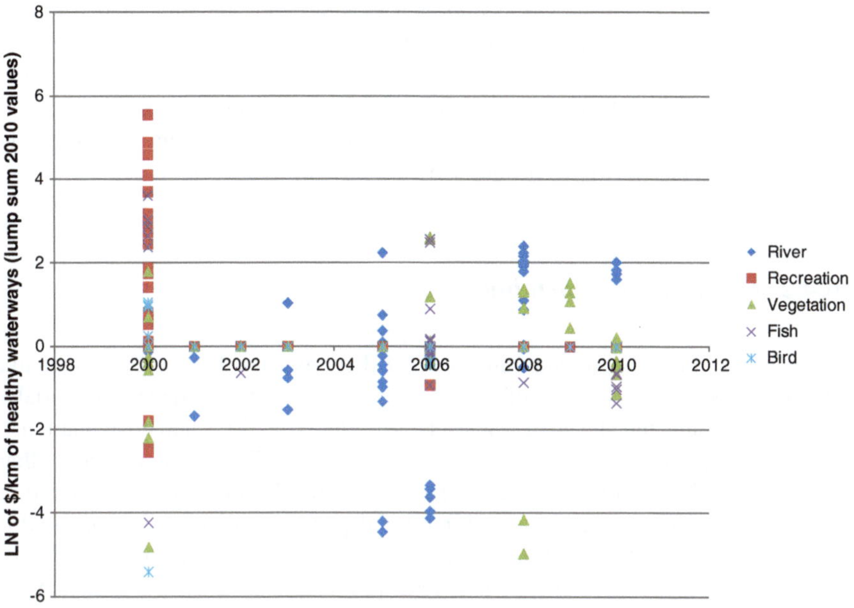

Fig. 16.1 Scatter plot of LN of WTP per km of healthy waterways (2010 values)

estimates were pooled into a single model, with independent variables added to the data set to represent differences among studies.

Substantial variation was also identified in the way that changes in the environmental good were described in the source studies. Where the good was defined in terms of river health, we identified changes in terms of absolute values (kilometers of waterways in good health) or percentage values (percentage of waterways in good health). This enabled changes to be compared across studies. The same variation occurred across the indicator variants of river health description, where variables such as vegetation, fish and birds were described in both absolute and percentage terms across studies. For the meta-analysis, absolute values were chosen as the consistent descriptive standard. Implicit prices from studies where changes were described in percentage terms were converted to values per kilometer using the length of the river system as the base.[2]

Value estimates from the selected studies were not directly comparable because of differences in attribute description, payment streams, and study year. Three steps were required to transform values from the individual case studies into a consistent estimate of WTP per kilometer of waterways in good health. To address description differences, values for percentage changes were transformed into absolute values by multiplying percentage changes by river length. To address variations in payment

[2]Many studies included this information as part of the framing to survey respondents. Where the information was not included in studies, the data was sourced from Norris et al. (2001).

streams, all WTP estimates were converted to lump sum amounts, using a 10 % discount rate. As the studies had been collected over an 11-year period between 2000 and 2010, WTP estimates needed to be converted to real values in a consistent year. To achieve this, the Consumer Price Index for Australia was used to bring all payment estimates into 2010 dollar equivalents. The resulting values of the dependent variable (in log form) are shown below in Fig. 16.1.

16.4.2 Explanatory Variables

The graphical analysis presented in Fig. 16.1 demonstrates that there is substantial variation in WTP per kilometer of waterways in good health, the dependent variable of interest. While the mean of values is $7.15/hh/km, the standard deviation is $25.46/hh/km. The maximum observation of $260.72/hh/km is more than 57,000 times the minimum value of $0.004/hh/km. To minimize this variation, three key strategies were employed. First, the dependent variable was converted to log form prior to the MA. Second, six extreme values were identified (from the log form of the dependent variable) and removed from the data set, reducing the maximum observation to $24.09/hh/km. Third, three different forms of weighting the dependent variable have been tested.

The other possible strategy to account for the heterogeneity in the dependent variable is to expand the set of independent moderators. For example, there are a number of methodological, framing and design variations among the CM experiments that may explain the heterogeneity of the remaining value estimates. These are reviewed below.

16.4.2.1 Site Differences

Site differences could be expected to explain part of the variation in WTP values. Differences include variations among catchments, where factors such as size (river length), location (state) and type (inland versus coastal) may influence how respondents view tradeoffs. Many studies could be identified within two key catchments: the Murray Darling river system draining parts of Queensland, New South Wales, Victoria and South Australia, and the Fitzroy River system in central Queensland.

16.4.2.2 Population Differences

Values may vary across populations and with population characteristics. There is some evidence from individual case studies that values differ according to whether the population sample comes from inside or outside catchments (Bennett et al. 2008a, b; Kragt et al. 2007; Morrison and Bennett 2004; van Bueren and Bennett 2004) and when state capital versus local populations are sampled (Bennett et al. 2008a, b; Kragt et al. 2007; Mazur and Bennett 2009; Morrison and Bennett 2004; Rolfe et al. 2002). There is also consistent evidence within the studies that key socio-demographic characteristics such as age, gender and household income influence WTP amounts.

16.4.2.3 Framing of the Tradeoffs

Several differences were identified across the studies in terms of the way that tradeoffs were framed to choice respondents. All the experiments were consistent in terms of presenting a status quo or constant base option plus two or more improvement options, together with a cost attribute. Most studies presented the information in absolute terms (kilometers of healthy waterways under different policy options), but one study (van Bueren and Bennett 2004) framed the information in terms of marginal changes, and one study (Windle and Rolfe 2006) presented both absolute and marginal levels together.

Differences in WTP per kilometer of improvement may also be driven by marginal effects. The total length of river systems that were assessed varied from 209,118 km (Australian total) to 44 km (Moorabool River), whereas the percentages in good condition ranged from a low of about 5 % for the Clarence River (Morrison and Bennett 2004) and the Goulburn River to 65 % for the Georges River (Kragt and Bennett 2009a, b). It is possible that respondents considered this information when identifying their values per each one kilometer improvement.

There were also differences in the way that condition trends were depicted, with the future base lower than current condition in 57 % of the experiments and equal to current condition in the remainder. Concerns about losses, in a form of endowment effect, may cause respondents to support options that avoid future losses more strongly than options that only add to the current status. There were also differences in the total range of improvement levels offered, from a low of 2 % of total river length (Mazur and Bennett 2009, 2010) to a high of 100 % of total river (Morrison and Bennett 2004). Where the amounts of level changes are proportionally higher, respondents may find improvement options more attractive.

16.4.2.4 Framing of Payment Mechanisms

A number of different payment mechanisms were applied in the different studies, with most using some form of local rates or levies to identify how payment would be collected. Some studies present a mix of payment vehicles, where respondents were informed that the higher costs would be generated by a mix of higher taxes, rates, charges and consumer costs. About half of the studies involved annual costs over a number of years, with 20-year time frames being the most common.

16.4.2.5 Data Collection

There was some variation in survey collection techniques, with 53 % of the samples collected through mail surveys and 47 % collected through dropoff/pickup techniques. The mean sample size was 232 respondents (standard deviation = 141), while the mean response rate was 40.6 % (standard deviation = 17.8 %).

16.4.2.6 Presentation of Levels

Differences were also identified across studies in the way that levels were presented in the choice sets. Tradeoffs were described only in absolute numbers (i.e. kilometers of waterways) in 36 % of the experiments, only in percentage terms in 10 % of the experiments, and with the use of symbols in 38 % of the surveys. Other formats included the joint use of absolute numbers and percentage terms (15 % of surveys) and the joint use of absolute numbers and symbols (6 % of surveys).

16.4.2.7 Choice Set Dimensions

There was limited variation in the dimensions of choice sets used in the experiments. All experiments used three alternatives (including one as a base), apart from one experiment which had five choice alternatives. The latter was also the only labeled experiment. Five attributes were used in 82 % of the experiments, with four attributes used in the remainder. Five choice sets per questionnaire were applied in 72 % of experiments, with six choice sets in 18 % and eight choice sets in 9 %.

16.4.2.8 Analysis of Results

The statistical models used in the data analysis were generally confined to three main approaches when only the models used to predict results were considered. Conditional logit models were employed for 38 % of the studies, nested logit models for 52 %, and random parameters logit models for 9 %. Reported model fits in terms of adjusted rho-square values ranged from a low of 0.034 to a high of 0.41. Forty-one percent of the studies had been published in refereed journal articles or book chapters, with the remainder in the grey literature as conference papers and research reports. There was potential for individual researcher effects, with three key researchers involved across the pool of studies. Jeff Bennett (Australian National University) was involved in 13 of the studies, Mark Morrison (Charles Sturt University) in two (with 36 separate experiments), and John Rolfe (Central Queensland University) in six.

16.5 Results

The analysis of relationships between the implicit prices (WTP per kilometer of healthy waterways) and the potential explanatory variables was conducted with least squares regressions, taking into account the panel format and heteroskedasticity in the data. The meta-model used to predict the marginal rate of substitution (implicit price) between income and a healthy waterways attribute can be described more generally as follows:

$$\ln(MWTP_i) = \sum_j \beta_j X_{ij} = \beta_1 X_{i1} + \beta_2 X_{i2} + \beta_3 X_{i3} + \cdots + \beta_J X_{iJ} + \varepsilon_i \qquad (16.1)$$

where $\ln(MWTP_i)$ is the vector containing the natural log of the implicit price found in study i and X_{ij} represents the design matrix for the covariates, consisting of amenity characteristics, population characteristics, and study and methodology characteristics, with the latter capturing variations in tradeoff framing, payment mechanisms, data collection, level description, choice set design, and data analysis. The estimates for the regression coefficients β_j (j = 1…J) are obtained through maximum likelihood (ML) techniques.

Four types of models were estimated. The first was a simple ordinary least squares (OLS) model using all relevant variables. This was repeated with a random effects (RE) model (to address the panel nature of the data). All variables were included in these models to identify the significant moderators in the regression. The other four RE models all included a correction for the heteroskedasticity in the data (using the White corrector), but the first did not include any weighting for study size or quality. The last three models tested different weightings for surrogates of study quality: sample size, response rate and model fit (the latter indicated by adjusted R-square statistics).

16.5.1 General MA Model

An example of a general MA is demonstrated in the models presented in Table 16.2. These models analyze the 148 observations in the data set (after six extreme values had been removed), where the source studies include value estimates for river health that had been scoped in different ways. To address these variations in the amenity of interest, additional regressors were included in the models to describe whether the dependent variable was defined in terms of recreation, vegetation, or fish/bird health. To improve the performance of the models with the heteroskedasticity corrector, only significant variables were included in these latter models; the potential for omitted variable bias should be recognized when comparing the different models.

The models for the full data set all exhibit relatively high explanatory power, as shown by the R-square statistics. The use of weighted models with a White corrector for heteroskedasticity reduces model fit slightly over the simple OLS and RE models.

Table 16.2 Meta-analysis regression models (full sample of 148 observations)

	Simple OLS	Random effects models		Random effects model with heteroskadasticity (White) corrector		
Weighting used	None	None	None	Sample size	Response rate	Adjusted R-square
	Coef.	Coef.	Coef.	Coef.	Coef.	Coef.
Constant	22.006***	3.943	1.228	8.605***	7.890***	3.327***
Years from 2000	−0.062	0.033				
DV for local population	−0.141	0.221		0.372**		
DV for annual Payment	2.760***	2.666***	1.375***	2.659***	2.436***	2.095***
DV for rate or levy payment vehicle	−1.827**	−0.985	−1.266***		−0.967**	−1.647***
DV for Murray Darling	−0.281	−0.737***	−0.849***	−0.668***	−0.918***	−0.785***
LN of river length	−0.979***	−0.026***	−0.027***	−0.031***	−0.028***	−0.027***
Percent good condition now	−0.029***	−0.017***	−0.021***	−0.030***	−0.029***	−0.017***
DV if future base is lower	−0.389**	0.202				0.299***
Range of attribute levels relative to river length (%)	−0.035***	−0.021**	−0.023**	−0.039***	−0.034***	−0.022*
DV if years to improvement not specified	4.493***	3.542***	2.466***	2.583***	3.820***	3.438***
DV if level presented as absolute amounts	−1.467**	−1.500	−2.731***	−1.994***	−1.470**	−2.563***
DV if level presented as marginal amounts	−5.569***	−8.930***	−9.388***	−9.835***	−8.570***	−9.043***
Percent good condition now	−0.029***	−0.017***	−0.021***	−0.030***	−0.029***	−0.017***
DV if future base is lower	−0.389**	0.202				0.299***
Range of attribute levels relative to river length (%)	−0.035***	−0.021**	−0.023**	−0.039***	−0.034***	−0.022*
DV if years to improvement not specified	4.493***	3.542***	2.466***	2.583***	3.820***	3.438***
DV if level presented as absolute amounts	−1.467**	−1.500	−2.731***	−1.994***	−1.470**	−2.563***
DV if level presented as marginal amounts	−5.569***	−8.930***	−9.388***	−9.835***	−8.570***	−9.043***

(continued)

Table 16.2 (continued)

	Simple OLS	Random effects models		Random effects model with heteroskadasticity (White) corrector		
Weighting used	None	None	None	Sample size	Response rate	Adjusted R-square
	Coef.	Coef.	Coef.	Coef.	Coef.	Coef.
DV if levels shown only as % amounts	−1.328***	0.826				
DV if levels are shown as symbols	0.854	3.198***	2.193***	2.530***	2.601***	2.402***
DV if mail survey	−0.254	0.860		0.637***		
Total number of surveys collected	0.000	0.000				
Response rate	0.002	−0.020*		−0.015**		
DV if improvement options are labeled	−2.187***	−0.111				
Number of attributes	0.246	−0.927		−1.669***	−1.377***	
DV if MNL model estimated	−0.009	−0.113				
Percent male in sample	−0.003	−0.017				−0.034**
Average age in sample	−0.010	0.021				
Log of average income in sample	−0.370*	−0.199			−0.269**	−0.319***
DV if published	−0.216	0.794*	0.761**	0.632*	0.492*	1.091***
DV if dep. var. not initially assessed in kms	−2.821***	−2.870***	−2.662***	−2.373***	−2.613***	−2.712***
DV of river health in terms of recreation use	3.544***	3.890***	4.784***	4.476***	4.652***	5.013***
DV of river health in terms of healthy vegetation	1.485	2.626**	3.143***	2.764***	2.987***	3.732***
DV of river health in terms of fish or bird stocks	1.344	2.449**	2.891***	2.462***	2.983***	3.385***
Number of observations	148	148	148	148	148	148
OLS-adjusted R-Squared	0.867	0.825	0.796	0.816	0.817	0.801
Random effects model R-Squared		0.825	0.789	0.801	0.805	0.806
Probability RE model is different to OLS	NA	0.072	0.114	0.212	0.125	0.090

(continued)

Table 16.2 (continued)

	Simple OLS	Random effects models		Random effects model with heteroskadasticity (White) corrector		
Weighting used	None	None	None	Sample size	Response rate	Adjusted R-square
	Coef.	Coef.	Coef.	Coef.	Coef.	Coef.
Standard deviation of dependent variable	1.818	1.818	1.818	1.447	1.880	1.821
Predicted value (per 1 km of river health improvement) (at means of indep. variables)	\$5.463	\$5.453	\$5.315	\$3.008	\$5.930	\$5.713

DV dependent variable; * $p < 0.10$; ** $p < 0.05$; *** $p < 0.01$

Both the response rate and the adjusted R-square variables generate slightly higher model fits when they are used to weight the sample compared to using sample size.

To aid in the comparative analysis, the value predicted by each model for one kilometer of rivers in good health has been estimated, using the mean values for each moderator to derive the sum of the products ($X\beta$). A correction for the log transformation bias ($\sigma^2/2$) has been added to the sum of the products before the exponential transformation (antilog) has been taken. It is notable that there is limited variation in value prediction from the different models from \$5.32/km to \$5.93/km, apart from the random effects model weighted by sample size, which has a prediction value of \$3.01/km. There is a difference between predicted values for the dependent variable and the mean of the unlogged (trimmed) dependent variable at \$3.42/km. This difference is due to the fact that the dependent variable is the log of the implicit price rather than the implicit price itself. Hence, the mean model prediction is not expected to match the mean (untransformed) implicit price, due to the intervening log transformation.

The analysis finds that many regressors used to describe the river health values were significant in explaining values. Location in the Murray Darling basin (the largest in Australia) reduced value estimates, whereas values were also lower for rivers that had larger areas currently in good condition. Consistent with declining marginal utility, respondents had lower values per kilometer of river health for larger systems and for larger changes in improvement (the latter indicated by the ratio of attribute level ranges to the length of river in the study). If the condition of the river was expected to keep declining (the future base was set lower than current condition), this had a positive influence on WTP in one of the models.

Population characteristics had more limited significance. Age was not significant in any of the models, while Percent Male was significant in one model (negative effect), Income in two models (negative effect) and Local Population in one model (positive effect).

The framing and presentation of the tradeoffs in the choice experiments were shown to be important explanatory variables. Coefficients were positive (values would be higher) when studies did not specify how long it would take for environmental improvements to occur,[3] when attributes with multiple descriptors had at least one descriptor showing absolute changes in levels, and when attributes were described by symbols. Coefficients were negative (values would be lower) when attribute levels were presented only as absolute amounts or as marginal changes, or when the amount of river in question was not defined in kilometers.

As expected, the form of the payment vehicle had a significant influence. The use of annual payments was a positive predictor of value estimates compared to lump sum payments, whereas the use of compulsory mechanisms such as rates and levies had a negative influence. The types of models used in the estimation were not identified as important, but higher values were associated with studies that were published compared to studies in the grey literature.

Other design and collection factors had limited influence. Sample size was not significant in any of the models, but response rate was significant and negative in three. The latter result may indicate that as a more comprehensive sample was collected, the heterogeneity of responses increased. The number of attributes in choice sets was a negative influence in one model, while the use of mail surveys for collection was positive in another.

In common with many MA studies, the small data set limited the number of attributes that were statistically significant. Some variables that appeared to be significant in bi-variate tests, such as author effects and the year of the study, did not maintain significance in the larger models, indicating that the effects were driven by design and study factors. The limited sample size may also explain the difficulty of including both weighting and heteroskedasticity corrections together. Langrange Multiplier tests indicate that standard random effects model is preferred over the OLS models at the 7 % level, and that the random effects model with a weighting for model fits and the White correction for heteroskedasticity was preferred at the 9 % level; however the other random effects models were not significantly different from the OLS models at the 10 % level.

16.5.2 Targeted MA Model

A targeted modeling approach further restricted the dataset only to the 51 observations that directly assessed river health, excluding those observations that defined river health in terms of terms of recreation, vegetation, or fish/bird health. This helps to test whether reducing the size of the data set by targeting a narrower protocol for the dependent variable of interest and reducing the number of required explanatory variables reduces or improves the accuracy of the resulting function. This issue of whether the focus of an MA should be general or targeted is a familiar problem to MA practitioners (e.g., Moeltner et al. 2007; Stapler and Johnston 2009).

---

[3]This result has to be treated with caution, as this was a characteristic of only one study (Morrison and Bennett 2004), and may be driven by other study characteristics.

Table 16.3 Meta-analysis regression models (subsample for river health only)

	Simple OLS	Random effects model with heteroskadasticity (White) corrector			
Weighting used	None	None	Sample Size	Response Rate	Adjusted R-Square
	Coef.	Coef.	Coef.	Coef.	Coef.
Constant	−0.816	−11.441*	−12.613**	−10.702*	−15.925***
Years from 2000	0.222***	0.203***	0.214***	0.199***	0.216***
DV for annual payment	4.314***	4.328***	4.248***	4.310***	4.349***
DV for Murray Darling	−0.601***	−0.615***	−0.623***	−0.594***	−0.532***
River length	0.000***	0.000***	0.000***	0.000***	0.000***
Percent good condition now	−0.033***	−0.029***	−0.032***	−0.030***	−0.025***
DV if levels shown only as % amounts	3.938***	3.838***	3.737***	3.888***	4.034***
Response rate	0.029**	0.035***	0.045***	0.030***	0.041***
Average age in sample	−0.080**	−0.069**	−0.090***	−0.062*	−0.066***
Log of average income in sample		2.034*	2.395**	1.887	2.860***
DV if published	−0.640*	−0.673**	−0.831*	−0.577*	−0.708***
DV if dep. var. not initially assessed in kms	−3.763***	−3.674***	−3.569***	−3.653***	−3.681***
Number of observations	51	51	51	51	51
OLS-adjusted R-Squared	0.943	0.945	0.932	0.942	0.956
Random effects model R-Squared		0.957	0.951	0.956	0.944
Probability RE model is different to OLS	NA	0.110	0.111	0.113	0.089
Standard deviation of dependent variable	1.906	1.818	1.628	1.915	1.890
Predicted value (per 1 km of river health improvement) (at means of indep. variables)	\$5.817	\$5.818	\$3.534	\$5.970	\$5.441

DV dependent variable; * $p < 0.10$; ** $p < 0.05$; *** $p < 0.01$

Restricting the data set to the more narrowly focused amenity has some effect on mean values for the dependent variable. The “river-only” data set has lower average values of \$2.83/km compared to \$3.42/km for the full data set, as well as lower variance in the sample (standard deviation of \$3.31/km compared to \$5.59/km). This indicates that expanding the scope of studies in MA to increase the number of data points comes at the cost of additional systematic influences on values, as well as increased heterogeneity. Results for the targeted MA model are shown in Table 16.3. Note that the predicted mean values continue to be higher than the actual mean value of the data set because the log transformation of the dependent variable.

Tests for whether the dependent variable of interest should be targeted or general can be made in three ways by comparing the models reported in Tables 16.2 and 16.3. First, the model fits for the more narrowly defined amenity appear to be improved (higher adjusted R-square values) in this case study, possibly because of lower heterogeneity in the dependent variable. However, not as many explanatory variables are significant because of the smaller data set as well as invariance in some variables over the reduced pool of sampled studies. This means that the use of the expanded (general) data set produces more detailed transfer functions. The third test is to compare the predicted values, where the difference in value predictions ranges between 6–17 % of the values predicted by the corresponding general model. This suggests that there is little difference in predictive performance between the two approaches.

A potential next step in the analysis could be to conduct additional evaluations of model robustness and specification choices. Because these methods are the primary focus of Chap. 17, they are not discussed here. Additional analyses could also be conducted to evaluate the benefit transfer accuracy of the various MA models (e.g., in terms of transfer error) and compare this accuracy to that achievable through various unit value transfers. Results of similar types of evaluations in other case studies are provided in Chap. 14.

16.6 Conclusions

Meta-analysis, like other tools used for benefit transfer, involves tradeoffs and compromises. These include assumptions and adjustments required to analyse results from a heterogeneous set of primary studies within a single statistical model used to predict values for unstudied policy sites. The case study presented in this chapter is useful for demonstrating both the advantages and challenges of an MA for benefit transfer. Among the advantages is the capacity to synthesize information from a multiple studies. Here, values from 148 choice experiments conducted over a 10-year period across Australia were synthesized into a single, systematic value function that can be applied for benefit transfers. As noted above, this required a number of adjustments and assumptions to account for heterogeneity in valued commodities, sites, populations and methodologies among the primary studies. The resulting benefit transfer function is capable of predicting values for a variety of different sites, populations and purposes. The use of this function for benefit transfer also ameliorates problems related to the selection of individual sites or studies for unit value transfer. This is a particularly relevant advantage in the present case, given the wide disparity in unit values found across our sample of primary studies.

In performing the MA, however, a number of challenges had to be addressed. The first was to identify a sufficient number of suitable source studies. Restricting the selection only to choice experiments limited methodological variations, whereas limiting studies only to Australia and from the year 2000 onwards minimized other sources of heterogeneity. In addition, six studies (4 %) were trimmed from the analysis because of extreme values.

The second challenge was to use a consistent definition of the dependent variable. This was particularly important given study variations discussed above. In this case, only source studies that measured river health in some way and that could be assessed in terms of kilometers of rivers in good condition were selected, even if there were other variations in the amenity of interest across case studies. To address this, moderator variables were used to identify whether the river health was aligned with a particular purpose or amenity such as recreation, vegetation, and fish or birds; the subsequent modeling showed that these moderators were particularly influential. These protocols exemplify the type of choices faced in the specification of nearly all MAs in the valuation literature.

The third challenge was to identify the various site, population and methodological variations that could have influenced the value of the dependent variable across different studies. The specification of the amenity and the way that it was framed to respondents was particularly important in the subsequent regression analysis, whereas population and methodological variables were less influential. Despite the highest value of the dependent variable being 9,300 times more than the lowest value (after extreme values were removed), the regressions explained more than 80 % of the heterogeneity in this dependent variable. Values were sensitive to the amount of the amenity of interest, the proportion in good condition, and the amount of improvements that were being offered. This result suggests that value transfers that do not adjust for these differences may generate misleading results and high-error transfers.

Fourth, the case study also demonstrated the application of statistical treatments to address three key issues in this type of MA: the panel nature of the data set, the need for weighting of some or all observations, and issues of hetroskedasticity. In this study the predictions of the dependent variable do not vary greatly across the different modelling approaches, but this may not always be the case. The three potential weightings that were tested did not have a major impact on value estimates. It is notable that weighting by sample size produced the worst-performing model; this may reflect that sample size by itself does not necessarily indicate how representative a sample may be or the subsequent statistical performance.

Finally, this study also provides empirical insight into the degree of commodity consistency that should be required within valuation metadata (i.e., how narrowly focused the selection of source studies should be). Here, relaxing the definition of the dependent variable of interest to increase the pool of available studies does enable more detailed benefit transfer functions, but impacts on prediction accuracy are limited. We also caution that the amenity of interest must still be defined narrowly enough for the studies to be comparable, and that broadening the pool of studies is likely to introduce greater unexplained heterogeneity into the data sets.

Acknowledgments Thanks go to Jeff Bennett, Russell Blamey, Mark Morrison and Jill Windle for helping to provide relevant data for the meta-analysis. The time for Roy Brouwer to assemble the data was funded by the CSIRO National Research Flagship Water for a Healthy Country.

References

Barrio, M., & Loureiro, M. (2010). A meta-analysis of contingent valuation forest studies. *Ecological Economics, 69*, 1023–1030.

Bateman, I. J., & Jones, A. P. (2003). Contrasting conventional with multi-level modelling approaches to meta-analysis: Expectation consistency in UK woodland recreation values. *Land Economics, 79*, 239–258.

Bateman, I. J., Brouwer, R., Ferrini, S., Schaafsma, M., Barton, D. N., Dubgaard, A., et al. (2011). Making benefit transfers work: Deriving and testing principles for value transfers for similar and dissimilar sites using a case study of the non-market benefits of water quality improvements across Europe. *Environmental and Resource Economics, 50*, 365–387.

Bennett, J., Dumsday, R. & Gillespie, R. (2008a). *Analysing options for the Red Gum forests along the Murray River*. Paper presented at the 52nd Australian Agricultural and Resource Economics Society, Canberra, February.

Bennett, J., Dumsday, R., Howell, G., Lloyd, C., Sturgess, N., & Van Raalte, L. (2008b). The economic value of improved environmental health in Victorian Rivers. *Australasian Journal of Environmental Management, 15*, 138–148.

Bergstrom, J. C., & Taylor, L. O. (2006). Using meta-analysis for benefits transfer: Theory and practice. *Ecological Economics, 60*, 351–360.

Boyle, K., & Bergstrom, J. (1992). Benefit transfer studies: Myths, pragmatism and idealism. *Water Resources Research, 28*, 657–663.

Brander, L. M., Florax, R. J. G. M., & Vermaat, J. (2006). The empirics of wetland valuation: A comprehensive summary and a meta-analysis of the literature. *Environmental and Resource Economics, 33*, 223–250.

Brander, L. M., Van Beukering, P., & Cesar, H. S. J. (2007). The recreational value of coral reefs: A meta-analysis. *Ecological Economics, 63*, 209–218.

Brouwer, R. (2000). Environmental value transfer: State of the art and future prospects. *Ecological Economics, 32*, 137–152.

Brouwer, R., Langford, I. H., Bateman, I. J., & Turner, R. K. (1999). A meta-analysis of wetland contingent valuation studies. *Regional Environmental Change, 1*, 47–57.

Christensen, P. (2003). *Topics in meta-analysis: A literature survey*. Oslo: Institute of Transport Economics.

Colombo, S., Calatrava-Requena, J., & Hanley, N. (2007). Testing choice experiment for benefit transfer with preference heterogeneity. *American Journal of Agricultural Economics, 89*, 135–151.

Dekker, T., Brouwer, R., Hofkes, M., & Moeltner, K. (2011). The effect of risk context on the value of a statistical life: A Bayesian meta-model. *Environmental and Resource Economics, 49*, 597–624.

Glass, G. V. (1976). Primary, secondary, and meta-analysis of research. *Educational Researcher, 5*, 3–8.

Glass, G. V., McGaw, B., & Smith, M. L. (1981). *Meta-analysis in social research*. Beverly Hills, CA: Sage.

Hatton MacDonald, D., & Morrison, M. (2010). Untangling differences in values from internet and mail stated preference studies. Paper presented at the 4th World Congress on Environmental Economics, Montreal, Canada, June 28–July 2.

Johnston, R. J., & Moeltner, K. (2014). Meta-modeling and benefit transfer: The empirical relevance of source-consistency in welfare measures. *Environmental and Resource Economics, 59*, 337–361. doi:10.1007/s10640-013-9730-3.

Johnston, R. J., & Rosenberger, R. S. (2010). Methods, trends and controversies in contemporary benefit transfer. *Journal of Economic Surveys, 24*, 479–510.

Johnston, R. J., & Thomassin, P. J. (2010). Willingness to pay for water quality improvements in the United States and Canada: Considering possibilities for international meta-analysis and benefit transfer. *Agricultural and Resource Economics Review, 39*, 114–131.

Johnston, R. J., Besedin, E. Y., & Wardwell, R. F. (2003). Modeling relationships between use and nonuse values for surface water quality: A meta-analysis. *Water Resources Research, 39*, 1363–1372.

Johnston, R. J., Besedin, E. Y., Iovanna, R., Miller, C., Wardwell, R., & Ranson, M. (2005). Systematic variation in willingness to pay for aquatic resource improvements and implications for benefit transfer: A meta-analysis. *Canadian Journal of Agricultural Economics, 53*, 221–248.

Johnston, R. J., Besedin, E., & Ranson, M. H. (2006). Characterising the effects of valuation methodology in function-based benefits transfer. *Ecological Economics, 60*, 407–419.

Kaul, S., Boyle, K. J., Kuminoff, N., Parmeter, C. F., & Pope, J. C. (2013). What can we learn from benefit transfer errors? Evidence from 20 years of research on convergent validity. *Journal of Environmental Economics and Management, 66*, 90–104.

Kragt, M. & Bennett, J.W. (2009a). *Using choice experiments to value river and estuary health in Tasmania with individual preference heterogeneity.* Paper presented at 53rd Australian Agricultural and Resource Economics Society Conference, Cairns.

Kragt, M. & Bennett, J.W. (2009b). *Using choice experiments to value river and estuary health in Tasmania with individual preference heterogeneity.* Canberra: Environmental Economics Research Hub Research Report 16, Australian National University.

Kragt, M. & Bennett, J.W. (2010). *The impacts of attribute level framing and changing cost levels on choice experiments value estimates.* Paper presented at the 54th Australian Agricultural and Resource Economics Society Conference, Adelaide.

Kragt, M., Bennett, J., Lloyd, C., & Dumsday, R. (2007). *Comparing choice models of river health improvement for the Goulburn River.* Paper presented at the 51st Australian Agricultural and Resource Economics Conference, 13–16 February, Queenstown.

Lindhjem, H., & Navrud, S. (2008). How reliable are meta-analyses for international benefit transfers? *Ecological Economics, 66*, 425–435.

Londoño, L. M., & Johnston, R. J. (2012). Enhancing the reliability of benefit transfer over heterogeneous sites: A meta-analysis of international coral reef values. *Ecological Economics, 78*, 80–89.

Loomis, J., & Rosenberger, R. (2006). Reducing barriers in future benefit transfers: Needed improvements in primary study design and reporting. *Ecological Economics, 60*, 343–350.

Mazur, K.M. (2011). *Choice modelling in the development of natural resource management strategies in NSW catchments.* Ph.D thesis, Canberra: Australian National University.

Mazur, K. & Bennett, J.W. (2009). *Location differences in communities' preferences for environmental improvements in selected NSW catchments: A choice modelling approach.* Paper presented at 53rd Australian Agricultural and Resource Economics Society Conference, Cairns, 10–13 February.

Mazur, K. & Bennett, J.W. (2010). *Framing for incentive compatibility in choice modelling*. Paper presented at 54th Australian Agricultural and Resource Economics Society Conference, Adelaide, 9–12 February.

Moeltner, K., Boyle, K., & Paterson, R. W. (2007). Meta-analysis and benefit transfer for resource valuation-addressing classical challenges with Bayesian modelling. *Journal of Environmental Economics and Management, 53*, 250–269.

Morrison, M., & Bennett, J. (2004). Valuing New South Wales rivers for use in benefits transfer. *Australian Journal of Agricultural and Resource Economics, 48*, 591–611.

Morrison, M., & Bennett, J. (2006). Valuing New South Wales rivers for use in benefits transfer. In J. Rolfe & J. Bennett (Eds.), *Choice modelling and the transfer of environmental values.* Cheltenham: Edward Elgar.

Nelson, J. P., & Kennedy, P. E. (2009). The use (and abuse) of meta-analysis in environmental and resource economics: An assessment. *Environmental and Resource Economics, 29*, 323–336.

Norris, R. H., Young, B., Liston, P., Bauer, N., Davies, N., Dyer, F., et al. (2001). *The assessment of river condition: An audit of the ecological condition of Australian Rivers.* Canberra: National Land and Water Resources Audit.

Rolfe, J. (2006). Theoretical issues in using choice modelling data for benefit transfer. In J. Rolfe & J. Bennett (Eds.), *Choice modelling and the transfer of environmental values* (pp. 28–53). Cheltenham: Edward Elgar.

Rolfe, J., & Bennett, J. (2009). The impact of offering two versus three alternatives in choice modelling experiments. *Ecological Economics, 68*, 1140–1148.

Rolfe, J., & Brouwer, R. (2013). Design effects in a meta-analysis of river health choice experiments in Australia. *Journal of Choice Modelling, 5*, 81–97.

Rolfe, J., & Windle, J. (2003). Valuing the protection of aboriginal cultural heritage sites. *The Economic Record, 79*, s85–s95.

Rolfe, J., Loch, A., & Bennett, J. (2002). Tests of benefits transfer across sites and population in the Fitzroy Basin. In Valuing floodplain development in the Fitzroy Basin Research Report No. 4. Rockhampton: Central Queensland University.

Rosenberger, R. S., & Loomis, J. (2000). Using meta-analysis for benefit transfer: In-sample convergent validity tests for an outdoor recreation database. *Water Resources Research, 36*, 1097–1107.

Rosenberger, R. S., & Stanley, T. D. (2006). Measurement, generalization and publication: Sources of error in benefit transfers and their management. *Ecological Economics, 60*, 372–378.

Shrestha, R., Rosenberger, R., & Loomis, J. (2007). Benefit transfer using meta-analysis in recreation economic valuation. In S. Navrud & R. Ready (Eds.), *Environmental value transfer: Issues and methods* (pp. 161–177). Dordrecht: Springer.

Smith, V. K., & Osbourne, L. (1996). Do contingent valuation estimates pass the scope test? A meta-analysis. *Economics Letters, 32*, 67–272.

Smith, V. K., & Pattanayak, S. K. (2002). Is meta-analysis a Noah's Ark for non-market valuation? *Environmental and Resource Economics, 22*, 271–296.

Stapler, R. W., & Johnston, R. J. (2009). Meta-analysis, benefit transfer, and methodological covariates: Implications for transfer error. *Environmental and Resource Economics, 42*, 227–246.

Stanley, T. D., Doucouliagos, H., Giles, M., Heckemeyer, J. H., Johnston, R. J., Laroche, P., et al. (2013). Meta-analysis of economics reporting guidelines. *Journal of Economic Surveys, 27*, 390–394.

Swait, J., & Louviere, J. J. (1993). The role of the scale parameter in the estimation and comparison of multinomial logit models. *Journal of Marketing Research, 30*, 305–314.

Van Bueren, M., & Bennett, J. (2004). Towards the development of a transferable set of value estimates for environmental attributes. *Australian Journal of Agricultural and Resource Economics, 48*, 1–32.

Van Houtven, G., Powers, J., & Pattanayak, S. K. (2007). Valuing water quality improvements in the United States using meta-analysis: Is the glass half-full or half-empty for national policy analysis? *Resource and Energy Economics, 29*, 206–228.

Windle, J., & Rolfe, J. (2004). Assessing values for estuary protection with choice modelling using different payment mechanisms, in *Valuing floodplain development in the Fitzroy basin*, Research Report No. 10. Rockhanpton: Central Queensland University.

Windle, J., & Rolfe, J. (2006). Non market values for improved NRM outcomes in Queensland. Research Report No. 2 in the non-market valuation component of AGSIP Project #13. Rockhampton: Central Queensland University.

Woodward, R. T., & Wui, Y.-S. (2001). The economic value of wetland services: A meta-analysis. *Ecological Economics, 37*, 257–270.

Zandersen, M., & Tol, R. (2009). A meta-analysis of forest recreation values in Europe. *Journal of Forest Economics, 15*, 109–130.

Chapter 17
Meta-analysis: Econometric Advances and New Perspectives Toward Data Synthesis and Robustness

Kevin J. Boyle, Sapna Kaul and Christopher F. Parmeter

Abstract This chapter outlines statistical and econometric procedures that can be applied to the analysis of meta-data. Particular attention is paid to ensuring robustness of the insights from a meta-regression. Specific detail is paid to the fine econometric details to sharpen the insights of practitioners when deciding which tools to use for a meta-analysis for literature assessment of policy prescriptions.

Keywords Meta-analysis · Horizontal robustness · Vertical robustness · Nonparametric regression analysis · Bandwidth selection · Nonparametric specification test · Fixed effects meta-regression · Random effects meta-regression

17.1 Introduction

Any analyst conducting a meta-analysis must face two realities. First, the available studies and observations are not random draws from any known sampling frame. Second, the analyst must make careful and informed decisions regarding the inclusion/exclusion of individual studies, observations and explanatory variables. Given these constraints of sample availability and the attendant influence of analyst judgment, a natural question to ask is whether meta-analysis findings are robust. Although there are a variety of empirical tools that one might use to address this

K.J. Boyle (✉)
Virginia Tech, Blacksburg, VA, USA
e-mail: kjboyle@vt.edu

S. Kaul
University of Utah, Salt Lake City, UT, USA
e-mail: sapna.kaul@hci.utah.edu

C.F. Parmeter
University of Miami, Coral Gables, FL, USA
e-mail: cparmeter@bus.miami.edu

R.J. Johnston et al. (eds.), *Benefit Transfer of Environmental and Resource Values*,
The Economics of Non-Market Goods and Resources 14,
DOI 10.1007/978-94-017-9930-0_17

question, published meta-analyses rarely include extensive evaluation of whether and how results are (or are not) sensitive to included/excluded information and model specification choices.

Consider meta-analysis as a research assessment tool. We can think of using a meta-analysis to advance a field of study in two dimensions. First, we can identify the key findings of a literature where the collective insights from multiple studies provide stylized facts. Second, we can uncover areas where more research should be devoted within a literature because the existing empirical studies do not collectively support or refute an outcome. These alternative avenues may offer new insights into long-standing problems. Further, meta-analyses can shed light into how research in a given field is being conducted, which has its own benefits because it can offer new perspectives and criticisms to how problems should or should not be tackled. The importance of these uses of meta-analysis highlights the need to consider the robustness of empirical insights that arise from a meta-analysis.

The goal of this chapter is to illustrate a variety of approaches through which meta-analysts can better evaluate the robustness of their results. We discuss several dimensions of meta-analysis where robustness checks can be applied to gauge the influence of meta-data and analyst choices on insights drawn from meta-regression results. These robustness checks apply to either the horizontal or vertical dimensions of the meta-data. That is, deciding what studies or observations to include in a meta-analysis entails adding or subtracting observations (rows) from the meta-data; whereas deciding which descriptive features of the empirical studies to include as regressors entails adding or subtracting variables (columns). There are at least four related issues underlying meta-data construction:

- Sample selection: is the set of empirical studies available to the analyst subject to a systematic data-generating mechanism that results in some empirical studies being conducted and others not being conducted?
- Influential studies: are individual studies systematically different from the mass of the data from other studies in the meta-data?
- Influential observations: are individual observations systematically different from the mass of the other observations in the meta-data?
- Influential variables: do the variables available and chosen as regressors by the analyst affect estimation results?

Any and all of these dimensions can impact the meta-regression results. However, the presence of sample selection, or influential studies, observations or variables does not undermine the validity of a meta-data or meta-regression per se as these issues are not unique to meta-analysis and can occur in any empirical analysis. Their presence indicates areas where future research may be warranted. Influential studies and observations suggest further research on evaluating differences within and across studies and exploring the importance of study-specific heterogeneity in interpreting and applying meta-analysis findings. Influential variables provide insight on variables of interest in future primary studies and how study characteristics should be documented. Thus, one should not jump to the conclusion that the presence or absence of robustness implies desirable or undesirable outcomes. Rather, any specific result must be taken in context and the robustness checks described below can provide important insights for advancing any empirical literature.

Beyond focusing on the construction of the meta-data, there also remains the issue of the appropriate tools to estimate the meta-equation itself. The meta-equation is important because it governs the ability to uncover key insights from a literature. For example, meta-analysis has been deployed to determine the empirical methods that result in accurate benefit transfers. Consider the hypothetical situation where a researcher is deciding between using studies that have either deployed travel-cost models or contingent valuation for conducting a benefit transfer. Further, suppose the policy issue is a quality change. A traditional meta-equation would be additively separable in all variables in the meta-data. However, if an interaction existed between the study method and the type of policy change that was being analyzed, then the researcher may overlook a key subset of the literature that could assist with constructing a more accurate benefit transfer.

This situation describes a clear limitation of how current meta-equations are specified in the literature. Recent research has introduced the use of nonparametric methods to the analysis of meta-data to assist in this search for robust conclusions. A key issue is that a majority of meta-data variables may be discrete in nature and construction of **all potential** interactions in a meta-equation will quickly exhaust degrees of freedom and cause the empty cell phenomenon. Nonparametric methods are ideally suited to these challenges because they can leverage information in nearby, non-empty cells to combat the lack of information that exists with empty cells.

Given meta-analysis' role in providing objective insights into a given literature, it is only natural to turn the light on meta-analysis practices themselves and judge the robustness of their findings. In the remainder of this chapter we shall discuss recent econometric advances in the inspection of meta-data, the estimation of meta-equations, and the synthesis of results. These econometric procedures provide practitioners with a bevy of tools to discern the robustness of the insights gleaned from meta-analysis. We begin with a description of robustness checks based on an a priori parametric specification of a meta-equation to cement ideas. Next our discussion turns to the robustness of the meta equation itself, reviewing a recent set of methods perfectly suited to meta-data—smooth, discrete nonparametric regression. Given the nascent empirical underpinnings of smooth discrete nonparametric regression we devote a large portion of our review on the construction, intuition and implementation of this estimator. Finally, we discuss an array of alternative empirical approaches that can further be deployed in meta-analysis to determine the overall robustness of the key insights stemming from the meta-data.

17.2 Parametric Meta-analysis

17.2.1 Sample Selection

Sample selection in meta-analysis occurs when the set of empirical studies arises from some underlying data-generating mechanism that causes some studies to be funded and conducted and others to not occur. This selection process could occur for a variety of reasons that include coarseness of funding availability, investigator research interests, etc. The traditional sample selection process follows Heckman (1979) where there would be one observation per study and the analyst investigates why some study/observation combinations exist and others do not.

In statistical nomenclature, primary studies included in a meta-analysis may not represent a random sample from the potential population of studies. That is, while we wish to draw information from the universe of all potential studies of effect size, we observe information only from studies that were conducted **and** chosen for inclusion in the meta-analysis. Thus, we need to evaluate factors that lead to a study being undertaken and the factors that influence inclusion in the meta-data. Further, one can focus on how these two sets of factors differ.

Empirical studies are often funded when there is a policy question of interest. If a topic is of greater interest in some states than in others, it may follow that state funding has been leveraged to fund studies in the states where there is greater interest. For example, there may be a greater interest in water scarcity in southwestern states such as Arizona and New Mexico than in the New England states of Connecticut and Massachusetts. Thus, the sample selection would have water valuation studies occurring in states with higher relative scarcity and perhaps higher values for water. For study selection, studies published in peer-reviewed journals are likely to be present in any electronic literature search, but this may not be the case with studies that are documented only as "gray" literature reports.

Hoehn (2006) was the first to consider sample selection in meta-regression, looking across states for studies done on wetland valuation for a jurisdictional bias in the prioritization. Rosenberger and Johnston (2009) provide a detailed analysis of the many variants of selection that can result in biases in a meta-analysis. One of their forms of selection is research priority selection, where researchers may tend to focus on resources or events with larger outcomes. This is sample selection in the spirit of Hoehn (2006). Boyle et al. (2013) also follow this cross-state, jurisdictional approach to consider sample selection in a meta-analysis of benefit transfers for sport fishing valuation across the U.S.

A traditional meta-analysis follows a reduced-form, linear specification:

$$y_i = x_i\beta + \varepsilon, \tag{17.1}$$

where the dependent variable, y_i represents the effect size, and the vector of independent variables, x_i includes study specific characteristics. The parameter vector,

β measures the effects of study characteristics on effect size, and ε is a random error. In meta-analysis applications, the traditional Heckman framework needs to be expanded to allow for unbalanced panels of data; each study can generate multiple observations and the number of observations per study can vary.[1] Thus, the traditional cross-section sample selection correction will not work. Instead, a pseudo-panel approach needs to be adopted.

This more general valuation model is,

$$y_{ic} = x_{ic}\beta + \gamma\lambda_c + \omega_{ic}, \tag{17.2}$$

where i indexes values within the selection cluster (e.g., county, state and country), which may have more than a single study conducted, c indexes the identified clusters over which studies were conducted, $\{n_1, \ldots, n_C\}, N = \sum_{c=1}^{C} n_c$ is the overall sample size and λ_c is the inverse Mill's ratio (IMR) for cluster c derived from a selection equation assuming normally distributed errors. The n_c value estimates in the cth cluster share the same IMR. To estimate this model we have to replace the unknown λ_c with an estimate. This leads to an unbalanced panel meta-regression

$$\begin{aligned} y_{ic} &= x_{ic}\beta + \gamma\lambda_c + \gamma(\hat{\lambda}_c - \hat{\lambda}_c) + \omega_{ic} \\ &= x_{ic}\beta + \gamma\hat{\lambda}_c + \gamma(\lambda_c - \hat{\lambda}_c) + \omega_{ic} \\ &= x_{ic}\beta + \gamma\hat{\lambda}_c + \varepsilon_{ic}. \end{aligned} \tag{17.3}$$

The key distinction with this model is that ε_{ic} is correlated within each cluster since the *same* IMR is used for each value estimate. Thus, even though ε_{ic} is uncorrelated across clusters, we have to account for within cluster correlation when we estimate our first-stage selection equation.

To implement the panel meta-regression we follow procedure 3.2 of Wooldridge (1995), which we repeat here with updated notation.

1. Estimate the probability that at least one value estimate is observed in a selection cluster, $P(s_c = 1|z_c) = \Phi(z_c\delta)$, using data for all of the *potential* clusters where a study could have been conducted. $\Phi(\cdot)$ is the standard normal cumulative distribution function, z_c are the cluster-specific covariates that are used to determine if a study is conducted within a cluster and δ are the effects of the selection regressors. The explanatory variables do not have to coincide with the study-specific variables that enter our meta-equation, x_c, and typically would be different regressors.

[1] An important question is what level of aggregation should be considered for identification of sample selection. The state designations used by Hoehn (2006) and Boyle et al. (2013) arise from the availability of data. This may or may not be the best approach to identifying sample selection. We do not address this important question in this chapter, but leave it for future research.

2. For $s_c = 1$ compute $\hat{\lambda}_c = \lambda(z_c\hat{\delta})$. Here $\lambda(\cdot) = \phi(\cdot)/\Phi(\cdot)$, where $\phi(\cdot)$ is the standard normal density function.
3. Estimate Eq. (17.3) via ordinary least squares.[2]
4. Test $H_0 : \gamma = 0$ using the ordinary t-statistic from the regression in (17.3). If heteroscedasticity is a concern, then some form of heteroscedasticity and/or cluster-robust standard error should be used instead.

If one rejects the null hypothesis, then selection is present and the standard errors stemming from (17.3) are invalid.

To construct a variance-covariance matrix that is robust to the correlation introduced by replacing λ with an estimate in (17.3) we follow a modification of procedure 4.2 in Wooldridge (1995), which we reproduce here using the notation introduced above.

1. Estimate $P(s_c = 1|z_c) = \Phi(z_c\delta)$ using data for all of the clusters. This is the same step for the procedure to test for selection described above.
2. For $s_c = 1$ compute $\hat{\lambda}_c = \lambda(z_c\hat{\delta})$. Here $\lambda(\cdot) = \phi(\cdot)/\Phi(\cdot)$ where $\phi(\cdot)$ is the standard normal density function, which is the same as (2) above.
3. Estimate

$$y_{ic} = \hat{w}_{ic}\theta + \varepsilon_{ic} = x_c\psi + x_{ic}\beta + \gamma\hat{\lambda}_c + \varepsilon_{ic} \quad (17.4)$$

   via pooled OLS, where $\theta = (\psi,\ \beta,\ \gamma)$, $\hat{w}_{ic} = (1, x_{1c}, \ldots, x_{n_c c}, x_{ic}, \hat{\lambda}_c)$ and $x_c = (x_{1c}, \ldots, x_{n_c c})$
4. Now, let $\{1, \ldots, \bar{C}\}$ denote the set of clusters where at least one study was conducted. Further, n_c represents the number of estimates produced in the cth cluster. An estimate of the asymptotic variance-covariance matrix of $\hat{\theta}$ can be constructed from the pooled OLS residuals, $\hat{\varepsilon}_{ic} = y_{ic} - \hat{w}_{ic}\hat{\theta}$, via the formula $N^{-1}\hat{A}^{-1}\hat{B}\hat{A}^{-1}$, where

$$\hat{A} = N^{-1}\sum_{c=1}^{\bar{C}}\sum_{i=1}^{n_c}\hat{w}'_{ic}\hat{w}_{ic} \quad (17.5)$$

   and

$$\hat{B} = N^{-1}\sum_{c=1}^{\bar{C}}\hat{p}_c\hat{p}'_c, \quad (17.6)$$

[2]Note that we do not use the between correction employed in Wooldridge's actual procedure. This is due to the fact that he suggests estimating a probit for each time period, in doing so this would remove the inverse Mill's ratio from the regression.

with the vector $\hat{p}_c$ defined as $\hat{p}_c = \hat{q}_c - \hat{D}\hat{r}_c$ where $\hat{q}_c = \sum_{i=1}^{n_c} \hat{w}'_{ic}\hat{\varepsilon}_{ic}$,

$$\hat{D} = -N^{-1}\sum_{c=1}^{\bar{C}}\sum_{i=1}^{n_c}\hat{w}'_{ic}\hat{\theta}'\left(z_c\hat{\beta}\hat{\lambda}_c(z_c\hat{\beta} + \hat{\lambda}_c)\right)'$$

and the vector r_c has length n_c with common elements

$$\hat{M}_c^{-1}\left(\Phi(z_c\hat{\beta})\left[1 - \Phi(z_c\hat{\beta})\right]\right)^{-1}\phi(z_c\hat{\beta})z'_c\left[1 - \Phi(z_c\hat{\beta})\right]$$

where $\hat{M}_c = \left(\left(\Phi(z_c\hat{\beta})\left[1 - \Phi(z_c\hat{\beta})\right]\right)^{-1}\phi(z_c\hat{\beta})^2 z'_c z_c\right)$.

This procedure will produce asymptotically correct standard errors for the meta-regression estimators.[3]

While Hoehn (2006) and Boyle et al. (2013) both investigate selection within their meta-analysis, they reach different conclusions. Hoehn (2006) finds selection to be a statistically and economically significant feature within the meta-data on wetland valuation while Boyle et al. (2013) cannot reject the null hypothesis that selection is not present in their meta-data on sport fishing valuation. Ghermandi and Nunes (2013) note that in their meta-analysis of coastal recreation values, a research priority bias (sample selectivity) may exist, but they do not attempt to test for the presence of selection. In contrast, Ghermandi et al. (2010) uncover the likelihood of sample selection in the meta-analysis of provision of services from wetland ecosystems by noting that a majority of the studies in their meta-data stem from latitudes lower than 45°N. Despite these mixed results, existing empirical evidence supports conducting investigations for sample selection in meta-analysis where possible. However, it must be kept in mind that investigations of sample selection are dependent on available data, and test data may not represent the dimension where sample selection occurred. Thus, sample selection results always need to be interpreted with caution.

17.2.2 Value and Study (Horizontal) Robustness

Horizontal robustness involves investigating whether the inclusion or exclusion of individual effect sizes or effect sizes grouped by a particular study affect meta-

[3]With a small sample it may prove useful to deploy a bootstrap procedure. However, the bootstrapping mechanism is not exactly clear since there are two key issues, the selection that is present and the fact that most likely the meta-analyst will have an unbalanced panel in the value-estimate dimension. One approach could be to re-sample based on individual values as opposed to studies. However, this treats values from common studies as independent.

regression parameter estimates (Johnston and Rosenberger 2010; Boyle et al. 2013).[4] A variety of metrics can be constructed to compare meta-regression estimates with specific rows (or groups of rows) of the meta-data removed. A simple metric is to compute the absolute change in the meta-regression coefficient estimates across each effect size with a single observation removed (with replacement) from the estimation procedure. We focus on changes to specific parameter estimates, β_j, in the meta-equation

$$\Delta_{\beta_j} = \left| \frac{\hat{\beta}_j^{(-t)} - \hat{\beta}_j}{\hat{\beta}_j} \right|, \quad \forall t, \tag{17.7}$$

where $\hat{\beta}_j$ is the parameter estimate for the *j*th regressor including all observations and $\hat{\beta}_j^{(-t)}$ is the parameter estimate for the *j*th regressor when the *t*th valuation observation or study (groups of observations) is removed from the estimation. To characterize robustness one could then use these $n(ns_c)$ absolute parameter changes to calculate either the median or mean absolute deviations (MADs).

Consider a given observation or study that results in $\Delta_{\beta_j} = 0.5$. This would imply that the presence of this single observation (study) leads to a 50 % change in the effect size. While this might lead one to question if the estimated meta-equation is robust, the real insight is that it provides a clear signal to consider why this observation (study) has a larger effect size than the mass of the meta-data. Recall that during the construction of the meta-data the analyst imposes any number of *priors* (investigator decisions) on which studies/observations belong in the meta-data. One can use Δ_{β_j} as a guide to assess *ex post* features of the meta-data that warrant further scrutiny. Perhaps this particular observation (study) corresponds to a study that has a much smaller or much larger resource change than the mass of the meta-data, and this robustness check suggests that the meta-equation, as currently specified, is not capturing this nonlinearity. Thus, Δ_{β_j} can give insight into the effect that particular observations (studies) have on estimated effects sizes, which are the key pieces of information stemming from the meta-regression. An unfortunate consequence of using Δ_{β_j} to assess influential observations is that it does not explicitly recognize the random nature of the data. That is, we can only learn how large/small a change in the estimated effect sizes might occur, but not the related roles of influence and leverage. For those unfamiliar with leverage and influence, leverage refers to the ability of a single point to impact parameter estimates by virtue of its placement in regressor

[4]Individual effect sizes can be grouped according to a number of criteria. Here we consider one obvious group, by study, because observations for a particular study could be unique due to the application, design and conduct of the study, data analysis, and effect size reporting. We do not imply that other data groupings should not be investigated if such analyses seem appropriate for any specific meta-analysis.

space, while influence refers to the ability of a single point to impact parameter estimates by virtue of its placement in regressand space. To actually meaningfully impact parameters estimates a point must have both influence and leverage.

To provide econometric intuition for Δ_{β_j} we follow both Johnston and Rosenberger (2010) and Boyle et al. (2013) and discuss leverage/influence metrics when a single value is left out of the meta-data. This is followed by a discussion of the econometric steps needed when we elect to leave a study (as opposed to an observation) out of our meta-data.

Collecting all of our right-hand-side variables into the matrix X, our initial OLS estimates are constructed as

$$\hat{\beta} = (X'X)^{-1}X'y. \tag{17.8}$$

The full vector of leave-one-observation-out estimates can then be constructed as

$$\hat{\beta}^{(-t)} = \hat{\beta} - (X'X)^{-1}X_i'\hat{\omega}_i/(1 - h_{ii}), \tag{17.9}$$

where $\hat{\omega}_i$ is the ith residual from the initial OLS estimation, X_i' is the ith column of X', and h_{ii} is the ith diagonal element from the so called hat matrix $X(X'X)^{-1}X'$.

Two elements of the right-hand-side of (17.9) must be considered, $\hat{\omega}_i$ and h_{ii}. This can be seen by rearranging (17.9) to show the difference between the leave-one-out estimator and the actual estimator, which is $\hat{\beta} - \hat{\beta}^{(-t)} = (X'X)^{-1}X_i'\hat{\omega}_i/(1 - h_{ii})$. As $h_{ii} \rightarrow 0$ the ith observation has no influence on $\hat{\beta}$. In essence, the ith observation lies exactly on the line constructed using $\hat{\beta}$ so its removal does nothing to change the resultant estimate. However, as $h_{ii} \rightarrow 1$ the ith observation completely controls $\hat{\beta}$, i.e. the difference approaches ∞. A general rule-of-thumb is that if $h_{ii} \gg 2k/n$ (here k is the number of regressors in the regression and n is the number of observations used to estimate the equation) then the ith point possesses leverage (Belsley et al. 1980).

While h_{ii} is used to gauge the influence of a particular point, a large h_{ii} does not necessarily imply that an observation has any influence on our parameter estimates, only that it has the potential to influence our estimates. The reason for this is simple. In (17.9) notice that the leave-one-observation-out estimate (or index) also depends on the ith residual, error variable. If the residual were very small, then even if a point had high leverage, it would not be able to act on this influence to dictate the new estimate. An effect-size observation must have both in order to actually influence the parameter estimates. This is the reason that the simple calculation in (17.8) should not be used as a sole judge to evaluate robustness.

A simple measure of influence, which combines influence and leverage, is Cook's index. This index constructs a variance weighted version of $\hat{\beta} - \hat{\beta}^{(-t)}$, defined as

$$D_i = \frac{(\hat{\beta} - \hat{\beta}^{(-t)})' X' X (\hat{\beta} - \hat{\beta}^{(-t)})}{ks^2} = \frac{h_{ii}\hat{\omega}_i^2}{s^2(1 - h_{ii})^2 k}, \tag{17.10}$$

where s^2 is the standard error estimate of the residuals. Cook and Weisberg (1982) suggest using $D_i > 1$ as a threshold for an observation being influential. Note that we can also write

$$D_i = \frac{h_{ii}}{(1 - h_{ii})^2} \frac{\hat{\omega}_i^2}{s^2 k}, \tag{17.11}$$

which decomposes the index into two components; the first represents the potential influence of an observation of the meta-data and the second is a standardized residual from the initial meta-regression that is the leverage component.[5] Composite measures, such as D_i, are appealing because they explicitly account for both leverage and influence.[6]

A consequence of looking at such leave-one-observation-out diagnostics, however, is that they suffer from the well known masking effect. That is, by focusing attention on a single observation at a time, a given process may fail to uncover a subset of influential observations. Unfortunately, the methods described above tend to be less computationally easy to obtain when considering subsets of the data and rarely yield simple and intuitive closed form solutions. Moreover, while we have N leave-one-observation-out calculations, the number of potential data groupings increases dramatically with the sample size. Consider 100 total observations in which 3 observations are left out at a time; there would be a total of $\binom{100}{3} = 161{,}700$ subsets to consider.

Given the number of potential data groupings, tractability requires some way of simplifying the task of selecting data groupings. Here we propose reducing the number of groupings to subsets of the data with a natural relationship, such as effect sizes from the same study since observations from the same study may follow a common data generation mechanism.[7] For the leave-one-study-out analysis it can be shown that the difference between the initial meta-regression estimates and those that result when one omits a study is composed of a linear combination of the elements of the hat matrix. To simplify our notation we let $X_{(-t)}$ denote our matrix

[5] An alternative measure of overall influence of an observation is the change in the OLS residuals from leaving a single observation out. This index can be calculated as $DF_i = \hat{\omega}_i - \hat{\omega}_{(i)} = \frac{h_{ii}\hat{\omega}_i}{1-h_{ii}} = D_i \frac{s^2(1-h_{ii})k}{\hat{\omega}_i}$, where $\hat{\omega}_{(i)}$ is the ith residual, constructed leaving the ith observation out of the analysis. We elect to use D_i since in general DF_i and D_i yield similar results but D_i has a more intuitive interpretation of Eq. (17.11).

[6] See Davidson and MacKinnon (2004, Sect. 2.6) for more details on influence and leverage.

[7] This approach is common in the time-series literature where n_k consecutive observations would be left out (Poirretti 2003).

of covariates with the rows corresponding to the ith study being removed, and $y_{(-t)}$ denote the vector of corresponding effect sizes. We also let $X_{(t)}$ and $y_{(t)}$ denote the rows of X and y for the tth study. Our leave-one-study-out estimates are now written as[8]:

$$\hat{\beta}^{(-t)} = \left(X'_{(-t)}X_{(-t)}\right)^{-1} X'_{(-t)}y_{(-t)} = M_{(t)}\hat{\beta} + Q_{(t)}y_{(t)}, \tag{17.12}$$

where

$$M_{(t)} = I + R_{(t)}$$
$$Q_{(t)} = M_{(t)}(X'X)^{-1}X'_{(t)}$$

where $R_{(t)} = X'_{(t)}(X'X)^{-1}X_{(t)}$ and $P_{(t)} = X_{(t)}(X'X)^{-1}X'_{(t)}$. This decomposition shows that our leave-one-study-out estimator is a linear combination of the initial estimator and the effect sizes for the tth study. One point is clear, when leaving more than one observation out at a time there are now several sources of leverage and influence. The additional leverage/influence stems primarily from the fact that two points individually may have very little leverage (small h_{ii}), but they may have a large joint leverage, which can cause the new parameter estimates to deviate from the actual estimates. If the points, individually or jointly, have no influence then the leverage they possess will have no means by which to affect parameter estimates.

The leave-one-study-out residuals, defined as $\tilde{\omega}_{(t)}$, can be shown to be[9]

$$\tilde{\omega}_{(t)} = y_{(t)} - X_{(t)}\hat{\beta}_{(-t)} = \left(I - P^{(t)}\right)^{-1} \hat{\omega}_{(t)}, \tag{17.13}$$

where $\hat{\omega}_{(t)}$ are the residuals from the tth study when **all** the observations are included. To our knowledge this result has not previously been discussed in the literature and marks an important generalization of existing metrics that rely on leave-one-observation-out residuals. Here we see the importance of the strength of all the observations within a study, as captured by $P^{(t)}$.

From the derivations above we suggest a commonly used approach to evaluate, statistically, whether a study has a significant influence on meta-analysis parameter estimates. This approach recognizes that the presence of a study can have an influence on parameter estimates that is akin to testing for parameter equality using a Chow type test. That is, we propose the statistic[10]

[8]See Appendix A for a derivation.

[9]See Appendix B for full derivation.

[10]A heteroscedasticity robust version of this statistic could also be constructed consistent with the concerns of Nelson and Kennedy (2009).

$$F = \frac{\left(RSS_m - \left(RSS_t + RSS_{(-t)}\right)\right)/k}{\left(RSS_t + RSS_{(-t)}\right)/(n - 2k)}, \tag{17.14}$$

where RSS_m the residual sum of squares from the original meta-regression, RSS_t is the residual sum of squares using only the observations corresponding to study t, and $RSS_{(-t)}$ is the residual sum of squares obtained by running the meta-regression omitting the values from the tth study. Here, k represents the number of parameters estimated in the baseline meta-regression and n is the total number of observations. F is distributed $F(k, n - 2k)$ and can be used to easily determine if a set of value estimates is driving differences in parameter estimates when we omit the tth study. We note that in certain settings, including the one presented here, some studies will contain fewer observations than there are parameters ($n_t < k$). In this case the appropriate F statistic can be shown to be

$$F = \frac{\left(RSS_m - RSS_{(-t)}\right)/n_t}{RSS_{(-t)}/\left(n_{(-t)} - k\right)}, \tag{17.15}$$

where n_t is the number of observations from the tth study and $n_{(-t)} = n - n_t$. Additionally, while the F statistic proposed here tests whether the entire set of estimated parameters differs in the meta-regression when a study is removed, one could also look at individual coefficients as well, using a simple paired t-test to address parameter estimates of the greatest importance for research of policy assessments.

The calculations reported in (17.7) and (17.13)—and related tests in (17.14) and (17.15)—solely quantify/test the magnitude of the effect on the meta-regression parameter estimates by omitting observations from the meta-data. This is important information, but does not provide a decision rule for research evaluation or policy analyses. That is, these calculations/test statistics do not suggest a plan of action when an observation or study is found to have a large influence on parameter estimates; it is up to the discretion of the researcher to make the decision on influence. The above assessments, however, can help identify when meta-data specification choices are likely to have influential consequences that may be relevant for research evaluation or policy analysis.

17.2.3 Variable Robustness

Another approach to assess robustness is to consider the variables to be included in the meta-regression. Two key study features influence what variables are included in meta-regressions: (1) the information available in the documentation of original studies and (2) the analyst's choice of what regressors to include in the meta-regression.[11] We refer to this as vertical robustness because it investigates the columns within the meta-data.

Vertical robustness can be handled in a variety of manners. Here we consider the use of extreme bounds analysis (EBA), championed by Leamer (1983).[12] EBA assesses the robustness of parameter estimates by looking at changes in statistical significance and sign of a parameter estimate when specific explanatory variables (data columns) are excluded from the estimation. Leamer's key insight is that parameter estimates chosen by researchers often reflect a priori assumptions that can influence estimation results; i.e., the results typically presented are those that fit best with prior assumptions of the analyst, not necessarily those that reflect the underlying data generation processes. By repeatedly estimating models with individual regressors or groups of regressors omitted, one can gauge the influence on the remaining parameter estimates in the meta-regression, and specifically those parameter estimates that might be most relevant for research assessment or policy analysis.

To place the meta-regression within the confines of EBA, we partition our meta-regression variables into two sets. The first set includes the particular variable (or set of variables) of interest within the meta-equation, x_{ic}^{A}; those that are of interest for research assessment or to support policy analysis. The second set includes all other variables, x_{ic}^{P} yielding

$$y_{ic} = x_{ic}^{A}\beta^{A} + x_{ic}^{P}\beta^{P} + \omega_{ic}. \tag{17.16}$$

This setup allows us to determine the influence of the x_{ic}^{P} covariates on the estimate of β^{A}; assuming the x_{ic}^{A} variables are always included in the estimation.[13] To accomplish this, we vary the variables which enter x_{ic}^{P} in the meta-equation to quantify effects on estimates of β^{A}. Even for moderate sets of covariates within the meta-data, this can result in a large number of models to estimate. For example, if P contains 11 potential regressors, there are $2^{11} = 2048$ potential subsets of P to investigate, if we consider every possible linear combination of the variables to determine the effects of our estimates for β^{A}.

Define $\hat{\beta}^{A\tilde{P}}$ to be the estimator of the meta-regression using $x^{\tilde{P}}$, $\tilde{P} \subseteq P$ and $\hat{\sigma}_{A\tilde{P}}$ the estimator of the standard error of $\hat{\beta}^{A\tilde{P}}$. The lower extreme bound can be defined as $\underline{\beta}_{j}^{EBA} = \min_{\tilde{P}\in P}\left(\hat{\beta}_{j}^{A\tilde{P}} - 2\hat{\sigma}_{j}^{A\tilde{P}}\right)$ and the upper extreme bound can be defined as $\overline{\beta}_{j}^{EBA} = \max_{\tilde{P}\in P}\left(\hat{\beta}_{j}^{A\tilde{P}} + 2\hat{\sigma}_{j}^{A\tilde{P}}\right)$. Within EBA, a variable x^{j}, $j \in A$, is deemed robust if

[11]Another issue to consider is the measurement of the regressors. Measurement is affected by documentation in original studies and choices made by the analyst. We do not address issues of variable measurement here, but this is another area of fruitful future research.

[12]A model-averaging approach is also feasible. Chapter 22 provides more discussion of these methods.

[13]We have used P and A to signify "potentially included" and "always included," respectively.

$0 \notin \left(\underline{\beta}_j^{EBA}, \overline{\beta}_j^{EBA}\right)$ for all i, both the upper and lower extreme bounds must have the same sign for this to hold.

If this criterion is not met then the parameter estimates of interest are not deemed robust. Unfortunately, these criteria often classify few, if any, parameter estimates as robust (Sala-I-Martin 1997). Given that the estimator is itself a random variable with an underlying distribution, it is possible that this distribution straddles 0, ensuring that for any finite sample there will be some positive probability of having an estimate of the wrong sign even if this probability is very small, e.g., less than 1, 5 or 10 %. To address this concern within the confines of EBA[14] Boyle et al. (2013) suggest using the extreme decile range of the parameter estimates (plus/minus two standard deviations) to assess robustness. While ad hoc, this rule-of-thumb approach provides reasonable bounds for which to assess robustness by asking if the mass of the data support a conclusion of robustness. This is much in the same spirit of classical hypothesis testing where tests are done not at the 0 % level but at the 1, 5 or 10 % levels to balance type I and type II errors.

EBA is now viewed as a precursor to the now common model-averaging paradigms[15] (see Moeltner et al. 2007) in which one searches over a set of models to determine which variables have the highest probability of being in the final model. This was partly in response to the overly restrictive criteria needed to place confidence in parameter estimates. Here, we are interested in traditional models in which the analyst chooses the set of regressors and wishes to evaluate the effect of changes in the conditioning set (excluded regressors) on parameters estimates for the variables of interest. For example, in a structural model for meta-equation, certain economic variables (A) would be expected in the model and there may be other taste and characteristic variables (P) that may be associated with effect size that would not be explicitly specified in the equation. EBA, with the modification for identifying robust regressors, provides insights into what one learns from a meta-regression conditioned on variables that appear in the meta-regression.

Prior to carrying out either a horizontal or vertical robustness check of the meta-regression, the practitioner must construct the meta-data. This is not innocuous. Which studies and which variables enter the meta-data will have far-reaching impacts on what is learned from the meta-analysis. A recent literature on the construction of optimal meta-data has emerged to tackle to this important issue.

17.2.4 Optimal Meta-data

Recall that the construction of the meta-data is from a pool of S total studies on the subject of interest and each study produces n_s observations. However, reporting

[14]Other authors have used entirely different methods, such as model averaging.

[15]Approaches such as these are discussed in more detail in Chaps. 21 and 22.

differences across studies may make the collection of all *N* observations difficult. This stems from the fact that certain regressors cannot be collected from particular studies. Moeltner et al. (2007) refer to this as the "N versus K" issue. Thus, meta-analysts trade off the final sample size (*N*) for the meta-analysis to ensure that enough regressors (*K*) are accounted for so that concerns over omitted variables are alleviated. A small sample (likely stemming from a focus on maximizing *K*) can also lead to a lack of precision of the meta-regression estimates.

From the standpoint of assessing the literature in question, increasing *N* runs the risk of including studies that may not be similar to the majority of studies. However, both the horizontal and vertical robustness checks described above, along with the procedures detailed in Moeltner et al. (2007) provide a framework for evaluating how a single study compares with the remaining studies in terms of parameter homogeneity.

Moeltner and Rosenberger (2008) provide an excellent discussion of these issues surrounding the construction of the meta-data, building on the work of Moeltner et al. (2007). One could think of extending Moeltner and Rosenberger's (2008) partitioning of the data, which they call the data space, using the horizontal/vertical robustness checks described above to assess the impact of the alternative data sources. To be specific, Moeltner and Rosenberger (2008) wish to use meta-analysis to conduct a benefit transfer for the expected benefits to anglers of coldwater fisheries under different expectations of catch rates. Their construction of the meta-data results in four potential data spaces. Their baseline data space contains only four studies, with a total of 29 observations for coldwater fisheries in rivers (the policy context at hand). They augment the meta-data by including observations for both warm and coldwater river fisheries, which add another five studies and 39 observations; then coldwater fisheries in rivers and still water, adding nine studies and 50 observations; and finally, they add 12 studies conducted over both warm and cold water fisheries in both rivers and still water, adding in 112 observations. This clearly illustrates the scope issue of collecting meta-data.

Another example is the meta-analysis work on international coral reef values (Brander et al. 2007; Londoño and Johnston 2012). Londoño and Johnston (2012) pay specific attention to the construction of their meta-data to ensure robustness. Specifically, they note that their focus for the meta-data is studies in which willingness-to-pay is linked directly/explicitly to recreational activities/access to coral reefs whereas Brander et al. (2007) include studies that have more opaque links to coral reef valuations in their meta-data. Further, given the relatively few studies conducted on coral reef valuation within a given country, both meta-analyses pool studies across countries. Overall, Londoño and Johnston (2012) have a more refined scope for their meta-data than Brander et al. (2007). This is evidenced in Londoño and Johnston's (2012) transfer error exercise where they find lower out of sample transfer errors than Brander et al. (2007).

By starting from the baseline, optimal scope data, the horizontal, study specific robustness checks can be applied to discern the impact that increasing the data space has on the results. This exercise could be further iterated to drop observations within a data source (as opposed to a single study) and then combinations of the data spaces. Concurrently, vertical robustness can address the influence of adding (or deleting) regressors as studies are added (observations dropped). Thus, horizontal and vertical robustness can provide important insights with respect to the "N versus K" trade off in any meta-analysis. In general, when issues about the scope of the meta-data exist, meta-analysts should provide checks to readers on the sensitivity/robustness of their findings to the levels of scope of their data.

17.3 Nonparametric Meta-analysis

The issues and procedures discussed above were based on a linear specification of meta-regression equation. When the underlying meta-equation is not linear, a traditional parametric analysis can be misleading. Further, many regressors are coded as binary variables and the "empty cell" phenomenon can actually lead to inefficient parametric estimates, even with a correctly specified meta-equation if the meta-data sample is small relative to the number of cells (potential binary variable interactions) that compose the data. Nonparametric regression methods can be deployed to mitigate both of these issues.

Following Manski's (2007) "bottom up" approach, coordinated nonparametric and parametric estimation methods can provide an additional robustness check to evaluate the insights from a meta-regression. By starting with the less restrictive nonparametric analysis and moving to parametric estimation, where linearly and separability are imposed, we can glean more insights from the meta-data. Parametric estimation allows very specific insights, which may not hold if the imposed assumptions are inappropriate. Nonparametric estimation does not suffer this restriction, but insights are more general, given the lack of prior assumptions imposed on the model. If both estimation approaches provide similar estimation results, then the analyst can be confident that that the findings are not linked to the assumptions imposed by parametric estimation.

To understand the role that nonparametric estimation can play in investigating the robustness of insights from meta-equations, we provide simplified descriptions of the approach below.

17.3.1 A Simple Nonparametric Model with Discrete Data

Consider a simple meta-regression setting with a single quantitative variable, x, that takes values in $\{0, 1\}$:

$$y_i = m(x_i) + \varepsilon_i, \tag{17.17}$$

where ε_i is the random component of the meta-analysis. While we have purposely left the meta function, $m(\cdot)$ unspecified, given the discrete nature of x_i, we could just have easily specified the meta function as $\beta_0 + \beta_1 x_i$. The standard way to estimate this model is using ordinary least squares (OLS). However, in this setting OLS is equivalent to the frequency estimator. To see this, define $\bar{y}_1$ and $\bar{y}_0$ to be the mean of y when $x_i = 1$ and $x_i = 0$, respectively. Then, the frequency estimator of β_0 and β_1 is

$$\hat{\beta}_0 = \bar{y}_0; \quad \hat{\beta}_1 = \bar{y}_1 - \bar{y}_0. \tag{17.18}$$

In this relatively simple setting a standard parametric approach is equivalent to a nonparametric approach, here $\hat{m}(0) = \hat{\beta}_0$ and $\hat{m}(1) = \hat{\beta}_1$.

At this point it may seem that with discrete covariates there would be no difference from a parametric approach, however, as the number of discrete variables increases, one can immediately see the need for nonparametric analysis. First, consider the case with two discrete variables, each taking values in $\{0, 1\}$. In this case there are four cells which y would need to be averaged over to discern the impacts of a specific quantitative setting, $\{(0, 0), (0, 1), (1, 0), (1, 1)\}$. As we continue to add more variables to our analysis the number of potential cells increases. For a parametric analysis to capture the cell-specific effects, a large number of interaction variables would be required for the estimation and a sufficient number of observations in each cell is needed to produce reliable estimates. In many meta-analysis settings, it is likely that the number of cells will be greater than or roughly equivalent to the number of available observations. For example, in Kaul et al. (2013), there are 1071 observations, but $5 \times 2^7 = 640$ potential cells, which is less than two observations per cell on average. This presents serious issues for parametric estimation.

To avoid issues with sparse cells in a meta-regression, a common approach is to assume additive separability amongst the categories. This is equivalent to the linear in parameters meta-regression specifications that dominate the environmental and resource economics literature.[16] Thus, for two distinct quantitative variables, x_1 and x_2, a parametric model imposing this constraint would be

$$y_i = \beta_0 + \beta_1 1\{x_{1i} = 1\} + \beta_2 1\{x_{2i} = 1\} + \varepsilon_i. \tag{17.19}$$

This approach avoids the empty cell issue since each distinct dimension of cells (two versus the four identified here) is spread over n observations. However, an unfortunate side effect of the additive separability assumption is that the estimated effects are implicitly assumed to be constant over alternatives for the other variables, interaction effects are ignored. That is, β_1 is constant, whether x_{2i} is 0 or 1. This may not be a reasonable assumption depending upon the specific application of the analysis. Consider an example here. Assume that we have meta-data on

[16]An alternative, but important issue is how the meta-analyst chooses to specify the categories across regressors within the meta-data. The meta-analyst could elect to have fewer categories to minimize the number of cells. However, this draws into question the commonality of categories and their impact on effect sizes. This is another investigator assumption that we do not address in this chapter, but it warrants empirical investigation to determine the consequence on estimated meta-equations.

benefit transfer validity based on contingent-valuation and choice modeling studies. Further let x_1 represent if a value transfer was performed in benefit transfer and x_2 represent the primary study survey response rate. It is possible that the impact of a value transfer on transfer validity will depend on the response rates that could potentially lead to higher or lower value estimates. The additively-separable model is agnostic to this potential confound.

To avoid the unfortunate side effects of assuming additive separability as well as avoiding sparse cell issues associated with the construction of all possible variable interactions, we describe a smoothing approach to meta-regression that was introduced in Kaul et al. (2013). This approach is nonparametric in nature, is asymptotically equivalent to the frequency approach, and has been shown in repeated simulations to outperform both ad hoc parametric specifications as well as the frequency estimator (Hall et al. 2007; Henderson et al. 2012a, b).

17.3.2 Smoothing Discrete Data

To understand how the smoothing of discrete data takes place, consider the frequency approach to estimating cell means. In the bivariate, binary regressor case we would have

$$\bar{y}_{(0,0)} = \frac{\sum_{i=1}^{n} y_i \cdot 1\{x_{1i} = 0\}1\{x_{2i} = 0\}}{\sum_{i=1}^{n} 1\{x_{1i} = 0\}1\{x_{2i} = 0\}}, \quad \bar{y}_{(0,1)} = \frac{\sum_{i=1}^{n} y_i \cdot 1\{x_{1i} = 0\}1\{x_{2i} = 1\}}{\sum_{i=1}^{n} 1\{x_{1i} = 0\}1\{x_{2i} = 1\}}$$
$$\bar{y}_{(1,0)} = \frac{\sum_{i=1}^{n} y_i \cdot 1\{x_{1i} = 1\}1\{x_{2i} = 0\}}{\sum_{i=1}^{n} 1\{x_{1i} = 1\}1\{x_{2i} = 0\}}, \quad \bar{y}_{(1,1)} = \frac{\sum_{i=1}^{n} y_i \cdot 1\{x_{1i} = 1\}1\{x_{2i} = 1\}}{\sum_{i=1}^{n} 1\{x_{1i} = 1\}1\{x_{2i} = 1\}}$$

The frequency estimator partitions the data into homogenous cells and takes averages of y within each cell. Effectively, no information from other cells is used when this averaging takes place. We can think of the frequency estimator as a smoothing estimator that assigns 0 weight to nearby cells. This weighting scheme is too rigid with respect to the homogeneity of cells. The idea behind the smoothing approach is to replace the indicator functions in the cell mean estimators above with weight functions that allow observations in neighboring cells to be used in the averaging. That is, we allow for the possibility of non-zero weight to be assigned to nearby cells.

While it may seem counter intuitive to smooth discrete data, by leveraging information in nearby cells the finite sample properties of smooth discrete estimators can substantially dominate their unsmooth counterparts. To begin, we introduce smoothing functions to smooth two types of discrete data, unordered and ordered. A discrete unordered variable, for example, could be used to classify the type of empirical analysis conducted for each study that contributes outcomes to our meta-analysis (e.g., revealed versus stated preference) while an ordered discrete variable might represent an improvement in quality valued within a study.

We assume that we have a total of q discrete variables with q^u unordered discrete variables and q^o ordered discrete variables ($q = q^o + q^u$). Our unordered smoothing function (commonly known as a kernel) is given by

$$\ell^u\left(x_i^u, x^u, \lambda^u\right) = (\lambda^u)^{1\{x_i^u \neq x^u\}}, \tag{17.20}$$

where λ^u is our smoothing parameter for unordered data. The exponent, $1\{x_i^u \neq x^u\}$ assigns weight based on whether x_i^u is different from the evaluation point, x^u. When these two points are the same, a weight of 1 is given to the observation; when these two points are different, a weight of $\lambda^u < 1$ is given. Note that the difference between x_i^u and x^u does not matter, only that the two points are not identical. Further, observe that when $\lambda^u = 0$, no weight is given to points that differ from the point of interest, x^u. In this sense, the smoothing method contains the frequency approach, which would not give weight to differing points. In contrast, when $\lambda^u = 1$, all points receive equal weight. In this case there is no reason to include this set of categories since different points cannot be distinguished. Thus, insight can be gained only when $0 < \lambda^u < 1$.

The ordered kernel (or smoothing) function is given by

$$\ell^o\left(x_i^o, x^o, \lambda^o\right) = (\lambda^o)^{|x_i^o - x^o|}, \tag{17.21}$$

where λ^o is the smoothing parameter. Here, the weighting of nearby cells decreases as the point of interest becomes further from the comparison point. Moreover, we see the two key features of the ordered kernel function that its unordered counterpart possessed, when $\lambda^o = 0$ we essentially have the frequency method (no smoothing is conducted) and when $\lambda^o = 1$ all points, no matter how disparate, receive equal weighting, suggesting that this set of categories should not be included.

The ordered and unordered kernel functions will replace the indicator functions in the cell mean estimator. To allow even greater flexibility we allow each individual component of x^u and x^o to be smoothed differently. This requires introducing an individual smoothing parameter for each discrete regressor. Thus, for $x = (x_1^u, x_2^u, \ldots, x_{q^u}^u, x_1^o, x_2^o, \ldots, x_{q^o}^o)$, we will have a vector of smoothing parameters, $\lambda = (\lambda_1^u, \lambda_2^u, \ldots, \lambda_{q^u}^u, \lambda_1^o, \lambda_2^o, \ldots, \lambda_{q^o}^o)$. Finally, we introduce the product kernel, which will take the place of the multiplicative indicator functions appearing in the cell mean estimator. The product kernel is

$$\begin{aligned} L_{\lambda,ix} &= L^u_{\lambda^u}\left(x_i^u, x^u\right) L^o_{\lambda^o}\left(x_i^o, x^o\right) = \prod_{s=1}^{q_u} l^u\left(x_{is}^u, x_s^u, \lambda_s^u\right) \prod_{s=1}^{q_o} l^o\left(x_{is}^o, x_s^o, \lambda_s^o\right) \\ &= \prod_{s=1}^{q_u} \left(\lambda_s^u\right)^{1\{x_{is}^u \neq x_s^u\}} \prod_{s=1}^{q_u} \left(\lambda_s^o\right)^{|x_{is}^o - x_s^o|}. \end{aligned}$$

Our discrete kernel regression estimator is

$$\hat{m}(x) = \frac{\sum_{i=1}^{n} y_i L_{\lambda,ix}}{\sum_{i=1}^{n} L_{\lambda,ix}}. \tag{17.22}$$

when $\lambda = 0$ for each component of x, the kernel regression estimator is identical to the cell mean estimator. We delay discussion on the selection of λ until Sect. 17.3.4.[17]

To gain some intuition as to why smoothing is likely to improve finite sample estimation, consider the case in which a single ordered discrete variable exists that has five cells $\{0, 1, 2, 3, 4\}$. This is a natural simplification of any empirical setting. Suppose that **no observations** exist for category 2. In this case we can still present function estimates using the smoothing approach, whereas the frequency method would be unable to produce an estimate. Our nonparametric categorical regression estimator for category 2 would be

$$\hat{m}(2) = \frac{\sum_{i=1}^{n} y_i \ell^o(x_i^o, 2, \lambda^o)}{\sum_{i=1}^{n} \ell^o(x_i^o, 2, \lambda^o)} = \frac{\sum_{|x_i-2|=1} y_i \lambda^o + \sum_{|x_i-2|=2} y_i (\lambda^o)^2}{\sum_{|x_i-2|=1} \lambda^o + \sum_{|x_i-2|=2} (\lambda^o)^2}, \tag{17.23}$$

where λ^o is referred to as the bandwidth and is used to smooth over the five cells. This shows that the nonparametric smoothing estimator is a weighted average of

[17]While it is likely that the majority of the meta-data will be discrete in nature, it is certainly reasonable to have variables that are continuous in nature, such as the sample size used in an empirical analysis, or the standard deviation of the estimate of interest. In this case we can smooth over both the discrete and continuous variables in the meta data with only slightly more complex notation. Our kernel weighting takes the same approach, introducing a weight function geared towards continuous data to pair with our product kernel for discrete data. For a single continuous covariate, a kernel function takes the form $h^{-1}k\left(\frac{x_i - x}{h}\right)$, where h is our smoothing parameter and $k(\cdot)$ is commonly chosen to be a probability density function, such as the normal probability density, $k\left(\frac{x_i - x}{h}\right) = \left(\sqrt{2\pi}\right)^{-1} e^{\left(-(x_i - x)^2/(2h^2)\right)}$. To be as general as possible we now allow for q regressors with q^c continuous regressors, q^u unordered discrete variables and q^o ordered discrete variables ($q = q^c + q^o + q^u$) Our generalized product kernel in the mixed, continuous-discrete data setting is

$$\begin{aligned} W_{h\lambda,ix} &= K_h\left(x_i^c, x^c\right) L^u_{\lambda^u}\left(x_i^u, x^u\right) L^o_{\lambda^o}\left(x_i^o, x^o\right) \\ &= \prod_{s=1}^{q_c} k\left(\frac{x_{is}^c - x_s^c}{h_s}\right) \prod_{s=1}^{q_u} \left(\lambda_s^u\right)^{1\left(x_{is}^u \neq x_s^u\right)} \prod_{s=1}^{q_u} \left(\lambda_s^o\right)^{\left|x_{is}^o - x_s^o\right|}. \end{aligned}$$

The smoothed meta-function is estimated as

$$\hat{m}(x) = \frac{\sum_{i=1}^{n} y_i W_{h\lambda,ix}}{\sum_{i=1}^{n} W_{h\lambda,ix}}.$$

nearby cells (0, 1, 3, 4), provided that $\lambda^o \neq 0$. The weights depend upon the distance the cell is from the cell of interest, $|x_i - 2|$. The frequency approach could not produce an estimate for this case given the fact that no observation was equal to a 2.

This is a simplified version of the sparse cell problem that the frequency estimator is likely to encounter in applied meta-analysis research. Regressors in meta-analyses are often binary variables that describe study specific characteristics and not all combinations of these characteristics are represented in the meta-data. In fact, with equations often having four or more binary regressors it is very likely that some cells will be empty or have few observations. By smoothing over the categories, the nonparametric approach introduces a flexible and efficient estimator.

Even though the smoothing estimator avoids many adverse side effects of the parametric and frequency estimators, there are several performance issues which need to be mentioned. When irrelevant discrete covariates are included in the analysis (similar to including additional regressors in a standard parametric analysis) this produces a bias in the smoothed estimator (Ouyang et al. 2009). This bias does not exist with the standard frequency estimator. Additionally, the speed at which the estimator converges to its limiting distribution is slower[18] than in the setting where only the correct covariates are included. This is important for a meta-regression since it is not always clear at the onset what features of empirical analyses influence the outcome of interest and should be included as regressors in the meta-equation.

17.3.3 Intuition for the Smoothed Estimator

A simple approach to highlighting and visualizing the differences between a nonparametric meta-analysis and a parametric meta-analysis is to deploy a 45° plot where the y and x axes are the same (see Henderson et al. 2012a). For both estimators, plot the estimated parameter estimates against themselves (which will occur along a 45° line). The insight comes from the dispersion of the nonparametric effects versus the single point estimate for the parametric estimation.

Figure 17.1 illustrates example 45° plots for the case of a single binary variable, but with different dispersion (heterogeneity) of the estimated effects. The left panel, labeled heterogeneous, depicts a setting where the parametric and nonparametric estimates provide different insights about the impact of the binary variable. The nonparametric estimates vary between −1 and 1 and have roughly equal numbers of effects on both sides of zero. In contrast, the parametric estimate (given by the + sign) is positive and 0.06. In this case we would conclude that there is heterogeneity that is obscured by using a single regressor parametric meta-equation. Alternatively, in the right panel, labeled homogeneous, we see that even though the nonparametric

[18]This speed of convergence is important given the dramatically different behavior of the estimator in the presence of irrelevant discrete variables.

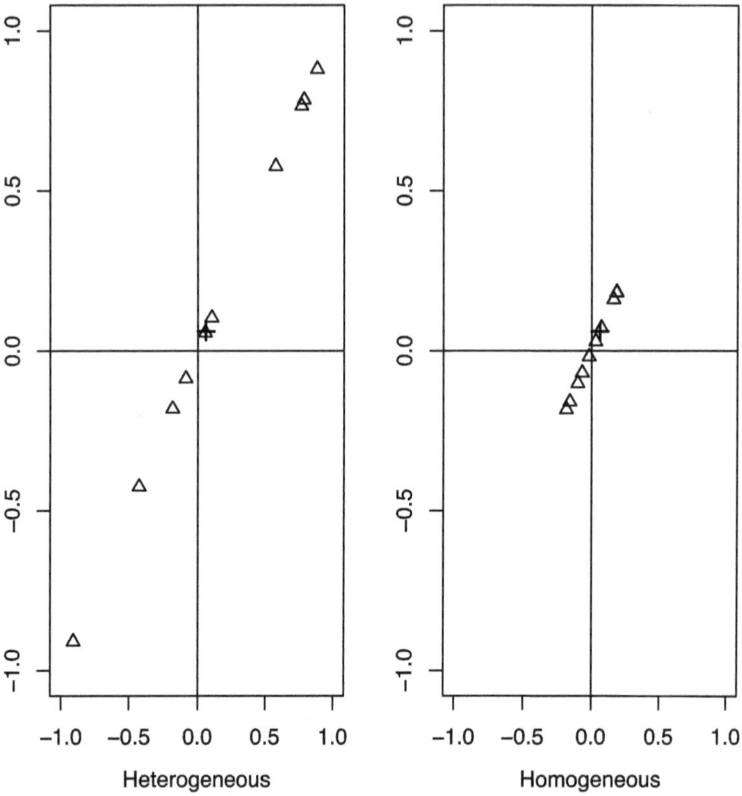

Fig. 17.1 *Illustration of the 45° plot* The Δ s represent the counterfactual estimates from the nonparametric estimator while the +s are the parametric, single cell, meta-regression estimates

estimates still straddle zero, they vary only from about −0.2 to 0.2 and are much closer to the parametric estimate. By using the 45° plots for all of the effects sizes, it is easy to discern where the heterogeneity exists in the meta-data that would be missed only by considering a single parametric meta-regression parameter estimate. This may not be of such concern in panel (b) where the nonparametric and parametric analyses provide similar insights; yet the parametric estimator still does not pick up the heterogeneity in the negative domain. Confidence intervals can also be added to the plots to uncover statistically significant heterogeneity that exists in the meta-data.

Consider an example here. Assume that we are conducting a meta-analysis of existing benefit-transfer, convergent-validity studies. The effect size is the percent error from convergent validity studies of benefit transfers. Further, we want to examine the impact of the variable BASELINEΔ that represents the percent change in value from baseline for some quality change on transfer errors. Both the nonparametric and parametric models may indicate that the variable, BASELINEΔ has a positive effect on transfer errors. This is the "face-value" robustness result.

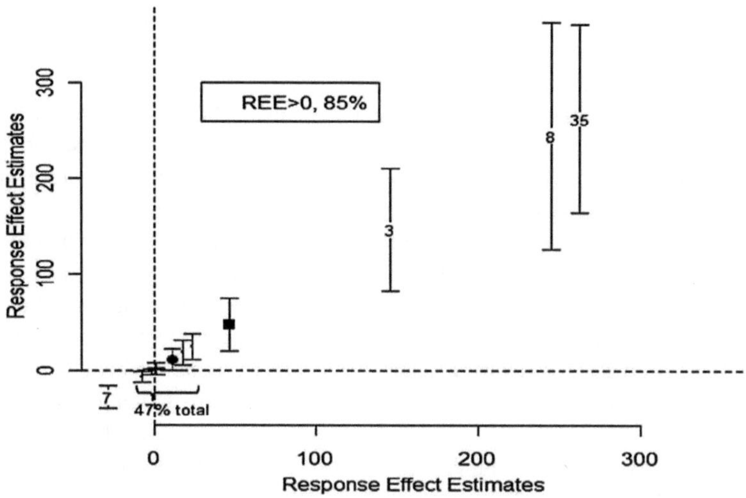

Fig. 17.2 *Benefit transfer error response effects for BASELINEΔ*: The independent variable is percent benefit transfer error and the dependent variable is BASELINEΔ that represents the percent change in value from baseline for some quality change. The figure plots WLS point estimates and nonparametric response effect estimates (*REE*), and 95 % confidence intervals for each. The *square* and *circle* represent the WLS point estimates based on the data with and without outliers, respectively. The *horizontal bars* above and below denote 95 % confidence intervals. The numbers indicate the share of the response effects at the point where the number is located. Clustering of REEs occurs because all of the independent variables are binary

However, when the response effects from the nonparametric analysis are plotted on a 45° plot with whisker plots,[19] a richer story arises (Fig. 17.2). The majority (85 %) of the nonparametric response effects are positive, indicating higher transfer errors associated with BASELINEΔ, which is the confirmation of the parametric coefficient that is significant and positive. There are three other results that could not be gleaned from a parametric estimate:

- a small number of the response effects are negative, not positive, and the largest negative effect size is significantly different from zero;
- nearly 50 % of the response effects are small and clustered near 0 (zero), no error; and
- the largest spikes of response effects are very large and are in excess of 200 % errors.

[19]The whisker plots mimic the 45° plots earlier, except confidence intervals are added in the form of lines extending vertically from each estimate such that the length of the lines is the length of the confidence interval. Any confidence level can be used and when 0 is contained in the interval the point estimate is statistically insignificant for that response effect.

Hence, the average effect revealed by the parametric analysis masks a more nuanced and practically relevant underlying pattern in generalization errors for research evaluation or prospective policy analysis.

These nuanced insights suggest that the analyst should seek further understanding of why some studies produce the opposite (negative) response effect from the mass of the data and others produce very large errors. Part of these insights might arise from unpacking the combination of other regressor values that are contained in each cell observation for the nonparametric estimator. From a literature evaluation perspective these results should prompt more research to understand from theoretical/methodological perspectives why there are two clusters of response effects that tell different stories. From a policy perspective, consider an analyst at the U.S. EPA making a decision using a meta-analysis-based benefit transfer to support a regulatory impact analysis. Imagine that only a value estimate was available for transfer. From Kaul et al.'s (2013) robust meta-analysis results, the analyst would need to know if their transfer is in the low-error type or the high-error type of applications to make an informed policy recommendation.

17.3.4 Smoothing Parameter Selection

Given the importance of the smoothing parameters (λs) in the performance of the smooth mixed data nonparametric regression estimator, it is advised to select them using a data-driven procedure. The most common approach to obtain bandwidths is to minimize the least squares cross-validation function

$$LSCV(\lambda^u, \lambda^o) = \sum_{i=1}^{n} (y_i - \widehat{m}_{-i}(x_i))^2,$$

where $\widehat{m}_{-i}(x_i)$ is the leave-one-out estimator, defined as

$$\widehat{m}_{-i}(x_i) = \frac{\sum_{j=1, j\neq i}^{n} y_j L_{\lambda, jx_i}}{\sum_{j=1, j\neq i}^{n} L_{\lambda, jx_i}}.$$

The intuition underlying this method of bandwidth selection is to estimate the unknown function at each point *without using that point*, and then predict what that point's observed response is. The set of bandwidths which yield the lowest average squared prediction error are deemed the best. A variety of other criterion to select the bandwidths have been proposed (such as improved Akaike Information Criterion; see Hurvich et al. 1998), but the LSCV criterion is the most popular amongst practitioners (Henderson et al. 2012b; Delgado et al. 2013).

Aside from selecting the bandwidths, the LSCV criterion also provides bandwidths that provide interesting empirical interpretations. First, notice what happens if any of the discrete bandwidths are 1. In this case the weighting does not depend on x_i meaning that the weighting function can be brought outside of the summation side in both the numerator and the denominator, so that this variable is removed from the construction of the estimator. This is also true for any continuous variable with a bandwidth $h_s = \infty$, though once h_s is more than a few standard deviations this effect takes hold. More generally, if we partition regressors of each type $x = (x^u, x^o)$ into $x^r = (x^{u_r}, x^{o_r})$ and $x^{ir} = (x^{u_{ir}}, x^{o_{ir}})$,[20] representing those variables of each type which are relevant to the smoothing of the estimator and those that are irrelevant to the smoothing of the estimator, respectively, then we can write our estimator as:

$$\widehat{m}(x) = \frac{\sum_{i=1}^{n} y_i L_{\lambda,ix}}{\sum_{i=1}^{n} L_{\lambda,ix}} = \frac{\sum_{i=1}^{n} y_i L_{\lambda,ix^r}}{\sum_{i=1}^{n} L_{\lambda,ix^r}} = \widehat{m}(x^r),$$

reducing the dimensionality of the problem. The ability to automatically remove irrelevant discrete variables is important for the performance of the smoothed estimator. When only discrete variables are present, the presence of irrelevant discrete regressors impedes the performance of the smoothed discrete regression estimator, in the sense that a correctly specified parametric model will now be more accurate than the smooth discrete regression estimator. In the case where only relevant discrete regressors are included the smoothed discrete regression estimator has the same level of accuracy as a correctly specified parametric model (see Li and Racine 2007).

Another specification issue that the meta-analyst must tackle is how to include the categories in the analysis. A simple example should suffice. Suppose there are two binary categories in the meta-data, one for competing methods used to construct the effect-size estimates, and another for the type of policy change conducted with those effect-size estimates. The traditional setting would be to include one categorical variable for each specific variable. However, an alternative approach would be to construct each specific cell and include one variable in the meta-regression. In this case we would have an unordered categorical variable that takes values in $\{0, 1, 2, 3\}$, where 0 signifies (0, 0), 1 signifies (0, 1), 2 signifies (1, 0) and 3 signifies (1, 1). While both approaches have merit, the condensed approach does not allow determination of the relevancy of a specific variable. That is, since we only have a single bandwidth, we can only say if the combined variable is irrelevant, not if specific categories within that variable are irrelevant. Thus, it may be prudent to keep the meta-data as general as possible to allow direct insights into these types of questions.

[20]Or, with mixed continuous-discrete data from $x = (x^c, x^u, x^o)$ into $x^r = (x^{c_r}, x^{u_r}, x^{o_r})$ and $x^{ir} = (x^{c_{ir}}, x^{u_{ir}}, x^{o_{ir}})$.

17.3.5 A Test for Correct Parametric Specification

Given the ability to smooth over a large array of discrete covariates, a natural question to ask is whether a parametric meta-function is correctly specified. This is an important component to assessing specification robustness of the meta-equation. It is not simply enough to find heterogeneous marginal effects from the smooth regression estimator. Given the common small samples in meta-analyses, rigorous statistical testing should be undertaken to ensure that this heterogeneity is not an artifact of the sample size. Here we detail the implementation of a test which can be used in the presence of mixed data to test for correct specification of the meta-function (Hsiao et al. 2007).

Consider a pre-specified parametric meta-function, perhaps linear, which we define as $m(x, \beta)$, where β is a k dimensional parameter vector.[21] Our null hypothesis is that the parametric model is correctly specified. Formally, we wish to the test the null that

$$H_0 : P[E(y|x) = m(x, \beta)] = 1$$

for some β versus the alternative that

$$H_1 : P[E(y|x) = m(x, \beta)] \neq 1$$

for any β.

Hsiao et al. (2007) propose using

$$\widehat{I}_n = n^{-2} \sum_{i=1}^{n} \sum_{j=1, j \neq i}^{n} \widehat{\varepsilon}_i \widehat{\varepsilon}_j L_{\lambda, ji},$$

to test H_0, where $\widehat{\varepsilon}_i = y_i - m(x_i, \hat{\beta})$ is the parametric residual for the ith observation. The intuition underlying this test is that if the meta-function is correctly specified, then the residuals should have no relationship with the covariates. Given that we are not requiring a specific parametric model in our alternative hypothesis, we use kernel smoothing to cover all potential forms with which $\widehat{\varepsilon}_i$ may depend on x_i. This gives the specification test power in all directions away from the null.

This test has an asymptotically normal distribution with variance

$$\widehat{\sigma}_n^2 = \frac{2}{n(n-1)} \sum_{i=1}^{n} \sum_{j=1, j \neq i}^{n} \widehat{\varepsilon}_i^2 \widehat{\varepsilon}_j^2 L_{\lambda, ji}^2,$$

---

[21] In this general setup q need not be equal to p. For example, including an unordered discrete variable with three different values $\{0, 1, 2\}$ would require two dummy variables in a parametric framework but only a single categorical regressor in our nonparametric framework.

where $L^2_{\lambda,ji}$ represents the product of the squares of each of the individual kernel functions. Given these results, our test statistic is constructed as

$$\widehat{T}_n = n\frac{\widehat{I}_n}{\widehat{\sigma}_n}.$$

Hsiao et al. (2007) show that $\widehat{T}_n$ converges to the standard normal distribution under the null.[22]

In the applied nonparametric literature, it is common to forego asymptotic approximations and use bootstrap methods to construct critical values for the test statistic. A straightforward bootstrap approach, allowing for heteroscedasticity of the error terms is:

1. Compute the test statistic $\widehat{T}_n$ for the original sample of $\{x_1, x_2, \ldots, x_n\}$ and $\{y_1, y_2, \ldots, y_n\}$.
2. For each observation i, construct the centered bootstrapped residual ε_i^*, where $\varepsilon_i^* = \frac{1-\sqrt{5}}{2}\left(\widehat{\varepsilon}_i - \overline{\widehat{\varepsilon}}\right)$ with probability $p = \frac{1+\sqrt{5}}{2\sqrt{5}}$ and $\varepsilon_i^* = \frac{1+\sqrt{5}}{2}\left(\widehat{\varepsilon}_i - \overline{\widehat{\varepsilon}}\right)$ with probability $1 - p$.[23]Then construct the bootstrapped left-hand-variable as $y_i^* = m\left(x_i, \widehat{\beta}\right) + \widehat{\varepsilon}_i^*$ for $i = 1, 2, \ldots, n$. Call $\left\{y_i^*, x_i\right\}_{i=1}^n$ the bootstrap sample. Note that the resampling is on the residuals, holding the covariates in the meta-data fixed. Thus, the only variable which changes in the bootstrapped meta-data is the effect size.
3. Calculate $\widehat{T}_n^*$ where $\widehat{T}_n^*$ is calculated the same way as $\widehat{T}_n$ except that y_i is replaced by y_i^*.
4. Repeat steps 2–3 a large number (B) of times and then construct the sampling distribution of the bootstrapped test statistics. We reject the null that the parametric is correctly specified if the estimated test statistic $\widehat{T}_n$ is greater than the upper α-percentile of the bootstrapped test statistics.

17.3.6 Why Use a Nonparametric Approach?

The smoothing approach has the ability to uncover structure that a traditional, additively separable parametric setup would miss. If these cross product terms are important for accurate assessment of effect size estimation then a nonparametric setup seems appropriate. Further, if one were to deploy least-squares cross-

---

[22]See Hsiao et al. (2007) for a description of the test statistic when continuous data are present as well.

[23]This bootstrap procedure ensures that the first three moments of the bootstrapped residuals are identical to the first three moments of the actual residuals.

validation to estimate the bandwidths, a direct comparison between the smoothed estimator and the cell estimator is possible, given that the estimated bandwidths may be 0. Thus, the frequency approach is fully nested in the smooth approach when using smoothing methods.

Given the visual aids to help display the variety of results provided with a nonparametric analysis, coupled with the relative ease of specification testing, we recommend deploying a nonparametric meta-analysis to assess the robustness of the assumed functional form of the meta-regression. Further, in the presence of discrete meta-regressors, the smoothed estimator does not suffer from the curse of dimensionality that is a common critique of nonparametric methods. Although we must worry about the inclusion of irrelevant meta-regressors, this adversely impacts parametric estimation as well.

17.4 Exploiting the Panel Nature of the Meta-data

Meta-data often comprise a panel because multiple observations can be gleaned from each individual study. A common approach to estimating the meta-regression with pooled data is to deploy a random or fixed effects estimator. For example, Rose and Stanley (2005) conduct a meta-analysis to infer the effect of a common currency union on international trade. They find considerable differences between the coefficient estimates of fixed and random effects models indicating heterogeneity in their meta-data across panel segments.

Estimation of population effect sizes via different methods can indicate the extent of heterogeneity in the data. Before estimating the mean population-effect size, some assumptions about the true data-generating process, like all other aspects of meta-analysis estimation procedures, need to be made. If observations on effect size are assumed to be homogeneous (i.e. true effect across studies is the same), then the variation in observed effect size arises from sampling error. However, if the true effect size is assumed to be different for all studies, the total variation in observed effect size is the sum of within and between study variation. Given these alternative perspectives, which may not be obvious from the data, robustness checks are also needed when exploiting the panel nature of meta-data.

In the former case, a fixed effects model is used to estimate the population effect size while the latter requires using the random effects models to estimate the mean effect size. Hedges and Vevea (1998) argue that the decision of using fixed versus random effects models essentially depends on the *nature of desired inference*. If the inferences are made only for the observed set of studies (conditional inference), then a fixed effects model is appropriate. However, for unconditional inferences about the overall population of studies a random effects model should be used. For example, if the nature of inference demands estimation of an unconditional effect size (i.e. that satisfies external validity for the pool of studies not conducted or studies that will be undertaken in the future) then a random effects model is favorable. Below, we discuss the fixed and random effects models where only one effect size is provided by each study. We then extend our discussion to panel stratification in meta-analyses where each study may provide multiple observations.

Assume that the effect size is normally distributed as $y_i \sim N(\bar{y}, \sigma_i^2)$ for i = 1, …, S, where S is the number of studies. In the fixed effects model, the variance minimizing weighted mean effect is estimated as:

$$\hat{\bar{y}} = \frac{\sum_{i=1}^{S} w_i y_i}{\sum_{i=1}^{S} w_i} \tag{17.24}$$

where w_i are weights assigned to each individual study.[24] With multiple studies, weights can be computed by estimating the within study variance as $w_i = 1/\sigma_i^2$, for i = 1, …, S. The intuition behind this weighting scheme is that studies with smaller variance are likely to be more precise and should be allocated more weight. The variance of the common effect size is the inverse of sum of weights i.e. $\hat{\sigma}^2 = \frac{1}{\sum_{i=1}^{S} w_i}$.

A constant true effect size can be assumed only when the studies are nearly identical in their data generating process and estimation procedures. However, this is not observed in many economics based meta-analyses. With heterogeneity, the weighted mean is estimated in the same fashion as (17.24); however, the weights are different. The study specific weights are estimated as $w_i = 1/(v_i^2 + \sigma_i^2)$, where v_i^2 is the between-study variation and σ_i^2 is the within-study variation. Der Simonian and Liard (1986) introduced the moment-based estimate of between-study variation

[24]Weighting the dependent and independent variables is commonly used to account for heterogeneity. In environmental economics, weights are often based on the sample size of primary studies, since the number of estimates to report is an arbitrary decision of each primary investigator. Thus, the observations on y and X used to estimate Eq. (17.1) are down-weighted by the number of observations on effect size provided by each study. This procedure helps to mitigate the influence of studies that potentially influence the results by providing more observations than others studies (Nelson and Kennedy 2009). Bergstrom and Taylor (2006) discuss weighting by the variance of the effect size estimate or weighting by the effect size significance probabilities. These approaches contain assumptions. First, sample size weighting implies that each study should have equal weight and it is the observations that are of importance. This may not be the case in studies with split samples that effectively carry out multiple independent experiments where any single experiment could have been published independently. Second, weighting by variance implies that precision is the key feature. However, focusing on variance overlooks accuracy; it is possible to have a very precise measure that is not accurate. Third, weighting by p-values implies that significant effect sizes carry the most important information. This overlooks the consideration that insignificant effect sizes may be real and should be given equal consideration in the analysis. Thus, while there are logical reasons for using weights in meta-analyses, these considerations are not without concerns. Thus, it is important to investigate estimation with and without weights, and perhaps using a variety of plausible weighting schemes to evaluate the robustness of parameter estimates.

$$\hat{v}^2 = \begin{cases} \frac{Q-(s-1)}{c} & \text{if } Q \geq s-1 \\ 0 & \text{otherwise,} \end{cases}$$

where Q and c are defined as

$$Q = \sum_{i}^{s} w_i y_i^2 - \frac{\sum_i^s w_i y_i^2}{\sum_i^s w_i}, c = \sum_i^s w_i - \frac{w_i^2}{\sum_i^s w_i}. \tag{17.25}$$

Divergence in the estimated population effect size via the fixed and random effects models suggests potential heterogeneity. To further investigate this issue, statistical tests of heterogeneity can be employed. These tests are based on the logic of comparing between and within variations in effect size. Consider the null hypothesis of equal true effect size across studies. It is common to estimate a fixed effects model if this null hypothesis is rejected. The following test statistics can be used for hypothesis testing

1. One of the most commonly used statistics in science-related meta-analyses is the Q statistic that equals $\sum_{i=1}^{s}\left((y_i - \hat{\bar{y}})/\sqrt{w_i}\right)^2$, where $\bar{y}$ is the mean of the effect size. This test examines whether the total variation is due to within study variation. Under the null hypothesis of the same true effect size, the Q statistics follows a χ^2 distribution with degrees of freedom equal to $s - 1$ (Der Simonian and Liard 1986). The drawback of using the Q statistic is that it is known to have low power when the number of studies is low (Higgins and Thompson 2002).
2. The H statistic equals $Q/s - 1$, and measures the potential amount of heterogeneity. The degrees of freedom for Q statistic is $s - 1$ and $E(Q) = s - 1$. Homogeneity of effect size is indicated when $H = 1$. Higgins and Thompson (2002) point out that no universal rules can be placed on the accepted value of H for claiming mild, moderate or severe heterogeneity. However, values above 1.5 indicate potential heterogeneity and demand further investigation.
3. The I^2 statistic is defined as $I^2 = 100 \cdot (Q - (s - 1))/Q = (H^2 - 1)/H^2$. This test compares observed heterogeneity and total variation. Low values of I^2 reflect no observed heterogeneity. Large values of I^2 indicate increasing heterogeneity.

Both H and I^2 statistics are more powerful than Q statistic as they do not depend on the number of studies (Higgins and Thompson 2002). These simple tests help in investigating heterogeneity in effect sizes and can be easily employed in economics as additional robustness checks. Heterogeneity in effect size should be followed by steps to control for study-specific effects using a random or fixed effect specification of the meta-regression. Prior to estimating these models, the analyst should resolve stratification of the meta-data, which exploits the panel features. Rosenberger and Loomis (2000) discuss a variety of stratification alternatives for meta-data. Each of these additional stratification approaches can provide further checks on the robustness of the meta-regression estimates.

The panel data approach is important for evaluating the robustness of the parameter estimates from a meta-regression. There are several reasons for this. First, in the presence of fixed effects, coefficient estimates may be biased, leading to statistically poor estimates of the desired effects the meta-regression is intended to provide. Second, in the presence of random effects there may exist both within and across strata variance effects that could make inference on the meta-regression coefficient estimates inaccurate. Both of these effects need to be considered. Also it is important to recognize the distinction between incorporating fixed or random effects into the panel. Including fixed effects acknowledges that there are strata specific effects that are not represented by regressors in the estimated equation and are correlated with the regressors in the meta-regression, whereas accounting for random effects suggests that the regressors are uncorrelated with the latent strata specific effects. An alternative to applying common panel methods is the multilevel modeling approach of Bateman and Jones (2003) and Brouwer et al. (1999). The multilevel modeling is similar to the random effects specification of a panel regression model.

Given the variety of strata that can be thought of as generating a hierarchical structure to the meta-data, it is important to test across various levels of stratification. Rosenberger and Loomis (2000) developed levels of panel stratification to account for the panel nature of the meta-data while allowing exploitation of the panel structure. Their additional stratification levels of the meta-data include the year when the studies were conducted (time stratification), the researchers who conducted the studies (researcher stratification) or the explicit features of the data common to numerous studies (data structure stratification). Alternatively, the multilevel model meta-regression of Bateman and Jones (2003) stratifies the data by author whereas Florax et al. (2005) stratify their meta-data on valuation of pesticide risk exposure by estimation method.

A common symptom of the panel structure for meta-data is that many of the stratification levels may contain a single observation. For example, if stratification is at the author level, a majority of the studies may have unique authors while a few studies may be conducted by the same author. Thus, accounting for intra-panel correlations is not feasible. Rosenberger and Loomis' (2000) data structure stratification proceeds by stratification of the meta-data into single studies that provide a single estimate for a recreation activity; studies that provide multiple estimates but across different target populations; and studies that provide multiple estimates for the same target population. This stratification results in all strata having at least two observations, so that panel effects could be discerned within the meta-data.

Two tests of panel stratification of the meta-data are provided in Rosenberger and Loomis (2000). They provide a Lagrange multiplier test to determine if panel effects are present and a Hausman test (assuming panel effects exist) to test between fixed (correlated) or random (uncorrelated) effects. The recent work of Guggenberger (2010) suggests careful implementation of Hausman tests in panel data analysis. It is common to report inference results from the random effects model when the Hausman test fails to reject the null of uncorrelated panel effects. However, this induces a pre-test bias. Recall that the fixed effects estimator is consistent even under the random effects setting, so it is advised that inference results are presented from the fixed effect model even when a Hausman test determines random effects to be the appropriate specification. The same warning should apply to the testing framework of Rosenberger and Loomis (2000) who only use the Hausman test *after* rejecting the null of no panel effects. For assessing robustness to panel stratification, both tests of Rosenberger and Loomis (2000) should be conducted and any inference from the meta-regression should include both fixed and random effects t-statistics to avoid likely pre-test biases.

17.5 Conclusion

Meta-analysis is a prudent empirical approach to summarize key findings in a literature. The results stemming from meta-regressions can be used in two ways. The first is a normative use to assess the status of an empirical literature and use the results to identify questions and gaps in the literature that could be addressed by future studies. The second is a positive use to make predictive prescriptions for decision making, e.g., in non-market valuation predicting value estimates (benefit transfer) to support cost-benefit analyses of public programs. The former application is relatively straightforward, while the latter is more controversial. In either case, as with all applied research, the robustness of the key findings should be scrutinized with respect to modeling decisions. Robustness checks can address the strengths of research insights and the criticisms of prescriptive uses of estimation results. The format of meta-data offers an array of avenues to assess the robustness (or lack thereof) of insights from the meta-regression.

This chapter discusses a number of econometric approaches that can be employed to investigate the robustness of meta-analysis regression estimates. These robustness procedures address sample selection, horizontal and vertical robustness of the meta-data, specification of the meta-equation via nonparametric estimation, and methods to control for heterogeneity that arise from the panel nature of most meta-data. While the methods outlined here are by no means exhaustive, they highlight some of the techniques that are available to undertake robustness analyses for building credible insights.

As stated at the outset of this chapter, we do not interpret robustness to imply good/bad or reliable/unreliable outcomes or any other normative statement about the empirical outcomes of any meta-analysis. Rather, we view the robustness as informational; whether the presence/absence of robustness is good, neutral or bad news is context-specific. Given meta-analysis' role in highlighting the features of a literature it seems only natural that the meta-analyst would examine what findings from the meta-analysis itself are significantly robust.

Sample selection, or research priority, effects can be difficult to deal with in a meta-analysis because the dimensionality of selection must be specified and then requires the acquisition of a secondary dataset to investigate selectivity. Regarding horizontal robustness, while the discussion has focused on observations and studies, one could also think of using alternative data structures to analyze meta-regression estimates. Regardless of how selection and horizontal robustness of the meta-data are defined, the approaches we describe are appropriate for investigating robustness.

Vertical robustness can be investigated in two dimensions. First, we can examine the effects of the inclusion/exclusion of regressors on estimated results. Second, we can investigate the effects of model specifications by exploring nonparametric estimation techniques. Our discussion of nonparametric smoothing of discrete data was necessarily long in order to provide the appropriate intuition to give meta-analysts, new to this literature, confidence in applying these methods in future studies. The empty cell phenomenon is likely to remain for the foreseeable future, until bodies of literature in environmental economics expand to fill in the gaps. Further, even with more studies, if new methods and new datasets are used then the number of cells available is likely to expand as well. The discrete smoothing methods of nonparametric estimation can be used in conjunction with parametric methods to identify the robustness of regressor-specific response effects. This is perhaps where the greatest insights can be extracted from the meta-data from the robustness checks we have described.

We also do not view the issues of robustness in this chapter as the end. As meta-analysis evolves, new methods and concerns over the collection, construction and analysis of the meta-data undoubtedly will evolve. Not all of the robustness checks we outline are universally relevant in all applications, but some variant of them is always relevant. One thing that we can be certain of is that the meta-analysts need to be cognizant of the robustness of their findings before making general recommendations for the literature at hand or to the decision makers.

Appendix A: Derivation of (17.12)

We have

$$\begin{aligned}
\hat{\beta}^{(-t)} &= \left(X'_{(-t)}X_{(-t)}\right)^{-1} X'_{(-t)}y_{(-t)} \\
&= \left(X'X - X'_{(t)}X_{(t)}\right)^{-1}\left(X'y - X'_{(t)}y_{(t)}\right) \\
&= \left[(X'X)^{-1} + (X'X)^{-1}X'_{(t)}\left(I - X_{(t)}(X'X)^{-1}X'_{(t)}\right)^{-1}X_{(t)}(X'X)^{-1}\right]\left(X'y - X'_{(t)}y_{(t)}\right) \\
&= \left(I + (X'X)^{-1}X'_{(t)}\left(I - P^{(t)}\right)^{-1}X_{(t)}\right)\hat{\beta} - (X'X)^{-1}X'_{(t)}\left(I + \left(I - P^{(t)}\right)^{-1}P^{(t)}\right)y_{(t)} \\
&= M_{(t)}\hat{\beta} + Q_{(t)}y_{(t)},
\end{aligned}$$

where $P^{(t)} = X_{(t)}(X'X)^{-1}X'_{(t)}$. The first equality follows by definition of the OLS estimator, the second by our definitions for $X_{(-t)}$, $X_{(t)}$, $y_{(-t)}$ and $y_{(t)}$, and the third by Racine (1997, Eq. 9).

Appendix B: Derivation of (17.13)

$$\begin{aligned}
\tilde{\omega}_{(t)} &= y_{(t)} - X_{(t)}\hat{\beta}^{(-t)} = y_{(t)} - X_{(t)}\hat{\beta} - P^{(t)}\left(I - P^{(t)}\right)^{-1}X_{(t)}\hat{\beta} + P^{(t)}y_{(t)} + P^{(t)}\left(I - P^{(t)}\right)^{-1}P^{(t)}v_{(t)} \\
&= y_{(t)} - X_{(t)}\hat{\beta} - P^{(t)}\left(I - P^{(t)}\right)^{-1}X_{(t)}\hat{\beta} + P^{(t)}v_{(t)} - P^{(t)}X_{(t)}\hat{\beta} + P^{(t)}X_{(t)}\hat{\beta} \\
&\quad + P^{(t)}\left(I - P^{(t)}\right)^{-1}P^{(t)}y_{(t)} - P^{(t)}\left(I - P^{(t)}\right)^{-1}P^{(t)}X_{(t)}\hat{\beta} \\
&\quad + P^{(t)}\left(I - P^{(t)}\right)^{-1}P^{(t)}X_{(t)}\hat{\beta} \\
&= \hat{\omega}_{(t)} + P^{(t)}\hat{\omega}_{(t)} + P^{(t)}\left(I - P^{(t)}\right)^{-1}P^{(t)}\hat{\omega}_{(t)} \\
&\quad + \left[P^{(t)}\left(I - P^{(t)}\right)^{-1} + P^{(t)} + P^{(t)}\left(I - P^{(t)}\right)^{-1}P^{(t)}\right]X_{(t)}\hat{\beta} \\
&= \left[I + P^{(t)} + P^{(t)}\left(I - P^{(t)}\right)^{-1}P^{(t)}\right]\hat{\omega}_{(t)} \\
&\quad + \left[P^{(t)}\left(I - P^{(t)}\right)^{-1} + P^{(t)} + P^{(t)}\left(I - P^{(t)}\right)^{-1}P^{(t)}\right]X_{(t)}\hat{\beta} \\
&= \left(I - P^{(t)}\right)^{-1}\hat{\omega}_{(t)}.
\end{aligned}$$

The last equality here follows from the fact that

$$I + P^{(t)} + P^{(t)}\left(I - P^{(t)}\right)^{-1}P^{(t)} = \left(I - P^{(t)}\right)^{-1}$$

$$P^{(t)}\left(I - P^{(t)}\right)^{-1} + P^{(t)} + P^{(t)}\left(I - P^{(t)}\right)^{-1}P^{(t)} = 0,$$

both of which can be discerned from the matrix equality $A - A(A+B)^{-1}A = B - B(A+B)^{-1}B$, when $(A + B)^{-1}$ exists (let $A = I$ and $B = -P^{(t)}$).

References

Bateman, I. J., & Jones, A. P. (2003). Contrasting conventional with multi-level modeling approaches to meta-analysis: Expectation consistency in U.K. woodland recreation values. *Land Economics, 79*(2), 235–258.

Belsley, D., Kuh, E., & Welsch, R. E. (1980). *Regression diagnostics*. New York: Wiley.

Bergstrom, J. C., & Taylor, L. O. (2006). Using meta-analysis for benefit transfers: Theory and practice. *Ecological Economics, 60*(2), 351–360.

Boyle, K. J., Parmeter, C. F., Boehlert, B. B., & Paterson, R. W. (2013). Due diligence in meta-analyses to support benefit transfers. *Environmental and Resource Economics, 55*(3), 357–386.

Brander, L., van Beukering, P., & Cesar, H. (2007). The recreational value of coral reefs: A meta-analysis. *Ecological Economics, 63*(1), 209–218.

Brouwer, R., Langford, H., Bateman, I. J., & Turner, R. K. (1999). A meta-analysis of wetland contingent valuation studies. *Regional Environmental Change, 1*(1), 47–57.

Cook, R. D., & Weisberg, S. (1982). *Residuals and influence in regression*. New York: Chapman and Hall.
Davidson, R., & MacKinnon, J. G. (2004). *Econometric Theory and Methods*. New York: Oxford University Press.
Delgado, M. S., Henderson, D. J., & Parmeter, C. F. (2013). Does education matter for economic growth? *Oxford Bulletin of Economics and Statistics, 76*(3), 334–359.
Der Simonian, R., & Kacker, R. (2007). Random-effects model for meta-analysis of clinical trials: an update. *Contemporary Clinical Trials, 28*, 105–114.
Der Simonian, R., & Laird, N. (1986). Meta-analysis in clinical trials. *Controlled Clinical Trails, 7*, 177–188.
Florax, R. J. G. M., Travisi, C. M., & Nijkamp, P. (2005). A meta-analysis of the willingness to pay for reductions in pesticide risk exposure. *Review of Agricultural Economics, 32*(4), 441–467.
Ghermandi, A., & Nunes, P. A. L. D. (2013). A global map of coastal recreation values: Results from a spatially explicit meta-analysis. *Ecological Economics, 86*, 1–15.
Ghermandi, A., van den Bergh, J. C. J. M., Brander, L. M., de Groot, H. L. F., & Nunes, P. A. L. D. (2010). Values of natural and human-made wetlands: A meta-analysis. *Water Resources Research, 46*, 1–12.
Guggenberger, P. (2010). The impact of a Hausman pretest of the size of a hypothesis test: The case. *Journal of Econometrics, 156*, 337–343.
Hall, P., Li, Q., & Racine, J. (2007). Nonparametric estimation of regression functions in the presence of irrelevant regressors. *The Review of Economics and Statistics, 89*(4), 784–789.
Heckman, J. J. (1979). Sample selection bias as a specification error. *Econometrica, 47*(1), 153–161.
Hedges, L. V., & Vevea, J. L. (1998). Fixed-and random-effects models in meta-analysis. *Psychological Methods, 3*(4), 486–504.
Henderson, D. J., Kumbhakar, S. C., & Parmeter, C. F. (2012a). A simple method to visualize results in nonlinear regression models. *Economics Letters, 117*(3), 578–581.
Henderson, D. J., Papageorgiou, C., & Parmeter, C. F. (2012b). Growth empirics without parameters. *The Economic Journal, 122*(559), 125–154.
Higgins, J. P. T., & Thompson, S. G. (2002). Quantifying heterogeneity in a meta-analysis. *Statistics in Medicine, 21*, 1539–1558.
Hoehn, J. P. (2006). Methods to address selection effects in the meta-regression transfer of ecosystem values. *Ecological Economics, 60*(2), 389–398.
Hsiao, C., Li, Q., & Racine, J. (2007). A consistent model specification test with mixed discrete and continuous data. *Journal of Econometrics, 140*(2), 802–826.
Hurvich, C. M., Simonoff, J. S., & Tsai, C. (1998). Smoothing parameter selection in nonparametric regression using an improved Akaike information criterion. *Journal of the Royal Statistical Society, 60*, 271–293.
Johnston, R. J., & Rosenberger, R. S. (2010). Methods, trends and controversies in contemporary benefit transfer. *Journal of Economic Surveys, 24*, 479–510.
Kaul, S., Boyle, K. J., Kuminoff, N. V., Parmeter, C. F., & Pope, J. C. (2013). What can we learn from benefit transfer errors? Evidence from 20 years of research on convergent validity. *Journal of Environmental Economics and Management, 66*(1), 90–104.
Li, Q., & Racine, J. S. (2007). *Nonparametric econometrics: Theory and practice*. Princeton, NJ: Princeton University Press.
Londoño, L. M., & Johnston, R. J. (2012). Enhancing the reliability of benefit transfer over heterogeneous sites: A meta-analysis of international coral reef values. *Ecological Economics, 78*(1), 80–89.
Leamer, E. (1983). Lets take con out of econometrics. *American Economic Review, 73*(1), 31–43.
Manski, C. F. (2007). *Identification for prediction and decision*. Cambridge: Harvard University Press.

Moeltner, K., Boyle, K. J., & Paterson, R. W. (2007). Meta-analysis and benefit transfer for resource valuation: Addressing classical challenges with Bayesian modeling. *Journal of Environmental Economics and Management, 53*(2), 250–269.

Moeltner, K., & Rosenberger, R. S. (2008). Predicting resource policy outcomes via meta-regression: Data space, model space, and the quest for 'Optimal Scope'. *The B.E. Journal of Economic Analysis and Policy, 8*(1), 1–31.

Nelson, J. P., & Kennedy, P. E. (2009). The use (and abuse) of meta-analysis in environmental and natural resource economics: an assessment. *Environmental and Resource Economics, 42*, 345–377.

Ouyang, D., Li, Q., & Racine, J. (2009). Nonparametric estimation of regression functions with discrete regressors. *Econometric Theory, 25*(1), 1–42.

Proietti, T. (2003). Leave-K-out diagnostics in state-space models. *Journal of Time Series Analysis, 214*(2), 221–236.

Racine, J. S. (1997). Feasible cross-validatory model selection for general stationary processes. *Journal of Applied Econometrics, 12*, 169–179.

Rose, A. K., & Stanley, T. D. (2005). A meta-analysis of the effect of common currencies on international trade. *Journal of Economic Surveys, 19*(3), 347–375.

Rosenberger, R. S., & Johnston, R. J. (2009). Selection effects in meta-analysis and benefit transfer: Avoiding unintended consequences. *Land Economics, 85*(3), 410–428.

Rosenberger, R. S., & Loomis, J. B. (2000). Panel stratification in meta-analysis of economic studies: an investigation of its effects in the recreation valuation literature. *Journal of Agricultural and Applied Economics, 32*(3), 459–470.

Sala-I-Martin, X. X. (1997). I just ran two million regressions. *The American Economic Review, 87*, 178–183.

Wooldridge, J. M. (1995). Selection corrections for models under conditional mean independence assumptions. *Journal of Econometrics, 68*(1), 115–132.

Part IV
Spatial and Geographical Considerations

Chapter 18
Spatial and Geographical Aspects of Benefit Transfer

Marije Schaafsma

Abstract This chapter discusses methodological issues associated with the spatial nature of values for environmental goods and services and implications for valuation and benefit transfer (BT). It is designed to complement Chap. 13, which discusses the importance of a spatial framework for analysis of ecosystem services, and Chap. 20, which demonstrates the use of geographic information systems (GIS) for value mapping and BT. The chapter provides a broad perspective on the potential causes of spatial heterogeneity in environmental values, with particular attention to the relevance of this heterogeneity for stated preference (SP) valuation and the transfer of resulting welfare estimates. This includes discussions of spatial variability in the provision of ecosystem services, distance decay and substitution effects, and additional spatial patterns in willingness to pay (WTP). The chapter also suggests different ways that spatial variations such as these may be accommodated to support reliable benefit transfer.

Keywords Spatial heterogeneity · Distance decay · Substitution · Geographical · Scale · Ecosystem service

18.1 Introduction

It is now widely acknowledged that spatial context is one of the key issues in ecosystem service assessment, both in biophysical and economic models. Many recent conceptual papers stress the importance of spatial context in environmental valuation (Fisher and Turner 2008; TEEB 2010; Turner et al. 2010). The development of large-scale assessment of environmental goods and services and their

M. Schaafsma (✉)
Geography and Environment & Centre for Biological Sciences,
University of Southampton, Southampton, UK
e-mail: m.schaafsm@soton.ac.uk

R.J. Johnston et al. (eds.), *Benefit Transfer of Environmental and Resource Values*,
The Economics of Non-Market Goods and Resources 14,
DOI 10.1007/978-94-017-9930-0_18

valuation drives the need for spatially explicit, transferable value functions (Bateman et al. 2011b).

This chapter discusses some of the main types and causes of spatial heterogeneity in environmental valuation, aiming to answer the questions: Which spatial factors cause differences in individual and per unit area willingness to pay (WTP) values (e.g., per hectare) between policy and study sites? How can studies and models be improved to ensure that WTP estimates reflect spatial heterogeneity? And how can benefit transfer errors be reduced by accounting for spatial heterogeneity? The chapter focuses on stated preference (SP) studies, but supplements these with relevant examples from the revealed preference (RP) literature. Although SP studies typically express results in WTP per respondent or household, some benefit transfer (BT) studies convert these into WTP per hectare to account for differences in area size when transferring values to a new study site.[1]

The chapter starts from the notion that different people living in different places will often express different WTP values for otherwise identical environmental changes. Mean value transfers may work well when the study and policy sites are very similar in terms of population and site characteristics, but function transfers tend to be more accurate for dissimilar sites (Bateman et al. 2011a). To estimate more valid and reliable WTP values, one would ideally develop a WTP function that is specified after testing a set of variables that can accommodate spatial variation in WTP and are supported by theory. However, a review of the published SP literature shows that heterogeneity in the spatial distribution of the benefits of ecosystem services is often ignored or addressed in a very simplified manner in the design of surveys, model specification, and value aggregation. This has important consequences for the potential transferability of study results, and the transfer errors that are likely to result.

In this chapter, various spatial and geographical aspects relevant to environmental valuation studies in general, and BT more specific, will be discussed.[2] First, spatial factors and heterogeneity in (biophysical) environmental impacts that affect WTP values are highlighted. WTP values are expected to be location-dependent when the provision of ecosystem goods and services to which they relate is also size-, scale- and location-dependent. An example is water quality provision where upstream water management may affect downstream quality levels. Second, according to economic theory, distance, as a proxy for travel costs and availability of substitutes, is an expected driver of WTP and is clearly spatially defined. Third, substitutes are rarely randomly distributed across space (e.g., Termansen et al. 2008) and differences in substitute availability among respondents in different areas are expected to lead to differences in individual WTP across regions. This leads to a

[1]These studies typically assume that the relationship between size, the production of ecosystem services and the associated WTP benefits is linear, i.e., fixed economic and biophysical values per hectare. This assumption is likely to be erroneous, as will be discussed in this chapter, but nevertheless applied, driven by demand for easily applicable estimates of economic value and insufficient knowledge about the "real" functional form of the size-value relationship.

[2]As discussed by Ferrini et al. in Chap. 13, both biophysical and economic values can be transferred.

discussion of distance decay, a term used to describe the decrease in individual WTP as distance between the good and the individual's home increases (Sutherland and Walsh 1985). Fourth, other causes of spatial heterogeneity in WTP values for environmental goods may be caused by differences in location-characteristics and the surrounding landscape of these sites (Johnston et al. 2002; Willis and Garrod 1993), differences in socio-demographic or other population characteristics, including spatial perception factors (e.g. sense of place, see for example Brody et al. 2004), and population density.

Each of these issues will be discussed in turn, first with regard to implications for primary study valuation, and then with regard to accommodation within benefit transfer. The chapter also highlights some of the primary literature in each area, as well as important areas for future research.

18.2 Spatial Heterogeneity of Environmental Impacts

As noted by Johnston and Rosenberger (2010), the valuation literature is dominated by works exploring novel methodologies rather than those focused on the provision of high quality, well-reported, and transferable empirical estimates. As a result, differences across the spatial, ecological and biophysical characteristics of sites and the level of provision of environmental goods and services sometimes receive less attention. However, it is important to consider that the provision levels of goods and services of sites are often location-specific, and understanding of this spatial variation is necessary to properly define the environmental goods and services under valuation.

Spatially differentiated biogeochemical characteristics of sites, such as temperature, rainfall, wind, sea currents and altitude, but also historical human use and environmental degradation, can affect the level of ecosystem or environmental service provision. In Chap. 13, Ferrini et al. provide an example of spatial heterogeneity in values for carbon sequestration. Considering other examples, house prices will be affected by air pollution only if they are located downwind from the source (Cameron 2006). Mangrove values decrease with distance from the seaward edge (i.e., increase with distance from land), because the outer patches of mangroves are most effective in reducing the impacts of storm waves (Barbier 2012). Other environmental effects and their values do not decay continuously and monotonically over space, but are affected by borders and barriers of the natural or political environment (Ferrini and Fezzi 2012; Perrings and Hannon 2001).

Ecological processes also operate over a wide range of spatial scales (e.g., site, landscape, global level), and processes may be connected across scales through feedback relationships. This leads to non-linear systems. Understanding the complexity of the system helps to identify the areas where beneficiaries of an environmental change can be found. Although some services are beneficial on a national scale, other services create utility only on a local or regional scale. Ecosystem scales do not necessarily overlap with the scale of the beneficiaries, or the scale at which the political system is organized. For example, the catchment (or watershed) scale is typically recommended as the appropriate level of analysis for freshwater ecosystems. However, beneficiaries may live well beyond the geographical boundaries of catchments, whereas relevant governing institutions can be found at smaller scales.

Ecological connections among sites and their ecosystem functions and processes may cause interdependencies in the provision of environmental goods and services. When improving water quality upstream, this is likely to have spill-over effects downstream following dispersion patterns of the [avoided] pollution. The identification of beneficiaries of an environmental change hence depends critically on the scope of the ecosystem services and impact hereon. Similarly, when environmental improvements lead to increased wildlife numbers, dispersal may also increase wildlife abundance at related sites. Although recreation demand models reflect choices among multiple sites, they rarely reflect spatial relationships in ecological production functions among the sites. Newbold and Massey (2010) develop a modeling framework which combines site choice models with species distribution models, but research in this area is limited and no applications within the SP literature exist.

There also exist ecological non-linearities in the relation between size of the study site and the level of ecosystem service provision. For example, some habitats need to be of a minimum size to avoid ecological collapse, and the risk of collapse increases exponentially with the decline in habitat (Barbier 2012). Similarly, some wildlife species have minimum area requirements, or require the ability to transit between habitat patches (Bauer et al. 2010). In such cases, simple benefit transfers based on a mean per hectare value from a large to a small area should take into account that the smaller site may not be able to provide the same service level, and the potential importance of connectivity.

Another example is that the shape of a mangrove forest determines its flood protection function. Mangroves need to be of a minimum width to withstand waves, and will have human benefits only if there is human capital to be protected. In the three diagrams in Fig. 18.1, the total area of mangroves is similar, but the value of

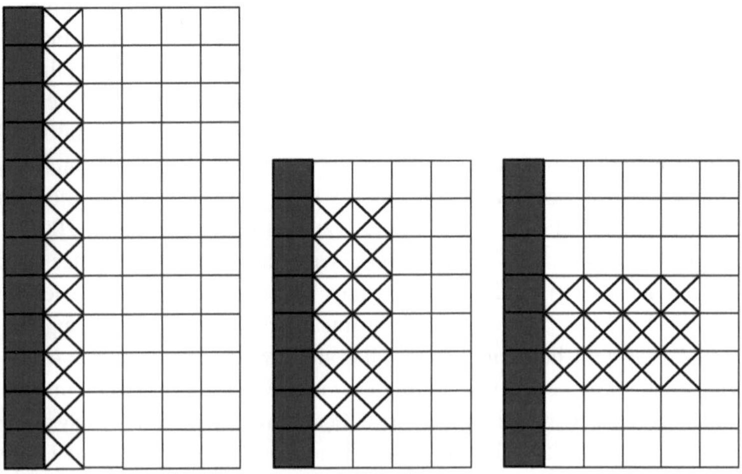

Fig. 18.1 Shape matters for flood protection value

protection service is different and follows a non-linear relationship. Grey squares represent built capital, crossed squares represent mangroves along the coastline and white squares represent the sea. If the minimum width of mangroves is two units to protect coastal properties from storm waves, the middle configuration of mangrove units has the highest flood protection value. In the left-hand picture, the mangroves are not wide enough to protect the properties, and the right-hand pictures reflects a sub-optimal configuration, as only three property units are protected. The relationship between width and marginal WTP is thus non-linear.

Management of natural areas can also involve tradeoffs where optimizing the provision of one ecosystem service at a site comes at the expense of other ecosystem services at the same site. Alternatively, management actions aimed to optimize one ecosystem service may simultaneously improve other services, leading to synergies in the overall provision of ecosystem services. Such interactions have to be described clearly and properly in SP valuation scenarios in order to avoid ambiguity about the scope of the study and the definition of the good under valuation.

This section is by no means exhaustive, but highlights some of the main spatial and geographical factors that affect the level of service provision by sites. Differences among sites in size and shape, their interaction with other ecosystems, and their position relative to a specific population of beneficiaries may cause significant differences in the extent to which they are able to deliver beneficial goods and services. The effectiveness of environmental policies to remediate environmental pressures and impacts may also be reduced. For benefit transfer studies, it is important to understand these differences in order to accurately adjust economic values between study and policy sites, and to avoid unforeseen sources of generalization error.

18.3 Substitutability and Complementarity

Substitutes are goods that can satisfy the same need or fulfill the same consumption goal: the decision maker chooses among goods that are substitutable in terms of their utility provision, at least in the margin. For example, different nature and recreation sites can serve as substitutes for each other in terms of the functions and associated goods and services they provide and associated use and nonuse values.[3] Substitution effects refer to the effect of changes in the presence or characteristics (price and quality) of substitutes on the WTP for (a change in the characteristics of) the alternative of interest (Koelemeijer and Oppewal 1999). Substitution rates depend on the level of (perceived) similarity of sites in their provision of goods and

[3]The economic concept of dependency between sites in terms of complementarity and substitution effects may be different from "ecological substitution" (Mitsch and Gosselink 2000). Sites of different habitat types may not be able to provide the same services in ecological terms, but may provide the same level of utility.

services, including spatial site characteristics, such as size or distance to beneficiaries. Complementarity occurs when goods are consumed jointly, for instance when improving access to one site increases the demand for, and thereby the value attached to, other sites.

The availability of substitutes or complements influences scarcity conditions and thereby public WTP for a good (Carson et al. 1998). According to economic theory, for example, marginal utility declines with every additional unit obtained: WTP for an additional hectare of a nature type of which supply is already high should be lower than for an additional hectare of a scarce nature type. WTP estimates should therefore vary with the size of the good. When transferring WTP estimates per hectare, it is often ignored that economic valuation studies produce marginal values that are non-linear and relative to the existing size of the area or good. Failure to account for these patterns during benefit transfer can lead to large generalization errors.

In BT studies, moreover, policy and study sites are unlikely to be surrounded by the same number of substitutes with the same quality; the availability of substitutes often varies over regions (Jørgensen et al. 2013). This can lead to differences in mean WTP estimates. Unlike many RP studies, most SP studies do not consider substitution effects.[4] Moreover, substitution effects that arise if the characteristics of multiple sites change simultaneously are given little attention in the valuation literature (Carson et al. 2001; DeShazo et al. 2009). This is surprising not only from a theoretical perspective, as substitutability plays a central role in microeconomic utility theory, but also from a practical point of view, considering that most national environmental policies are large-scale projects. Hence, these projects will not only affect the supply at a single study site, but also at surrounding substitute sites.

The majority of SP studies focus on the valuation of a single site. Critics of single-site contingent valuation (CV) studies (not including choice experiments, which are discussed below) point at the relatively large contribution that some people are willing to make in light of the large number of available substitutes (Arrow et al. 1993). If respondents' attention is focused on a single site only and drawn away from possible substitutes, studies may fail to account for changes in the availability or characteristics of relevant alternatives. Similarly, disregarding complementary sites may cause underestimation of WTP values.

Failure to account for substitution effects has consequences for the reliability and validity of using WTP estimates in both aggregation and benefits transfer procedures. Empirical CV studies have found that the sum of the value of goods measured individually is often higher than the value measured for all goods at once.[5] In case of substitution or diminishing marginal utility, this summation problem is theoretically valid (Hoehn and Randall 1989). For instance, respondents in an area with several lakes that are polluted are likely to value cleaning up the first lake

[4]An exception is the substitutability among attributes within choice experiments.

[5]As noted by Carson (2012), this finding is not unique to CV. It is also observed in market and experimental settings.

"on offer" more than cleaning up the second lake, because the first lake can be a substitute for the second, and the respondent has a budget limitation which reduces the money available for cleaning up the second and subsequent lakes.[6,7] In this case, valuing substitute goods separately and then adding up their individual values without any adjustment will overstate the "true" value, as every respondent will treat the lake they are asked to value as if it were the first and only good. When the WTP estimate for improving the first lake is transferred to the other lakes, summing these values will produce unreliable aggregate WTP estimates. Substitutability and complementarity imply that the marginal WTP for changes at one site depends on the availability and characteristics of relevant alternatives. Therefore, transferring WTP values from a study site with few substitutes to a policy site with many available alternatives is likely to result in high transfer errors, even if the relevant population of beneficiaries and other site characteristics are comparable.

The effectiveness of a simple sentence in a SP questionnaire to remind respondents of substitutes, as recommended by Arrow et al. (1993), has been questioned (Kotchen and Reiling 1999; Loomis et al. 1994; Whitehead and Blomquist 1999). Cummings et al. (1994) and Neill (1995) argue that to elicit WTP estimates that reflect substitution effects, respondents should be asked to value substitutes and study sites simultaneously. Alternatively, surveys should provide at least some description of available substitutes, using pictures, maps or text. The recent literature shows that it is becoming more common to include maps in the survey materials of SP studies to depict the study site (e.g., Bateman and Langford 1997; Johnston et al. 2002; Rollins and Lyke 1998), but additional information about substitute sites is usually not provided (but see Brouwer et al. 2010).

A small number of CV studies include multi-program scenarios in which different goods are valued simultaneously to test for substitutability and complementarity effects. Respondents are presented programs at different locations and asked to value the study site as well as its alternatives. These studies are limited to estimating the effect of the inclusion of different locations in alternative policy scenarios (availability effects). They do not estimate to what extent the utility of one alternative responds to changes in the characteristics of another. The empirical results of these studies are mixed in terms of the resulting substitution and complementarity effects. In a CV study on five different environmental programs, for example, Hoehn and Loomis (1993) find lower WTP values for combinations of

[6]In CV studies, sequencing and scope-sensitivity effects are sometimes attributed to substitution (Carson et al. 2001). However, the resulting magnitude of the scope effect of these studies does not give a valid indicator of substitution effects (Banerjee and Murphy 2005). This is because a priori information about the substitutability of the goods in question is missing and the magnitude of these scope effects has been argued to be too large to be interpreted validly as substitution effects (Bateman et al. 2004a).

[7]In some cases, SP studies find differences in WTP values for goods valued independently and grouped that cannot be explained by economic theory; this is usually referred to by SP critics as framing, part-whole or embedding bias (Carson et al. 1998; Freeman 2003; Hoehn and Loomis 1993).

these programs than for an individual program, implying substitution effects. The results in Cummings et al. (1994) from a CV study about multiple policy programs in the U.S. also suggest substitution between environmental and non-environmental goods. However, Hailu et al. (2000) find that the WTP for a combination of packages is higher than the WTP for the packages separately, reflected through a positive interaction term between the programs, suggesting that goods in the same region are considered complementary goods.

The problem with the CV method with respect to estimating substitution effects is that the number of valuation questions that can be included in surveys and the possibilities to create variation in the scenario, for instance via the number and characteristics of alternative sites, are limited. The effect of on-site characteristics, such as size or recreational facilities, on site-selection behavior and WTP can be assessed only by changing these characteristics in the valuation scenario, or evaluating site selection behavior across a large number of sites that differ in these characteristics (see, for example, Scarpa et al. 2000).

Choice experiments (CE) provide more flexibility to estimate substitution effects, because the CE design can include different alternatives, varying their characteristics over the choice tasks (Boxall et al. 1996; Rolfe et al. 2002). CE data are analyzed using the same random utility model (RUM) framework as recreation demand and travel cost studies. RUM-based travel cost studies analyze choices among multiple destinations with different site characteristics (Parsons 2003). Hence, if there are multiple relevant sites apart from the main study site, a CE could be a suitable technique to assess substitution effects of changes in the price, quality or availability of substitutes by designing the experiment as a site choice study. However, in the existing literature almost all multiple site studies, which focus on choices among sites as a function of site access and other site characteristics, such as water quality changes, are based on RP data (Kaoru 1995; Needelman and Kealy 1995; Parsons and Massey 2003). Few CE studies in the SP environmental valuation literature have focused on substitution effects among sites, one example being the CE study of Rolfe et al. (2002), in which respondents are asked to choose between rainforests in different continents across the world.

Information about tradeoffs among sites can be obtained directly only by asking people to choose among sites and including the distance to those sites in the analysis. However, *a posteriori* options to assess effects of alternatives that are not included in the good under valuation, i.e. not in the choice set, are also possible. The WTP function can be extended with a variable that reflects the presence of substitutes, including their number (e.g. Brown and Duffield 1995) or size (e.g. Brander et al. 2012; Jones et al. 2010; Pate and Loomis 1997; Sen et al. 2013). Variables reflecting the quantity of a good can be used as a proxy to account for the scarcity of a good, but do not provide information about substitution behavior in terms of the effect of changes in the price or quality of the available substitutes on WTP. Alternatively, the cost of visiting other sites can be used. Assuming that the distance to substitutes is a valid proxy for the WTP of the substitute sites of the good under valuation, Bateman et al. (2011a) include the distance to other sites in the value function of sites that undergo water quality improvement works and find that respondents with more alternatives nearby hold lower WTP values, as theoretically expected. The accuracy of WTP estimates in primary SP studies could potentially increase by including such relative scarcity indicators in the WTP model, and make them more suitable for BT studies.

A challenge in estimating substitution effects is defining the relevant choice set of alternatives, i.e. the number of substitutes that can and should be included in the design and analysis. Choice set specification is important as it can affect parameter estimates (DeShazo and Fermo 2002; Pellegrini et al. 1997) and resulting economic welfare estimates (Parsons and Hauber 1998). In most spatial choice studies, there are many possible alternatives (Cummings et al. 1994) and the boundaries of the study area are difficult to define. Large choice sets can be problematic for modeling in RP studies, whereas CE studies must limit the choice set size even further to reduce task complexity for respondents. In addition, the relevant set of substitutes may vary across goods and sampled populations (Bhat and Zhao 2002; Boyle and Bergstrom 1999). Schaafsma and Brouwer (2013) test whether varying the size of the choice affects WTP values, but do not find significant effects when respondents are asked to evaluate smaller subsets of options before evaluating larger sets. The choice set composition problem affects not only the results of primary studies, but is also relevant to BT studies using function transfer, as the substitutes for the policy site need to be identified as well. Ideally, the same selection process is used for the BT exercise, so the process should be transferable.

18.4 Distance Decay

Distance decay implies that the further the respondent lives away from the site under valuation, the less (s)he is willing to pay for conserving or improving this site. This effect is expected to be most prominent in the WTP for environmental goods with mainly recreational or other use values. The use of distance decay functions in SP research primarily mirrors the RP valuation work based on travel cost studies. The main theoretical expectation regarding the effect of distance on WTP is that as distance increases, travel cost to the site increases, and in turn demand for the site decreases. Although this effect is expected to be found for both the policy and the study site, and therefore important to be reflected in benefit transfer studies, primary SP studies do not always account for distance decay in the WTP for the study site, and moreover, the distance decay effect is not necessarily the same for the policy site. Bateman et al. (2006) discusses implications of these issues for BT.

One of the main reasons to account for distance decay is to determine the size of the economic "market" of the environmental good. This market reflects the geographic area where the affected population lives over which (non-zero) WTP values can be aggregated to calculate the total welfare change to society (Loomis 1996). Simply put, it is the geographic area over which WTP is greater than zero. Bateman et al. (2002) propose distance decay as a validity check and its examination as a minimum requirement for SP studies. But only about 35 SP studies, which vary widely in the environmental goods, countries and populations they cover, account for the effect of distance on WTP. Approximately 85 % of these studies find significant distance effects (Schaafsma 2010, updated).

In addition to the direct travel cost effect, a number of alternative explanations for distance decay have been suggested. In SP studies, the demand function is not estimated based on the number of trips, as in travel cost studies of observed behaviors. Nevertheless, distance is often inversely correlated with visitation rates, length of residency, information and knowledge about the good under valuation. These variables are common explanatory variables in many SP studies (e.g., Bateman et al. 2005; Concu 2005; Pate and Loomis 1997). As people live closer to the study site, they are more likely to know more about the site, either via their own visits or those by friends and family (Sutherland and Walsh 1985).

Although distance decay of use values has a clear theoretical explanation, there is no theoretical expectation for declining nonuse values as distance increases. Empirical results in the SP literature are mixed and partly obscured because distinguishing pure nonuse from use values is difficult and often mixed with discussions of differences between users and nonusers. No distance decay effect among nonusers was found for national parks (Barrick and Beazley 1990) or symbolic species that people rarely see, such as seals in the Netherlands (Bulte et al. 2005). The uniqueness of these goods may call for a protection status, leading to widely spread knowledge about the good, and implies that there are likely to be few substitutes. For salmon, also a symbolic species, Loomis (1996, 2000) finds a weak distance decay effect (i.e. "flat" distance decay function) and Pate and Loomis (1997) do not find any distance decay effect. However, in contrast to theoretical expectations, some studies find distance decay in nonuse values held by users (Sutherland and Walsh 1985), while other studies find distance decay in WTP-estimates of nonusers (Bateman et al. 2000; Hanley et al. 2003), which may reflect option values expressed by current nonusers (Bateman et al. 2006).

Sampling biases may also obscure or bias distance decay estimates (Bateman et al. 2006). Studies using samples that are not based on the economic market area but on a smaller area near the environmental good may produce distance decay effects that cannot be extrapolated beyond the sampling area. To ensure that a distance decay effect can be estimated reliably to demarcate the geographical boundaries of the market, a sampling strategy should be applied that is stratified according not only to population density and socio-demographics, but also to distance and substitute availability variables. Accurate estimation of distance decay effects requires large sampling areas, which cover, where possible, the population that holds a WTP close to zero. This may also ensure that, when aggregating or transferring WTP, the function does not have to be applied to areas farther away that are not covered by the original sample (i.e. extrapolated beyond parameter space). Therefore, when selecting primary studies for BT, the sampling strategy should also be evaluated.

The distance between an environmental good and human populations is also important for studies that transfer the aggregated individual WTP value for sites or values per hectare (e.g., Eade and Moran 1996; Brander et al. 2012) from densely to sparsely populated areas. Environmental goods and services are valued only when there are human beneficiaries present to enjoy them. Aggregate WTP estimates of locally or regionally important goods and services should therefore reflect differences in population density.

Economic theory does not provide unambiguous guidance regarding the functional form of distance decay functions (Ferrini and Fezzi 2012), but those empirical studies that account for distance decay mostly use exponential and log linear functions. The little empirical evidence and wide variation in findings regarding explanatory factors have a number of implications for benefit transfer studies. For simple mean WTP transfers, no rules of thumb can be defined to identify a minimum or maximum range from the policy site over which the mean WTP estimates from study sites should be applied. The existing empirical studies find widely ranging market sizes. Moreover, testing the assumption put forward by Hanley et al. (2003) that distance decay differs between users and nonusers and use and nonuse values, Schaafsma et al. (2013) find that achieving a similar environmental objective may be associated with different values and distance decay functions when the environmental goods and services are provided by different types of sites. This implies that function transfer exercises using a primary study that includes a simple distance decay effect may also lead to transfer errors, especially when the type of value (use or nonuse), the road network, or the availability of substitutes differs.

Substitutability and distance effects are interdependent (see Fig. 18.2). Substitute sites are usually not randomly distributed over space in terms of quantity and quality. As distance from the site or the geographical scale of the study increases, the number of substitutes is likely to increase too. The availability of substitutes is one of the possible causes of distance decay. Similarly, distances between alternatives and between alternatives and respondents influence the substitutability of sites in the same geographical market.

Fig. 18.2 Combined effect of distance and substitutes

The first graphs shows declining marginal utility in the provision level of environmental goods: an additional unit of an environmental good is valued lower than the previous unit as the number of goods (quantity) expands, reflecting substitution. The second graph shows distance decay. The third graph shows the combined effect of distance and substitutes (i.e. the first two graphs) when substitute availability increases with distance.

The distribution of alternative locations over space and their possible substitutability or complementarity (e.g. in the case of multiple-site trips) is expected to cause variation in substitutability and distance decay across regions (Moran 1999). This is depicted in Fig. 18.3, in which the gray-scaled circles around the location of interest L1 reflect WTP-categories from the highest values (darker) to the lowest values (lighter). The rectangles represent different lakes (L1, L2, L3) providing environmental goods and services. The grey scales in the figure reflect different WTP categories: higher values are reflected by darker, and lower values by lighter grey. On the left-hand side of the figure, the distance decay function is equal in all directions, whereas on the right-hand side, substitute sites cause a higher distance decay effect, and therefore lower WTP values and smaller markets, in the direction of other sites, such as L2 and L3. Thus, the figure shows that distance decay functions will not typically be uniform, but depend on the location of the substitutes. Accounting for such variation can improve the accuracy of WTP estimates, but requires a two-dimensional analysis.

Schaafsma et al. (2013) test the assumptions of continuous, spatially homogeneous distance decay, using the geographical expansion method first applied by Cameron (2006) in an hedonic pricing study. The results of a CE on water quality improvements in the Netherlands show clear heterogeneity in WTP across space. These differences would lead to different results in aggregation for some areas and would therefore have considerable policy consequences were the values to be used in price or tax incentives. Schaafsma et al. (2012) account for heterogeneity in distance decay by subdividing the sample of respondents into different compass directions (northeast, southwest, etc.) and including the directions as dummy variables in the WTP model. The mapped findings reveal a WTP pattern that is

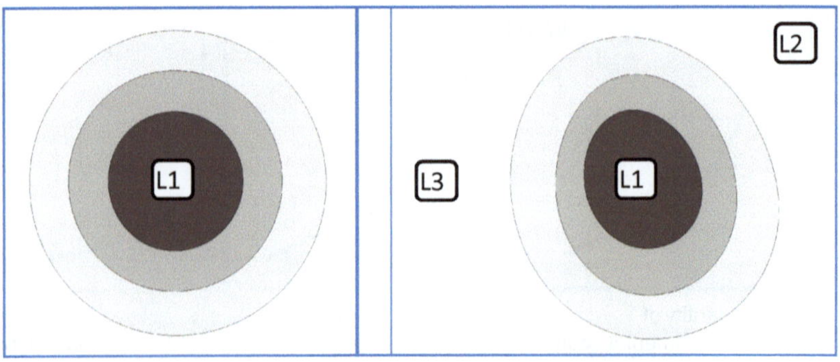

Fig. 18.3 Distance decay without (*left*) and with (*right*) substitution effects

sensitive to the location of the available substitutes relative to the site of interest. However, such dummy variables are not suitable for transfer to other policy areas because the spatial distribution of substitute sites at the policy site may be different.

WTP models of choice studies, where respondents choose between sites, would ideally allow for substitution patterns flexible enough to control for the spatial distribution of alternatives. Error components models can allow for correlation in the unobserved variation across geographically nearby locations by adding error components to the utility functions of locations near one another (Herriges and Phaneuf 2002). For example, Termansen et al. (2008) apply this approach in a recreation choice study using revealed preference data and find a significant improvement in model fit.

18.5 Other Spatial Factors and Variation in Individual WTP

Apart from distance and substitute effects, there are differences in the geographical context of policy and study sites that may lead to differences in individual WTP across space. For example, the surrounding landscape can affect public preferences for a site. Liekens et al. (2013) show that a nature site is valued higher when sharing a boundary with another nature site, whereas sites surrounded by industrial areas are valued lower. They develop a transferable value function that accounts for the effect of different types of surrounding land use on WTP for nature areas, in addition to distance and size. In a CE study by Johnston et al. (2002), respondents show sensitivity to spatial attributes and the spatial configuration of alternative land use scenarios, even when these are not mentioned explicitly in the text.

In another example, Brouwer et al. (2010) show that preferences for water quality improvements differ across regions. Taking this regional heterogeneity into account results in lower transfer errors (Martin-Ortega et al. 2012). Whenever goods have local importance due to a cultural or other association associated with political or social areas, WTP is likely to fall beyond associated borders (e.g., national borders). Such discontinuous patterns in WTP for "local" goods are not only related to distance, but may also reflect other underlying preferences such as a sense of place or belonging, a "sense of ownership" (Bateman et al. 2004b) or "spatial identity" (Hanley et al. 2003). These studies include a dummy indicator for zones (in kilometer ranges around the asset or for administrative zones) to indicate whether the respondent is a resident of the country, province or state in which the good is located.

Alternatively, latent class models that cluster respondents by spatial unit (e.g. region) can also be used to reveal differences in preferences in WTP across space. This approach is applied, for example, in Garrod et al. (2012), who show that respondents in the UK tend to prefer agri-environmental projects in landscapes nearest to where they live. Transferring the values of a habitat type where this type is abundant to another area where this type is less common may thus result in transfer errors, even when other characteristics of the study and policy sites are similar (similar to results found by Brouwer and Spaninks 1999).

To reveal spatial heteroskedasticity (i.e., a spatial pattern in the variance of the model), spatial analysis techniques have been used in combination with SP data. As an indicator of remaining spatial heterogeneity, Campbell et al. (2008) use the Moran's I statistic, which is a two-dimensional measure of spatial autocorrelation, accounting for correlation between nearby locations in space. Their findings suggest that benefits of landscape improvements are not spatially uniform, but instead clustered in homogenous regions. However, Meyerhoff (2011) finds only weak spatial autocorrelation among individual WTP estimates in a CE study on preferences for wind turbines when estimating Moran's I statistics. Johnston and Ramachandran (2014) include an indicator of local spatial association in a mixed logit model to analyze the spatial distribution of the individual-specific parameter estimates in a study of migratory fish passage restoration on Rhode Island. They find that the implicit prices have local "hot spots", where respondents have significantly higher WTP compared to other areas, a pattern which cannot be revealed using distance decay analysis. These analyses, particularly the latter one, can provide relevant information to policymakers and help reveal where voters may support environmental projects. However, the results of these statistical analyses are site-specific and not directly useable in BT studies. As long as the underlying drivers of these local hot spots in WTP are not identified, it is very difficult if not impossible to identify similar hot spots in new policy sites. This is an important area for future research.

18.6 Concluding Remarks

Understanding the spatial nature of the environmental services and the spatial distribution of the associated benefits is paramount for reliable estimation of both individual and total WTP and the identification of the relevant population of beneficiaries. Such understanding will improve the reliability of original value estimates for study sites, but is also required to inform the adjustments necessary for valid and accurate BT.

This chapter has reviewed the SP literature to assess how SP studies account for spatial and geographic effects on WTP. As the SP literature has focused extensively on the problems and biases associated with WTP responses related to the valuation of single goods or sites, relatively little attention has been paid to the spatial context in which environmental goods are often embedded. When available, primary studies that account for distance and substitution effects should be selected for BT purposes.

It has not become the norm yet in the scientific SP literature to parameterize substitution and distance effects in WTP models. Broader modeling of these effects could significantly improve the potential transferability of WTP values, as additional control could be included for variation across policy and study sites for the availability and quality of substitutes. The variation in empirical findings of distance decay effects suggests that when this effect is expected to be present in WTP for goods with use values, distance decay functions will vary across (policy and study) sites and populations. One of the main gaps in the SP literature is evaluating the effects of substitutes on WTP. Providing information about the good under valuation, its substitutes and their locations, and including questions about respondents' use and perception of these alternatives, may increase respondents' consideration of substitutes and thereby the chances of detecting significant substitution effects in primary studies. That would give BT studies the possibility to control for differences in supply and scarcity between study and policy sites, thereby potentially improving transfer accuracy. However, the literature has yet to demonstrate whether the incorporation of information on substitutes enhances transfer accuracy in practice. This is an important area for future work.

These substitution effects could be especially relevant for large-scale studies in which environmental changes at multiple sites are evaluated. At the same time, the complexity of ecosystem analysis may increase as the scale of analysis increases and involve non-linearities in ecosystem provision. In such cases, the applicability of using fixed values per hectare and the results of small-scale and single-site primary studies for BT purposes has to be evaluated. Often, such practices will lead to significant generalization error.

The increased focus on spatial factors in ecosystem assessment and intensified application of geographic information systems (GIS) in valuation studies (see Chap. 20) may shift more attention (and capacity) toward the assessment of spatial heterogeneity and location-specific values. As this chapter has argued, there is scope to improve primary valuation studies in terms of how spatial factors are included in questionnaire design, the selection of valuation methods, the definition of the good(s) or services under valuation, the sampling strategy and modelling. Estimating transferable and spatially explicit functions should become a primary objective for studies aiming to provide policy relevant results.

References

Arrow, K., Solow, R., Portney, P. R., Leamer, E. E., Radner, R., & Schuman, H. (1993). Report of the NOAA panel on contingent valuation. *Federal Register, 58*(10), 4601–4614.

Banerjee, S., & Murphy, J. H. (2005). The scope test revisited. *Applied Economics Letters, 12*, 613–617.

Barbier, E. B. (2012). A spatial model of coastal ecosystem services. *Ecological Economics, 78*, 70–79.

Barrick, K., & Beazley, R. (1990). Magnitude and distribution of option value for the Washakie Wilderness, northwest Wyoming, USA. *Environmental Management, 14*, 367–380.

Bateman, I. J., & Langford, I. H. (1997). Non-users' willingness to pay for a National Park: An application and critique of the contingent valuation method. *Regional Studies, 31*, 571–582.

Bateman, I. J., Langford, I. H., Nishikawa, N., & Lake, I. (2000). The Axford debate revisited: A case study illustrating different approaches to the aggregation of benefits data. *Journal of Environmental Planning and Management, 43*, 291–302.

Bateman, I. J., Carson, R., Day, B., Hanemann, W. M., Hanley, N., Hett, T., et al. (2002). *Economic valuation with stated preference techniques: A manual*. Cheltenham, UK: Edward Elgar.

Bateman, I. J., Cole, M., Cooper, P., Georgiou, S., Hadley, D., & Poe, G. L. (2004a). On visible choice sets and scope sensitivity. *Journal of Environmental Economics and Management, 47*, 71–93.

Bateman, I. J., Cooper, P., Georgiou, S., Navrud, S., Poe, G. L., Ready, R., Riera, P., Ryan, M., & Vossler, C. A. (2004b). *Scope sensitivity tests for preference robustness: An empirical examination of economic expectations regarding the economic valuation of policies for reducing acidity in remote mountain lakes*. CSERGE Working Paper EDM-04-03. Norwich, UK: University of East Anglia.

Bateman, I. J., Georgiou, S. & Lake, I. R. (2005). *The aggregation of environmental benefit values: A spatially sensitive valuation function approach*. CSERGE Working Paper EDM 05-04. Norwich, UK: University of East Anglia.

Bateman, I. J., Day, B. H., Georgiou, S., & Lake, I. (2006). The aggregation of environmental benefit values: Welfare measures, distance decay and total WTP. *Ecological Economics, 60*, 450–460.

Bateman, I. J., Brouwer, R., Ferrini, S., Schaafsma, M., Barton, D., Dubgaard, A., et al. (2011a). Making benefit transfers work: Deriving and testing principles for value transfers for similar and dissimilar sites using a case study of the non-market benefits of water quality improvements across Europe. *Environmental and Resource Economics, 50*, 365–387.

Bateman, I. J., Mace, G. M., Fezzi, C., Atkinson, G., & Turner, R. K. (2011b). Economic analysis for ecosystem service assessments. *Environment and Resource Economics, 48*, 177–218.

Bauer, D. M., Swallow, S. K., & Paton, P. W. C. (2010). Cost-effective species conservation in exurban communities: A spatial analysis. *Resource and Energy Economics, 32*, 180–202.

Bhat, C. R., & Zhao, H. (2002). The spatial analysis of activity stop generation. *Transportation Research Part B: Methodological, 36*, 557–575.

Boxall, P. C., Adamowicz, W. L., Swait, J., Williams, M., & Louviere, J. (1996). A comparison of stated preference methods for environmental valuation. *Ecological Economics, 18*, 243–253.

Boyle, K. J., & Bergstrom, J. C. (1999). Doubt, doubts, and doubters: The genesis of a new research agenda? In I. J. Bateman & K. G. Willis (Eds.), *Valuing environmental preferences: Theory and practice of the contingent valuation method in the US, EU, and developing countries* (pp. 183–206). Oxford: Oxford University Press.

Brander, L. M., Brauer, I., Gerdes, H., Ghermandi, A., Kuik, O., Markandya, A., et al. (2012). Using meta-analysis and GIS for value transfer and scaling up: Valuing climate change induced losses of European wetlands. *Environmental and Resource Economics, 52*, 395–413.

Brody, S. D., Highfield, W., & Alston, L. (2004). Does location matter? Measuring environmental perceptions of creeks in two San Antonio watersheds. *Environment and Behavior, 36*, 229–250.

Brouwer, R., & Spaninks, F. A. (1999). The validity of environmental benefits transfer: Further empirical testing. *Environmental and Resource Economics, 14*, 95–117.

Brouwer, R., Martin-Ortega, J., & Berbel, J. (2010). Spatial preference heterogeneity: A choice experiment. *Land Economics, 86*, 552–568.

Brown, T. C., & Duffield, J. W. (1995). Testing part-whole valuation effects in contingent valuation of instream flow protection. *Water Resources Research, 31*, 2341–2351.

Bulte, E., Gerking, S., List, J. A., & de Zeeuw, A. (2005). The effect of varying the causes of environmental problems on stated WTP values: Evidence from a field study. *Journal of Environmental Economics and Management, 49*, 330–342.

Cameron, T. A. (2006). Directional heterogeneity in distance profiles in hedonic property value models. *Journal of Environmental Economics and Management, 51*, 26–45.

Campbell, D., Hutchinson, W. G., & Scarpa, R. (2008). Incorporating discontinuous preferences into the analysis of discrete choice experiments. *Environmental and Resources Economics, 41*, 401–417.

Carson, R. T. (2012). Contingent valuation: A practical alternative when prices aren't available. *Journal of Economic Perspectives, 26*, 27–42.

Carson, R. T., Flores, N. E., & Hanemann, W. M. (1998). Sequencing and valuing public goods. *Journal of Environmental Economics and Management, 36*, 314–323.

Carson, R. T., Flores, N. E., & Meade, N. F. (2001). Contingent valuation: Controversies and evidence. *Environmental and Resource Economics, 19*, 173–210.

Concu, G. B. (2005). *Are non-use values distance-independent? Identifying the market area using a choice modelling experiment*. Murray Darling Program Working Paper M05_6. Brisbane: University of Queensland.

Cummings, R. G., Ganderton, P. T., & McGuckin, T. (1994). Substitution effects in CVM values. *American Journal of Agricultural Economics, 76*, 205–214.

DeShazo, J. R., & Fermo, G. (2002). Designing choice sets for stated preference methods: The effects of complexity on choice consistency. *Journal of Environmental Economics and Management, 44*, 123–143.

DeShazo, J. R., Cameron, T. A., & Saenz, M. (2009). The effect of consumers' real-world choice sets on inferences from stated preference surveys. *Environmental and Resource Economics, 42*, 319–343.

Eade, J. D. O., & Moran, D. (1996). Spatial economic valuation: benefits transfer using geographical information systems. *Journal of Environmental Management, 48*, 97–110.

Ferrini, S., & Fezzi, C. (2012). Generalized additive models for nonmarket valuation via revealed or stated preference methods. *Land Economics, 88*, 613–633.

Fisher, B., & Turner, R. K. (2008). Ecosystem services: Classification for valuation. *Biological Conservation, 141*, 1167–1169.

Freeman, A. M, III (2003). *The measurement of environmental and resource values: Theory and methods*. Washington, DC: Resources for the Future Press.

Garrod, G., Ruto, E., Willis, K., & Powe, N. (2012). Heterogeniety of preferences for the benefits of environmental stewardship: A latent class approach. *Ecological Economics, 76*, 104–111.

Hailu, A., Adamowicz, W. L., & Boxall, P. C. (2000). Complements, substitutes, budget constraints and valuation. *Environmental and Resource Economics, 16*, 51–68.

Hanley, N., Schlapfer, F., & Spurgeon, J. (2003). Aggregating the benefits of environmental improvements: Distance-decay functions for use and non-use values. *Journal of Environmental Management, 68*, 297–304.

Herriges, J. A., & Phaneuf, D. J. (2002). Inducing patterns of correlation and substitution in repeated logit models of recreation demand. *American Journal of Agricultural Economics, 84*, 1076–1090.

Hoehn, J. P., & Loomis, J. B. (1993). Substitution effects in the valuation of multiple environmental programs. *Journal of Environmental Economics and Management, 25*, 56–75.

Hoehn, J. P., & Randall, A. (1989). Too many proposals pass the benefit cost test. *The American Economic Review*, 544–551.

Johnston, R. J., & Ramachandran, M. (2014). Modeling spatial patchiness and hot spots in stated preference willingness to pay. *Environmental and Resource Economics*, 59, 363–387. Online version available, doi:10.1007/s10640-013-9731-2.

Johnston, R. J., & Rosenberger, R. S. (2010). Methods, trends and controversies in contemporary benefit transfer. *Journal of Economic Surveys, 24*, 479–510.

Johnston, R. J., Swallow, S. K., & Bauer, D. M. (2002). Spatial factors and stated preference values for public goods: Considerations for rural land use. *Land Economics, 78*, 481–500.

Jones, A., Wright, J., Bateman, I. J., & Schaafsma, M. (2010). Estimating arrival numbers for informal recreation: A geographical approach and case study of British woodlands. *Sustainability, 2*, 684–701.

Jørgensen, S. L., Olsen, S. B., Ladenburg, J., Martinsen, L., Svenningsen, S. R., & Hasler, B. (2013). Spatially induced disparities in users' and non-users' WTP for water quality improvements—Testing the effect of multiple substitutes and distance decay. *Ecological Economics, 92*, 58–66.

Kaoru, Y. (1995). Measuring marine recreation benefits of water-quality improvements by the nested random utility model. *Resource and Energy Economics, 17*, 119–136.

Koelemeijer, K., & Oppewal, H. (1999). Assessing the effects of assortment and ambience: A choice experimental approach. *Journal of Retailing, 75*, 319–345.

Kotchen, M. J., & Reiling, S. D. (1999). Do reminders of substitutes and budget constraints influence contingent valuation estimates? Another comment. *Land Economics, 75*, 478–482.

Liekens, I., Schaafsma, M., De Nocker, L., Broekx, S., Staes, J., Aertsens, J., & Brouwer, R. (2013). Developing a value function for nature development and land use policy in Flanders, Belgium. *Land Use Policy, 30*, 549–559.

Loomis, J. (1996). How large is the extent of the market for public goods: Evidence from a nationwide contingent valuation survey. *Applied Economics, 28*, 779–782.

Loomis, J. (2000). Vertically summing public good demand curves: An empirical comparison of economic versus political jurisdictions. *Land Economics, 76*, 312–321.

Loomis, J., Gonzalez-Caban, A., & Gregory, R. (1994). Do reminders of substitutes and budget constraints influence contingent valuation estimates? *Land Economics, 70*, 499–506.

Martin-Ortega, J., Brouwer, R., Ojea, E., & Berbel, J. (2012). Benefit transfer and spatial heterogeneity of preferences for water quality improvements. *Journal of Environmental Management, 106*, 22–29.

Meyerhoff, J. (2011). Do turbines in the vicinity of respondents' residences influence choices among programmes for future wind power generation? In *International Choice Modelling Conference 2011*, Leeds, UK.

Mitsch, W. J., & Gosselink, J. G. (2000). The value of wetlands: Importance of scale and landscape setting. *Ecological Economics, 35*, 25–33.

Moran, D. (1999). After Axford: What lessons for environmental valuation in the UK? CSERGE Working Paper PA-1999-01. Norwich, UK: University of East Anglia.

Needelman, M. S., & Kealy, M. J. (1995). Recreational swimming benefits of New Hampshire lake water quality policies: An application of a repeated discrete choice model. *Agricultural and Resource Economics Review, 24*, 78–87.

Neill, H. R. (1995). The context for substitutes in CVM studies: Some empirical observations. *Journal of Environmental Economics and Management, 29*, 393–397.

Newbold, S. C., & Massey, D. M. (2010). Recreation demand estimation and valuation in spatially connected systems. *Resource and Energy Economics, 32*, 222–240.

Parsons, G. R. (2003). The travel cost model. In P. A. Champ, K. J. Boyle, & T. C. Brown (Eds.), *A primer for nonmarket valuation* (pp. 269–329). Dordrecht: Kluwer Academic Publishers.

Parsons, G. R., & Hauber, A. B. (1998). Spatial boundaries and choice set definition in a random utility model of recreation demand. *Land Economics, 74*, 32–48.

Parsons, G. R., & Massey, D. M. (2003). A RUM model of beach recreation. In N. Hanley, W. D. Shaw, R. E. Wright, & J. C. Coleman (Eds.), *The new economics of outdoor recreation*. Cheltenham, UK: Edward Elgar.

Pate, J., & Loomis, J. (1997). The effect of distance on willingness to pay values: A case study of wetlands and salmon in California. *Ecological Economics, 20*, 199–207.

Pellegrini, P. A., Fotheringham, A. S., & Lin, G. (1997). An empirical evaluation of parameter sensitivity to choice set definition in shopping destination choice models. *Papers in Regional Science, 76*, 257–284.

Perrings, C., & Hannon, B. (2001). An introduction to spatial discounting. *Journal of Regional Science, 41*, 23–38.

Rolfe, J., Bennett, J., & Louviere, J. (2002). Stated values and reminders of substitute goods: Testing for framing effects with choice modelling. *Australian Journal of Agricultural and Resource Economics, 46*, 1–20.

Rollins, K., & Lyke, A. (1998). The case for diminishing marginal existence values. *Journal of Environmental Economics and Management, 36*, 324–344.

Scarpa, R., Hutchinson, W. G., Chilton, S. M., & Buongiorno, J. (2000). Importance of forest attributes in the willingness to pay for recreation: A contingent valuation study of Irish forests. *Forest Policy and Economics, 1*, 315–329.

Schaafsma, M. (2010). Spatial effects in stated preference studies for environmental valuation. Ph. D. Thesis,Vrije Universiteit Amsterdam.

Schaafsma, M., & Brouwer, R. (2013). Testing geographical framing and substitution effects in spatial choice experiments. *Journal of Choice Modelling, 8*, 32–48.

Schaafsma, M., Brouwer, R., & Rose, J. (2012). Directional heterogeneity in WTP models for environmental valuation. *Ecological Economics, 79*, 21–31.

Schaafsma, M., Brouwer, R., Gilbert, A., van den Bergh, J. C. J. M., & Wagtendonk, A. (2013). Estimation of distance-decay functions to account for substitution and spatial heterogeneity in stated preference research. *Land Economics, 89*, 514–553.

Sen A., Darnell A., Crowe A., Bateman, I. J., Munday, P., & Foden J. (2013). Economic assessment of the recreational value of ecosystems in Great Britain. Forthcoming in *Environmental and Resource Economics.*

Sutherland, R. J., & Walsh, R. G. (1985). Effect of distance on the preservation value of water quality. *Land Economics, 61*, 281–291.

TEEB. (2010). Integrating the ecological and economic dimensions in biodiversity and ecosystem service valuation. Chapter 1 in *The economics of ecosystems and biodiversity: The ecological and economic foundations.* http://www.teebweb.org.

Termansen, M., Zandersen, M., & McClean, C. J. (2008). Spatial substitution patterns in forest recreation. *Regional Science and Urban Economics, 38*, 81–97.

Turner, R. K., Morse-Jones, S., & Fisher, B. (2010). Ecosystem valuation: A sequential decision support system and quality assessment issues. *Annals of the New York Academy of Sciences, 1185*, 79–101.

Whitehead, J. C., & Blomquist, G. C. (1999). Do reminders of substitutes and budget constraints influence contingent valuation estimates? Reply to another comment. *Land Economics, 75*, 483–484.

Willis, K. G., & Garrod, G. D. (1993). Valuing landscape: A contingent valuation approach. *Journal of Environmental Management, 37*, 1–22.

Chapter 19
Reliability of Meta-analytic Benefit Transfers of International Value of Statistical Life Estimates: Tests and Illustrations

Henrik Lindhjem and Ståle Navrud

Abstract If there are no applicable domestic studies, there are many ways to utilize the international literature to conduct benefit transfer (BT). In the health economics literature simple unit transfer methods, rather than function-based methods, are the most commonly used. In this chapter we utilize a large database of Value of a Statistical Life (VSL) estimates, derived from stated preference studies worldwide, to investigate the reliability of meta-analytic BT (MA-BT) and compare this method with simple unit transfers in a case study illustration. Meta-regression analysis is a way to estimate how different policy-relevant factors affect VSL and is thought to improve accuracy in BT. We discuss in particular how different quality criteria to screen available studies and VSL estimates may influence BT accuracy. Results show that quality screened MA-BT models give lower transfer errors, and in the case study example MA-BT methods achieve accuracy gains over the use of unit transfer methods. However, the unscreened MA-BT method achieved around the same accuracy as the best unit transfers based on quality screened data. Hence, transfer accuracy may in some contexts depend as much on the quality of the underlying data as on the BT method itself.

Keywords Meta-analysis · Benefit transfer · Value of statistical life · Quality screened data

H. Lindhjem (✉)
Norwegian Institute for Nature Research (NINA), Vista Analysis, Oslo, Norway
e-mail: henrik.lindhjem@nina.no

S. Navrud
School of Economics and Business, Norwegian University of Life Sciences, Akershus, Norway
e-mail: stale.navrud@nmbu.no

R.J. Johnston et al. (eds.), *Benefit Transfer of Environmental and Resource Values*, The Economics of Non-Market Goods and Resources 14,
DOI 10.1007/978-94-017-9930-0_19

19.1 Introduction

Many countries have a relatively small body of national valuation literature to use as a basis for benefit transfer (BT). A solution to this problem is to expand the information base to include relevant international valuation literature of acceptable quality, and then use a BT method to derive and transfer a suitable welfare estimate to the national context in question. For environmental goods, such as water quality, wetlands, coral reefs, forest conservation benefits etc., meta-analytic benefit transfer (MA-BT) has become an increasingly common method, at least for academic investigations of reliability (e.g., Brander et al. 2012; Johnston and Thomassin 2010; Lindhjem and Navrud 2008; Londoño and Johnston 2012; Stapler and Johnston 2009). Although the evidence is mixed, for the international context at least, the emerging consensus seems to be that such function-based transfers outperform unit value transfers (Kaul et al. 2012; Rosenberger and Stanley 2006).

For some reason, and as pointed out by Johnston and Rosenberger (2010), the health economics literature deriving value of statistical life (VSL) estimates from stated or revealed preference studies seems to be more doubtful of the potential reliability gains from function-based benefit transfer. Both the academic literature and many international agencies, such as the World Bank, still emphasize simple unit value transfer, typically adjusted only by gross domestic product (GDP) differences between countries. To our knowledge, there are very few examples of function-based transfers in the health economics literature. Brouwer and Bateman (2005) is a first application of a standard function-based transfer of willingness to pay (WTP) for health risk reductions (rather than VSL directly), while Dekker et al. (2011) uses a Bayesian meta-model of VSL for BT. Preceding Dekker et al., OECD initiated a project compiling a large database of VSL estimates from stated preference studies worldwide, resulting in several preliminary reports and two final publications using meta-analysis (see Lindhjem et al. 2011; OECD 2012).[1]

This chapter utilizes the OECD database of VSL estimates to investigate two issues that in our opinion are important, and so far under-appreciated, in MA-BT research in general, and in the health economics literature in particular. It is generally agreed that the reliability of any function-based transfer depends on the quality of the primary research upon which it is based. However, very few studies try to operationalize measures of quality and investigate the impact of data screening on BT accuracy. In the literature, it is typically unclear which quality criteria have been used to include or exclude estimates from a meta-database, and the effect of such often subjective choices remains hidden. Finding appropriate criteria to determine which studies and estimates are acceptable for inclusion in meta-analysis when used as the basis for policy formulation has been emphasized by the Environmental Economics Advisory Committee of the US EPA's Science Advisory Board (Morgan and Cropper 2007). Using the different quality criteria

[1]See full dataset and publications at http://www.oecd.org/env/tools-evaluation/vsl.htm.

discussed in Lindhjem et al. (2011), the first issue we investigate is whether quality screened meta-regression models perform better when used for BT.

The second issue we investigate and illustrate is how the different meta-regression models perform in comparison with different types of naïve or simple unit transfers. Although we do not conduct a comprehensive and definitive analysis, for this dataset of around 850 VSL estimates we provide an illustrative example, useful for practitioners, to compare transfer accuracy between methods.

The outline of the chapter is as follows. The next section explains two ways to define benefit transfer accuracy or reliability, one relative and one absolute. We then explain briefly the testing procedure and logic of the illustration we use to compare reliability or accuracy between different MA-BT applications and unit value transfer methods. We use the terms "reliability" and "accuracy" interchangeably for the extent to which a transferred estimate is of a similar value to a pre-determined, benchmark value (true value), representing a policy context. Section 19.3 presents the meta-dataset of VSL estimates, the variables used in the meta-regressions, the quality screening criteria and the estimated models that are later used for BT. Section 19.4 then first conducts some simple tests of the reliability of using these models for BT, before going through a comprehensive illustration comparing results with unit value transfers. Section 19.5 concludes and points to some further research needs. Although this study is explorative and not comprehensive in judging the use of MA-BT, we find that the accuracy of the quality screened MA-BT methods tested is higher than for the various unit value transfers.

19.2 Reliability Tests and Testing Procedures for VSL Transfers

19.2.1 Defining VSL and Reliability and Transfer Error

The value of statistical life (VSL) is a summary measure of the willingness to pay (WTP) for a mortality risk reduction, and a central input into the calculation of benefits of policies that save lives. Risk reduction benefits are computed as VSL × L, where L is the expected number of lives saved by the policy. VSL is the marginal value of a reduction in the risk of dying (prematurely), and is therefore defined as the rate at which people are prepared to trade off income for risk reduction:

$$VSL = \frac{\partial WTP}{\partial R} \tag{19.1}$$

where R is the risk of dying. The VSL can also be described as the total WTP by a group of N people experiencing a uniform reduction of $1/N$ in their risk of dying. Consider, for example, a group of 10,000 individuals, and assume that each of them is willing to pay USD 50 to reduce his, or her, own risk of dying by 1 in 10,000.

The VSL for the population group implied by this WTP is USD 50/0.0001, or USD 500,000.

To measure the reliability and validity of BT, it is common to use the concept of (relative) transfer error (TE), which for VSL normally would be defined as:

$$TE_R = \frac{|VSL_T - VSL_B|}{VSL_B} * 100\% \tag{19.2}$$

where T = transferred (predicted) value, either from a function or unit value transfer and, B = estimate of the true (but unknown) value ("benchmark") at national policy site. TE is most commonly defined in percentage terms and measures by how many percent the estimated and transferred value "miss" the true value for a particular policy context, assuming that one could know what this true value is. Convergent validity studies testing transfer errors often use a benchmark value for this true value, for example the VSL estimate from an accurate or robust study, and then test how different BT techniques perform when predicting this value.

BT validity assessments have shifted focus from testing whether the two estimates are statistically indistinguishable to the concept of reliability for policy use, which requires that TE is relatively small but not necessarily zero (Czajkowski and Scasny 2010; Navrud and Ready 2007). This is because the most appropriate null hypothesis is that TE is larger than zero since theory predicts that values for environmental and other benefits should vary between contexts for many reasons (Kristofersson and Navrud 2005). There is no agreement on maximum TE levels for BT to be reliable for different policy applications, though 20 and 40 % have been suggested (Kristofersson and Navrud 2007). The required accuracy would also depend on the use (Navrud and Pruckner 1997), and sensitivity of decisions to differences in the size of estimated benefits, e.g., whether costs and benefits are close.

While relative TE is the most commonly used measure of transfer reliability or accuracy, it may not always be the most appropriate measure. For example, in cost-benefit analysis it is the absolute, rather than the relative difference, that matters to the cost-benefit ratio. When assessing the relative TEs for MA-BT models, small differences between benchmark and predicted values where low values are involved translate into large transfer errors in relative terms (Lindhjem and Navrud 2008). We therefore also use an absolute measure of TE, in which we do not distinguish between missing the true value on the positive or negative side:

$$TE_A = |VSL_T - VSL_B| \tag{19.3}$$

19.2.2 Testing Procedure and Illustration

The OECD database of VSL estimates was constructed to conduct the meta-regression analyses that are used in this chapter, both as the basis for BT and for the measurement of transfer reliability. First, we use a data splitting technique in which *N* different MA-BT functions were estimated utilizing $N - 1$ of the VSL data for each run, since the *N*th VSL estimate the model predicts is taken out. This *N*th VSL estimate represents for each run the true value, and is the benchmark used to assess how well each MA model predicts. Then the overall mean and median TE_R and TE_A for all the *N* model runs are calculated, with the former sometimes termed the mean and median Absolute Percentage Error (Brander et al. 2006). Simple and more comprehensive versions of the meta-regressions, for different types of quality screened datasets, are used for the BT tests. We will return to the model details below.

We also use a simple case study illustration to more closely mimic an actual BT situation. A single VSL estimate is drawn randomly from one study to represent a benchmark, unknown VSL value for a specific mortality risk reduction policy under assessment. This is assumed to be the true VSL. The next step is then to use the other studies to transfer a best VSL estimate to that policy context using different BT techniques, and compare the TE. The procedure we use here builds on Chapter 4 in OECD (2012) and Lindhjem and Navrud (2008). Before presenting the results of these BT tests, we briefly describe the underlying dataset and MA models used for BT.

19.3 Meta-analysis Data and Regression Models

19.3.1 Metadata and Variable Definitions

The OECD database of VSL estimates contains about 850 estimates of sample mean adult VSL estimates from Stated Preference (SP) surveys conducted in 38 countries around the world, with final entries made in 2010. It covers SP surveys[2] of WTP for mortality risk reductions in the environmental, health, or traffic context in the period from 1970 to 2008. All VSL estimates were adjusted for inflation to 2005 values in national currencies and then converted to 2005 U.S. dollars, using purchasing power parity (PPP) factors. Information was then extracted from the studies and coded into a meta-dataset containing more than 50 variables. After extensive preliminary analysis, the final variable set was chosen for this study, as given in Table 19.1, divided into three variable categories: (1) risk valuation context, (2) methodological choices, and (3) income and survey year (see also Lindhjem et al. 2011). The third column (Sign) gives the expected relationship with VSL, and the final column gives the mean and standard deviation (SD).

The mortality risk change presented to respondents in the SP surveys was normalized to an annual risk change to ensure commensurability. The risk change affects a private individual, his or her household, or the general public. There are

[2]Values are mostly sourced from the contingent valuation method, but also from other SP approaches like choice experiments (CE).

Table 19.1 Meta-analysis variables, expected VSL relationships and descriptive statistics

Variable	Description	Sign	Mean (SD)[b]
Dependent variable			
lnvsl	Natural logarithm of sample mean VSL in PPP-adjusted USD 2005 (mean, annual WTP divided by annual risk change, PPP-adjusted based on AIC[a])		14.50[c] (1.59)
Risk valuation context variables			
lnrchrisk	Continuous: Log of change in mortality risk on an annual basis per 1000 (normalized per year from study info)	0	−8.48[d] (2.13)
public	Binary: 1 if public good; 0 if private (risk affects only the individual asked or his/her household)	±	0.30 (0.46)
envir	Binary: 1 if environment-related risk change; 0 if health-related	?	0.24 (0.42)
traffic	Binary: 1 if traffic-related risk change; 0 if health-related	?	0.30 (0.45)
latent	Binary: 1 if risk change occurs after a certain time; 0 if the risk change is immediate	±	0.14 (0.35)
cancerrisk	Binary: 1 if reference to cancer risk in survey; 0 if not	+	0.13 (0.34)
household	Binary: 1 if WTP is stated on behalf the household; 0 if WTP is only for the individual asked	+	0.29 (0.45)
Methodological variables			
noexplan	Binary: 1 if no visual tool or specific explanation of the risk change was used in survey; 0 if otherwise	+/?	0.14 (0.33)
turnbull	Binary: 1 if WTP was estimated using Turnbull, nonparametric method; 0 parametric method	–	0.04 (0.20)
Income and survey year			
lngdp	Continuous: Log of GDP/capita, USD 2005, PPP-adjusted based on AIC[a]	+	9.65 (0.86)
lnyear	Continuous: Log of year of data collection, adjusted to start at log2 for earliest survey included from 1970	±	3.41 (0.32)

Source Reproduced from Lindhjem et al. (2011)
[a]*PPP*—purchasing power parity. *AIC*—actual individual consumption
[b]Mean and standard deviation (SD) are for overview purposes and sake of brevity given only for the whole, unscreened data set of 856 estimates
[c]This translates into a mean VSL of around 1.98 million USD 2005
[d]625 estimates contain information about the risk change valued

two variables controlling for risk context (environment and traffic, while health is the hidden category). We also control for a potentially positive "cancer premium" effect based on the fear of cancer and a theoretically ambiguous effect of risk latency.[3] The relationship between the size of the risk change and WTP should from

[3]People are known to discount the future at a positive rate. But their utility will also vary at different ages in ways that can make WTP higher to reduce future mortality risks than to reduce immediate risks (Hammitt and Liu 2004).

theory be positive and approximately proportional (Hammitt 2000). VSL should therefore largely be unaffected by the change in risk, at least for small changes and for low baseline risks. However, primary stated preference studies typically find that people's WTP is quite insensitive to the size of the risk change; i.e., they fail internal and/or external scope tests (Hammitt and Graham 1999). This means that questions involving smaller risk changes tend to result in higher VSL estimates, which we also find (see below). We use the information about scope sensitivity in screening the data, as one could argue that those studies that find scope sensitivity may have been more successful in explaining risk changes to respondents. This explanation may have involved indicating probabilities using square grids or other tools and making sure respondents are trained in understanding probabilities before embarking on the valuation task.

Of socioeconomic and other variables, only GDP per capita (adjusted using actual PPP correction in the same way as the VSL estimates) and the year of the survey (for one of the meta-regressions), were retained. Most studies report mean (household or individual) income from the total sample, but not for subsamples from which many of our estimates typically are derived. In order not to lose these estimates, we use GDP per capita instead as a proxy for individual income.[4] Further details of the dataset and discussion of the expected relationships of variables with VSL are given in Lindhjem et al. (2011) and OECD (2012).

19.3.2 Meta-regression Approach

The following meta-regression model, based on fairly standard practice in the MA literature, is used:

$$lnvsl_{si} = \beta_0 + \beta_1 lngdp_{si} + \sum_k \beta_k X_{si}(k) + \varepsilon_{si} \tag{19.4}$$

where lnvsl_{si} is the natural logarithm of VSL for estimate i from survey group s; $lngdp_{si}$ is the natural logarithm of per capita GDP, and X_{si} is a vector of other explanatory variables, as explained in Table 19.1. This model is estimated using ordinary least squares (OLS). Since the number of estimates varies widely across survey groups s, the OLS is weighted by the reciprocal of the number of estimates in each group, so as to weight each survey group equally (rather than giving equal weight to each individual VSL estimate) (see for example Mrozek and Taylor 2002).[5]

[4]As noted in Lindhjem et al. (2011), the correlation between log of GDP and log of reported sample income was found to be very high, so GDP per capita is a good proxy for this purpose.

[5]In Lindhjem et al. (2011), precision weights based on the standard deviation was also used. This is the weighting scheme recommended by USEPA (2006), but it is difficult to apply in practice, since many studies do not report the necessary information. For simplicity, and not to lose too many estimates, we do not do this here.

Further, a "cluster" option is used for estimating robust standard errors, in order to account for the correlation between different estimates within the same survey group. Clustered OLS (with or without some kind of weighting) is still the most common approach in the MA literature, although other models are also common (see for example the review by Nelson and Kennedy 2009). The log-log model we use has the advantage that the estimated coefficients for GDP per capita and the risk change have natural interpretations as elasticities. Note that a "risk change elasticity" of −1 implies that WTP is independent of the risk change, indicating preferences that are completely insensitive to scope. An elasticity equal to zero implies that WTP increases proportionally with the risk reduction, as predicted by theory.[6]

19.3.3 Quality Screening of Data and Meta-regression Models

19.3.3.1 Screening of the Metadata

In MA generally, it is controversial to screen out studies based on quality, as there is no general agreement about what constitutes quality in general or quality for a specific purpose. Hence, many meta-analysts recommend to "err on the side of inclusion" (Stanley and Jarrell 2005). Still, for MA-BT in policy applications, there are good reasons to explore quality screening criteria and their effects on results. This is also the recommendation by Morgan and Cropper (2007). The reason is that good studies, on average, provide better information that is closer to the "truth" in some sense. We introduce some quality screening criteria discussed in Lindhjem et al. (2011) one by one and provide meta-regression results that will be utilized in later sections for the BT tests.

19.3.3.2 Full Dataset and First-Level Screening

We start by reporting results for the full data set where no screening criteria were applied, for the sake of comparison. In Lindhjem et al. (2011), five regression models were run, gradually increasing the number of explanatory variables for each dataset. For our purpose here we simplify by including first the simplest model, retaining only log of GDP per capita and a constant, and then the comprehensive model using all variables in Table 19.1 (Models I and II). The first model is included to evaluate whether introducing the full range of explanatory variables will increase precision in BT. Note that the risk change variable is not included here, since many studies do not report this information.

[6]An elasticity above zero would also be possible if WTP increases more than proportionally with the risk reduction. This is, however, less commonly observed.

For the next subset of the data, the size of the risk reduction reported in the surveys is included as an explanatory variable (Models III and IV). Some estimates are lost, as this information is not always reported, but something is gained as the model is more appropriate. In addition to leaving out studies that do not report the risk change valued as basis for deriving VSL, we use two additional screening criteria:

- Subsamples smaller than 100 observations and main survey samples less than 200 observations are omitted.[7]
- Samples that are not representative of a broad population are omitted.[8]

Compared with the full, unscreened data set, this data set is likely to be of higher quality. The number of observations has been cut in half. For this dataset we also show the results for both a simple and a comprehensive model. Results for all four regression models are displayed in Table 19.2.

Increasing the number of variables increases the explained variation (i.e., R-square) from 40 to 53 % for the full dataset and from 70 to 83 % for the screened dataset. GDP per capita is highly significant for all four models, yielding an elasticity of GDP to VSL of between 0.9 and 1.3. Elasticities derived from individual stated preference surveys (rather than MA studies) are often below unity, while newer studies show elasticities equal to unity or above (supporting findings from Viscusi 2010). When controlling for the risk change, the elasticity drops below unity.[9] Lindhjem et al. (2011) find that controlling for the risk change in the regressions helps explain around half of the reduction in the income elasticity. This means that some of the effect on VSL is not due to the increase in GDP capita but to the fact that surveys conducted in higher-income countries tend to present lower risk changes for respondents to value.

Model II show that traffic risks are valued higher, as are risks related to cancer. These coefficients are no longer significant in the screened dataset, where environmental and public risks changes instead become significant. Surveys that do not carefully inform respondents about the magnitude of the risk change, by using visual tools or appropriate explanations (variable "noexplan"), tend to provide higher estimated VSLs for both Models II and IV.

The coefficient on the risk change variable is between −0.57 and −0.45 for the two models. This means that the respondents' WTP is not very sensitive to the size of the risk change, and that WTP does not increase in proportion to the risk reduction, as predicted by theory. As seen from the definition of VSL in (1.1), VSL therefore decreases when the risk change increases. This finding can be seen as a potential problem for both policy and research, as using lower risk change levels in

---

[7]Often a main survey has a range of subsamples with smaller sample sizes.

[8]This may be particular types of workers, commuters or similar special groups.

[9]Caution should be exercised when interpreting the income elasticity, as the potential effect of real income growth on VSL over the long time period the data cover is not easily adjusted in this type of MA. Adjustment of VSL and GDP per capita to a common base year is the most commonly used method.

Table 19.2 Meta-regression results for full dataset (Models I and II) and first-level screened dataset (Models III and IV)

	Model I	Model II	Model III	Model IV
lngdp	1.296* (0.206)	1.168* (0.212)	0.865* (0.189)	0.774* (0.195)
turnbull		0.0322 (0.494)		−0.0951 (0.684)
envir		0.264 (0.348)		−0.622*** (0.338)
traffic		0.617** (0.308)		−0.331 (0.239)
public		−0.415 (0.356)		−0.887* (0.252)
household		−0.165 (0.279)		0.0305 (0.228)
cancerrisk		0.948* (0.356)		0.485 (0.308)
latent		−0.415 (0.392)		−0.327 (0.371)
noexplan		1.026* (0.320)		0.712* (0.223)
lnchrisk			−0.446* (0.0911)	−0.575* (0.0862)
Constant		2.046 (1.997)	2.768 (2.097)	1.942 (2.390)
Estimates	852	852	397	397
R-squared	0.403	0.529	0.703	0.829
Root mean squared error	1.361	1.213	0.905	0.694

Robust standard errors in parentheses
*$p < 0.01$, **$p < 0.05$, ***$p < 0.1$

the surveys would give higher VSL estimates. We therefore also investigate a subset of the data in the next section that come from surveys that pass external and internal scope tests (see explanation below).

19.3.3.3 Stricter Data Screening

In this section two approaches to improving the quality of data by further screening are tested. We start from the previous dataset of 397 estimates which includes the risk reduction variable, population representative and reasonable-sized survey samples. First, we screen out all estimates that come from studies that do not pass both external and internal scope tests. When mean WTP is found in statistical tests to be significantly higher for respondents faced with risk change "A" compared with risk change "B," and A > B, the test is normally interpreted as "passed" in the literature. The test is called external if the different sized risks are asked in separate samples, while it is internal (and a weaker test) if the same respondent is given different risk levels.[10] As shown in Lindhjem et al. (2011) it is only the strict requirement that surveys should pass both tests that manages to reduce the "risk change elasticity" substantially, down to between −0.3 and −0.25 (and the latter is not significant) in Models V and VI in Table 19.3.[11] Note that only 79 estimates remain from the dataset of 397 observations. In Model VI, the only other significant coefficients are GDP and latency. The explained variation increases from 62 to 76 % when including the full range of explanatory variables.

The second screening criterion, used in Lindhjem et al. (2011), that we test here is limiting the data to estimates that are derived from survey questionnaires (and associated data collection procedures) that use an accepted "good practice" approach. Although there is no general agreement on this, the careful survey initially developed by Krupnick, Alberini and co-authors has been tried and tested in several countries (Alberini et al. 2004; Krupnick et al. 2002, 2004).[12] The idea is that if more of the methodological variation can be reduced, the effects of other variables more relevant for BT and policy use will be revealed more clearly. Another advantage is that different versions of the survey have been used in several countries, ensuring variety in some of the policy-relevant variables (such as income). Using this "good practice" screening criterion, the number of estimates drops from 397 to 169. These 169 estimates are all drawn from the use of small variations of the same survey instrument. The log of survey year was added to the regression model. We include only the full model, not the simple version, for this dataset as an example for comparison.

Results for this dataset are given for Model VII in Table 19.3. This model explains approximately 81 % of the variation in the VSL estimates. Both the risk change and GDP per capita again are highly significant. Hence, the careful survey approach by Krupnick et al. (2002) does not manage to remove people's sensitivity to the risk change. VSL tends to be lower for risk reductions that are latent and for estimation procedures using the lower-bound nonparametric Turnbull estimator. It can also be noted that newer surveys tend to give higher VSL estimates, the reasons for which are unclear.

We will in the next section investigate how these models perform in BT, compared to the data subject to less strict or no screening.

[10]Note that we do not apply the stricter requirement that WTP should be proportional to the risk change, i.e., that WTP_A/WTP_B and A/B should be equal. Information about proportionality is not always reported in studies conducting scope tests.

[11]Lindhjem et al. (2011) also compare results between those studies that do pass scope test or only one of the external or internal tests.

[12]This questionnaire values health risk reductions only (with no reference to specific causes of the risk); using a 1000-square grid for displaying and training respondents to understand the magnitude of the risk changes. The variables "household," "envir," "traffic," "cancerrisk," "public," and "noexplan" drop out, as the values of these are the same for all estimates.

Table 19.3 Meta-regression results for full dataset (Models I and II) and first-level screened dataset (Models III and IV)

	Model V	Model VI	Model VII
lngdp	0.283 (0.161)	0.336** (0.134)	0.435* (0.0811)
lnchrisk	−0.298*	−0.245 (0.135)	−0.507** (0.166)
turnbull	(0.0571)	(0.135) −0.476	(0.166) −0.591***
envir		(0.299) 0.130	(0.249)
traffic		(0.294) −0.190	
public		(0.197) −0.0143	
household		0.0845 (0.332)	
cancerrisk		0.0188 (0.125)	
latent		−0.695* (0.245)	−0.227* (0.0536)
lnyear			4.222** (1.254)
Constant	8.981* (1.641)	9.192* (1.659)	−8.903 (4.924)
Estimates	79	79	169
R-squared	0.615	0.756	0.815
Root mean squared error	0.539	0.451	0.359

Robust standard errors in parentheses
$^{*}p < 0.01$, $^{**}p < 0.05$, $^{***}p < 0.1$

19.4 Testing the MA Models for Benefit Transfer

In this section we conduct the BT tests explained in Sect. 19.2.2 based on the meta-regression models estimated in Sect. 19.3 above.

19.4.1 Out-of-Sample Transfers for Full Dataset and First-Level Screening

As described above, we first use the meta-regression models to predict one VSL estimate in turn from the remaining BT values, and complete this procedure for all estimates. Results are first displayed graphically, i.e., the predicted values (zigzag line in the figures) and the VSL estimates that are predicted (rising graph in the figures) are compared in ascending order from the lowest to the highest VSL estimates in the dataset. The difference represents the (log of the) absolute error for each VSL estimate. The graphic depiction of this exercise for the unscreened data in Models I and II are given in Fig. 19.1, which shows that Model II predicts marginally better. However, both models seem to miss the benchmark values, especially at the extremities of the data.

We conduct the same procedure for Models III and IV, where the dataset has undergone first-level screening and where the risk change variable is included. It can be seen from comparing the two panels in Fig. 19.2 that these models come closer in their predictions, and that including the comprehensive set of variables in Model IV, rather than just GDP and the risk change, removes some of the high predictions that increase overall TE.

We then calculate the measures for mean and median TE_R and TE_A, and display the results in Table 19.4. The mean TE_R is more than 200 % for Model I and decreases to around 100 % for the screened data in Model III and IV. Median TE_R is surprisingly similar for all four models, though the rank between them is as expected. Measuring TE_A, the transfers on average miss by more than 4.8 million USD for Models I and II, and less than half that for the quality screened models at 1.7 and 1.4 million USD for the last two models, respectively. Measuring TE_A, the precision increases (especially considering medians) relatively more between the

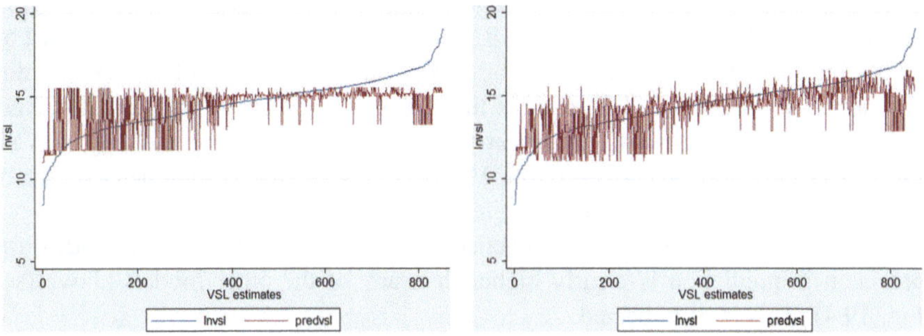

Fig. 19.1 LnVSL (*increasing line*) and predicted (transferred) lnVSL (*zig-zag line*) from Model I (*left*) and Model II (*right*) of the unscreened data

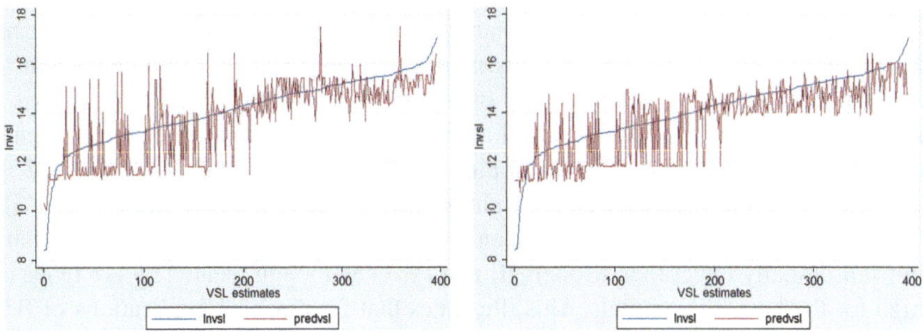

Fig. 19.2 LnVSL (*increasing line*) and predicted (transferred) lnVSL (*zig-zag line*) from Model III (*left*) and Model IV (*right*) of the first-level screened data

Table 19.4 Transfer errors from out-of-sample predictions, full dataset and first-level screening

Model	Mean TE_R (%)	Median TE_R (%)	Mean TE_A (USD)	Median TE_A (USD)
I	258	64.8	4,821,439	1,402,022
II	133.7	67.2	4,385,816	1,094,655
III	103.5	57	1,711,435	707,672
IV	96.7	55.9	1,381,858	811,920

two pairs of models, compared to considering TE_R alone. This justifies the consideration of TE_A in BT assessments as an additional measure of accuracy.

Although there is no convincing theoretical or empirical justification for doing so, trimming away the 2.5 % highest and lowest VSL estimates may yield some (relatively small) precision gains over the ones given in Table 19.4, as shown in OECD (2012). We do not test this further here.

19.4.2 Out-of-Sample Transfers for Stricter Data Screening

We repeat the same procedure for the estimates from scope-sensitive surveys and for the "good practice" survey. Results for the two models using the scope-sensitive dataset are displayed visually in Fig. 19.3. It is difficult to judge BT accuracy just by inspecting these figures and comparing them with the models in Sect. 19.4.1, due amongst other reasons to difference in the scales and number of transfers involved. For Model V, it seems that it under-predicts substantially for the higher VSL estimates. This gap is narrowed for Model VI, when the comprehensive set of variables is included (right panel in Fig. 19.3).

For the data from the "good practice" survey of Krupnick et al. (2002), the precision in prediction is clearly higher than any of the other models above (see Fig. 19.4). This is as expected.

We display the results for mean and median TE_R and TE_A in Table 19.5. The mean TE_R is a decent 52 % for Model V and decreases even further (to 43 %) for the scope—sensitive model with the full set of explanatory variables. The mean TE_R for Model VII confirms the visual inspection above, with highest accuracy at a TE_R of 26 %. Median TE_R is somewhat lower than the mean, but not as significant as for the results in Table 19.4. This means that the latter three models seem to under-predict almost as often as they overshoot. Interestingly, Model VII shows a precision gain over Models V and VI as measured by TE_A that is relatively higher than the reduction in TE_R. For example, a reduction from mean 43 % to 22.2 % yields an absolute precision gain of almost 75 % from USD 1.1 million to USD 276,235. It can also be noted that mean and median TE_A for Models V and VI are not substantially lower than Models III and IV, even though mean TE_R are twice as high for the latter two models. This illustrates that for practical applications of BT, the emphasis on relative TE may be misleading. The more relevant question is if the absolute values of TE are acceptable for different policy assessments.

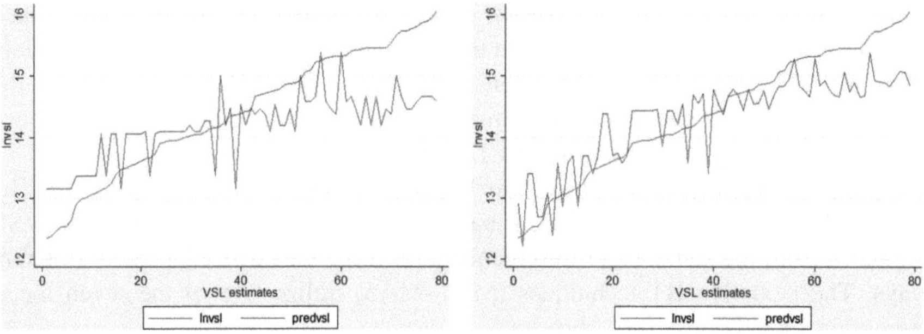

Fig. 19.3 LnVSL (*increasing line*) and predicted (transferred) lnVSL (*zig-zag line*) from Model V (*left*) and Model VI (*right*) of the scope-sensitive data

Fig. 19.4 LnVSL (*increasing line*) and predicted (transferred) lnVSL (*zig-zag line*) from Model VII, the "good practice" survey data

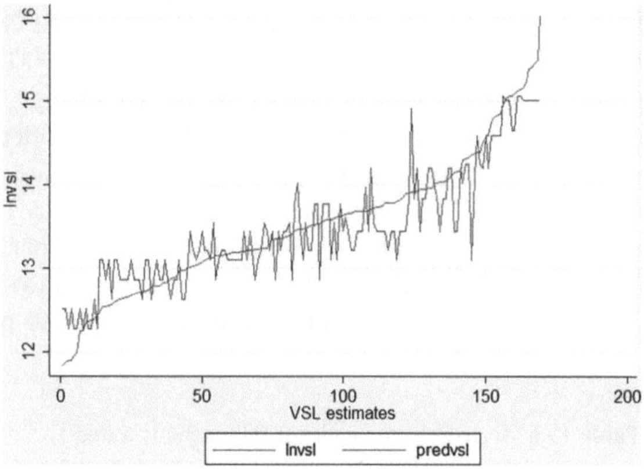

Table 19.5 Transfer errors from out-of-sample predictions, stricter screening

Model	Mean TE_R (%)	Median TE_R (%)	Mean TE_A (USD)	Median TE_A (USD)
V	52.3	51.6	1,432,451	566,752
VI	43	27.1	1,113,542	714,265
VII	26.2	22	276,235	147,942

19.4.3 Comparing Accuracy of BT Methods: An Illustration

19.4.3.1 Choice of "Benchmark Value" and BT Methods for Comparison

To more closely replicate an actual BT situation, we draw a single VSL estimate randomly from one study to represent a benchmark, unknown VSL value for a policy under assessment. The next step is to use the other studies to transfer a best

VSL estimate to that policy context, based on simple and more sophisticated MA-BT techniques. We then calculate transfer errors. This illustration is not meant to give the definite answer about transfer accuracy, but rather to show how the methods can be applied and how quality screening procedures may influence results. The example is based on OECD (2012).

Table 19.6 provides a menu of possible BT methods an analyst could use. The first six BT techniques (N1–N6) are based on naïve unit transfers of mean VSL estimates from the screened or unscreened part of the data and adjusted in different ways. The next five BT techniques (MA1–MA5) utilize five of the seven meta-regression models discussed above.

We decided to choose a study from Japan as the source for a benchmark value to be approximated through BT techniques (Itaoka et al. 2007). The study used the "good practice" questionnaire developed by Krupnick and colleagues, as explained above, and should represent a good quality estimate of VSL. The study reports several estimates and a VSL value of USD 2,795,978 was chosen randomly as our benchmark value. The study valued a 1 in 10,000 risk change related to health (rather than environment or traffic); the risk change was assumed to be immediate (not latent), chronic and private (affects the respondent and his household only) and was explained to respondents using a 1000 square grid. Further, the survey was conducted in 1999, using self-administration on a PC, asking WTP for mortality risk reduction through a dichotomous choice format. Before demonstrating the use of BT methods in the following section, we removed from the data the 30 other VSL estimates from the same study from Japan to mimic a real BT situation.

Table 19.6 Common international BT methods tested

Number		BT method for VSL	Description/model used
N1	Naïve unit BT	Mean of most similar studies	Adjusted by GDP Income elasticity set to unity
N2		Raw mean of unscreened studies	Same as above
N3		Mean of first-level screened studies	Same as above
N4		Mean of first-level screened studies with same risk change	Same as above
N5		Mean of first-level screened studies with same risk change	Same as above, except income elasticity set to 0.8
N6		Mean of OECD countries, first-level screening	Adjusted by GDP Income elasticity set to unity
N7		Mean of similar "good practice" studies	Adjusted by GDP Income elasticity set to unity
MA1	Meta-analytic BT	Unscreened	Full model II, Table 19.2
MA2		First-level screening	Simple model III, Table 19.2
MA3		First-level screening	Full model IV, Table 19.2
MA4		Scope-sensitive studies	Full model VI, Table 19.3
MA5		Similar "good practice" studies	Model VII, Table 19.3

19.4.3.2 N1—VSL Estimate from Most Similar International Studies

If a suitable domestic study that has valued a similar risk change does not exist (as is the case for our Japanese example), an option is to choose a similar international study. It is not straightforward to decide which "similarity criteria" should be applied (and in which order), as the analyst will typically not find one, unique study that matches all the risk and population characteristics that define the policy context of interest.

One place to start is to look for studies that value the same risk reduction, since we know this affects the VSL. This reduces the number of potential VSL estimates from the full dataset of 821 (when all the Japanese estimates have been removed) to 84 eligible estimates. Further, if we think that the type of risk should be the same ("health"), this leaves 68 potential estimates. Of these, 54 estimates are for chronic, private and immediate risk changes. Selecting with the remaining variables from Table 19.1, that the risk change affects the individual (rather than the household) and is not related to cancer, finally leaves 47 candidate VSL estimates. This search can go on until a sufficiently similar study is found. However, it would be difficult to decide which variables should be used to judge similarity, in which order, and when to stop the screening process.

Weighing the mean so that each study counts equally (rather than each estimate) the mean VSL of the final 47 estimates is USD 5,729,686. Adjusting for GDP per capita differences of Japan and the weighted average GDP per capita of the 47 international estimates finally yields a transferred estimate of **USD 6,171,170**, more than double the benchmark value.

19.4.3.3 N2—Raw Mean of Unscreened Studies

A simpler method than picking a single study or do a detailed matching of variable characteristics with the policy context, as we did above, would instead be to take a raw mean of VSL estimates of all collected studies and adjust with GDP differences. A weighted mean VSL for this procedure is **USD 7,778,766**.

19.4.3.4 N3—Mean of First-Level Screened Studies

Instead of taking the raw, uncritical mean of all VSL estimates, a quality screening could first be conducted. Using the first-level screening discussed in Sect. 19.3, the number of estimates is reduced from 852 to 397. AIC-adjusted GDP per capita for Japan for this year was USD 20,438, while the weighted mean of the GDP per capita for the sample was USD 17,860. Assuming an income elasticity of VSL of 1 for simplicity, leaves a simple, income adjusted transferred VSL estimate to Japan of **USD 3,558,376**.

19.4.3.5 N4—Mean of First-Level Screened Studies with Same Risk Change, Elasticity = 1

Doing the same exercise as for N3, but including only studies that have the same risk reduction as the Japanese study of 1/10,000, reduces the number of available estimates to 34. Performing a GDP/capita adjustment yields a transferred VSL estimate for Japan of **USD 3,710,774**, when the income elasticity is set to unity.

19.4.3.6 N5—Mean of First-Level Screened Studies with Same Risk Change, Elasticity = 0.8

There may be reasons to use lower income elasticity than 1, as found in some of the meta-regression models (see for example recommendations in OECD 2012). If we instead use 0.8 for income elasticity, as recommended by OECD (2012), the relative difference between GDP/capita will be raised to the power of 0.8 before being multiplied with the unadjusted mean VSL.[13] There may be reasons to use an elasticity below unity for some BT applications (see OECD 2012 for a discussion). This procedure yields a transferred estimate of **USD 3,800,408.**

19.4.3.7 N6—Mean of OECD Countries

In the examples above, we have not made any considerations about whether people in low-income countries may have other risk preferences or if there are other reasons for excluding estimates from such countries. Since the country we want to transfer to is Japan, it may make sense to limit studies to higher income countries, such as OECD countries. This reduces the number of estimates to 603. Adding first-level screening leaves 233 estimates. Mean, income-adjusted VSL from these is **USD 3,525,709**, when the income elasticity is set to unity.

19.4.3.8 N7—Mean of Similar "Good Practice" Studies

Taking the mean of the estimates using the "good practice" approach to VSL valuation implied by the questionnaire developed by Krupnick et al. (2002) yields a VSL estimate of **USD 1,761,986**, based on 150 estimates.

[13]The formula is VSL_T = mean VSL_I*(GDP_T/mean GDP_I)^e, where T is transfer country, I is estimates from international studies, and e is the elasticity of GDP/capita.

19.4.3.9 MA1—MA-BT, Unscreened Data

If a meta-regression analysis was carried out with no concern regarding screening based on criteria of quality, one could take as a starting point Model II in Table 19.2. When removing the Japanese estimates, the estimated meta-regression function of this model is:

$$\begin{aligned}\ln\mathrm{VSL} = {} & 2.707 + 1.177 * \ \mathrm{lngdp} + 0.216 * \mathrm{envir} + 0.595 * \ \mathrm{traffic} \\ & -0.431 * \mathrm{public} \ -0.180 * \mathrm{household} + 1.007 * \ \mathrm{cancerrisk} \\ & - 0.395 * \mathrm{latent} \ + 1.010 * \mathrm{noexplan} \ + 0.0310 \ * \mathrm{Turnbull}\end{aligned} \quad (19.5)$$

First, we use this equation to estimate and transfer a VSL value to the policy context in Japan. Since the methodological values are unknown at the policy site (in reality), common practice is to set the values of the methodological variables equal to some best practice value. In this case, it is good practice to use thorough explanation in explaining risk changes (hence "noexplan" is set to zero). Similarly, since the Turnbull approach typically yields a lower bound on VSL, this variable is also set to zero. The issue of whether variables that are not significant should be excluded (normally in BT they are not) is disregarded here. Further, since the risk is related to health, for an individual (not a household), a private risk program, immediate and not related to cancer, all these variables are set to zero. That leaves the following simple equation:

$$\ln\mathrm{VSL} = 2.707 + 1.177 * \mathrm{lngdp} \quad (19.6)$$

Inserting log of the GDP per capita for Japan of USD 20,438 and taking the antilog (inverse) of lnVSL yields an estimate of VSL of **USD 1,773,960.**[14]

19.4.3.10 MA2—MA-BT, First-Level Screening, Simple Model of GDP and Risk Change

Instead of using the unscreened model above, we apply the first-level screening of estimates and use the simplest model first (i.e., Model III of Table 1.2). Inserting values for log of the risk change (1/10,000) and GDP per capita yields an estimated VSL of **USD 2,244,547**.

[14]To make the calculations simpler and more transparent for this example, no correction is made for so-called "econometric error" when converting from log form (cf. Bockstael and Strand 1987). Several authors in the literature follow the same approach (e.g., Stapler and Johnston 2009) and such correction may have only a relatively small impact on the estimated VSL values when considering the overall sensitivity of results in this example.

19.4.3.11 MA3—MA-BT, First-Level Screening, Full Model

We now do the same as the above, except use the full model (i.e., Model IV of Table 19.2) with all variables, and then insert policy context values for each of them. This yields a transferred VSL estimate of **USD 2,822,560**. Hence, controlling for a wider range of variables than the risk change shifts the transferred value by ca. USD 600,000.

19.4.3.12 MA4—MA-BT, Scope-Sensitive Studies, Full Model

If we are very strict, only including VSL estimates that come from studies that pass both internal and external scope tests, the number of estimates for MA is reduced to 69 (since the Japanese estimates are removed). Re-estimating Model VI inserting values for the risk change and GDP per capita for Japan, yields a transferred VSL estimate of **USD 2,986,626**.

19.4.3.13 MA5—MA-BT, Similar "Good Practice" Studies

We now use the data from only those studies that use the good practice survey questionnaire, i.e., Model VII from Table 19.3. Conducting the same procedure as above gives an estimate of VSL of **USD 2,206,645**. Compared to the MA-BT examples above, this model also includes the year of data collection. In the same way as for the previous MA-BT functions, all other variables except GDP per capita, the risk change and study year, were set to zero to fit the policy context to which the estimate is transferred.

19.4.3.14 Summary of BT Illustration

The estimated VSL (and corresponding TE_R) values from the examples are displayed in Fig. 19.5. The left, vertical axis shows the values in USD while the right axis shows the TE_R. The horizontal axis displays the 12 different BT methods used, where the last five are the MA-BT methods. The dotted, horizontal line represents the benchmark value from the Japanese study.

It can be seen from the figure that the simple, naïve BT methods N1 (mean of most similar studies) and N2 (raw mean of unscreened studies) give much higher transfer errors than the other methods (up to 180 % for N2). More elaborate unit transfers of mean VSL in methods N3–N7 produce precision levels that approximate acceptable levels (below 40 %), similar to the first MA-BT method (MA1—unscreened data). Interestingly, all the unit transfer methods overshoot the benchmark value, while the MA-BT methods generally miss on the low side.

When introducing screening criteria, the accuracy of the BT increases, yielding the lowest error in this example for the first-level screened MA-BT model, with the

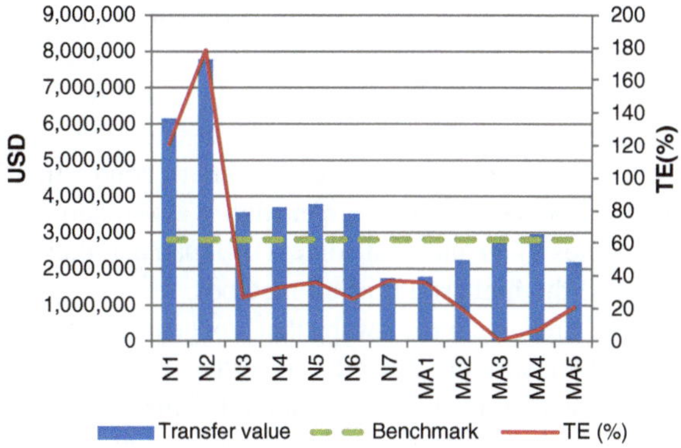

Fig. 19.5 Comparison of simple methods with meta-analytic BT for an example scenario

comprehensive set of explanatory variables (MA3). The TE_R for all the quality screened MA-BT models are below 20 % and within acceptable levels for most policy uses. Note that there is no difference between judging the BT methods by TE_R or TE_A in this example, since the benchmark value is held constant throughout the comparison.

19.5 Conclusions and Further Research

This chapter has conducted reliability tests of meta-analytic benefit transfer models and provided a simple illustrative example comparing such BT methods with simpler, unit based transfers, often used in practice by consultants and various policy agencies. We investigated in particular how quality screening in meta-analysis affects precision in MA-BT and how precision compares between MA-BT and unit transfer methods.

The first accuracy test was carried out for three main types of quality screening criteria applied to the data. It is clear from these tests that the transfer errors decrease both when introducing more quality screening and when including more explanatory variables. Considering TE in an absolute sense (i.e., measured in monetary terms), not just percentage-wise, gives a more accurate picture of BT precision, as well as providing more suitable estimates for cost-benefit analysis.

The second accuracy test compared different types of meta-regression models. An estimate of VSL from Japan was randomly picked to represent an unknown, true VSL value at a policy site or context, and then different BT techniques were used to derive a VSL value from the international literature that could be transferred to the Japanese context, since no other relevant studies were found from this country. Seven simple unit transfer methods were compared with five MA-BT methods. Though no general conclusions can be drawn based on this example, it demonstrated what is generally thought to be true but rarely tested, that MA-BT may be able to increase precision over unit transfer methods. All the quality screened MA-BT methods gave relative TE of less than 20 %, which should be acceptable for most policy applications. Somewhat surprisingly, perhaps, five of seven unit transfer methods yielded a precision level better or equal to the MA-BT method that was based on the full, unscreened dataset (TE below 40 %). In other words, conducting a simple quality screening of a dataset and then transferring a GDP-adjusted mean VSL yielded equally precise transfers as conducting MA-BT on a complete dataset that has not been quality screened. It is also notable that matching an international set of studies to the national context of Japan and then performing a simple unit transfer yielded the highest TE of all methods.

We are of course aware that no general conclusions can be drawn from this simple illustration. An area of further research would be to do a comprehensive assessment and comparison for the BT methods for all VSL estimates in the dataset, similar to the tests conducted in for example Johnston and Thomassin (2010) and Lindhjem and Navrud (2008). Another avenue of research would be to further investigate different types of data quality screening criteria and their impact on BT precision for different types of BT methods. Our simple tests and example illustrates that the choice of BT method is not an easy one. If a MA-BT approach is used, the choice of quality screening procedure (and other methodological choices) will influence the results. It is, however, comforting that the demonstration of sensibly quality screened MA-BT applications tested here did give reliability gains over the simple unit transfermethods. Whether the considerable efforts needed to conduct a comprehensive and high-quality MA-BT justifies these gains in accuracy, depends on the level of precision needed, an issue worthy of more systematic investigation in the future.

Acknowledgments We would like to acknowledge important contributions by Nils Axel Braathen of the Environment Directorate, OECD and Vincent Biausque of the National Institute of Statistics and Economic Studies, France.

References

Alberini, A., Cropper, M., Krupnick, A., & Simon, N. (2004). Does the value of a statistical life vary with age and health status? Evidence from the United States and Canada. *Journal of Environmental Economics and Management, 48*, 769–792.

Bockstael, N. E., & Strand, E. (1987). The effect of common sources of regression error on benefit estimates. *Land Economics, 63*, 11–20.

Brander, L. M., Florax, R. J. G. M., & Verrmaat, J. E. (2006). The empirics of wetland valuation: A comprehensive summary and a meta-analysis of the literature. *Environmental and Resource Economics, 33*, 223–250.

Brander, L. M., Brauer, I., Gerdes, H., Ghermandi, A., Kuik, O., Markandya, A., et al. (2012). Using meta-analysis and GIS for value transfer and scaling up: Valuing climate change induced losses of European wetlands. *Environmental and Resource Economics, 52*, 395–413.

Brouwer, R., & Bateman, I. J. (2005). Benefits transfer of willingness to pay estimates and functions for health-risk reductions: A cross-country study. *Journal of Health Economics, 24*, 591–611.

Czajkowski, M., & Scasny, M. (2010). Study on benefit transfer in an international setting. How to improve welfare estimates in the case of the countries' income heterogeneity? *Ecological Economics, 69*, 2409–2416.

Dekker, T., Brouwer, R., Hofkes, M., & Moeltner, K. (2011). The effect of risk context on the value of a statistical life: A Bayesian meta-model. *Environmental and Resource Economics, 49*, 597–624.

Hammitt, J. K. (2000). Valuing mortality risk: Theory and practice. *Environmental Science and Technology, 34*, 1396–1400.

Hammitt, J. K., & Graham, J. D. (1999). Willingness to pay for health protection: Inadequate sensitivity to probability? *Journal of Risk and Uncertainty, 18*, 33–62.

Hammitt, J. K., & Liu, J. T. (2004). Effects of disease type and latency on the value of mortality risk. *Journal of Risk and Uncertainty, 28*, 73–95.

Itaoka, K., Krupnick, A., Akai, M., & Alberini, A. (2007). Age, health, and the willingness to pay for mortality risk reductions: A contingent valuation survey of Shizuoka, Japan, residents. *Environmental Economics and Policy Studies, 8*, 211–237.

Johnston, R. J., & Rosenberger, R. S. (2010). Methods, trends and controversies in contemporary benefit transfer. *Journal of Economic Surveys, 24*, 479–510.

Johnston, R. J., & Thomassin, P. J. (2010). Willingness to pay for water quality improvements in the United States and Canada: Considering possibilities for international meta-analysis and benefit transfer. *Agricultural and Resource Economics Review, 39*, 114–131.

Kaul, S., Boyle, K. J., Kuminoff, N. V., Parmeter, C. F., & Pope, J. C. (2012). What can we learn from benefit transfer errors? Evidence from 20 years of research on convergent validity. *Journal of Environmental Economics and Management, 66*, 90–104.

Kristofersson, D., & Navrud, S. (2005). Validity tests of benefit transfer—are we performing the wrong tests? *Environmental and Resource Economics, 30*, 279–286.

Kristofersson, D., & Navrud, S. (2007). Can use and non-use values be transferred across countries? In S. Navrud & R. Ready (Eds.), *Environmental value transfer: Issues and methods* (pp. 207–226). Dordrecht: Springer.

Krupnick, A., Alberini, A., Cropper, M., Simon, N., O'Brien, B., Goeree, R., & Heintzelman, M. (2002). Age, health and the willingness to pay for mortality risk reductions: A contingent valuation survey of Ontario residents. *Journal of Risk and Uncertainty, 24*, 161–186.

Lindhjem, H., & Navrud, S. (2008). How reliable are meta-analyses for international benefit transfers? *Ecological Economics, 66*, 425–435.

Lindhjem, H., Navrud, S., Braathen, N. A., & Biausque, V. (2011). Valuing mortality risk reductions from environmental, transport and health policies: A global meta-analysis of stated preference studies. *Risk Analysis, 31*, 1381–1407.

Londoño, L. M., & Johnston, R. J. (2012). Enhancing the reliability of benefit transfer over heterogeneous sites: A meta-analysis of international coral reef values. *Ecological Economics, 78*, 80–89.

Morgan, M. G., & Cropper, M. (2007). SAB Advisory on EPA's Issues in Valuing Mortality Risk Reduction. Memorandum from the Chair, Science Advisory Board, and the Chair, Environmental Economics Advisory Committee, to EPA Administrator Stephen L. Johnson. EPA-SAB-08-001.

Mrozek, J. R., & Taylor, L. O. (2002). What determines the value of life? A meta-analysis. *Journal of Policy Analysis and Management, 21*, 253–270.

Navrud, S., & Pruckner, G. (1997). Environmental valuation—To use or not to use? A comparative study of the United States and Europe. *Environmental and Resource Economics, 10*, 1–26.

Navrud, S., & Ready, R. (2007). Lessons learned for environmental value transfer. In S. Navrud & R. Ready (Eds.), *Environmental value transfer: Issues and methods* (pp. 283–290). Dordrecht: Springer.

Nelson, J. P., & Kennedy, P. E. (2009). The use (and abuse) of meta-analysis in environmental and natural resource economics: An assessment. *Environmental and Resource Economics, 42*, 345–377.

OECD (Office of Economic Cooperation and Development) (2012). *Mortality risk valuation in environment, health and transport policies*. OECD Publishing. doi:10.1787/9789264130807-en.
Rosenberger, R., & Stanley, T. D. (2006). Measurement, generalization, and publication: Sources of error in benefit transfers and their management. *Ecological Economics, 60*, 372–378.
Stanley, T. D., & Jarrell, S. D. (2005). Meta-regression analysis: A quantitative method of literature surveys. *Journal of Economic Surveys, 19*, 299–308.
Stapler, R. W., & Johnston, R. J. (2009). Meta-analysis, benefit transfer, and methodological covariates: Implications for transfer error. *Environmental and Resource Economics, 42*, 227–246.
US EPA (2006). Report of the EPA Work Group on VSL Meta-Analysis, Report NCEE–0494. National Center for Environmental Economics, US EPA. http://yosemite.epa.gov/ee/epa/eerm.nsf/vwRepNumLookup/EE-0494?OpenDocument.
Viscusi, W. K. (2010). The heterogeneity of the value of statistical life: Introduction and overview. *Journal of Risk and Uncertainty, 40*, 1–13.

Chapter 20
GIS-Based Mapping of Ecosystem Services: The Case of Coral Reefs

Luke M. Brander, Florian V. Eppink, Philipp Schägner, Pieter J.H. van Beukering and Alfred Wagtendonk

Abstract This chapter illustrates the process of mapping ecosystem service values with an application to coral reef recreational values in Southeast Asia. The case study provides an estimate of the value of reef-related recreation foregone, due to the decline in coral reef area in Southeast Asia, under a baseline scenario for the period 2000–2050. This value is estimated by combining a visitor model, meta-analytic value function and spatial data on individual coral reef ecosystems to produce site-specific values. Values are mapped in order to communicate the spatial variability in the value of coral reef degradation. Although the aggregated change in the value of reef-related recreation due to ecosystem degradation is not high, there is substantial spatial variation in welfare losses, which is potentially useful information for targeting conservation efforts.

Keywords Value mapping · GIS · Meta-analysis · Coral reefs · Recreation

L.M. Brander (✉) · P.J.H. van Beukering · A. Wagtendonk
Department of Environmental Economics, Institute for Environmental Studies,
Vrije Universiteit, Amsterdam, The Netherlands
e-mail: lukebrander@gmail.com

P.J.H. van Beukering
e-mail: pieter.van.beukering@vu.nl

A. Wagtendonk
e-mail: alfred.wagtendonk@vu.nl

L.M. Brander
Kadoorie Institute, University of Hong Kong, Pokfulam, Hong Kong

F.V. Eppink
Landcare Research, Auckland, New Zealand
e-mail: eppinkf@landcareresearch.co.nz

P. Schägner
Department of Landscape Architecture and Environmental Planning,
Technical University Berlin, Berlin, Germany
e-mail: philipp.schaegner@gmail.com

R.J. Johnston et al. (eds.), *Benefit Transfer of Environmental and Resource Values*,
The Economics of Non-Market Goods and Resources 14,
DOI 10.1007/978-94-017-9930-0_20

20.1 Introduction

The framework of ecosystem services (ES) is widely used for understanding and communicating the links between ecosystems and human well-being (Millennium Ecosystem Assessment 2005). Many studies aim to integrate ES assessments into decision-making processes (TEEB 2010; UK NEA 2011). The economic value (i.e., contribution to human welfare) of an ES is, as with any good or service, determined by its supply and demand. The supply side of an ES is largely determined by ecological processes and characteristics (e.g., functioning, fragmentation, productivity, resilience or climate) that may be influenced by human activities, either deliberately or inadvertently. The understanding and modeling of the supply of ES has largely been taken up by natural scientists (e.g., ecologists, geographers, hydrologists). The demand side of an ES is largely determined by the characteristics of human beneficiaries of the ES (population, preferences, distance to the resource, etc.) and modeling hereof has largely been taken up by economists. It has been recognized that the determinants of both the supply and demand of ES are spatially variable, which makes the assessment of ES values inherently a spatial analysis. In recent years, a growing body of literature has assessed ES spatially by producing digital maps either of ES supply or its value. With the development of advanced GIS technology, mapping of ES values has emerged and become an important research issue, in particular the mapping of monetary values for ES value (Bateman et al. 1999; Brainard 1999; Maes et al. 2013; Schägner et al. 2013; Troy and Wilson 2006). This literature therefore includes studies that produce graphical value maps as well as analyses that explicitly address spatial variability in values.

We define mapping of ES values as the valuation of ES in monetary terms across a relatively large geographical area that includes the examination of how values vary across space. Thereby, mapping of ES values reveals additional information as compared to traditional site-specific ES valuation, which is beneficial for designing spatially efficient policies and institutions for maintaining ES supply. Most often, this mapping involves some type of benefit transfer, in which values from one set of locations are used to project or approximate values in other areas.

To some extent, spatial issues have been disregarded in environmental and resource economics, including ES valuation, but have attracted increasing attention with the emergence of advanced GIS technology in the 1990s (Bockstael 1996). The first studies to map ES values examined recreational values for Welsh forests (Bateman et al. 1995) and multiple ES across a protected area in Belize (Eade and Moran 1996). Since then, the number of publications mapping ES values has grown exponentially. Schägner et al. (2013) provide a review of the literature on mapping ES values and show that almost 60 % of such studies have been published after 2007. The methodologies applied in these studies differ widely, particularly with respect to how spatial variation in ES values is estimated. The precision and accuracy of mapped ES values have been questioned, and accordingly the utility for policy guidance. However, no consensus has been reached on which methods can and should be used to inform specific policy contexts (de Groot et al. 2010).

The purpose of this chapter is to develop and apply a method for mapping the value of the recreational use of ecosystems, based on a meta-analytic benefit function transfer. The chapter is organized as follows: Sect. 20.2 describes the methods that have been applied in the literature so far. Section 20.3 describes an application of value mapping to assess the welfare loss associated with coral reef degradation in Southeast Asia under a business-as-usual scenario for the period 2000–2050. This section contains details on the case study region, methodology, visitor model, meta-analytic value function, scenario for coral reef degradation and value maps. Section 20.4 provides conclusions on the results, methods and avenues for future research.

20.2 Methodologies for Mapping Ecosystem Service Values

The estimation of accurate ES values requires that models account for spatial heterogeneity in biophysical and socioeconomic conditions. The spatial perspective of variation in ES values is relatively new and has not been extensively researched (Schaafsma et al. 2012). Insufficient knowledge exists about how ES values differ across space and the spatial determinants of these values (Bateman et al. 2002; Bockstael 1996; de Groot et al. 2010; Plummer 2009; Schaafsma et al. 2013). Spatial factors that affect the supply of ecosystem services include, among others: ecosystem area (possibly characterized by a non-linear relationship and/or with thresholds), networks, fragmentation, and biodiversity. Spatial factors that affect demand for ecosystem services include: the number of beneficiaries, distance to the ecosystem, availability of substitutes, complements, and accessibility. See Bateman et al. (2002) and Hein et al. (2006) for more detailed discussions of spatial determinants of ecosystem service demand and supply.

Besides communication and visualization, value mapping makes site-specific ecosystem service values available on a large spatial scale. It allows decision makers to extract estimated values from a map or database for the locations or areas of policy interest in order to evaluate potential policy measures. New time-consuming primary valuation studies may therefore not be necessary.

Spatially explicit ES value maps have specific advantages for several types of policy applications including green accounting, land use policy evaluation, resource allocation and payments for ES. Green accounting includes information on environmental goods and services and/or natural capital in national accounts. Mapping of ES values allows the estimation of values at different spatial scales, and the aggregation of total ES values across the region of interest for inclusion in green accounts (TEEB 2010). For land use policy evaluation, the mapping of ES values allows for the evaluation of broad land use policies at a regional or even supra-national level. Typically, land uses are multi-functional and therefore provide multiple services. ES value mapping displays the tradeoffs and synergies in ES values that may result from land use change. For improving resource allocation, the mapping of ES values not only supports decisions on whether or not to implement a policy measure, it also informs where to implement a policy measure. It allows the identification of locations in order to minimize negative or maximize positive impacts on the provision of ecosystem service (Naidoo et al. 2008; Polasky et al. 2008). Regarding payments for ES, by making ES values spatially explicit, schemes can be designed to allow for more efficient and cost-effective incentives across providers. The levels of payments can then be more closely related to the value of services provided by different locations.

Methodologies used for mapping ecosystem service supply can be divided into five main categories (Eigenbrod et al. 2010; Schägner et al. 2013): (1) one-dimensional proxies for ecosystem services, such as land cover or land use (e.g., Costanza et al. 1997; Helian et al. 2011; Simonit and Perrings 2011); (2) non-validated models: ecological production functions based on likely causal combinations of explanatory variables, which are grounded in researcher or expert assumptions (e.g., Holzkämper and Seppelt 2007; Naidoo and Adamowicz 2005; Zhang et al. 2011); (3) validated models: ecological production functions, which are calibrated based on primary or secondary data on ecosystem service supply (e.g., Coiner et al. 2001; Mashayekhi et al. 2010); (4) representative samples of the study area: data on ecosystem service supply that is collected for the specific study area (e.g., Chen et al. 2009; Crossman et al. 2010); and (5) implicit modeling of ecosystem service supply within a value transfer function, i.e., the quantity of ecosystem service supply is modeled within the valuation of the ecosystem service using variables that capture supply-side factors (e.g., Brander et al. 2012; Costanza et al. 2008).

20.3 Application: Mapping Coral Reef Values in Southeast Asia

This section provides an illustration of the process of mapping ecosystem services values in an application to value changes in coral reef recreational values in Southeast Asia. The purpose of this case study is to illustrate the data, methods and results of a value mapping exercise.

20.3.1 Coral Reef Recreation, Threats and Values in Southeast Asia

Southeast Asia has the most extensive and diverse coral reefs in the world. They cover approximately 70,000 km^2, which is 28 % of the global total area of coral reef (Burke et al. 2011). Within the region, the Coral Triangle, which includes the reefs of Indonesia, the Philippines and Malaysia, contains 76 % of all known coral species and hosts 37 % of all known coral reef fish species. The coral reefs of Southeast Asia are highly productive ecosystems that provide a variety of valuable ecosystem services to local populations (Burke et al. 2011; UNEP 2006).

These ecosystem services include coastal protection, habitat and nursery functions for commercial and subsistence fisheries, recreational and tourism opportunities, and the existence of diverse natural ecosystems. In this case study we focus on the recreational and tourism uses of coral reefs.

Tourism is one of the largest and fastest growing industries in the world. In Southeast Asia, tourism accounted for 11.1 % of the region's GDP in 2012 and is forecast to grow at 5.8 % per annum over the coming decade (WTTC 2013). Reef-related tourism is expected to increase even more rapidly (Musa and Dimmock 2012). Recreational activities associated with coral reefs include diving, snorkeling, viewing from boats, and fishing. In addition, many beaches are protected by reefs or formed from coral material. Cesar et al. (2003) estimate the total global annual value of coral reef-based recreation and tourism at $9.6 billion.

Despite the provision of multiple valuable services, the coral reefs of Southeast Asia are the most threatened in the world (Burke et al. 2011). The threats are both local and global in origin and include non-sustainable fishing practices (Pet-Soede et al. 2000), sedimentation, pollution and waste, mining and dredging, damaging tourism practices, invasive alien species, climate change-related increases in temperature and sea level rise (Cesar 2000), and ocean acidification due to higher concentrations of CO_2 in the atmosphere (Veron et al. 2009). In addition, natural threats such as diseases and the occurrence of outbreaks of dominant (invasive) species are compounded by weakened ecosystem functioning (Burke et al. 2011).

Given the range and serious nature of threats to the ecological integrity of coral reefs, there is a need for more information on the value of welfare losses associated with a decline in the provision of ecosystem services (Millennium Ecosystem Assessment 2005). Information on the value of coral reef ecosystem services can be used in a number of different policy-making contexts, including the justification for establishing marine protected areas, determination of compensation payments for damage to coral reefs, setting of user fees for access to protected areas, cost-benefit analysis of conservation and restoration measures, and advocacy regarding the economic importance of properly functioning marine ecosystems (Van Beukering et al. 2007).

20.3.2 Outline of the Case Study Methodology

The aim of this case study is to provide an estimate of the loss in value of coral reef-related recreation resulting from the decline in coral reef area under a business-as-usual scenario for the period 2000–2050. In other words, it estimates one component of the cost of policy inaction from not adequately addressing the multiple threats facing coral reefs in the region.

The changes in coral reef-related recreation values are mapped in order to account for spatial variation in the determinants of value and present the results in a spatially explicit way, allowing for the identification of high impact locations. Following Sen et al. (2014), the selected methodology uses a combination of a validated model for visits to coral reefs and a meta-analytic value function to estimate the value per visit. An alternative approach would be to use a meta-analysis to estimate recreational values on a per hectare basis and implicitly model the number of visits to each hectare of an ecosystem within the value function. This is the approach used, for example, by Ghermandi and Nunes (2013) for estimating the recreational value of the world's coasts. Due to data limitations on recreational visit flows at a global scale with which to estimate a model of visits, they transfer values on a per hectare basis rather than per recreational visit.

The methodology involves the following steps:

1. Estimate a model of recreational visits to individual coral reef sites. The visitor model relates the number of visits per day to the site and context characteristics of each coral reef ecosystem such as degree of siltation or fishing damage.
2. Estimate a value function for coral reef recreation through a meta-analysis of existing monetary estimates. The value function relates the value per visitor day to the characteristics of the ecosystem and its surroundings.
3. Develop a database of coral reef ecosystems in Southeast Asia containing information on the variables included in the visitor model and value function estimated in steps 1 and 2.
4. Develop a baseline scenario for the change in the quality and spatial extent of coral reef ecosystems in Southeast Asia for the period 2000–2050. This baseline scenario is spatially variable to reflect variation in location-specific pressures on coral reef ecosystems.
5. Combine the models and data generated in steps 1 through 4 to produce estimates of the value of the loss in coral reef-related recreation under the baseline scenario. This approach allows the estimation of spatially variable, site-specific values that reflect the characteristics and context (e.g., pressure or threat) of each coral reef.

20.3.3 Visitor Model

In the first step of the analysis, we estimate a visitor model which explains variation in the number of visits by individual visitors to a given coral reef site per day. This is modeled as a function of several explanatory variables describing the characteristics of the ecosystem and its surroundings. We estimate the visitor model using a large sample survey for coral reef sites in Southeast Asia.[1] These data have a panel structure in that multiple observations of visitor numbers are taken for the same coral reef site at different points in time. Using a GIS, the visitor data are combined with additional information on spatially referenced variables obtained

[1]Reef Check is a volunteer survey program that has collected biophysical and visitor data at reef sites for more than 3000 survey sites in 80 countries globally since 1997 (see: www.reefcheck.org).

Table 20.1 Variables included in the visitor model for Southeast Asia

Variable	Variable definition	Mean	Standard deviation
Visitors	Number of visitors per day	16.216	15.396
Siltation	Dummy: 1 = siltation; 0 = none	0.717	0.451
Fishing damage	Dummy: 1 = fishing damage; 0 = none	0.290	0.454
Air temperature	Average air temperature (°C)	30.795	1.751
Area of coral cover	Area of coral cover (km^2)	11.351	38.553
Area of mangroves	Area of mangroves within 50 km (km^2)	32.298	79.124
Population	Population within 50 km	739,273	920,681
GCP	Gross cell product within 50 km (US$)	6732	4533

from multiple sources (including area of other ecosystems, population and economic activity in the vicinity of each coral reef site).

The dependent variable in the estimated regression model (y) is the number of visitors per day to a specific reef location. The explanatory variables are grouped in two matrices that include the site characteristics in Xs and context characteristics in Xc. Table 20.1 presents the list of variables included in the analysis with the mean and standard deviation of each.

The model fit was considerably improved, and heteroskedasticity mitigated, by using the natural logarithms of the area and context variables. Following Bateman and Jones (2003), Brander et al. (2007), and Brouwer et al. (1999), we use a multilevel modeling (MLM) approach to estimate the meta-regression.[2] MLM allows a relaxation of the common assumption of independent observations, and enables us to examine hierarchies within the data, such as similarity of observations for the same reef. The use of MLM provides an indication of where the assumption of independence may be invalid, and also improves the estimation of standard errors on parameter coefficients. The estimated model is given in Eq. 20.1:

$$y_{ij} = \alpha + \beta s X s_{ij} + \beta c X c_{ij} + u_j + e_{ij} \tag{20.1}$$

where the subscript i takes values from 1 to the number of observations of visits and subscript j takes values from 1 to the number of reefs. α is the constant term, u_j is a vector of residuals at the second (reef) level, e_{ij} is a vector of residuals at the first (observation) level, and the vectors β contain the estimated coefficients on the respective explanatory variables. In this equation, both u_j and e_{ij} are random quantities with means equal to zero. We assume that these variables are uncorrelated and also that they follow a Normal distribution so that it is sufficient to estimate their variances, σ_u^2 and σ_e^2 respectively (Rasbash et al. 2003). This type of model is also known as an error variance components model, given that the residual

[2]The software used is MLwiN version 2.0 (see Rasbash et al. 2003).

variance is partitioned into components corresponding to each level in the hierarchy. In our model, the level 2 residuals represent each reef's departure from the population mean, represented by the constant term, and the level 1 residuals reflect the conventional error variance at observation level. The estimated regression model is presented in Table 20.2.

As expected, the presence of siltation and damage due to dynamite fishing at a coral reef site reduces the number of visitors to that site. Air temperature is also found to have a statistically significant negative effect on the number of visitors at a coral reef site. This indicates that additional increases in temperature reduce the attractiveness of recreation locations. An optimal temperature or possible non-linear effects with temperature were examined by including a quadratic term in the regression model, but no statistically significant effects were found. The estimated coefficient on the area of coral cover at the site is positive but not quite statistically significant at the 10 % level. The area of mangroves within a 50 km radius of the coral reef site is found to have a positive and statistically significant effect on the number of visits. This suggests that there may be positive effects from the extent of other coastal ecosystems on the attractiveness of coral reef sites to visitors. This apparent complementarity between ecosystems possibly indicates the degree of naturalness of the site location. The size of the population living within a 50 km radius of a coral reef site is found to have a negative and statistically significant effect on the number of visitors. On one hand this result is somewhat surprising, since the population in the vicinity of a coral reef represents potential visitors.

Table 20.2 Estimated visitor model for Southeast Asia

Variable	Variable definition	Coefficient	Standard error
Constant	–	−37.301**	16.073
Siltation	Dummy: 1 = siltation; 0 = none	−5.866***	0.932
Dynamite fishing damage	Dummy: 1 = fishing damage; 0 = none	−7.036***	1.212
Air temperature	Air temperature (°C)	−0.569***	0.162
Area of coral cover	Natural log of area of coral cover (km^2)	1.027	0.638
Area of mangroves	Natural log of area of mangroves within 50 km (km^2)	0.685*	0.373
Population	Natural log of population within 50 km	−0.886*	0.467
GCP	Natural log of Gross Cell Product within 50 km (US$)	9.672***	1.373
Level 1 (observation) variance		145.509***	12.697
Level 2 (reef) variance		12.569***	0.927
−2*log likelihood		4447.873	
N		658	

$^{*}p \leq 0.10$, $^{**}p \leq 0.05$, $^{***}p \leq 0.01$

On the other hand, visitors to coral reefs are often not local residents. This may particularly be the case in developing countries for which a large proportion of coral reef visitors are international tourists. In this respect, visitor models for coral reefs may differ substantially from visitor models for other ecosystems, for which the size and proximity of the local population are important explanatory factors (Sen et al. 2014).

The negative effect of population in the vicinity of a coral reef site is interpreted here as the pressure and impact of urbanization and other types of development on the attractiveness of a coral reef to visitors. The estimated coefficient on gross cell product (GCP), which is a spatially disaggregated measure of economic activity equivalent to gross domestic product (GDP),[3] indicates that visitor rates are higher in regions with higher income levels. This variable does not necessarily represent the income of visitors themselves, given that visitors are often international tourists, but may reflect the availability and quality of infrastructure in a region. The estimated level 2 (reef-specific) variance indicates that there remains unexplained reef-specific variation in visitor numbers. Calculating the variance partition coefficient [12.569/(12.569 + 145.509) = 0.08] shows that approximately 8 % of residual variance in visitor numbers can be attributed to unobserved differences between reefs.[4]

20.3.4 Meta-Analytic Value Function for Reef Recreation

Following Brander et al. (2007) and Londoño and Johnston (2012), a meta-analysis of the coral reef valuation literature is used to estimate a value function for coral reef-related recreation. The coral reef value dataset used to estimate value functions for coral reef ecosystem services is an extension of the data described in Brander et al. (2007). These data have been expanded to include a number of recent coral reef valuation studies. We restrict this data set, however, to select only estimates obtained using contingent valuation or travel cost methods in order to ensure the theoretical validity of the welfare estimates (e.g., we excluded estimates that measure gross revenues). The restricted sample size is 74, of which 47 are contingent valuation estimates and 27 are travel cost estimates.

---

[3]The conceptual basis of gross cell product (GCP) is the same as gross domestic product (GDP) as developed in national income accounts. The basic measure of output is gross value added in a specific geographical region. Gross value added is defined as total production of market goods and services less purchases from other businesses. Under the principles of national economic accounting, GCP will aggregate up across all cells within a country to GDP (Nordhaus et al. 2006). This variable is correlated with population, but not perfectly.

[4]We test the influence of unobserved reef specific effects using a likelihood ratio test, for which the null hypothesis is that $\sigma_u^2 = 0$. We compare the estimated model with a model where σ_u^2 is constrained to equal zero, i.e., a single level model. The value of the likelihood ratio statistic is 5157.32−4447.87 = 709.442. Comparing this to a chi-squared distribution on 1 degree of freedom, we conclude that there are significant unobserved differences between reef sites.

Fig. 20.1 Location of coral reef recreation valuation study sites

The studies included in our analysis were published between the years 1992 and 2012. The geographic distribution of study sites is presented in Fig. 20.1. Southeast Asia is reasonably well represented in the data with 13 valuation estimates (17 % of the sample). The locations of the remaining estimates are the Caribbean (16 %), the United States (51 %),[5] Indian Ocean (13 %), and Australasia (3 %).

The data on the value of reef-related recreation are standardized to a common currency, year of value and units using PPP adjusted exchange rates and GDP deflators from the World Bank World Development Indicators.[6] The standardized values are expressed in US$ per visitor day in 2007 prices. This is the dependent variable in the meta-analytic regression model. The model is given in Eq. 20.2:

$$\ln(y_i) = a + b_S X_{Si} + b_R X_{Ri} + b_M X_{Mi} + u_i \tag{20.2}$$

The subscript i assumes values from 1 to 74 (number of observations), a is the constant term, b_S, b_R and b_M are the coefficients of the explanatory variables and u is a vector of residuals. The explanatory variables consist of three categories, giving characteristics of: (i) the study site X_S, (ii) the recreational activities valued X_R, and (iii) the valuation method used X_M. Table 20.3 presents the full list of variables included in the analysis, with the mean and standard deviation of each.

The meta-regression results are presented in Table 20.4. Following best practice, heteroskedasticity-consistent standard errors are estimated. However, the null hypothesis of homogenous variance of the residuals cannot be rejected by White's test for heteroskedasticity (White's statistic = 21.589). The adjusted R^2 statistic indicates that approximately 41 % of the variation in the dependent variable is explained by the explanatory variables, which is comparable with similar meta-

[5]Including Hawaii.

[6]http://data.worldbank.org/data-catalog/world-development-indicators.

Table 20.3 Variables included in the meta-analytic value function

Variable	Variable definition	Mean	Standard deviation
Value per visit	US$ per visitor day	73.86	171.66
Visits per day	Visits per day	196.83	388.23
Area of coral cover	Area of coral cover (km^2)	16.29	26.83
Caribbean	Dummy: 1 = Caribbean; 0 = other	0.16	0.37
Indian Ocean	Dummy: 1 = Indian Ocean; 0 = other	0.13	0.34
Southeast Asia	Dummy: 1 = SE Asia; 0 = other	0.17	0.38
Australia	Dummy: 1 = Australia; 0 = other	0.03	0.16
Diving	Dummy: 1 = diving; 0 = other	0.77	0.42
Snorkelling	Dummy: 1 = snorkelling; 0 = other	0.64	0.48
Fishing	Dummy: 1 = fishing; 0 = other	0.07	0.25
CVM	Dummy: 1 = CVM; 0 = other (travel cost method)	0.61	0.49

Table 20.4 Estimated meta-analytic value function

Variable	Variable definition	Coefficient	Standard error
Constant		3.871***	1.087
Visits per day	Natural log of visits per day	−0.434**	0.174
Area of coral cover	Natural log of area of coral cover (km^2)	0.451*	0.278
Caribbean	Dummy: 1 = Caribbean; 0 = other	1.482**	0.736
Indian Ocean	Dummy: 1 = Indian Ocean; 0 = other	2.932***	0.943
Southeast Asia	Dummy: 1 = Southeast Asia; 0 = other	1.456*	0.822
Australia	Dummy: 1 = Australia; 0 = other	0.065	1.087
Diving	Dummy: 1 = diving; 0 = other	−0.276	0.476
Snorkelling	Dummy: 1 = snorkelling; 0 = Other	−0.980**	0.446
Fishing	Dummy: 1 = recreational fishing; 0 = other	0.131	0.491
CVM	Dummy: 1 = contingent valuation; 0 = other	−1.949***	0.449
Adjusted R^2	0.41		
N	74		

*$p \leq 0.10$, **$p \leq 0.05$, ***$p \leq 0.01$

analyses of the ecosystem service valuation literature (e.g., Brander et al. 2007; Ghermandi et al. 2010).

The estimated model broadly fits prior expectations. The estimated coefficient on the number of visitors to a reef has a negative sign and is statistically significant, suggesting that visitors prefer less crowded coral reefs. The area of coral cover has a positive effect on the welfare derived from a recreational visit. Visitors have a preference for coral reefs with larger areas. Regarding the results on the regional indicators, reefs in the Indian Ocean, Caribbean and Southeast Asia are all found to provide significantly higher recreational values than reefs in the U.S. (the omitted category in the set of regional dummy variables). The values of recreational visits to Australian reefs are not statistically significantly different from visits to U.S. reefs. Regarding the dummy variables indicating the principal recreational activity that is valued, only the estimated coefficient for snorkeling is statistically significant and indicates that the value of this activity is lower than for others.[7]

Regarding valuation methods, we find that contingent valuation (CV) estimates are statistically significantly lower than estimates obtained using the travel cost (TC) method. From a theoretical perspective we might expect CV estimates to exceed TC estimates, given that the former may include some element of nonuse value in addition to the direct use value of a recreational visit. On the other hand, TC estimates for recreational visits that are part of a more complex multi-purpose trip, such as a vacation to a tropical island, may over-estimate the value of individual constituent activities (Armbrecht 2014). Empirical evidence with regard to the extent that these two methods produce similar results is somewhat ambiguous. Carson et al. (1996) review 83 valuation studies for quasi-public goods from which 616 comparisons of CV and revealed preference (RP) estimates are made. The sample mean CV/RP ratio is 0.89, with a 95 % confidence interval of 0.81–0.96 and a median of 0.75. Although the results from this study show that RP methods produce higher value estimates than CV, they also show that estimates from these two methods are within the same range. Mayor et al. (2007) compare TC and CV estimates specifically for recreational visits and find that the former tend to exceed the latter. Previous meta-analyses of the coral reef valuation literature have found similar results to those of the present study (Brander et al. 2007; Londoño and Johnston 2012).

20.3.5 Data and Scenario for Coral Reef Loss, 2000–2050

The next step in assessing the welfare change associated with the loss of coral reef area over the period 2000–2050 is to develop a database of coral reef ecosystems in Southeast Asia that contains information on the variables included in the visitor model and the meta-analytic value function. We then develop a baseline scenario for the change in the spatial extent of coral reef ecosystems in Southeast Asia for the period 2000–2050.

Individual ecosystem or patch-level data on coral reefs in Southeast Asia were obtained from the UNEP World Conservation Monitoring Centre (WCMC,

[7]The omitted category of reef-related recreation is a general category of "other" activities, including the viewing of coral reefs from boats. Our prior expectation is that the value of diving would be higher than other reef-related recreational activities. We do not, however, find evidence that the value of diving is different from recreational fishing or reef viewing. These activities can evidently also be of high recreational value.

described in Giri et al. 2011). For each of the 5290 coral reef patches in Southeast Asia that are included in the UNEP-WCMC database, we used a GIS to obtain information on the area of each coral reef and area of mangroves, population and gross cell product within 50 km.

We make use of the results of the *Reefs at Risk Revisited* assessment by the World Resources Institute (Burke et al. 2011) to define a baseline scenario for coral reef change for the period 2000–2050. This assessment provides a spatially explicit projection of the degree to which coral reefs are threatened. The threats included in the *Reefs at Risk Revisited* assessment are coastal development, watershed-based pollution, marine-based pollution and damage, overfishing and destructive fishing, thermal stress and ocean acidification. These local and global threats are combined into an integrated index representing the degree to which coral reefs are threatened. Threat levels are classified as low, medium, high, very high, or critical. The proportion of coral reefs in the low- or medium-threat categories declines over time, whereas the proportion of coral reefs that are highly, very highly or critically threatened increases dramatically. We used spatially differentiated change factors derived from the *Reefs at Risk Revisited* integrated threat data, combined with the patch-level data on coral reefs from the UNEP-WCMC, to calculate the change in area of each patch of coral reef for the period 2000–2050. The baseline loss of coral cover is presented in Fig. 20.2.

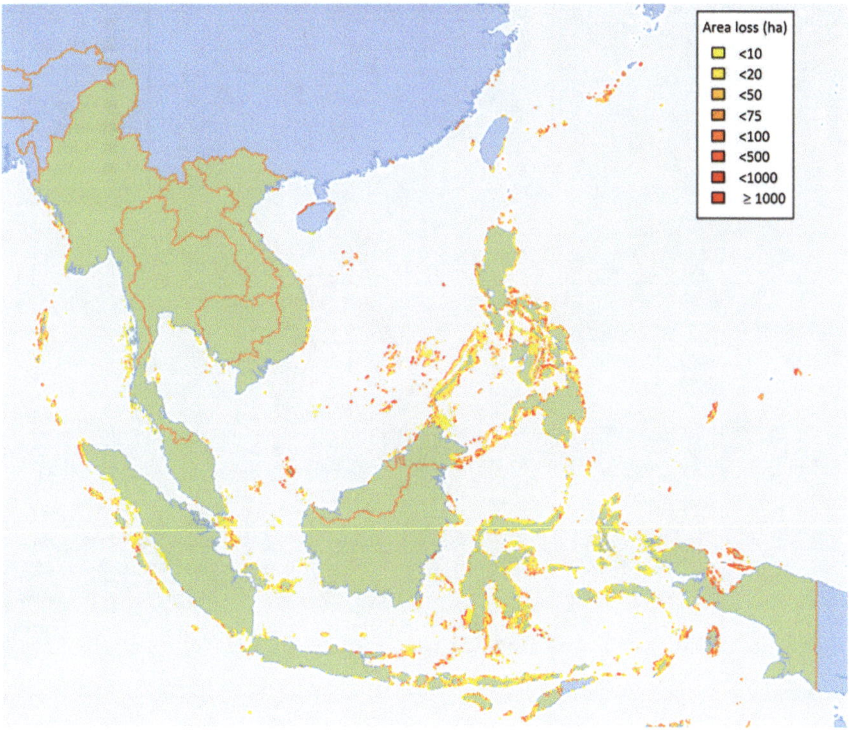

Fig. 20.2 Change in area of coral cover 2000–2050 in Southeast Asia

20.3.6 Results and Value Maps

The final step in the assessment is to combine the models and data generated in the previous steps to produce estimates of the value of the loss in coral reef-related recreation under the baseline scenario.

At the level of individual patches of coral reef, patch-specific parameter values are substituted into the visitor model to estimate the number of visitors to each site. Visitor numbers are estimated for the year 2050 by using the areas of coral cover and mangroves existing in 2000 (i.e., under a conservation scenario) and the projected areas in 2050 (i.e., the baseline scenario). The difference between these two scenarios gives the estimated site-specific change in visitor numbers due to ecosystem degradation. The change in visitor numbers is represented in Fig. 20.3 and is shown to be relatively insensitive to loss in coral cover. The average decrease in the annual visitation rate per site is only approximately 190 visitors. Nevertheless, there is substantial spatial variability across sites, due to both the underlying popularity of a site and the extent of change in the area of coral cover at that location. For example, the decrease in visitor numbers is shown to be higher for coral reefs on the east coast of Vietnam than for the west coast of Myanmar and Thailand.

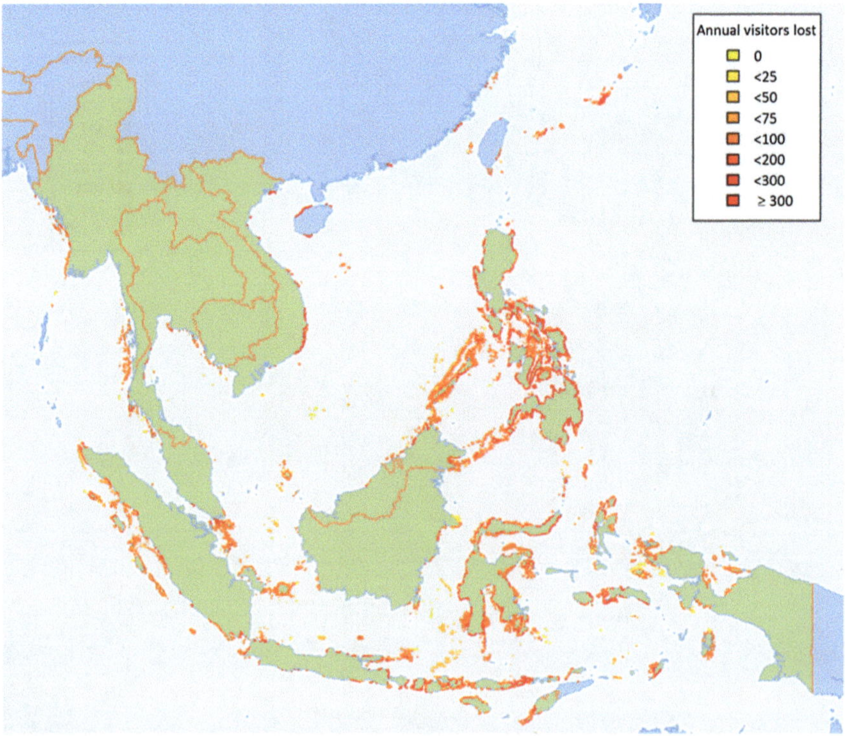

Fig. 20.3 Change in coral reef-related recreation visits per day in Southeast Asia

The value per visit to each site is computed by substituting patch-specific parameter values into the meta-analytic value function. This is done using pre- and post-change areas of coral cover and visitor numbers in order to estimate the value of a visit to each site before and after ecosystem service degradation.

Two components of the change in welfare due to ecosystem degradation are then computed. The first component is the loss in consumer surplus associated with the decrease in the number of visitors. This is computed as the decrease in visitors at each site multiplied by the pre-change value per visitor (i.e., the loss in value to those that no longer visit). The second component is the loss in consumer surplus associated with the decrease in value of visits that still take place (i.e., visitors may continue to visit a site but derive lower utility per visit from doing so). This is computed as the decrease in value per visit at each site multiplied by the number of visitors under the degradation scenario. Lower- and upper-bound values are calculated using the 95 % prediction intervals for each coral reef, which are computed using the method proposed by Osborne (2000). The prediction intervals provide an indication of the precision with which the estimated value function can predict out-of-sample values. The results are presented in Fig. 20.4 and in Table 20.5, aggregated to the country level. For Southeast Asia as a whole, the annual loss in consumer surplus from reef-

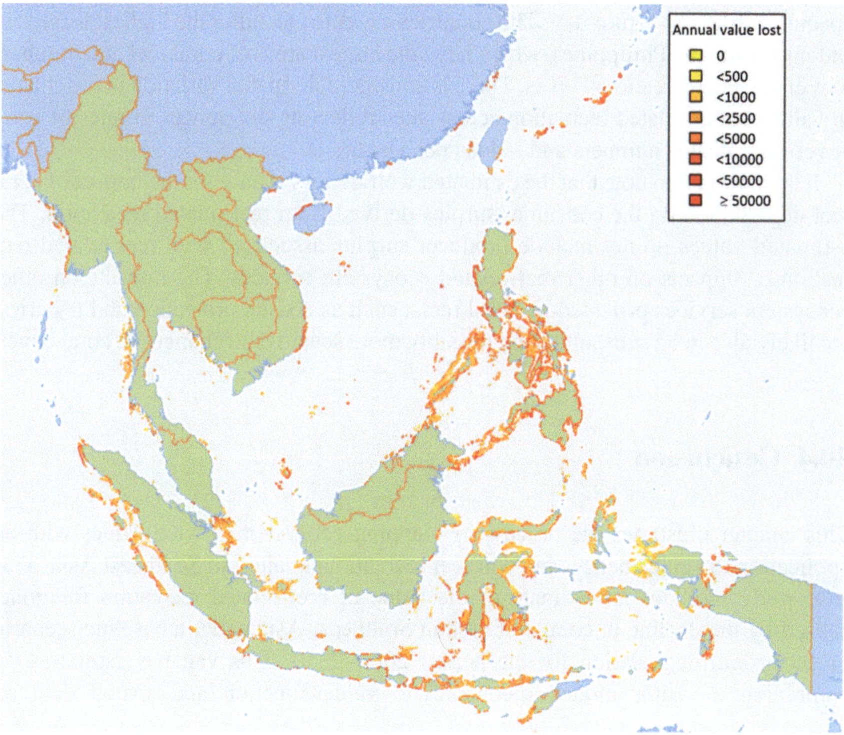

Fig. 20.4 Loss in the annual value of coral reef-elated recreation in 2050 due to policy inaction

Table 20.5 Change in consumer surplus of reef-related recreation in Southeast Asia caused by Ecosystem Degradation, 2050 (2007 US$)

Country	Value per visitor day	Total change in consumer surplus (000s)	Lower bound 95 % prediction interval (000s)	Upper bound 95 % prediction interval (000s)
Cambodia	11.20	−124	0	−1392
Indonesia	8.90	−59,468	−1099	−665,880
Malaysia	10.80	−3140	−280	−35,161
Myanmar	4.60	−2836	−253	−31,754
Philippines	6.50	−56,749	−5068	−635,440
Singapore	2.60	−176	−16	−1972
Thailand	5.80	−1936	−30	−21,680
Vietnam	4.00	−3577	−319	−40,058
Southeast Asia	6.80	−128,007	−2848	−1,433,337

related recreation in 2050 due to coral reef degradation is approximately US$ 120 million (with a 95 % prediction interval of US$ 3 million–1.4 billion). The 95 % prediction interval is very large and reflects the high uncertainty in estimating site-specific values per visitor day. The countries expected to suffer the highest losses are Indonesia and the Philippines, which have the largest areas of coral reef and numbers of reef-related recreational visits. There is considerable spatial variation in the change in value of reef-related recreation across sites reflecting differences in rates of coral cover loss, visitor numbers and values per visitor.

It is important to note that the estimated welfare loss is only for the impact of coral reef degradation on the consumer surplus derived from reef-related recreation. The estimated values do not include producer surplus associated with reef-related recreation or impacts on other reef-related ecosystem services. The impacts on other ecosystem services provided by coral reefs, such as coastal protection and fisheries, are likely also to be substantial and possibly more sensitive to changes in coral cover.

20.4 Conclusion

This chapter illustrates the process of mapping ecosystem service values with an application to value changes in coral reef recreational values in Southeast Asia. This case study provides an estimate of the value of reef-related recreation foregone, caused by the decline in coral reef area in Southeast Asia under a baseline scenario of ecosystem degradation for the period 2000–2050. This value is estimated by combining a visitor model, meta-analytic value function and spatial data on individual coral reef ecosystems to produce site-specific values. The case study illustrates the data, methods and results of a value-mapping exercise and allows several general conclusions to be drawn.

The estimated changes in visitors and values of reef-related recreation across Southeast Asia are not particularly high relative to their absolute values. Both visitation rates to coral reefs and values per visit are found to be relatively unresponsive to changes in the area of coral cover.[8] The aggregated loss of consumer surplus derived from reef-related recreation due to ecosystem degradation under the baseline scenario is therefore limited. The central estimate of annual loss in 2050 of US$ 128 million is not high, considering the size of the ecosystem providing the recreational services. The case study results do show, however, substantial spatial variation in the value of coral over loss. This information can potentially be used in economic analyses for targeting conservation efforts to specific locations. With additional information on the spatial variability of conservation costs, a spatially explicit cost-benefit analysis could be conducted to identify the location of conservation efforts in the region that would generate the highest returns. Such an analysis could be useful in locating new protected areas or planning new tourism developments.

There are several important limitations to the case study that are worth noting. There is a substantial challenge in obtaining reliable spatially disaggregated data on visitor numbers and characteristics with which to estimate a visitor model. The Reef Check data that we use in the case study application are focused primarily on the status of the reefs themselves, rather than on visitor numbers or visitor characteristics. We are therefore unable to include potentially important variables describing visitor characteristics in the model, such as recreational activity, income, origin and travel time. Future research should aim to collect such visitor-level data and include it in the estimation of visitor models. The lack of visitor-level data also restricts the options for including visitor characteristics in the meta-analytic value function, since it is necessary to have policy site data on each explanatory variable included in value function. Information on the income of visitors as a determinant of recreational value is again notably absent.

A second important limitation of the case study application is the restricted extent to which the supply of the ecosystem service is modeled. The supply side of reef-related recreation is essentially modeled implicitly in the visitor function, i.e., coral reefs supply recreational opportunities to the extent that people want to visit them. This approach may be defensible in the case of a cultural ecosystem service such as recreation, but still neglects other potentially important ecosystem characteristics that may determine the provision of the service, such as coral and fish diversity or water clarity. The method makes the analysis relatively simple but sidesteps the greater complexity involved in modeling the ecological functioning that underlies the supply of most ecosystem services. In general, accounting for spatial

[8]The regional mean proportional changes in visitor numbers and value per visit are −6 and −12.5 % for a −27 % change in the area of coral cover.

variability in ecosystem service values requires a closer integration of the biophysical assessment of ecosystem services into the valuation of ecosystem services. The disconnection between these steps in the ecosystem service assessment process remains challenging; future applications should attempt to better combine ecological and economic modeling of the determinants of ecosystem service values.

Third, the analysis of visitor behavior and recreational value does not account for the potential impact of changes to substitute (or perhaps complement) sites. The current model treats each site as independent, and does not allow for the possibility that simultaneous changes in the quality of multiple coral reef sites will influence visits and value in a way not captured through the aggregation of single-site estimates. To the extent that these cross-site effects are relevant, estimates may depart from those reported here.

References

Armbrecht, J. (2014). Use value of cultural experiences: A comparison of contingent valuation and travel cost. *Tourism Management*,. doi:10.1016/j.tourman.2013.11.010.

Bateman, I. J., Brainard, J. S. & Lovett, A. A. (1995). *Modelling woodland recreation demand using geographical information systems: A benefit transfer study (GEC 95-06)*. Norwich, UK: University of East Anglia, Centre for Social and Economic Research on the Global Environment.

Bateman, I. J., & Jones, A. P. (2003). Contrasting conventional with multi-level modeling approaches to meta-analysis: Expectation consistency in U.K woodland recreation values. *Land Economics, 79*, 235–258.

Bateman, I. J., Jones, A. P., Lovett, A. A., Lake, I. R., & Day, B. H. (2002). Applying geographical information systems (GIS) to environmental and resource economics. *Environmental and Resource Economics, 22*, 219–269.

Bateman, I. J., Lovett, A. A., & Brainard, J. S. (1999). Developing a methodology for benefit transfer using geographical information systems: Modelling demand for woodland recreation. *Regional Studies, 33*, 191–205.

Bockstael, N. E. (1996). Modelling economics and ecology: The importance of a spatial perspective. *American Journal of Agricultural Economics, 78*, 1168–1180.

Brainard, J. S. (1999). Integrating geographical information systems into travel cost analysis and benefit transfer. *International Journal of Geographical Information Science, 13*, 227–246.

Brander, L. M., Brauer, I., Gerdes, H., Ghermandi, A., Kuik, O., Markandya, A., et al. (2012). Using meta-analysis and GIS for value transfer and scaling up: Valuing climate change induced losses of European wetlands. *Environmental and Resource Economics, 52*, 395–413.

Brander, L. M., van Beukering, P., & Cesar, H. S. J. (2007). The recreational value of coral reefs: A meta-analysis. *Ecological Economics, 63*, 209–218.

Brouwer, R., Langford, I. H., Bateman, I. J., & Turner, R. K. (1999). A meta-analysis of wetland contingent valuation studies. *Regional Environmental Change, 1*, 47–57.

Burke, L., Reytar, K., Spalding, M., & Perry, A. (Eds.). (2011). *Reefs at risk revisited*. Washington, DC: World Resources Institute.

Carson, R. T., Flores, N. E., Martin, K. M., & Wright, J. L. (1996). Contingent valuation and revealed preference methodologies: Comparing the estimates for quasi-public goods. *Land Economics, 72*, 80–99.

Cesar, H. S. J. (2000). Coral reefs: Their functions, threats and economic value. In H. S. J. Cesar (Ed.), *Collected essays on the economics of coral reefs* (pp. 14–39). Sweden: Cordio.

Cesar, H. S. J., Burke, L., & Pet-Soede, L. (2003). *The economics of worldwide coral reef degradation*. The Netherlands: Cesar Environmental Economics Consulting, Arnhem and World Wildlife Fund, Zeist.

Chen, N., Li, H., & Wang, L. (2009). A GIS-based approach for mapping direct use value of ecosystem services at a county scale: Management implications. *Ecological Economics, 68*, 2768–2776.

Coiner, C., Wu, J., & Polasky, S. (2001). Economic and environmental implications of alternative landscape designs in the Walnut Creek Watershed of Iowa. *Ecological Economics, 38*, 119–139.

Costanza, R., d'Arge, R., de Groot, R., Farber, S., Grasso, M., Hannon, B., et al. (1997). The value of the world's ecosystem services and natural capital. *Nature, 387*, 253–260.

Costanza, R., Pérez-Maqueo, O., Martinez, M. L., Sutton, P., Anderson, S. J., & Mulder, K. (2008). The value of coastal wetlands for hurricane protection. *Ambio, 37*, 241–248.

Crossman, N. D., Connor, J. D., Bryan, B. A., Summers, D. M., & Ginnivan, J. (2010). Reconfiguring an irrigation landscape to improve provision of ecosystem services. *Ecological Economics, 69*, 1031–1042.

de Groot, R., Alkemade, R., Braat, L., Hein, L., & Willemen, L. (2010). Challenges in integrating the concept of ecosystem services and values in landscape planning, management and decision making. *Ecological Complexity, 7*, 260–272.

Eade, J. D. O., & Moran, D. (1996). Spatial economic valuation: Benefits transfer using geographical information systems. *Journal of Environmental Management, 48*, 97–110.

Eigenbrod, F., Armsworth, P. R., Anderson, B. J., Heinemeyer, A., Gillings, S., Roy, D. B., et al. (2010). The impact of proxy-based methods on mapping the distribution of ecosystem services. *Journal of Applied Ecology, 47*, 377–385.

Ghermandi, A., & Nunes, P. A. L. D. (2013). A global map of costal recreation values: Results from a spatially explicit meta-analysis. *Ecological Economics, 86*, 1–15.

Ghermandi, A., van den Bergh, J. C. J. M., Brander, L. M., de Groot, H. L. F., & Nunes, P. A. L. D. (2010). Values of natural and human-made wetlands: A meta-analysis. *Water Resources Research, 46*, 1–12.

Giri, C., Ochieng, E., Tieszen, L. L., Zhu, Z., Singh, A., Loveland, T., et al. (2011). Status and distribution of mangrove forests of the world using earth observation satellite data. *Global Ecology and Biogeography, 20*, 154–159.

Hein, L., van Koppen, K., de Groot, R. S., & van Ierland, E. C. (2006). Spatial scales, stakeholders and the valuation of ecosystem services. *Ecological Economics, 57*, 209–228.

Helian, L., Shilong, W., Guanglei, J., & Ling, Z. (2011). Changes in land use and ecosystem service values in Jinan, China. *Energy Procedia, 5*, 1109–1115.

Holzkämper, A., & Seppelt, R. (2007). Evaluating cost-effectiveness of conservation management actions in an agricultural landscape on a regional scale. *Biological Conservation, 136*, 117–127.

Londoño, L. M., & Johnston, R. J. (2012). Enhancing the reliability of benefit transfer over heterogeneous sites: A meta-analysis of international coral reef values. *Ecological Economics, 78*, 80–89.

Maes, J., Teller, A., Erhard, M., Liquete, C., Braat, L., & Berry, P., et al. (2013). *Mapping and assessment of ecosystems and their services: An analytical framework for ecosystem assessments under action 5 of the EU biodiversity strategy to 2020.* Luxembourg: Publications Office of the European Union.

Mashayekhi, Z., Panahi, M., Karami, M., Khalighi, S., & Malekian, A. (2010). Economic valuation of water storage function of forest ecosystems (case study: Zagros Forests, Iran). *Journal of Forestry Research, 21*, 293–300.

Mayor, K., Scott, S. & Tol, R. S. J. (2007). *Comparing the travel cost method and the contingent valuation method: An application of convergent validity theory to the recreational value of Irish forests.* Working Paper No. 190. Dublin: The Economic and Social Research Institute (ESRI).

Millennium Ecosystem Assessment. (2005). *Ecosystems and human well-being: General synthesis*. Washington, DC: Island Press.

Musa, G., & Dimmock, K. (2012). Scuba diving tourism: Introduction to special issue. *Tourism in Marine Environments Special Issue, 8*, 1–5.

Naidoo, R., & Adamowicz, W. L. (2005). Economic benefits of biodiversity exceed costs of conservation at an African rainforest reserve. *Proceedings of the National Academy of Sciences of the United States of America, 102*, 16712–16716.

Naidoo, R., Balmford, A., Costanza, R., Fisher, B., Green, R. E., Lehner, B., et al. (2008). Global mapping of ecosystem services and conservation priorities. *Proceedings of the National Academy of Sciences, 105*, 9495–9500.

Nordhaus, W., Azam, Q., Corderi, D., Hood, K., Victor, N. M., & Mohammed, M., et al. (2006). *The G-Econ database on gridded output: Methods and data*. Research Paper. New Haven, CT: Yale University.

Osborne, J. W. (2000). Prediction in multiple regression. *Practical Assessment Research and Evaluation, 7*(2), 1–9. (ISSN 1531-7714).

Pet-Soede, L., Cesar, H. S. J., & Pet, J. S. (2000). Blasting away: The economics of blast fishing on Indonesian coral reefs. In H. S. J. Cesar (Ed.), *Collected essays on the economics of coral reefs* (pp. 77–84). Sweden: Cordio.

Plummer, M. L. (2009). Assessing benefit transfer for the valuation of ecosystem services. *Frontiers in Ecology and the Environment, 7*, 38–45.

Polasky, S., Nelson, E., Camm, J., Csuti, B., Fackler, P., Lonsdorf, E., et al. (2008). Where to put things? Spatial land management to sustain biodiversity and economic returns. *Biological Conservation, 141*, 1505–1524.

Rasbash J., Steele, F., Browne, W., & Prosser, B. (2003). *A user's guide to MLwiN version 2.0*. London: Centre for Multilevel Modelling, Institute of Education, University of London.

Schaafsma, M., Brouwer, R., Gilbert, A., van den Bergh, J. C. J. M., & Wagtendonk, A. (2013). Estimation of distance-decay functions to account for substitution and spatial heterogeneity in stated preference research. *Land Economics, 89*, 514–537.

Schaafsma, M., Brouwer, R., & Rose, J. (2012). Directional heterogeneity in WTP models for environmental valuation. *Ecological Economics, 79*, 21–31.

Schägner, J. P., Brander, L. M., Maes, J., & Hartje, V. (2013). Mapping ecosystem services' values: Current practice and future prospects. *Ecosystem Services, 4*, 33–46.

Sen, A., Harwood, A. R., Bateman, I. J., Munday, P., Crowe, A., Brander, L. M., et al. (2014). Economic assessment of the recreational value of ecosystems: Methodological development and national and local application. *Environmental and Resource Economics, 57*, 233–249.

Simonit, S., & Perrings, C. (2011). Sustainability and the value of the regulating services: Wetlands and water quality in Lake Victoria. *Ecological Economics, 70*, 1189–1199.

TEEB. (2010). *The economics of ecosystems and biodiversity: Ecological and economic foundations*. London: Earthscan.

Troy, A., & Wilson, M. A. (2006). Mapping ecosystem services: Practical challenges and opportunities in linking GIS and value transfer. *Ecological Economics, 60*, 435–449.

Uk, N. E. A. (2011). *UK national ecosystem assessment: Understanding nature's value to society: Synthesis of the key findings*. Cambridge: UK National Ecosystem Assessment.

UNEP (United Nations Environment Programme) (2006). *Marine and coastal ecosystems and human well-being: Synthesis based on the findings of the millennium ecosystem assessment*. Nairobi: United Nations Environment Programme.

Van Beukering, P., Brander, L., Tompkins, E. & McKenzie, E. (2007). *Valuing the environment in small islands: An environmental economics toolkit.* Peterborough, UK: Joint Nature Conservation Committee (JNCC).

Veron, J. E. N., Hoegh-Guldberg, O., Lenton, T. M., Lough, J. M., Obura, D. O., Pearce-Kelly, P., et al. (2009). The coral reef crisis: The critical importance of <350 ppm CO_2. *Marine Pollution Bulletin, 58*, 1428–1436.

WTTC. (2013). *Travel and tourism economic impact 2013: Southeast Asia.* London: World Tourism and Travel Council.

Zhang, M., Zhang, C., Wang, K., Yue, Y., Xiangkun, Q., & Feide, F. (2011). Spatiotemporal variation of Karst ecosystem service values and its correlation and its correlation with environmental factors in Northwest Guangxi China. *Environmental Management, 48*, 1–12.

Part V
Bayesian Methods

Chapter 21
A Bayesian Model Averaging Approach to the Transfer of Subjective Well-Being Values of Air Quality

Carmelo J. León and Jorge E. Araña

Abstract This paper utilizes the subjective well being approach to benefit transfer of the welfare effects of reducing air pollution across a sample of citizens from several European countries. In addition, we compare the results of subjective well being utilizing the ordinal probit model approach with an alternative Bayesian Model Averaging (BMA) approach that selects the best model from the space of potential models. The results shows that the BMA approach significantly reduces transfer errors for air pollution externalities and provides a more robust strategy to modeling SWB data on pollution effects.

Keywords Air quality valuation · Subjective well-being · Bayesian model averaging · Benefit transfer

21.1 Introduction

Subjective well-being (SWB) is a valuation technique that has been increasingly utilized to measure the economic benefits of environmental quality (Bok 2010; Dolan et al. 2008; Ferrer-i-Carbonell and Gowdy 2007; Frey 2008; Frey and Stutzer 2002; Graham 2009; Kahneman et al. 2004; Welsch 2006; Welsch and Kühling 2010). The method involves asking individuals across society about what is the actual individual level of well-being (happiness, satisfaction) that they experience in their life at a moment of time or over a specified time period. By regressing the SWB across individual responses on the level of environmental quality and other socioeconomic

C.J. León · J.E. Araña (✉)
Instituto de Turismo y Desarrollo Económico Sostenible (TiDES),
Universidad de Las Palmas de Gran Canaria, Las Palmas, Spain
e-mail: jorge.arana@ulpgc.es

C.J. León
e-mail: cleon@daea.ulpgc.es

R.J. Johnston et al. (eds.), *Benefit Transfer of Environmental and Resource Values*,
The Economics of Non-Market Goods and Resources 14,
DOI 10.1007/978-94-017-9930-0_21

variables (including income and the level of environmental quality), an estimate of the monetary value of the level of environmental quality can be derived.

Although there is an extensive literature in economics highlighting a range of issues in making concise conclusions about the causes of SWB (Dolan et al. 2008), the approach has been used to value a wide range of public and environmental goods, either with individual (i.e. micro SWB) or aggregated data (i.e. macro SWB). Examples of these applications can be found in Becchetti et al. (2007), Frijters and van Praag (1998), and Rehdanz and Maddison (2005) for climate valuation; Di Tella and MacCulloch (2008) for the valuation of airport noise externalities; van Praag and Baarsma (2005) for assessing the costs of sulfur emissions; Dolan and Metcalfe (2008) for evaluating urban regeneration; and Luechinger and Raschky (2009) for preventing flood hazards.

The objective of this chapter is to propose an SWB benefit transfer (SWB-BT) technique that would enable researchers to produce estimates of SWB for the environmental quality of countries or regions that have not been covered with the existing studies. Among the primary research questions is the accuracy of the resulting out of sample SWB forecasts. To this end we utilize an innovative application of a Bayesian Model Averaging (BMA)[1] approach, which allows researchers to select the best econometric specification from the available data based on its predictive accuracy for new valuation contexts (Brock and Durlauf 2001; Brock et al. 2003; Sala-i-Martin 1997). Thus, the use of BMA enables the theoretical uncertainty existing among the factors determining individual SWB to be directly included in the estimation of the SWB equation.

This chapter focuses on the benefit transfer of air pollution externalities utilizing the SWB approach. Several papers have utilized SWB to value air quality. For instance, Welsch (2002, 2006, 2007) utilizes various cross-sections and panels of country-aggregated data to assess willingness to pay (WTP) for air quality improvements. In particular, Welsh (2006) focuses on the relationship between air pollution (nitrogen dioxide and lead) and SWB in 10 European countries, showing that air pollution plays an important role in predicting inter-country and inter-temporal differences in SWB. Di Tella and MacCulloch (2008) utilize aggregated country data to assess the relationship between SWB and per capita emissions of sulfur dioxide (SO_2), showing a negative relationship between these variables. Similarly, Luechinger (2010) finds a negative and significant relationship between SWB and national SO_2 concentrations in 13 European countries for a period of 15 years. In another study focusing on more disaggregated time series pollution data for regions in Germany, Luechinger (2009) reports a similar relationship between SWB and SO_2. In a recent paper, Levinson (2012) utilizes the SWB approach to derive welfare estimates of air pollution from particulate matter (PM10) in the United States, with environmental quality data obtained from the location where the happiness question was asked, rather than from regionally aggregated data.

[1]The Bayesian model average approach has been utilized in the context of environmental valuation by Koop and Tole (2004) and León-González and Scarpa (2008).

In the next section we outline the Bayesian Model Averaging approach to SWB-BT. This approach is a more flexible modeling strategy, departing from the basic SWB model that is estimated using an ordinal Probit model. The Bayesian Model Averaging approach allows us to select the model covariates that provide the best statistical performance to the data. Section 21.3 presents the data sources for the illustrated case study. Section 21.4 presents the results of the model, in which Bayesian Model Averaging is used to generate out-of-sample forecasts for the value of air quality in various European countries; these results are compared with those obtained with a standard ordinal Probit approach. Finally, Sect. 21.5 summarizes the main findings and implications of the results.

21.2 Valuing Air Quality Changes Based on Micro SWB Data

21.2.1 The Basic SWB Model

The standard micro SWB model involves a regression between the SWB or level of happiness of an individual and a set of variables that explain SWB, which include income and pollution or environmental quality. Thus, the objective is to prove the relationship between the respondents' stated level of SWB and a set of variables defining objective or subjective circumstances that might explain her level of SWB.

Since SWB is an ordinal variable, the basic approach shared by these applications is to run an Ordinary Least Squares (OLS) or an ordered Probit single regression of SWB/happiness data on relevant factors that are thought to determine individuals' well-being (e.g., income, lifestyle, etc.). That is,

$$H_{ij} = \alpha P_j + \gamma \ln Y_i + \beta X'_{ij} + \varepsilon_{ij} \tag{21.1}$$

where H_{ij} is the stated happiness of respondent i in location j. The variable P_j is the level of air quality experienced by individual i living at location j. The variable ($\ln Y_i$) represents the log of income.[2] The vector X_{ij} contains a set of other key aspects determining SWB, like demographic and local characteristics. ε_{ij} is a normally distributed error term with zero mean and unit variance. Once Eq. 21.1 is correctly estimated and identified, one can obtain the marginal rate of substitution (tradeoff) between air quality and income that will not alter the individual's perceived level of SWB. This can be defined as the marginal value of air quality improvements.

[2]The log specification of the income variable allows researchers to account for diminishing marginal utility of income on happiness (Levinson 2012). This assumption is supported by the extensive evidence estimating SWB functions, and it implies an increasing marginal WTP for air quality.

21.2.2 *Model Uncertainty and Bayesian Model Averaging for SWB Data*

Although most of the previous applications of SWB analysis employ the approach presented in the previous section, in practice there is uncertainty as to the key determinants of SWB, that is, which variables are relevant for each specific study. Iterative hypothesis testing presents the problem that each time a hypothesis test is carried out (i.e. whether or not a specific variable affects SWB), a possibility exists that a mistake has been made (i.e. the researcher rejects a better model for a not so good one). This possibility multiplies sequentially with each successive hypothesis test.

Nevertheless, even if a sequential hypothesis testing procedure does lead to the selection of the best model, standard decision theory implies that it is rarely desirable to present results for this model while ignoring all evidence from the not quite so good models. This point may explain the "common empirical wisdom that if one mines the data long enough one is bound to find something; however, one should not put too much trust in the finding." (Koop and Tole 2004, p. 33).

Given the problems in sequential hypothesis testing, a researcher may be tempted to include all potential variables in a regression. However, in general, the inclusion of irrelevant variables in an analysis will decrease the accuracy of the estimates and increase the difficulty of uncovering actual effects. In other words, if one starts running regressions combining various explanatory variables, variable x_1 will soon be found to be significant when the regression includes variables x_2 and x_3, but it becomes non-significant when x_4 is included. Since the "true" group of variables that should be included is not known, one is left with the question: ...what are the key factors that are really correlated with SWB, and therefore whose social value can be elicited through the use of an SWB analysis?

Such analysis leads to problems in interpreting the results of the relevant factors. Thus, the implicit use of *p*-values to select the relevant variables in this context leads to the variables with low values being discounted. Some authors demonstrate that the use of *p*-values following model selection can be dramatically misleading (Freedman 1983; Freedman et al. 1988). The most significant theoretical justifications for that bias are related to the fact that *p*-values ignore uncertainty about model form (Raftery 1995, 1996) and that the interpretation of the *p*-values when the set of potential variables is large is not the same as when there are only two alternative models (with and without the potential variable) (Miller 1990).

Bayesian model averaging (Hoeting et al. 1999) considers uncertainty in the selection of the variables explaining SWB. Hence, BMA allows researchers to estimate the probability that a specific variable x_1 affects SWB. As explained by Sala-i-Martin (1997, p. 182): ..."BMA allows us to depart from the standard zero-one labeling of variables as: significant" versus "non-significant". In fact, by mixing all potential explanatory models and considering the entire distribution of the estimators rather than a fixed point, BMA reports the exact probability that each lifestyle factor determines SWB".

Classical econometric methods (i.e. maximum likelihood) do not allow for the calculation of the probability that a model is the correct one. For this reason, many researchers use Bayesian methods to consider model uncertainty. Thus, the inclusion of uncertainty allows for the direct calculation of the probability of each potential variable being important in SWB, thereby representing a robust approach to test the hypotheses in this work. The mixed model that deals with uncertainty in variable selection and model specification is more accurate than any single model. Evidence of the superior predictive performance of BMA over other modeling approaches can be found, among others, in Fernandez et al. (2011), Koop and Tole (2004), Leon-Gonzalez and Scarpa (2008), and Raftery et al. (1997).

In spite all the theoretical advantages of BMA over more conventional regression analysis, empirical researchers always have the concern of whether or not using this approach makes a significant difference in economic terms rather than just on statistical terms (Ziliak and McCloskey 2008 for a nice review). In order to empirically compare the statistical and economic significance of using a BMA approach and the conventional ordinal Probit model for estimating determinants of SWB, a Monte Carlo simulation study was implemented. For the sake of simplicity, results are presented in Appendix A.

21.2.3 Welfare Estimation

The estimation of the relationship between air quality and SWB allows researchers to estimate the welfare impact of changes in air quality. This assumes that the SWB model correctly reflects the specification of a utility or welfare function. Therefore, totally differentiating the SWB equation leads to a monetary estimate of the change in SWB (utility or welfare) that is generated by a change in air quality, i.e. by calculating the quotient between the change in air quality and the marginal utility of income.

However, the estimation of the welfare change of air quality with models of individual SWB data differs somewhat from the estimation with aggregated data. The reason is that the questions on SWB and air quality perceptions are based on a 1–5 Likert scale, and therefore the appropriate model is ordinal and not linear. Therefore, the interpretation of the estimated coefficients with an ordinal scale model differs from the interpretation with linear regression models (Daykin and Moffatt 2002). The reason is that there is no natural function that reflects the conditional means of the model, since the dependent variable is simply a label that reflects the order in a non-quantitative level of the unobserved latent variable.

In order to calculate the impact of air quality on welfare in the ordinal scale model, it is necessary to relate the parameters of the model with the choice probabilities of each possible value of the scale, thus obtaining partial effects. That is, the marginal effect of a change in the level of air quality z_c on the probability of ordinal choice is:

$$\delta(z_c) = \frac{\partial \Pr(h = j/z)}{\partial z_c} = \left[f(\tau_{j-1} - \Gamma_c z_c^1) - f(\tau_j - \Gamma_c z_c^0)\right]\Gamma_c \tag{21.2}$$

where z_c is the covariate that represents the level of air quality, Γ_c is the parameter of this variable, z_c^0, z_c^1 are the initial and final values of the variable respectively, and τ_{j-1}, τ_j are the thresholds or cutoff points of the ordinal scale. Thus, in order to estimate the impact of a change in air quality on an individual's welfare we must substitute the value of the significant variables explaining SWB in Eq. 21.2, which can be approximated by the average value of these variables ($\overline{z}$), i.e.,

$$\frac{\partial \Pr(h = j/\overline{z})}{\partial z_c} = \left[f(\tau_{j-1} - \Gamma_c \overline{z}) - f(\tau_j - \Gamma_c \overline{z})\right]\Gamma_c \tag{21.3}$$

By conducting the same procedure to estimate the welfare impact of a change in the level of income (z_r), and by using the Delta method (Greene 2008, pp. 783–785), the monetary cost of air quality (CC) can be derived utilizing the following expression:

$$CC = \frac{\delta(z_c)}{\delta(z_r)} = \frac{\frac{\partial \Pr(h=j/z)}{\partial z_c} = \left[f(\tau_{j-1} - \Gamma_c z_c^1) - f(\tau_j - \Gamma_c z_c^0)\right]\Gamma_c}{\frac{\partial \Pr(h=j/z)}{\partial z_r} = \left[f(\tau_{j-1} - \Gamma_r z_r^1) - f(\tau_j - \Gamma_r z_r^0)\right]\Gamma_r} \tag{21.4}$$

where subscript r refers to the covariate of personal income.

21.3 The Data

The data were collected in 2010 with on-line sampling from the general public in France, Germany, Italy, Norway, Spain, Sweden and the United Kingdom. The sample size was 3830 individuals. The survey was designed to study the level of happiness of individuals in European countries, and was conducted by a professional firm following standard protocols for on-line surveying. Quota samples for age and sex were adopted in order to guarantee representative samples in these parameters. Internet sampling from a large panel of Internet users provides representative samples of the overall population if the web penetration in the population is larger than 50 % (ESOMAR 2011). Response rates were above 60 % in all countries. Participating subjects were incentivized with a €25 gift voucher for shopping in a major retail store in their respective countries. Non-respondents were followed up with two or three recall e-mails.

Respondents to the survey were asked a one-to-five Likert scale question on the state of happiness that they have in their current lives. The question was as follows: "How satisfied were you with your life in general in the last year, on a scale from 1 (not satisfied at all) to 5 (very satisfied)." A pretest with 418 individuals in all the countries considered in this study suggested that the questionnaire was generally well understood.

A common drawback of most previous literature using SWB to value air quality is the use of aggregate national or yearly measures of pollution.[3] It is well known that aggregating environmental quality across entire countries or regions masks much of its heterogeneity. In fact, it has been found that the standard deviation of particulate air pollution in the U.S. is twice as large if we look at daily observations within states instead of averages across states or years (Levinson 2012) and similar results within countries have been found in Europe (Welsch 2002).

This study addresses the issue by employing individual level SWB data and by collecting information regarding the level of environmental quality in the current location and at the same time as the survey was being implemented. For pollution information, we used data from the European Monitoring and Evaluation Programme (EMEP).[4] In particular, we employed the EMEP projection tool, which is a polar-stereographic projection that allows categorizing air quality levels with a cell-size resolution of 50 km × 50 km.

These secondary data and other socioeconomic characteristics of the respondents collected through the web-based survey were used as explanatory variables of the SWB responses in the econometric models. As discussed in the previous section, although the BMA accounts for all available information and model specifications (weighted by their ability to explain the sampling data set) the ordinal probit model considers only the variables and model specification that result in the best fit of the data. Also by incorporating socioeconomic background variables that have been identified by the past literature as important predictors of subjective well-being, plus time and region effects thereof, we aim to control for potential omitted variable bias.

21.4 Results

21.4.1 Estimated Models and WTP for Air Quality Improvements

The fundamental characteristic of the BMA approach is that it considers uncertainty in the selection of attributes determining SWB. This uncertainty is inherent to all studies of SWB since there can be several sets of explanatory variables that may be used to explain the data. The ordinal Probit model is an appropriate modeling specification for the type of ordinal data generated by the SWB question responses. Table 21.1 presents the estimation results for the ordinal Probit model and the BMA approach.

[3]Some very recent exceptions are Luechinger (2009) and Levinson (2012).

[4]http://www.ceip.at/webdab-emission-database.

Table 21.1 SWB estimation BMA and OLS results (standard errors in parentheses)

	Ordinal probit model	BMA	Percentiles posterior density		
	Mean (sd)	Mean	2.5 %	50.0 %	97.5 %
Constant	5.414*** (0.581)	3.125***	0.276	3.125	5.973
Age	0.289*** (0.068)	0.250***	0.199	0.250	0.301
Perceived HS	0.053 (0.033)	0.095***	0.080	0.095	0.109
Income	0.271 (0.205)	0.232***	0.157	0.232	0.307
Employed	0.141*** (0.059)	0.107**	0.071	0.107	0.142
Retired	−0.025*** (0.001)	−0.068**	−0.059	−0.068	−0.076
Self-employed	0.193*** (0.032)	0.108**	0.078	0.108	0.138
Gender	0.149*** (0.018)	0.169***	0.133	0.169	0.205
Married	2.135 (1.442)	1.486	−0.495	1.486	3.467
Kids	1.598** (0.787)	1.531	−0.628	1.531	3.690
Years of education	1.257 (1.122)	1.617	−2.026	1.617	5.260
SO_2	−0.118*** (0.052)	−0.090***	0.039	0.090	0.141
NO_2	−0.063*** (0.028)	−0.062***	0.014	0.062	0.118

* Significant at 0.10 level; ** significant at 0.05 level; *** significant at 0.01 level

In order to test whether individuals' SWB is affected by air pollution, an open research question is how to define the functional form of such a relationship. Most previous studies considered only one functional form (i.e. linear or logarithmic or with discontinuities).[5] We found that a linear functional form presented the best fit for the ordinal probit model. The BMA model, in contrast, considered different possible specifications.

After considering all potential variables in the model, BMA focuses only on the variables that are significantly correlated with SWB in a conventional statistical sense. That is, those variables for which the weighted cumulative distribution

[5]An example of discontinuous variables is presented in Menz and Welsch (2012). They found that a recoded pollution variable that takes value 0 if the air pollution level is lower than the critical annual mean value provided by the World Health Organization (WHO 2006) and the difference for values over such threshold resulted in a better fit for the CO_2, but not for the NO_2. The critical annual mean values were 20 ng/m for SO_2 and 40 g/m for NO_2.

Table 21.2 Purchasing power parity (PPP) converted WTP estimates based on BMA* (95 % confidence intervals in parentheses)

	NO_2		SO_2	
Country	E(WTP) 1 μg/m^3	E(WTP) 1 SD reduction	E(WTP) 1 μg/m^3	E(WTP) 1 SD reduction
Spain	107.41 [88.37, 126.45]	859 [671.90, 1046.70]	117.47 [95.61, 139.33]	857.56 [638.94, 1076.17]
Norway	114.82 [94.63, 135.01]	918.56 [630.37,1206.75]	191.50 [147.04, 235.96]	1397 [953.33, 1842.54]
Sweden	110.50 [89.85, 131.14]	883.99 [582.89, 1185.09]	197.13 [153.62, 240.64]	1439.05 [1003.96, 1874.14]
Italy	96.30 [75.26, 117.34]	770.41 [591.42, 949.39]	148.05 [91.32, 204.78]	1080.76 [513.48, 1648.03]
France	61.42 [48.65, 74.20]	491.38 [266.57, 716.19]	140.00 [114.70, 165.31]	1022.02 [769.97, 1275.07]
Germany	87.45 [69.46, 105.45]	699.64 [393.19, 1006.08]	176.21 [156.35, 196.07]	1286.33 [1087.73, 1484.94]
U.K.	50.31 [38.78, 61.84]	402.49 [242.86, 562.12]	104.60 [81.44, 127.76]	763.58 [532.01, 995.15]

*Euros in 2010 price levels

function (CDF) is larger than 0.95. The estimated weighted posterior median of the estimated coefficients for each variable are reported in Table 21.1, together with the 95 % posterior density values for each parameter. The table presents only the variables that appear to have a large probability of impacting SWB, i.e. that were significant in the regression analysis.[6] Thus, these variables were chosen to be included in the definitive model.

Table 21.2 shows the results of the BMA model's marginal WTP estimates for reducing SO_2 and NO_2 concentrations by 1 μm/m^3. The table also presents the results of the WTP estimates for reducing SO_2 and NO_2 concentrations by one standard deviation (SD). For NO_2 the largest value is obtained in Norway (€115.8 per 1 μg/m^3), followed by Sweden (€110.5) and Spain (€107.4). The lowest value is for the U.K. (€50.3). Therefore, there is a large dispersion between the highest and lowest values for this set of EU countries. These relative differences between the values across the different countries are similar for the measures of WTP for a one SD of NO_2 concentration.

The values of a unit reduction of SO_2 concentration are much larger than the value of a unit reduction of NO_2 for all countries except for Spain. The largest values of a unit reduction of SO_2 are found in Sweden (€197.1) and Norway (€191.5), followed by Germany (€176.2), Italy (€148.0) and France (€140.0). The U.K. again has the lowest value of a unit reduction of SO_2 with €104.6. The relative values are very similar, but much higher for a reduction of one SD of SO_2

[6]See Appendix for details of determining relevant covariates in the regression.

concentration level. Further, the relative dispersion in the values between the different countries is much lower for the reduction in one SD of SO_2 than for the reduction in a unit of SO_2.

21.4.2 Benefit Transfer Results

Benefit transfer involves predicting the value at an unstudied policy site from observations gathered at a set of study sites. In this paper, we test for the transferability of values by considering each of the countries in the data set as a potential policy site whose value might be transferred from the observations available from the rest of the countries. Thus, we compare the valuation result for each country obtained with the sampling data with the one that could be obtained by transferring the results based on the rest of the countries in the data set. There are several specific issues that need to be addressed in order to implement international benefit transfer studies, such as like currency conversion, differences in measurable attributes of the users, wealth versus income, differences in culture, extent of the market,[7] etc.

For the purposes of comparison, we evaluate the transfer of values based on the unit value transfer approach and those obtained with benefit function transfer. The *unit value* transfer approach involves transferring values from known study sites to some policy site (Luken et al. 1992; Ready and Navrud 2007; Bateman et al. 2011). In our case we transfer the elicited mean WTP obtained from the model estimations. On the other hand, the benefit function transfer is based on the specification of a valuation function utilizing explanatory variables that draw from the results and characteristics of a set of available studies (Nelson and Kennedy 2009). The latter can be approached either by considering the maximum likelihood (ML) estimates of the valuation function or the BMA approach that selects the most likely variables that specify the model.

The analysis was implemented using a jacknife methodology, that is, the predicted value for each country—both in the unit value (mean) as in the benefit function transfer (structural model)—was obtained by estimating the model for all the countries but the one in consideration. The same procedure was repeated for each country in the analysis. Then the transfer errors were calculated as the difference between the predicted and observed value for each country.

Table 21.3 presents the results of the transfer errors for all the countries considered as potential policy sites for the transfer of values, for both the SO_2 and NO_2 concentration reductions. In general, the lowest transfer errors are obtained when transferring values to Norway and Sweden. The highest transfer errors are obtained when transferring to Spain, with 83 % for the unit value transfer method for SO_2

[7]Some available reviews of the international benefit transfer issues are Shrestha and Loomis (2001), Ready et al. (2004), and Muthke and Holm-Mueller (2004), among others.

Table 21.3 Performance of value transfer methods (measured as percentage transfer errors)

Country	SO_2			NO_2		
	Unit value transfer	Value function transfer (ML)	Value function transfer (BMA)	Unit value transfer	Value function transfer (ML)	Value function transfer (BMA)
Norway	30.9	40.1	30.3	27.7	37.1	25.5
Sweden	34.7	41.0	30.5	30.2	37.1	27.5
Italy	48.3	47.1	45.2	48.0	42.7	40.4
France	39.1	38.0	36.4	41.3	41.3	31.7
Germany	44.9	43.8	41.6	42.5	47.4	37.8
U.K.	43.7	32.1	31.2	46.9	32.6	30.3
Spain	83.3	75.7	70.2	87.3	71.1	70.7
Average	46.4	45.4	40.8	46.3	44.2	37.7

and 87 % for NO_2. Overall, transfer errors are slightly lower for NO_2 than for SO_2. The unit value transfer approach leads to higher transfer errors than the value function transfer approach (either with ML or with BMA) for all countries except for Norway and Sweden, for both pollutants. Average transfer errors across countries are higher for the unit value transfer method (46 %) and for the ML value function transfer (45 % for SO_2 and 44 % for NO_2) than for the BMA value function transfer (41 % for SO_2 and 38 % for NO_2). This is the case also for the individual countries, although there are some differences in the performance of BMA across countries and pollutants.

Table 21.4 presents the changes in transfer errors obtained with the BMA approach against the unit value and the ML value function approaches. In the case of SO_2, the BMA approach reduces the average transfer error by 11 % against the unit value transfer method and by 10 % against the ML value transfer approach. For NO_2 the BMA approach reduces the average transfer error by 17 % against the unit value transfer method and by 16 % against the ML value transfer approach.

Table 21.4 Reduction in transfer errors (in percentage) of the BMA value function approach

Country	SO_2		NO_2	
	Unit value transfer	Value function transfer (ML)	Unit value transfer	Value function transfer (ML)
Norway	1.94	24.44	7.94	31.27
Sweden	12.10	25.61	8.94	25.88
Italy	6.42	4.03	15.83	5.39
France	6.91	4.21	23.24	23.24
Germany	7.35	5.02	11.06	20.25
U.K.	28.60	2.80	35.39	7.06
Spain	15.73	7.27	19.01	0.56
Average	11.29	10.48	17.34	16.23

Therefore, the BMA approach reduces on average the transfer errors for all countries between 10 and 17 %. The highest reduction in transfer errors with the BMA approach is obtained for the U.K. (−28 %) and Spain (−15 %) when compared against the unit value method in the reduction of SO_2, and also France (−23 %) in the reduction of NO_2. When compared against the ML value function transfer approach, the BMA leads to the highest reduction in transfer errors in the case of Sweden (−24 %) and Norway (−25 %) for SO_2. In the case of NO_2, the highest reductions of transfer errors with the BMA approach are obtained when compared with the value function transfer approach for Norway (−31 %), Sweden (−25 %), France (−23 %) and Germany (−20 %).

21.5 Conclusions

The SWB approach is increasingly utilized to value environmental and public goods. Since it has been increasingly applied to value air pollution and other externalities, there is a need to evaluate its performance in benefit transfer studies. In this paper we evaluated the application of SWB to benefit transfer for the case of air pollution externalities across a set of European countries. SWB was shown to be significantly related to the level of pollutants. This allowed us to derive economic values for these pollutants. The values are larger for the reduction of SO_2 than for the reduction of NO_2 for most individual countries. The value of NO_2 and SO_2 reductions are highest in Norway and Sweden and lowest in the U.K. These values are obtained by regressing for each specific country the individuals' SWB on the level of pollutants that individuals experience at a regional level as well as a set of socioeconomic characteristics.

The predictions of out of sample observations that are obtained with the traditional benefit transfer approaches (unit value transfer and value function transfer) are compared with those obtained with the BMA approach that selects the most likely model among the set of all potential models. That is, BMA selects the explanatory variables for SWB that provides the best fit to the data among the potential set of explanatory variables. The results show that SWB-BT based on BMA can successfully provide out of sample predictions on the economic value of the reduction of air pollutants, outperforming the traditional BT methods. These predictions do vary in terms of the type of pollutant and the country of interest. The largest reductions in prediction errors utilizing the BMA approach are obtained for Sweden and Norway. The average reduction in transfer errors across the set of EU countries in this study amounts to about 10 % for SO_2 and 17 % for NO_2.

Although the SWB approach seems to hold promise for benefits transfer applications, irrespective of applying it to SWB values or values directly elicited through stated preference methods, there also seems to be sufficient scope for further improvements in its application in out of sample predictions based on different modeling strategies like correcting for potential scale perception bias (León et al. 2013a, b; Araña and León 2012, 2013). The BMA approach provides researchers with a robust methodology that might increase the predictive power of benefit transfer by selecting the most likely model among all the potential models that can be utilized for specifying value transfer functions. An interesting avenue for further research is to extend BMA for other types of benefit transfer studies.

Appendix A Monte Carlo Simulation of BMA Versus Ordinal Probit

In this appendix we compare the performance of the ordinal Probit model against the BMA approach to estimate the social well-being (SWB) model. The empirical comparison of the two estimation approaches is not an easy task. The main issue comes from the fact that, while comparing alternative estimation strategies to specify the SWB model, one can know only whether they converge in terms of the mean parameters and/or whether or not the of one of them is higher or lower than the other one.

These results may be useful (although one may prefer to use a method that results in estimations with lowest variances), but they are very limited (since a lower variance for a biased estimation is likely to be worse that a high variance for an unbiased estimation). In other words, ideally researchers would like to choose among methodologies just by testing their to estimate the "true parameters" for the specific application at hand. Since one never knows what the true parameters are in any empirical comparisons, comparisons are always incomplete.

Thus, the comparison between the both approaches can be evaluated by Markov Chain Monte Carlo (MCMC). That is, we can compare both methodologies based on how close are the assumed true results to the true ones generated utilizing simulated data sets. The algorithm is as follows:

1. Let us consider that we know the true covariates that are relevant in determining SWB and the true value of their parameters, and that these are given by those presented in Table 21.5.
2. Let us collect a sample of 300 individuals from the SWB data set. Based on the true specification, we can estimate the true SWB for each individual.
3. True SWB levels and covariates are compared to the results of alternative estimation methods with the sample data in order to find the true relevant variables that determine SWB levels and their true level of parameters.
4. The sampling experiment is replicated 1000 times. For each sample their true SWB levels are evaluated based on the assumed true values of the parameters, and the Bayesian Model Averaging (BMA) and ordinal Probit models are estimated in order to compare their estimations with the true relevant covariates and parameter values.

After implementing the Monte Carlo experiments we have information for 1000 different studies (i.e. samples). For each one of them we know the final results of

Table 21.5 MC results for comparing BMA and ordinal probit model for SWB

Covariates	True mean values	Ordinal probit model		BMA	
		MSE	Percentage times covariate is defined as relevant	MSE	Percentage times covariate is defined as relevant
True relevant covariates					
Constant	3.1245	0.74	82.98	0.44	93.67
Age	0.2498	0.87	85.61	0.39	100.00
Perceived health status	0.0945	0.98	68.31	0.48	89.13
Income	0.2322	1.13	91.29	0.59	97.23
Employed	0.1067	0.92	72.01	0.44	91.35
Retired	−0.0676	0.81	84.23	0.32	95.63
Self-employed	0.2895	0.95	88.16	0.56	94.84
NO_2	−0.1076	0.79	88.53	0.41	91.62
SO_2	−0.1690	0.84	82.93	0.32	92.32
True non-relevant covariates					
Married	0	1.48	45.81	1.29	15.93
Kids	0	1.53	52.08	1.31	8.33
Years of education	0	1.61	37.55	1.22	4.07

the ordinal Probit model and BMA in terms of final relevant covariates and the estimated parameters value in each case. Table 21.5 presents results of the mean squared error (MSE) of each estimation methodology and the percentage of times that each covariate is assigned for each methodology. The criteria used to decide which variables were relevant or not was a p-value lower than 0.05 for the ordinal Probit approach and a posterior marginal probability containing a zero value lower than 5 %.

The results in Table 21.5 show that BMA improves performance over the ordinal Probit model by reducing estimated bias and reducing the percent of non-allocation of relevant variables to the SWB equation. BMA results in smaller MSE almost for all parameters and samples. In addition, the proportion of times that relevant variables are assigned as relevant by using BMA is much larger with ordinal Probit.

The performance of both estimation strategies for the group of non-relevant covariates is presented in the second part of Table B1, i.e. the covariates for which the true parameter was equaled to 0. It can be seen that the percentage of times that non-relevant covariates are considered as relevant for the ordinal Probit is significantly higher than for the BMA. Therefore, the BMA seems to be more robust to the non-inclusion of non-relevant attributes.

References

Araña, J. E., & León, C. J. (2012). Scale-perception bias in the valuation of environmental risks. *Applied Economics, 44*(20), 2607–2617.

Araña, J. E., & León, C. J. (2013). Correcting for scale perception bias in tourist satisfaction surveys. *Journal of Travel Research, 52*, 772–788.

Bateman, I. J., Brouwer, R., Ferrini, S., Schaafsma, M., Barton, D. N., Dubgaard, A., et al. (2011). Making benefit transfers work: Deriving and testing principles for value transfers for similar and dissimilar sites using a case study of the non-market benefits of water quality improvements across Europe. *Environmental and Resource Economics, 50*, 365–387.

Becchetti, L., Castriota, S., & Londono Bedoya, D. A. (2007). *Climate, happiness and the Kyoto protocol: Someone does not like it hot*. Departmental Working Paper No. 247, University of Rome Tor Vergata, Rome.

Bok, D. (2010). *The politics of happiness: What government can learn from the new research on well-being*. Princeton, NJ: Princeton University Press.

Brock, W., & Durlauf, S. (2001). Growth economics and reality. *World Bank Economic Review, 15*, 229–272.

Brock, W. A., Durlauf, S. N., & West, K. D. (2003). Policy evaluation in uncertain economic environments (with comments and discussion). *Brookings Papers of Economic Activity, 1*, 235–322.

Daykin, A. R., & Moffatt, P. G. (2002). Analyzing ordered responses: A review of the ordered probit model. *Understanding Statistics, 1*, 157–166.

Di Tella, R., & MacCulloch, R. (2008). Gross national happiness as an answer to the Easterlin paradox? *Journal of Development Economics, 86*, 22–42.

Dolan, P., & Metcalfe, R. (2008). *Valuing non-market goods: A comparison of preference-based and experience based approaches*. London: Tanaka Business School.

Dolan, P., Peasgood, T., & White, M. (2008). Do we really know what makes us happy? A review of the economic literature on the factors associated with subjective well-being. *Journal of Economic Psychology, 29*, 94–122.

ESOMAR. (2011). *Guideline for online research*. http://www.esomar.org/knowledge-and-standards/codes-and-guidelines/guideline-for-online-research.php.

Fernandez, C., Ley, E., & Steel, M. F. J. (2011). Model uncertainty in cross-country growth regressions. *Journal of Applied Econometrics, 16*, 563–576.

Ferrer-i-Carbonell, A., & Gowdy, J. M. (2007). Environmental degradation and happiness. *Ecological Economics, 60*, 509–516.

Freedman, D. A. (1983). A note on screening regression equations. *The American Statistician, 37*, 152–155.

Freedman, D. A., Navidi, W., & Peters, S. C. (1988). On the impact of variable selection in fitting regression equations. In T. K. Dijkstra (Ed.), *On model uncertainty and its statistical implications* (pp. 1–16). Berlin: Springer.

Frey, B. S. (2008). *Happiness: A revolution in economics*. Cambridge, MA: MIT Press.

Frey, B. S., & Stutzer, A. (2002). *Happiness and economics*. Princeton, NJ: Princeton University Press.

Frijters, P., & Van Praag, B. M. S. (1998). The effects of climate on welfare and well-being in Russia. *Climatic Change, 39*, 61–81.

Graham, C. (2009). *Happiness around the world: The paradox of happy peasants and miserable millionaires*. Oxford, UK: Oxford University Press.

Greene, William H. (2008). *Econometric analysis* (5th ed.). Upper Saddle River, NJ: Pearson/Prentice Hall.

Hoeting, J. A., Madigan, D., Raftery, A. E., & Volinsky, C. T. (1999). Bayesian model averaging: A tutorial. *Statistical Science, 14*, 382–417.

Kahneman, D., Krueger, A. B., Schkade, D. A., Schwarz, N., & Stone, A. A. (2004). Toward national well-being accounts. *American Economic Review, 94*, 429–434.

Koop, G., & Tole, L. (2004). Measuring the health effects of air pollution: To what extent can we really say that people are dying from bad air? *Journal of Environmental Economics and Management, 47*, 30–54.

León, C. J., Araña, J. E., & de León, J. (2013a). Valuing the social cost of corruption using subjective well being data and the technique of vignettes. *Applied Economics, 45*, 3863–3870.

León, C. J., Araña, J. E., & de León, J. (2013b). Correcting for scale perception bias in measuring corruption: An application to Chile and Spain. *Social Indicators Research, 114*, 977–995.

León-Gonzalez, R., & Scarpa, R. (2008). Improving multi-site benefit functions via Bayesian model averaging: A new approach to benefit transfer. *Journal of Environmental Economics and Management, 56*, 50–68.

Levinson, A. (2012). Valuing public goods using happiness data: The case of air pollution. *Journal of Public Economics, 96*, 869–880.

Luechinger, S. (2009). Valuing air quality using the life satisfaction approach. *Economic Journal, 119*, 482–515.

Luechinger, S. (2010). Life satisfaction and the transboundary problem. *Economics Letters, 107*, 4–6.

Luechinger, S., & Raschky, P. A. (2009). Valuing flood disasters using the life satisfaction approach. *Journal of Public Economics, 93*, 620–633.

Luken, R. A., Johnson, F. R., & Kibler, V. (1992). Benefits and costs of pulp and paper effluent controls under the clean-water-act. *Water Resources Research, 28*, 665–674.

Menz, T., & Welsch, H. (2012). Life-cycle and cohort effects in the valuation of air quality: Evidence from subjective well-being data. *Land Economics, 88*, 300–325.

Miller, A. J. (1990). *Subset selection in regression*. London: Chapman and Hall.

Muthke, T., & Holm-Mueller, K. (2004). National and international benefit transfer testing with a rigorous test procedure. *Environmental and Resource Economics, 29*, 323–336.

Nelson, J. P., & Kennedy, P. E. (2009). The use (and abuse) of meta-analysis in environmental and resource economics: An assessment. *Environmental and Resource Economics, 42*, 345–377.

Raftery, A. E. (1995). Bayesian model selection in social research (with discussion). In P. V. Marsden (Ed.), *Sociological methodology* (pp. 111–195). Cambridge, MA: Blackwell.

Raftery, A. E. (1996). Approximate Bayes factors and accounting for model uncertainty in generalised linear models. *Biometrika, 83*, 251–266.

Raftery, A. E., Madigan, D., & Hoeting, J. A. (1997). Bayesian model averaging for linear regression models. *Journal of the American Statistical Association, 92*, 179–191.

Ready, R., & Navrud, S. (2007). Morbidity value transfer. In S. Navrud & R. Ready (Eds.), *Environmental value transfer: Issues and methods* (pp. 77–88). Dordrecht: Springer.

Ready, R., Navrud, S., Day, B., Dubourg, R., Machado, F., Mourato, S., et al. (2004). Benefit transfer in Europe: How reliable are transfers between countries? *Environmental and Resource Economics, 29*, 67–82.

Rehdanz, K., & Maddison, D. (2005). Climate and happiness. *Ecological Economics, 52*, 111–125.

Sala-i-Martin, X. (1997). I just ran two million regressions. *American Economic Review, 87*, 178–183.

Shrestha, R. K., & Loomis, J. B. (2001). Testing a meta-analysis model for benefit transfer in international outdoor recreation. *Ecological Economics, 39*, 67–83.

Van Praag, B. M. S., & Baarsma, B. (2005). Using happiness surveys to value intangibles: The case of airport noise. *Economic Journal, 115*, 224–246.

Welsch, H. (2002). Preferences over prosperity and pollution: Environmental valuation based on happiness surveys. *Kyklos, 55*, 473–494.

Welsch, H. (2006). Environment and happiness: Valuation of air pollution using life satisfaction data. *Ecological Economics, 58*, 801–813.

Welsch, H. (2007). Environmental welfare analysis: A life satisfaction approach. *Ecological Economics, 62*, 544–551.

Welsch, H., & Kühling, J. (2010). Pro-environmental behavior and rational consumer choice: Evidence from surveys of life satisfaction. *Journal of Economic Psychology, 31*, 405–420.
WHO (2006). World Health Organization Air quality guidelines. Accesed on March 2015. http://whqlibdoc.who.int/hq/2006/WHO_SDE_PHE_OEH_06.02_eng.pdf
Ziliak, S. T., & McCloskey, D. N. (2008). *The cult of statistical significance: How the standard error costs us jobs, justice and lives*. Ann Arbor, MI: University of Michigan Press.

Chapter 22
Optimal Scope and Bayesian Model Search in Benefit Transfer

Klaus Moeltner

Abstract The practice of benefit transfer requires the choice of suitable sources of primary data. The existence of "Golden Rules" notwithstanding, there usually exists limited theoretical guidance for this crucial first step. This chapter illustrates how Bayesian econometric methods can be helpful in identifying the best combination of existing data sources for a given benefit transfer application. As shown in recent contributions, this "quest for optimal scope" can greatly enhance the robustness and efficiency of transfer estimates.

Keywords Data combination · Bayesian model search · Outdoor recreation · Willingness to pay

22.1 Optimal Scope in Benefit Transfer

As discussed in the preceding chapters of this book, benefit transfer (BT) requires first of all the identification of suitable sources of information or data to feed into the transfer estimate or function. Existing guidelines urge for commodity equivalence, population similarity, and similar baseline and change of environmental quality (e.g., Boyle and Bergstrom 1992; Brouwer 2000; Loomis and Rosenberger 2006; U.S. Environmental Protection Agency 2000). However, the strict adherence to this "Golden Rule" (GR) often leaves only a small sample of source observations to form the basis for the transfer. This negatively affects both the representativeness and the precision of the predicted outcome for the policy context.

Thus, the BT analyst is often faced with a fundamental dilemma: Make do with a handful of sources or relax one or more of these paradigms at the gain of more primary observations, i.e., a larger "information pool" to feed into the transfer

---

K. Moeltner (✉)
Department of Agricultural and Applied Economics, Virginia Tech, Blacksburg, VA, USA
e-mail: moeltner@vt.edu

R.J. Johnston et al. (eds.), *Benefit Transfer of Environmental and Resource Values*, The Economics of Non-Market Goods and Resources 14,
DOI 10.1007/978-94-017-9930-0_22

process. The latter approach is what Moeltner and Rosenberger (2008) call a broadening of *scope* of the BT context.[1]

Such a broadening-of-scope strategy is feasible and beneficial in many applications. As shown in Moeltner et al. (2009) different populations can exhibit very similar value distributions for a specific commodity. In their case, the willingness-to-pay (WTP) distributions for conserving a specific plot of farmland largely overlap for several mid-Atlantic communities, despite pronounced heterogeneity in underlying parameter estimates and population statistics. This suggests a broadening of scope along the population dimension, i.e., a pooling of primary observations across different populations, even though the BT context applies to only one of them.

Conversely, commodities that differ in policy-relevant dimensions can also share common value distributions within the same population, as illustrated in Moeltner and Rosenberger (2008). They find that pooling WTP estimates for running water fishing across coldwater and warmwater fisheries enhances the efficiency of benefit estimates for the coldwater context, the original BT target.

Moeltner and Rosenberger (2014) extend this concept further by examining whether information pooling might be feasible across *both* the commodity and population dimension. They consider 31 combinations of different outdoor activities and populations and find that many of them share common value distributions. Exploiting these information pools substantially enhances the efficiency of BT estimates, in the sense of a much reduced span of predictive confidence intervals.

If one accepts—or, better, embraces—the possibility of broadening scope in a given BT application, finding the best composition of source studies and observations can quickly become a formidable challenge. This chapter illustrates how Bayesian model search and model averaging techniques can provide guidance in finding the optimal information pool.

22.2 Optimal Scope and Model Uncertainty

In Bayesian estimation, a model m is defined by its likelihood function and the prior distribution of its parameters. We can generically express the former as $p(\mathbf{y}|\boldsymbol{\theta}, \mathbf{X}, m)$, and the latter as $p(\boldsymbol{\theta}|m)$, where $\mathbf{y}$ is a vector of outcomes (i.e. the "dependent variable" in regression analysis), $\mathbf{X}$ is an (optional) matrix of explanatory variables, and $\boldsymbol{\theta}$ comprises unknown parameters. If multiple models are considered, the analyst may also assign a separate prior model probability, $p(m)$.

[1]For this chapter we interpret the term "commodity" to encompass all relevant physical and environmental features at a given site, and the term "context" to capture the combined elements of site-specific commodity and population.

Using Bayes' Rule the model-specific *posterior* distribution for its parameters can be expressed as

$$p(\boldsymbol{\theta}|\mathbf{y},\mathbf{X},m)=\frac{p(\boldsymbol{\theta}|m)p(\mathbf{y}|\boldsymbol{\theta},\mathbf{X},m)}{p(\mathbf{y}|\mathbf{X},m)} \tag{22.1}$$

The term in the denominator of (22.1) is called "marginal likelihood". It can also be written as

$$p(\mathbf{y}|\mathbf{X},m)=\int_{\theta} p(\mathbf{y}|\boldsymbol{\theta},\mathbf{X},m)p(\boldsymbol{\theta}|m)\mathrm{d}\boldsymbol{\theta} \tag{22.2}$$

Intuitively, it describes what we would expect the data to look like, based on our parameter priors and likelihood function, before we collect it. It is not a direct function of $\boldsymbol{\theta}$, and can thus usually be ignored for most components of Bayesian analysis. However, the marginal likelihood is crucial for model comparison and model averaging. This becomes obvious when we derive an expression for the *posterior model probability* by re-applying Bayes' Rule:

$$p(m|\mathbf{y},\mathbf{X})=\frac{p(m)p(\mathbf{y}|\mathbf{X},m)}{p(\mathbf{y}|\mathbf{X})} \tag{22.3}$$

The numerator is the product of the *prior model probability* (often set to equal values across models in absence of strong priors) and the model-conditioned marginal likelihood from (22.2). We can now construct the *posterior odds ratio* for any two models as

$$\frac{p(m_1|\mathbf{y},\mathbf{X})}{p(m_2|\mathbf{y},\mathbf{X})}=\frac{p(m_1)p(\mathbf{y}|\mathbf{X},m_1)}{p(m_2)p(\mathbf{y}|\mathbf{X},m_2)} \tag{22.4}$$

Under equal model priors [i.e., $p(m_1)=p(m_2)$] this reduces to the *Bayes Factor* for model 1 versus 2, i.e.,

$$BF_{1,2}=\frac{p(\mathbf{y}|\mathbf{X},m_1)}{p(\mathbf{y}|\mathbf{X},m_2)} \tag{22.5}$$

which is simply the ratio of model-conditioned marginal likelihoods.

Similarly, when more than two models are under consideration marginal likelihoods can be used to derive posterior model probabilities, or *model weights*, by relating any model-specific marginal likelihood to the sum of marginal likelihoods across all models, i.e.:

$$p(m|\mathbf{y})=\frac{p(\mathbf{y}|\mathbf{X},m)p(m)}{\sum_{j=1}^{M}p(\mathbf{y}|\mathbf{X},j)p(j)} \tag{22.6}$$

where M is the *model space*, i.e., the full set of possible models considered by the analyst. If model priors are the same for all models, this further simplifies to

$$p(m|\mathbf{y},\mathbf{X}) = \frac{p(\mathbf{y}|\mathbf{X},m)}{\sum_{j=1}^{M} p(\mathbf{y}|\mathbf{X},j)} \tag{22.7}$$

A *model-averaged* posterior distribution for parameters θ can then be obtained via

$$p(\boldsymbol{\theta}|\mathbf{y},\mathbf{X}) = \sum_{m=1}^{M} p(\boldsymbol{\theta}|\mathbf{y},\mathbf{X},m)p(m|\mathbf{y},\mathbf{X}) \tag{22.8}$$

In practice this is implemented by first obtaining R draws of $p(\boldsymbol{\theta}|\mathbf{y},\mathbf{X},m)$ for each model, and then randomly drawing from these model-specific posteriors with relative frequency dictated by the computed model weights in (22.6) or (22.7). All subsequent inference, including posterior predictive densities, are then based on these model-weighted, or model-averaged, draws.

To relate this general framework of Bayesian model uncertainty to the question of optimal scope for BT, we need to distinguish between two types of evidence from existing primary sources: (1) All context-specific information as would be suggested by the GR of population and commodity equivalence, and (2) All other possibly relevant information that, via augmentation of scope, *might* improve BT accuracy and/or efficiency. For example, if the policy context is "small-game hunting in the northeast," an analyst following the GR would limit herself to source studies that cover this specific activity in this specific geographic region. In contrast, a researcher looking to augment scope may also consider other activities and regions with possibly related value distributions, such as "waterfowl hunting in the northeast," "big game hunting in the west," and so on.

Let the GR-subset of information be denoted as $\{\mathbf{y}_g, \mathbf{X}_g\}$ and the remaining available data as $\{\mathbf{y}_o, \mathbf{X}_o\}$, where subscript g stands for "Golden Rule," and subscript o for "other evidence." The generic terms $\mathbf{y}$ and $\mathbf{X}$ from the general exposition above are then obtained via the union of these subsets, i.e.,

$$\begin{aligned} \mathbf{y} &= \{\mathbf{y}_g \cup \mathbf{y}_o\} \\ \mathbf{X} &= \{\mathbf{X}_g \cup \mathbf{X}_o\} \\ \boldsymbol{\theta} &= \{\boldsymbol{\theta}_g \cup \boldsymbol{\theta}_o\}, \end{aligned} \tag{22.9}$$

where the last line indicates that parameters might differ depending on which subset of data is considered for estimation. In most applications the set of "other" evidence can be logically further subdivided into smaller components, say $\{\mathbf{y}_{o,j}, \mathbf{X}_{o,j}, \boldsymbol{\theta}_{o,j}\}, \quad j = 1 \ldots J$.[2] We can then translate the question of optimal scope

[2]The individual $\mathbf{y}_{o,j}$ vectors and $\mathbf{X}_{o,j}$ matrices can comprise individual-level observations or site- or study-specific aggregates, as would be the case in a typical meta-regression analysis. The

into one of *model selection* by considering the optimal subset of $\{\mathbf{y}_o, \mathbf{X}_o, \boldsymbol{\theta}_o\}$ that should be used in conjunction with $\mathbf{y}_g, \mathbf{X}_g$ and $\boldsymbol{\theta}_g$ to derive a benefit estimate or transfer function. In other words, we can define a model m as a specific partition of $\mathbf{y}$, $\mathbf{X}$ and $\boldsymbol{\theta}$ into BT-relevant and BT-irrelevant parts, i.e., into $\{\mathbf{y}_r, \mathbf{X}_r, \boldsymbol{\theta}_r\}$ and $\{\mathbf{y}_n, \mathbf{X}_n, \boldsymbol{\theta}_n\}$, where r indicates "relevant" and n stands for "non-relevant." This will be reflected in the likelihood function and parameter priors:

$$\begin{aligned} p(\mathbf{y}|\boldsymbol{\theta}, \mathbf{X}, m) &= p(\mathbf{y}_{r,m}|\boldsymbol{\theta}_{r,m}, \mathbf{X}_{r,m}) * p(\mathbf{y}_{n,m}|\boldsymbol{\theta}_{n,m}, \mathbf{X}_{n,m}), \quad \text{and} \\ p(\boldsymbol{\theta}|m) &= p(\boldsymbol{\theta}_{r,m}) * p(\boldsymbol{\theta}_{n,m}) \end{aligned} \tag{22.10}$$

In other words, priors and likelihood are multiplicatively separated into relevant and irrelevant components, where the definition of "relevance" is specific to a given model. The marginal likelihood, leading to posterior model probabilities, can then be written as

$$\begin{aligned} p(\mathbf{y}|\mathbf{X}, m) &= \int_{\boldsymbol{\theta}} p(\mathbf{y}|\boldsymbol{\theta}, \mathbf{X}, m) p(\boldsymbol{\theta}|m) \mathrm{d}\boldsymbol{\theta} = \int_{\boldsymbol{\theta}_{r,m}} p(\mathbf{y}_{r,m}|\boldsymbol{\theta}_{r,m}, \mathbf{X}_{r,m}) p(\boldsymbol{\theta}_{r,m}) \mathrm{d}\boldsymbol{\theta}_{r,m} \\ &\times \int_{\boldsymbol{\theta}_{n,m}} p(\mathbf{y}_{n,m}|\boldsymbol{\theta}_{n,m}, \mathbf{X}_{n,m}) p(\boldsymbol{\theta}_{n,m}) \mathrm{d}\boldsymbol{\theta}_{n,m} \end{aligned} \tag{22.11}$$

In most applications the optimal separation into relevant and irrelevant data parts will be largely an empirical question. This calls for an estimation approach that efficiently examines and compares different data combinations in a quest for optimal BT results. In a classical framework this would require an intractably long sequence of hypothesis tests, with the usual risks of propagating decision errors and other problems related to pretest estimators (e.g., Leamer 1983). In contrast, a Bayesian model search algorithm is ideally suited for this challenge.

The essential steps for Bayesian model search are as follows:

1. Define the full *model space M*, i.e., the total number of all feasible or permissible models. In the context of optimal scope, this translates into all possible separations of relevant and irrelevant data components.
2. Assign a *prior model probability* $p(m)$ to each possible case. This is often simply chosen as $\frac{1}{M}$.
3. Assign prior densities to all parameters $\boldsymbol{\theta}$ and for all models—several "convenient priors" are available for this step.
4. If M is "small enough," i.e., each individual model can be estimated within a reasonable time frame, derive *posterior model probabilities* $p(m|\mathbf{y})$ for each model. If there is overwhelming evidence for a single model, use that for

(Footnote 2 continued)
model search techniques discussed in this chapter can be applied to either situation, and even a mixture of data at different levels of aggregation.

subsequent analysis, i.e., for the construction of a transfer function of BT estimate. Else use the posterior model probabilities as discrete weights, and base inference on *model-averaged* results using (22.8).

5. If M is large, as will usually be the case, develop a search algorithm that efficiently travels through model space and identifies the most promising data combinations. Then proceed as in the previous step, with inference based on the most frequently chosen model, or a weighted average of all models that were visited at least once by the algorithm.

This avoids the classical pitfall of erroneously assigning zero probability to models that are at least to some extent supported by the data. In addition, it gives the analyst the opportunity to add additional available information for models or parameters that is not contained in the actual data via *informed priors*.[3]

Figure 22.1 illustrates the concept of Bayesian model averaging. It depicts a situation where two candidate models are ex ante considered by the analyst, i.e., $M = 2$, presumably with equal prior probabilities. To continue with our example from above, model 1 may have considered only data related to the GR context as relevant, i.e., "small game hunting in the northeast," and treated all other contexts as irrelevant. Model 2, in turn, may have combined the GR context with the "other" category, "waterfowl hunting in the northeast," to form the relevant information pool, and left only "big game hunting in the west" as irrelevant.

Each model in isolation generates a posterior distribution of WTP for the resource or amenity subject to valuation for the policy context, in this case "small-game hunting in the northeast." This is shown in the top panel of the figure. Model 1 produces a right-skewed distribution with an expected WTP of \$50.6. Model 2, in contrast, generates an approximately symmetric posterior distribution with expectation close to \$100. In addition to model-specific posterior distributions for the BT construct of interest, the Bayesian estimation also delivers posterior model probabilities, as discussed above. In rare cases these will favor a single model, which can then be selected for BT purposes. In the example at hand, the posterior weight for one or the other model would have to be truly overwhelming, say 0.99 or higher, to completely rule out the other possibility without seriously misrepresenting the model-unconditioned distribution of WTP, given the pronounced difference in location and shape of the posterior densities.

In less clear-cut cases, the Bayesian analyst will combine the posterior information from both models, properly weighted by their respective model probabilities. The higher the probability for one of the models, the more weight it receives in the averaged distribution. This is depicted in the center and top panels of the figure, which show cases where both models receive equal posterior weight (center), and where, respectively, model 2 receives 80 % of posterior evidence (bottom). The corresponding model-averaged distributions exhibit characteristics of both individual posteriors. The figure also illustrates that the model-*un*conditioned

[3]The benefits of informed priors within the context of BT are illustrated in Moeltner and Rosenberger (2014) and Moeltner and Woodward (2009).

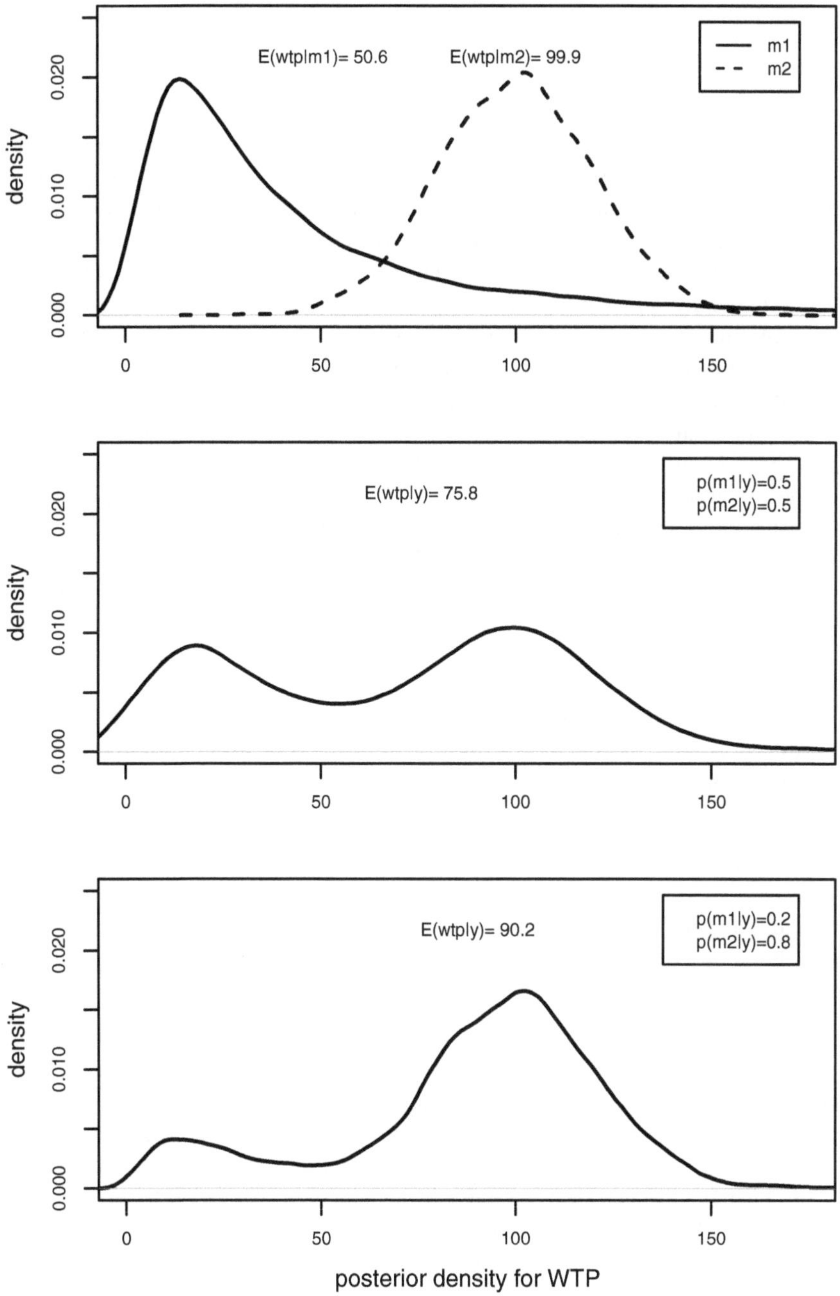

Fig. 22.1 Example for Bayesian model averaging

expectation of WTP ($76 in the first case, and $90.2 in the second) can be quite different from model-specific results. The same holds for the shape of the model-averaged posterior distribution, and thus for all other statistics of interest, such as the median or other percentiles, and the variance.

Since the model-specific Bayesian estimation produces a large number of draws from the posterior of WTP, model averaging can be conveniently accomplished by randomly selecting a share of these draws based on posterior model weights. For the current example, if 100,000 draws are available for each model, and posterior model weights are as in the bottom panel of the figure, we would randomly take 80,000 draws from $p(\text{wtp}|m_2)$, and combine them with 20,000 random draws from $p(\text{wtp}|m_1)$ for model-averaged inference and plotting. The details of these steps for specific applications are described in Moeltner and Rosenberger (2008) and Moeltner et al. (2009) (Technical Appendix).

The important thing to note is that the determination of "optimal scope" results in a clear identification of relevant sub-data only if the corresponding model receives overpowering posterior weight. In most cases, "optimal scope" actually translates into a probability-weighted mixture of different scopes. Continuing with the current example and the 20/80 weight split, this would be interpreted as "there is a 20 % chance that separating out the GR context for parameter estimation best describes the underlying data, while there is an 80 % chance that pooling parameters over small game hunting and waterfowl hunting in the northeast produces the best model fit."

The exact mechanism of determining which subsets of data might form information pools, i.e., the detailed nature of the Bayesian search "engine," depends largely on the following considerations:

1. Size of model space.
2. Existence of closed-form expressions for *analytical* model probabilities. This usually translates into the existence of an analytical solution to the marginal likelihood in (22.11).
3. Admissibility of *partial pooling*, i.e., pooling on only a subset of $\boldsymbol{\theta}_{r,m}$, leaving remaining parameters context-specific.

Each combination of these criteria implies different options and "best practices" for the Bayesian model search. As a general rule, the analytical derivation of model weights is critical when the model space is too large for the algorithm to produce exact *empirical* frequencies, i.e., when a large number of low-probability models is never "visited" by the search engine. In that case, the relative empirical frequencies of the visited models can be compared to their exact analytical counterparts, as illustrated in Moeltner and Rosenberger (2014). If they are close enough (e.g., as measured via their correlation), the analyst can be confident that the unvisited models are truly irrelevant. The drawback to the requirement of analytical model weights is that it imposes limitations on individual model specification, especially on the form of the likelihood function $p(\mathbf{y}|\boldsymbol{\theta}, \mathbf{X}, m)$.

In contrast, if the model space is small such that even highly unlikely models can be at least occasionally selected by the algorithm, the ability to compute analytical weights is of secondary consideration. The empirical model frequencies, i.e., model visitation count divided by the number of iterations, will be sufficiently accurate to form the basis for reliable posterior weights. This, in turn, allows for more flexibility in the types of statistical distributions that can be considered for $p(\mathbf{y}|\boldsymbol{\theta}, \mathbf{X}, m)$.

Finally, if partial pooling is of primary interest or relevance, model uncertainty is best introduced via mixture priors on individual parameters, as in Moeltner and Rosenberger (2008), rather than direct priors for each model. In most cases, this preempts the computation of analytical model weights and is thus more compatible with a "small model space" scenario.

The following section provides examples of existing BT applications that have employed Bayesian model search tools for guidance on optimal scope and information pools.

22.3 Bayesian Model Search and Benefit Transfer: Applications

22.3.1 Small Model Space and Partial Pooling

Moeltner and Rosenberger (2008) consider a situation where a BT function is sought to predict WTP/day for access to a running water/coldwater fishing site (i.e., a typical trout stream) with specific known characteristics $\mathbf{x}_p$, such as catch rate. Thus, only these characteristics can and will feature in an explicit BT function.

Accordingly, the standard GR-meta-regression approach would entail collecting all trustworthy estimates of WTP for running water/coldwater fishing available from existing source studies, estimating a secondary regression model, and combining the resulting estimated parameters $\hat{\boldsymbol{\theta}}_g$ with $\mathbf{x}_p$ to generate a transfer function and WTP estimates for the policy contexts. In Moeltner and Rosenberger's case, this would imply basing inference for the policy context, which is supposed to hold for the typical U.S. site, on only 29 observations from four studies.

The authors then consider a scope-augmentation of the dependent variable along various dimensions. For example, if a typical angler's WTP for running water/*warmwater* fishing is "similar enough" to that for running water/*coldwater* fishing, efficiency gains may be achieved by pooling all available data for both activities when deriving the transfer function. However, a "blind" pooling could produce misleading estimates if this motivating presumption is incorrect. Thus, Moeltner and Rosenberger (2008) pool the data sets, but allow ex ante for parameters to differ across the two groups via standard interaction terms. The task of the Bayesian search engine is then to determine the posterior probability for these interaction terms to be essentially zero, which would support information pooling. Even if only some of the interactions turn out to be irrelevant, the gain in sample size can still outweigh the cost of estimating additional "nuisance" parameters (the non-zero interactions) and lead to a more efficient, i.e., tighter posterior distribution of WTP for the policy context.

In terms of the preceding section, the authors estimate each scope-specific model in isolation, starting with the baseline context, and sequentially adding additional contexts from the "other" category. While this preempts a direct comparison of, or averaging over, scope-specific models, their approach allows for the possibility of partial pooling, i.e., a closer examination of which elements of $\boldsymbol{\theta}_{r,m}$ are truly shared by all given contexts—a maintained assumption in the general framework described above. This, in turn, leads to additional sub-models even within a given specification based on scope.

Moeltner and Rosenberger (2008) operationalize the algorithm by introducing mixture priors for the interaction terms, as suggested by Geweke (1996). Each interaction term is a priori specified to derive, with equal probability, from either a degenerate density at zero or a "healthy" density centered at zero, but with appreciable variance. The search engine than returns updated probabilities for the exact-zero case for each of these interactions. This information can then be used to compute empirical probabilities for each possible model, i.e., baseline case with all interactions set to zero, baseline plus one non-zero interaction, and so on. With model probabilities in hand, the authors then derive model-averaged posterior densities for each augmented data set under consideration.

This process is then repeated for different augmented data sets with differing degrees of deviation from the baseline. The authors find that all scope-augmented meta-regressions generate tighter WTP distributions than the baseline case. The biggest gain in posterior efficiency is achieved by using the fully augmented model, i.e., the specification with broadest scope, despite the numerous resulting interaction terms. This type of information gain would have been impossible to identify or utilize in a Classical econometric framework. Figure 22.2 shows the model-averaged posterior distributions for their BT scenario 2, for the baseline model and all four considered scope augmentations. The fully augmented version (dash-dotted line, labeled D3) has the smallest posterior standard deviation.

22.3.2 Large Model Space and Full Pooling

León-Gonzalez and Scarpa (2008) consider individual-level dichotomous-choice contingent valuation data for 42 forest destinations in the U.K. and Ireland. The probability of a "yes" response to a specific bid (access fee) was related to forest attributes, such as size, proportion of conifers, and congestion levels, as well as socio-demographic characteristics, such as income.

The authors argue that in a traditional BT context, aiming to predict WTP for a new site for which no primary visitation data are available, the site specific data of

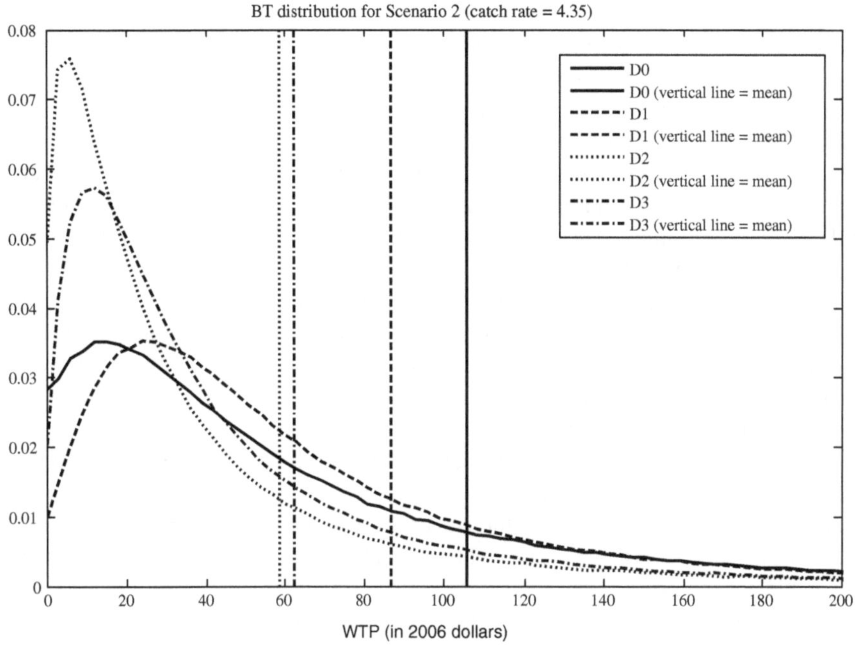

Fig. 22.2 Predicted WTP distributions for different levels of scope augmentation

these individual forests, or at least the subset that satisfies the GR of population similarity, would be pooled indiscriminately to estimate transfer function parameters. However, this could produce misleading predictions if parameters actually differ across sites, despite context equivalence. At the same time they note that rejecting poolability ex ante may imply ignoring transfer-relevant information, if two or more sites in fact do share the same model parameters.

This is an ideal setup for a large-scale Bayesian model search. It corresponds exactly to the general exposition in the preceding section. The entire set of 42 forests makes up the full data set $\{\mathbf{y}, \mathbf{X}, \boldsymbol{\theta}\}$. Declaring one of the forests, say the first, as the policy site makes that forest's information the GR set $\{\mathbf{y}_g, \mathbf{X}_g, \boldsymbol{\theta}_g\}$. The remaining sites form the "other evidence" category, i.e., $\{\mathbf{y}_{o,j}, \mathbf{X}_{o,j}, \boldsymbol{\theta}_{o,j}\}, \quad j = 2, 3, \ldots 42$. A specific model is then defined by the exact set of other forests that are added to the baseline data to form the BT-relevant information set $\{\mathbf{y}_{r,m}, \mathbf{X}_{r,m}, \boldsymbol{\theta}_{r,m}\}$.

Since, in theory, any pair, triplet, or larger set of forests can pool on all relevant parameters, the corresponding model space quickly escalates into the millions and billions—well beyond the reach of classical hypothesis tests based on pair-wise comparison. León-Gonzalez and Scarpa (2008) therefore employ a Markov-Chain Monte Carlo Model Composition (MC^3) algorithm that is designed to visit and compare many thousand models within a reasonable computational time frame.[4] At each iteration, the algorithm compares the current model, defined as a specific pooling pattern for the 42 sites, to a new "candidate" model, defined as the current model plus one minor alteration (one site leaves the pool or gets added to the pool). If the new model shows more promise based on a custom-tailored measure of comparative fit, it becomes the new current model. Else the algorithm remains at—or re-selects—the old model. This assures that better-fitting models are selected relatively more often. The resulting empirical model frequencies reflect these comparative probabilities.

The authors find that some sites exhibit close-to-perfect poolability (i.e., are always selected to fall into the pool of sites that share common parameters), whereas others are virtually always treated as independent (i.e., are never selected to fall into the pool of sites that share common parameters). There are also some in-between cases of sites that have pooling probabilities somewhere between zero and one. Two key results flow from this analysis: (1) Blindly pooling all or any arbitrary subset of sites results in misleading BT predictions, and (2) Exploiting the true pooling patterns made visible by the Bayesian search algorithm leads to substantial efficiency gains for BT, especially when a small sample of primary observations is available for the policy site.

Moeltner and Rosenberger (2014) adopt the León-Gonzalez and Scarpa (2008) framework to determine which subsets of recreational meta-data on WTP/day for various outdoor activities and U.S. regions might share common WTP distributions. Thus, as noted at the onset, they consider a scope augmentation along both the population and commodity dimension, i.e., across entire contexts. While León-Gonzalez and Scarpa (2008) examined pooling for the intercept and several slope parameters, Moeltner and Rosen-berger (2014) settle for pooling on a common mean and variance for the distribution of WTP.

They further extend the León-Gonzalez and Scarpa (2008) approach by allowing for the existence of *multiple* data pools. This is accomplished by first searching for a single dominant pool using the full meta-set, eliminating subsets that always pool with one another, and repeating the analysis to allow another pool to materialize. Overall, their analysis presents a promising picture for the potential of cross-context information borrowing for outdoor recreation. Of all 31 activity/region pairs in their meta-data, 26 pool at least occasionally with other combinations in one of the two estimation rounds, and several exhibit pooling probabilities with other contexts of close to 100 %.

Pairwise pooling probabilities across the 31 contexts can be conveyed at-a-glance via "heat maps", as shown in Fig. 22.3. The darker the shading of a given cell, the more often were the two corresponding contexts sorted into the common-parameters pool by the search algorithm, i.e., the higher is their posterior pooling

[4]For an accessible introduction to MC^3 techniques see Koop et al. (2007) (Chap. 16). The choice of parameter priors within an MC^3 context is discussed in Fernández et al. (2001).

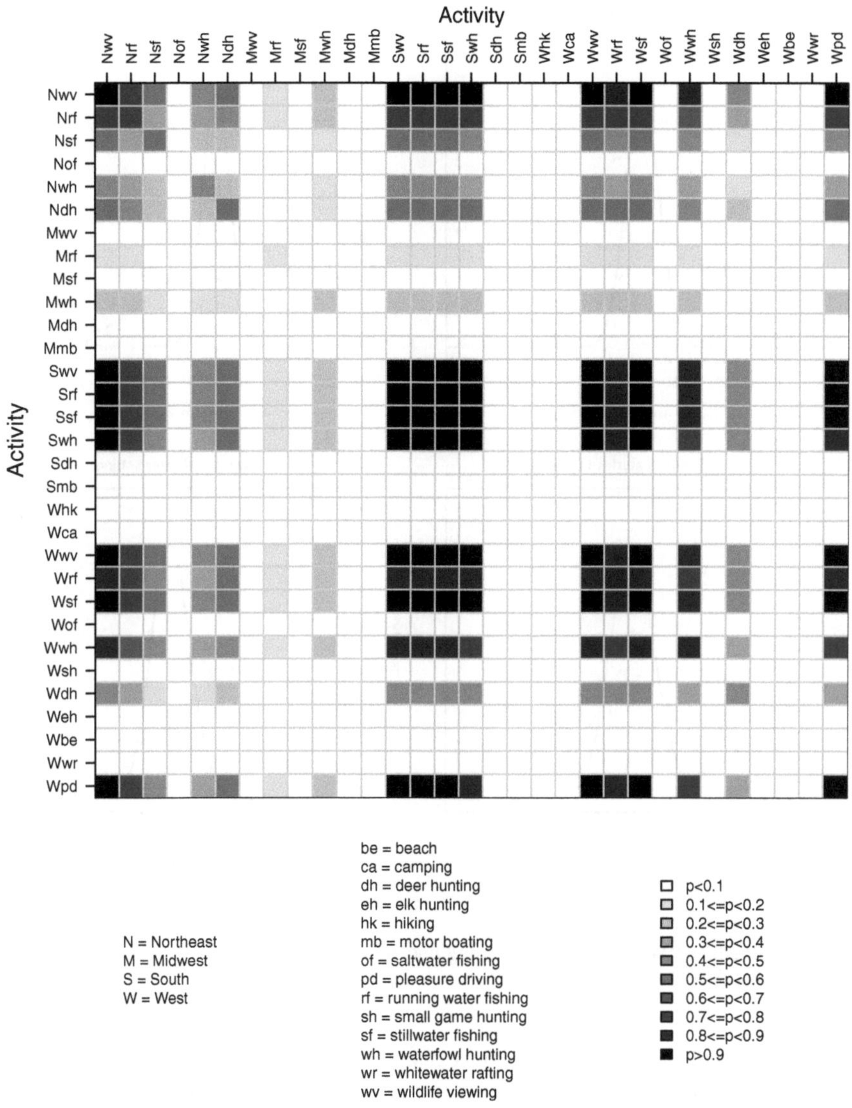

Fig. 22.3 Heat map for pairwise pooling probabilities. *Source* Moeltner and Rosenberger (2014)

probability (denoted as p in the figure legend). The diagonal of the heat maps captures the probability of inclusion in the pooled set for a single context.

For example, Fig. 22.3 illustrates that for three of the four regions considered in Moeltner and Rosenberger (2014), *wildlife viewing* pools at least moderately high with water-based fishing and hunting activities, which, in turn, pool with one another. This provides visual guidance as to which contexts have similar WTP distributions, and which ought to be processed independently.

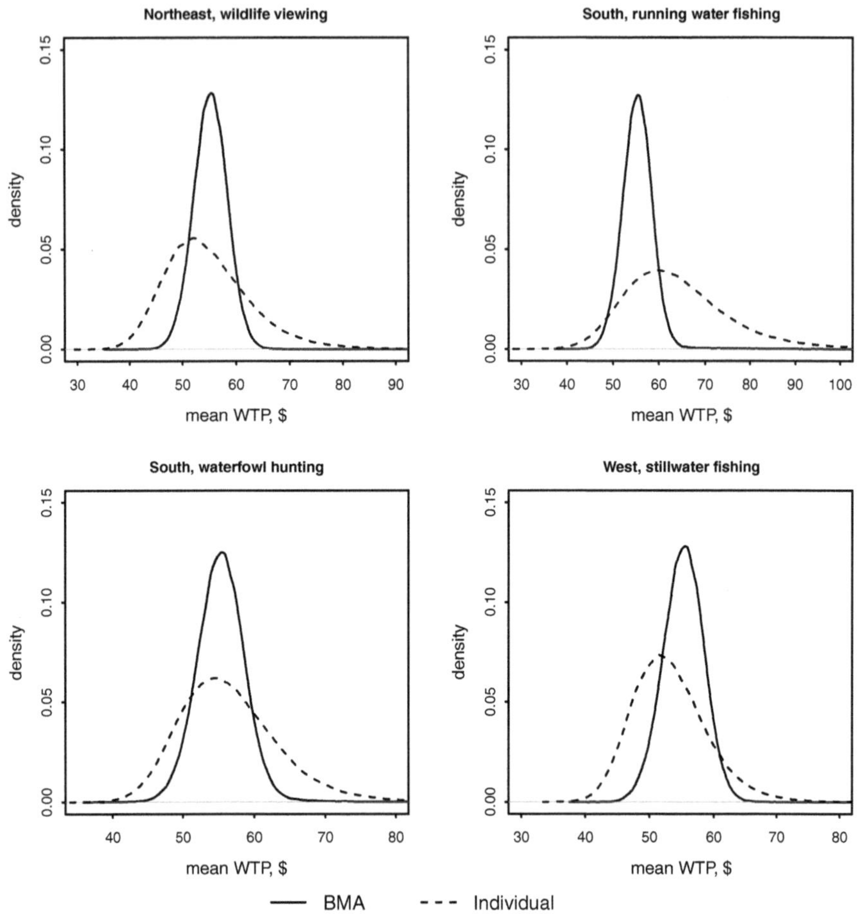

Fig. 22.4 Predicted WTP distributions: using pooled and context-specific data. *Source* Moeltner and Rosenberger (2014)

Exploiting these information pools results in substantial efficiency gains, i.e., tighter posterior distributions for benefit estimates associated with any of the highly pooled contexts. This is visualized in Fig. 22.4, which shows the posterior distribution of expected WTP for four contexts with posterior pooling probabilities of 95 % or higher.[5] In each case, the solid line represents the Bayesian Model-

[5]Since Moeltner and Rosenberger (2014) use a normal density function for log-WTP with mean β and variance σ^2, expected WTP for context j is computed as $E(y_j) = \exp\left(\beta_j + 0.5 * \sigma_j^2\right)$ for each draw of β_j and σ_j^2 flowing from the Bayesian model search algorithm. The resulting distribution of WTP, in dollar terms, can then be plotted using standard kernel density procedures.

Averaged, or pooled, result, and the dotted line the distribution obtained by using only the context-specific sub-data. Clearly, the solid lines describe a much tighter density than their dotted counterparts. This directly reflects the efficiency gain from recognizing shared WTP distributions.

Moeltner and Rosenberger's (2014) results are most useful when applied to a BT situation where a general, or aggregate value estimate is needed for a given activity and region, and when only a small sample of context-specific observations are available. Overall, they report efficiency gains in the form of reduced widths of 95 % credible intervals[6] for predicted benefits of 15–70 % for highly pooled contexts. In addition, their pooled results are based on a much larger underlying sample of studies and site-specific observations compared to the context-specific case. This boosts the representativeness of BT results, especially when they are supposed to hold at a regional or national level.

22.3.3 Towards Intelligent Model Weights

The applications discussed so far all produce model weights as a direct by-product of the Bayesian search algorithm. Moeltner et al. (2009) present an alternative approach that separates parameter estimation and the derivation of model weights into two steps. They independently estimate data from Choice Experiments on land preservation conducted in eight mid-Atlantic communities in a Bayesian framework. In each case, they derive the posterior predictive distribution (PPD) of WTP to preserve a specific plot of land, representing the policy context.

They then examine how well the resulting PPDs overlap with each other, i.e., how similar WTP distributions are across communities. The eight PPDs are depicted in Fig. 22.5. Clearly, there is substantial overlap across several locations, suggesting the potential for efficiency gains from combining community-specific information, i.e., from augmenting scope along the different-population dimension.

To explore this possibility Moeltner et al. (2009) sequentially declare each community as the hypothetical policy site. They then derive community-averaged transfer distributions by combining community-specific PPDs for the remaining locations with ex-post community weights. These weights can be constructed ad hoc (e.g., uniformly equal across locations), or based on an additional layer of community-level information, such as population density, housing stock, and open-space substitutes.

To compute these "intelligent weights," the authors regress an index of overlap of community-specific WTP densities based on round 1 results against community attributes. This yields a predicted overlap for the policy site by combining estimated regression coefficients with the community features of the target site. These weights

[6]In Bayesian analysis the 95 % credible interval comprises the area between the 2.5th and the 97.5th percentile of the posterior distribution for a parameter or predictive construct.

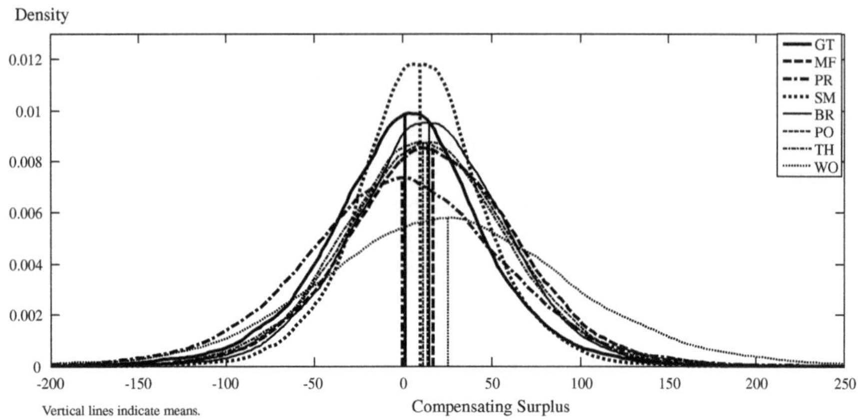

Fig. 22.5 Posterior distribution of WTP to preserve farmland, eight mid-Atlantic Communities. *Source* Moeltner et al. (2009)

are then calibrated to sum to one and used to obtain model-averaged draws of predicted, i.e., transferred WTP, following the technique described in Sect. 22.2.

The authors find that community-averaged predictive distributions that fit any of the target sites substantially better than the worst-case single-site transfer. They thus conclude that the community-averaged strategy is a more robust approach than a single-site transfer.

The Moeltner et al. (2009) two-step procedure will be preferable to the previously discussed self-contained or single-step strategies when the following conditions hold: (1) Each context in isolation provides a large enough data set such that small-sample issues are of secondary concern, (2) It is difficult to specify a tractable umbrella likelihood function that fits all individual contexts (for example due to different questionnaires used in choice elicitation), and (3) There is additional information outside the actual choice data available that might be informative for community "similarity," but would be difficult to introduce into the model via parameter or model priors. With the advent of advanced GIS technology and ever richer spatial data, the construction of such "intelligent weights," signifying *policy-relevant* spatial similarity across sites and thus the potential to form transfer partnerships, should take center stage in research on BT methodology in coming years.

22.4 Conclusion

This chapter provides an introduction on how Bayesian model search techniques can be employed to help the analyst choose the best set of source information for the derivation of transferred value distributions. While some headway as been made in designing these search tools, the search for optimal scope in BT is still in its infancy.

This is likely due to the almost universal acceptance of the GR paradigm—few BT researchers have to date looked beyond the "equal population, equal commodity" horizon. While a detailed theoretical discussion of the relevance and necessity of the GR is beyond the scope of this chapter we refer the reader to Moeltner and Rosenberger (2014), who argue that strict equivalence of all utility-theoretic components between a policy context and a candidate source for BT is not *necessary* to share a common value distribution for the policy-relevant commodity. Yet, it is the value distribution that is of primary interest to the policy maker, and not the underlying structural parameters.

Continued theoretical work is needed to formalize and extend this notion of *value equivalence* despite *parameter inequality*. This will likely lead to a relaxed definition of the "Golden Rule," and in turn encourage more BT researchers to embark on a "quest for optimal scope" in empirical work.

Acknowledgments The author wrote this chapter as a visiting professor at the University of Innsbruck, Austria, June 2012.

References

Boyle, K. J., & Bergstrom, J. C. (1992). Benefit transfer studies: Myths, pragmatism, and idealism. *Water Resources Research, 28*, 657–663.

Brouwer, R. (2000). Environmental value transfer: State of the art and future prospects. *Ecological Economics, 32*, 137–152.

Fernández, C., Ley, E., & Steel, M. F. J. (2001). Benchmark priors for Bayesian model averaging. *Journal of Econometrics, 100*, 381–427.

Geweke, J. (1996). Variable selection and model comparison in regression. *Bayesian statistics, 5*, 609–620.

Koop, G., Poirier, D., & Tobias, J. (2007). *Bayesian Econometric Methods*. Cambridge: Cambridge University Press.

Leamer, E. E. (1983). Let's take the con out of econometrics. *American Economic Review, 73*, 31–43.

León-Gonzalez, R., & Scarpa, R. (2008). Improving multi-site benefit functions via Bayesian model averaging: A new approach to benefit transfer. *Journal of Environmental Economics and Management, 56*, 50–68.

Loomis, J. B., & Rosenberger, R. S. (2006). Reducing barriers in future benefit transfers: Needed improvements in primary study design and reporting. *Ecological Economics, 60*, 343–350.

Moeltner, K., Johnston, R. J., Rosenberger, R. S., & Duke, J. M. (2009). Benefit transfer from multiple contingent experiments: A flexible two-step model combining individual choice data with community characteristics. *American Journal of Agricultural Economics, 91*, 1335–1342.

Moeltner, K., & Rosenberger, R. S. (2008). Meta-regression and benefit transfer: Data space, model space, and the quest for 'optimal scope'. *Journal of Economics Analytical and Pol, 8.*

Moeltner, K., & Rosenberger, R. S. (2014). Cross-context benefit transfer: A Bayesian search for information pools. *American Journal of Agricultural Economics, 96*, 488–496.

Moeltner, K., & Woodward, R. (2009). Meta-functional benefit transfer for wetland valuation: Making the most of small samples. *Environmental and Resource Economics, 41*, 89–109.

U.S. Environmental Protection Agency. (2000). *Guidelines for preparing economic analyses*. Technical Report EPA 240-R-00-003. United States Environmental Protection Agency, Office of Water.

Chapter 23
Structural Benefit Transfer Using Bayesian Econometrics

Daniel J. Phaneuf and George Van Houtven

Abstract In many instances, applying benefit transfer can be interpreted as an inherently Bayesian process. It typically requires the analyst to form beliefs (priors) about the values of interest, using evidence from the literature, and then update these beliefs with specific information about the policy site of interest. The analyst's benefit predictions are then based on this updated summary. Despite this methodological connection, relatively few benefit transfer studies have employed the Bayesian paradigm. In this chapter we describe a Bayesian approach using a structural benefit transfer model, meaning we use prior information and locally available data to estimate the parameters of a defined preference function. We demonstrate the approach through a recreation site choice application, which is based on (a) a prior distribution on marginal WTP for the recreation site attribute of interest (beach width); (b) a small amount of policy site choice micro data; and (c) an estimate of the aggregate proportion of times each alternative in the choice set is selected. Based on this experience, we conclude with observations regarding the advantages and challenges associated with the Bayesian approach.

Keywords Bayesian · Beach · Preference calibration · Site choice · Structural benefit transfer · Willingness to pay

---

D.J. Phaneuf (✉)
Department of Agricultural and Applied Economics,
University of Wisconsin, Madison, WI, USA
e-mail: dphaneuf@wisc.edu

G. Van Houtven
RTI International, Research Triangle Park, Durham, NC, USA
e-mail: gvh@rti.org

R.J. Johnston et al. (eds.), *Benefit Transfer of Environmental and Resource Values*,
The Economics of Non-Market Goods and Resources 14,
DOI 10.1007/978-94-017-9930-0_23

23.1 Introduction

Benefit transfer by definition is a pragmatic technique that seeks to make the most of existing information. All benefit transfer exercises, regardless of their sophistication, rely on judgments made by the analyst. The accuracy of transferred value predictions hinges heavily on these judgments. Indeed, one way of viewing benefit transfer is as a process is in which the analyst combines beliefs about the quantities needed for the analysis with observations of policy site data to produce a prediction of value. More specifically, the analyst's familiarity with the relevant literature allows her to summarize what is known about a value that is relevant for the policy decision. Her understanding of the specific conditions at the policy site helps her update this summary and interpret it in the context of the policy site. Based on this updated summary—which reflects both her reading of past literature and data gathering for the policy site—the analyst makes a prediction and assesses her confidence in the result. When described this way the process has a distinctly Bayesian feel. Nonetheless, only a small amount of research on benefit transfer methods has employed the Bayesian paradigm, and our sense is that explicitly Bayesian approaches are nearly non-existent in policy applications of benefit transfer. This chapter seeks to contribute to the research gap on the former, and proposes steps that might encourage a more decision-theory based approach in the latter.

Studies applying formal Bayesian methods in benefit transfer include Johnston and Moeltner (2014), Leon et al. (2002, 2003), León-Gonzalez and Scarpa (2008), Moeltner et al. (2007) and Moeltner and Rosenberger (2008). Among these, the Leon et al. studies investigate ways to augment policy-site specific dichotomous choice contingent valuation data with a prior distribution on willingness to pay (WTP) based on existing study results. In this chapter we extend this approach into a multinomial choice context, in which the objective is to estimate a preference function. Moeltner et al. (2007), Moeltner and Rosenberger (2008) and Johnston and Moeltner (2014), use the Bayesian paradigm in a meta-analysis (non-structural) context, with the last of these adapting the Bayesian model search algorithm first proposed by León-Gonzalez and Scarpa (2008). Our study is complementary to these, in that we contribute a structural analysis using the Bayesian paradigm.

An explicitly Bayesian approach to benefit transfer is appealing for at least three reasons. First, its reliance on a subjective interpretation of probability fits well with the notion of an expert expressing beliefs about policy-relevant phenomena. From a Bayesian perspective it is valid for an expert in non-market valuation to express her understanding of the value of an environmental commodity by stating her subjective belief about its probability distribution. Different experts can agree or disagree on the validity of each other's beliefs, and presumably the subjective distribution that is based on a better understanding of prior research and empirical evidence will be preferred. Second, the notion of having to explicitly construct a distribution to represent an understanding of existing literature and data imposes a certain discipline on the process. For example, to construct a summary of prior beliefs about a non-market value an analyst must be explicit about how the value is defined and must take concrete steps to transform disparate estimates in the literature into values that can be interpreted as realizations drawn from a single, specific distribution. While all transfer exercises need to at least implicitly complete this step, the Bayesian paradigm encourages a more careful documentation of the necessary assumptions and makes explicit the role of the expert's judgment in the process. Finally, the Bayesian process of using prior information together with new data is an intellectually appealing formalization of the pragmatic notion of evidence marshaling that characterizes benefit transfer.

In this chapter we illustrate a structural approach to benefit transfer (Smith et al. 2006; Van Houtven et al. 2011), meaning we are interested in using prior information and locally available data to specifically estimate the parameters of a defined preference function, rather than simply estimating a WTP value or reduced form benefit transfer function. In this application we use the random utility maximization (RUM) discrete choice framework as our organizing principle. The basic idea is to use existing studies to construct informative prior distributions for the parameters of the RUM utility function, obtain any locally available data for the policy site, and then combine the two information sources to estimate the policy site preference function. This preference function may then be used to generate the required benefit estimates. There are two defining features of our approach. First, we specify the RUM utility function in willingness to pay (WTP) space, so that the structural parameters can be interpreted as marginal WTP rather than marginal utility measures. This WTP specification is important in that it allows us to construct prior distributions over quantities that are comparable and have a clear economic meaning (recall that marginal utility estimates in RUM models are not comparable across studies, due to confounding with scale) (Swait and Louviere 1993). Second, we imagine an environment in which some policy-site data is available, perhaps from an existing secondary data source or via some limited data collection effort. Although this is a departure from the traditional environment in which the benefit transfer problem is considered, we argue that in many cases it is not an unrealistic situation.

The chapter proceeds as follows. In Sect. 23.2, we describe the economic model and the policy context in which it is relevant. Since our case study is a recreation site choice application, we couch the model in those terms, though we stress that the ideas apply to a wider range of applications. We describe the formal Bayesian framework that arises based on (a) a prior distribution on marginal WTP for the recreation site attribute of interest; (b) a small amount of policy site choice micro data; and (c) an estimate of the aggregate proportion of times each alternative in the choice set is selected. We follow with a statement of the posterior distribution that is the final source of inference for the transfer exercise, and then consider the technical steps needed to characterize the posterior distribution. Estimation of the model involves drawing realizations of the utility function parameters from their posterior distribution using Bayesian computational techniques. Although the steps involved with this are quite technical to describe, they employ standard methods that can packaged into flexible software platforms for use by non-experts.

In Sect. 23.3 we identify several strands of literature that are related to our analysis and help place our analytical methods in context. Then in Sect. 23.4 we present a case study examining beach visits in North Carolina and the value of maintaining beach width via nourishing. Section 23.5 concludes the chapter by assessing the research and data investment needs that are needed to operationalize our approach for policy purposes.

23.2 Model Overview

As noted above we use the RUM discrete choice framework as the basis for our analysis. This is attractive for two reasons. First, many of the transfer tasks undertaken by practitioners fit naturally into a space-based discrete choice environment. For example, when the policy site is a recreation destination it makes sense to characterize the relevant behavior as a choice of where to visit from among several options. In this case the benefits of a quality change are conveyed through the improved recreation experience at the policy site. A further example is neighborhood choice, in which households decide where to locate based in part on the bundle of local public goods conveyed by the location. Second, non-spatial policies can often be cast in a binary context comparing utility with and without the intervention. To illustrate, the benefits of a program to protect an endangered species arise from a comparison of people's well-being with and without the program. Thus we imagine a behavioral environment in which a person selects from among J alternatives with prices and levels of non-price (quality) attributes that vary across the alternatives.

Suppose the policy need centers on the value recreation that visitors have for a quality attribute at the policy site(s), which is denoted by q. Examples might include the value placed on additional beach width or water quality at a recreation site. There are J recreation sites a person can visit; we assume that a subset of these constitutes the policy sites and that the remainder is included to accurately reflect the range of available substitutes. A person i receives utility from a visit to site j according to

$$V_{ij} = -\alpha p_{ij} + \beta q_j + \gamma z_i q_j + \xi_j + \varepsilon_{ij}, \quad j = 1, \ldots, J, \tag{23.1}$$

where p_{ij} is the price person i pays if he selects j, q_j is the quality level at site j, and z_i is a policy relevant household characteristic that influences the value a person holds for q. By way of example, in the case study we examine below, p_{ij} is the travel cost of visiting a beach in North Carolina and q_j is the width of the beach. In the application we are interested in estimating the benefits of beach-width maintenance at a subset of the available destinations. The remaining terms in (23.1) are unobservable. The constant term ξ_j captures the unmeasured attributes of site j; these vary across sites but are assumed fixed across individuals. In our beach study this term would include, for example, the number of available parking spaces at beach j. The term ε_{ij} is the usual idiosyncratic component of preferences, and (α, β, γ) are utility function parameters. In (23.1) we have assumed that q_j and z_i are scalars for ease of exposition, although these terms could also be specified as vectors. Further, we have assumed that prices vary over people *and* sites and that quality varies only over sites. Such assumptions are typical in recreation demand studies. Other applications of the proposed model might have different patterns of variability—e.g., in a residential location application, the price of a neighborhood might be constant across people.

In a primary study of the same recreational behavior we would observe the choices, prices, and household characteristics for a sample of N people, and measures of the site characteristics for each of the J destinations. These data, together with an assumption on the distribution for ε_{ij}, permit estimation of the utility function parameters by maximum likelihood. With estimates of these parameters in hand it is relatively straightforward to predict the marginal WTP for q as a function of z as

$$MWTP(z_i) = \frac{\beta}{\alpha} + \frac{\gamma}{\alpha} z_i. \tag{23.2}$$

Similarly, for the linear-in-income model we have used, the WTP for a discrete change in q is

$$WTP = \frac{E\left(\max\left\{V_{ij}\left(q_j^1\right)\right\}\right) - E\left(\max\left\{V_{ij}\left(q_j^0\right)\right\}\right)}{\alpha}, \tag{23.3}$$

where superscripts 0 and 1 denote the baseline and new values for q_j, respectively. If ε_{ij} is assumed to have a type I extreme value distribution the familiar conditional logit model arises, in which

$$\Pr(j) = \frac{\exp(-\alpha p_{ij} + \beta q_j + \gamma z_i q_j + \xi_j)}{\sum_{k=1}^{J} \exp(-\alpha p_{ik} + \beta q_k + \gamma z_i q_k + \xi_k)} \tag{23.4}$$

and

$$E\left(\max\left\{V_{ij}(q_j)\right\}\right) = \ln\left[\sum_{j=1}^{J} \exp(-\alpha p_{ij} + \beta q_j + \gamma z_i q_j + \xi_j)\right]. \tag{23.5}$$

23.2.1 Benefit Transfer Context

In the benefit transfer context the objective is the same: to estimate the utility function parameters and use them to perform the WTP predictions shown in Eqs. (23.2) and (23.3). The way that we go about estimating the parameters, however, is less reliant on original data. To begin exploring these ideas it is useful to rewrite (23.1) as

$$\begin{aligned} V_{ij} &= -\alpha p_{ij} + \alpha\left(\frac{\beta}{\alpha} + \frac{\gamma}{\alpha} z_i\right) q_j + \xi_j + \varepsilon_{ij} \\ &= -\alpha p_{ij} + \alpha(\omega + \omega_z z_i) q_j + \xi_j + \varepsilon_{ij}. \end{aligned} \tag{23.6}$$

Though the utility function in (23.6) is equivalent to its counterpart in (23.1), its structural parameters have a different interpretation. In particular, ω is the population average marginal WTP for q, and ω_z is the deviation from the average based on the value of z. Train and Weeks (2005) refer to this formulation as the WTP space specification of the model because the parameters have an economic interpretation that is comparable across different models, studies, data sources, and econometric methods. Said another way, ω is not confounded with the scale of utility in the same way that marginal utilities such as β are. Thus ω and ω_z are quantities that can be inferred directly by examining the results of existing studies. This formulation of the conceptual model suggests that the first task in our Bayesian transfer exercise is to use the existing literature to elicit prior distributions for ω and ω_z. Note that we do not need to limit attention to RUM-based revealed preference recreation studies for this purpose. Any study providing information on how q conveys benefits in a recreation context is potentially informative.

Further progress can be made by rewriting the utility function again, so that it reads

$$V_{ij} = \delta_j - \alpha p_{ij} + \alpha\omega_z z_i q_j + \varepsilon_{ij}, \tag{23.7}$$

where

$$\begin{aligned} \delta_j &= \alpha\omega q_j + \xi_j, \quad j = 1, \ldots, J-1, \\ 0 &= \alpha\omega q_J + \xi_J, \end{aligned} \tag{23.8}$$

is the average (person-constant) utility for site j, and we have normalized $\delta_J = 0$ for identification. In this form we are able to see that site specific quantities other than q —in particular the unmeasured site attributes ξ_j—are important in determining the extent of substitutability among the sites in the choice set through their influence on the average utility δ_j. Accounting for unobserved site attributes via fixed effects/ alternative specific constants is now recognized as important in primary studies using discrete choice models (see, for example, Murdock 2006; Abidoye et al. 2010; Klaiber and Phaneuf 2010), and as such it makes sense to explicitly consider their role in benefit transfer.

We consider a context in which the J sites are partitioned into J_p policy sites and $J - J_p$ additional substitute sites; without loss of generality suppose the policy sites are indexed $j = 1, \ldots, p$ and that the non-policy sites are $j = p + 1, \ldots, J$. We assume the analyst has measured baseline quality q for all the sites in the choice set—both the policy and non-policy sites. Importantly, we assume that data are available describing the choices made by N individuals, where N can be small. We refer to these data as the micro choices, and discuss possible sources of these data later in the chapter. We also assume the analyst can observe or estimate the set of aggregate shares $s_1, \ldots, s_J$ that measure the proportion of total annual trips that occur at each of the J sites. We refer to these inputs as the macro shares data, and note that their realization might come about from secondary data or by assumption. For example, absent other information the analyst might simply assume an equal share of visits to each site, such that $s_j = 1/J$. One might also have access to gate counts or aggregate visitation data from existing sources. Finally, we assume that the analyst is able to use existing literature, her own experience, or access to other experts to research values of marginal WTP for q, perhaps as it varies with household characteristics z.

23.2.2 A Bayesian Model

In this section we present a formal Bayesian model that takes the discrete choice framework and transfer context as given, and derives the posterior distribution for the utility function parameters conditional on the prior assumptions and available data. The starting point is Eqs. (23.7) and (23.8). As usual we assume utility maximization so that the person selects site j if and only if $V_{ij} \geq V_{ik}$ for all $k \neq j$. We also assume that ε_{ij} is distributed independent extreme value so that the conditional logit model describes the micro choices. To these two standard assumptions we add the restriction that

$$s_j = \frac{1}{N}\sum_{i=1}^{N}\frac{\exp(\delta_j - \alpha p_{ij} + \alpha\omega_z z_i q_j)}{\sum_{k-1}^{J}\exp(\delta_k - \alpha p_{ik} + \alpha\omega_z z_i q_k)}, \quad j = 1,\ldots,J, \tag{23.9}$$

which implies that the average (across people) predicted probability that site j is selected is equal to the observed aggregate proportion of visits to site j. In a conditional logit model with a full set of alternative specific constants this relationship holds for the maximum likelihood estimates when s_j is the proportion of times in the sample that alternative j is chosen.[1] In our transfer context, however, s_j—obtained from auxiliary macro data—may not be the same as the micro sample frequency, and so we maintain (23.9) as a model assumption. We complete the model specification by assuming a normal probability distribution for unobserved site characteristics, such that $\xi_j \sim N(\mu, \sigma)$ for $j = 1, \ldots, J - 1$ and, via the normalization, $-\xi_J = \alpha\omega q_J$. Note that, under this formulation, μ is the average utility *difference* between the first $J - 1$ alternatives and the normalized alternative J, such

[1]More precisely, Eq. (23.9) is implied by the first order conditions that maximize the likelihood function when $J - 1$ unique alternative specific constants are included in the specification.

that across *all* sites ξ_j is now understood to have zero mean. To reduce notational clutter during our later derivations we rewrite (23.8) as

$$\begin{aligned}\delta_j &= \mu + \alpha\omega q_j + \xi_j,\\ &= W_j\tau + \xi_j, \quad j = 1,\ldots,J-1,\end{aligned} \tag{23.10}$$

where $W_j = (1, \alpha q_j)$ is a row vector, $\tau = (\mu, \omega)$ is a column vector.

The unknown parameters in the model are $(\alpha, \omega, \omega_z, \mu, \sigma)$; when referring to the full vector of these unknowns we use the notation $\theta = (\alpha, \omega, \omega_z, \mu, \sigma)$. Therefore, the objective of the analysis is to use prior information and locally available micro choice and macro share data to characterize the posterior distribution of θ. We denote the multivariate prior distribution for these unknown parameters by $\pi(\theta)$. A defining feature of our approach is that we use informative (marginal) priors for ω and ω_z, denoted by $\pi(\omega)$ and $\pi(\omega_z)$ respectively. For simplicity, we assume that ω and ω_z are independent. We discuss the specifics of how these parameters are defined in later sections. We assume that the prior distributions for α, μ and σ are independent and generally non-informative, so that[2]

$$\pi(\theta) = \pi(\omega)\cdot\pi(\omega_z)\cdot\pi(\alpha)\cdot\pi(\mu)\cdot\pi(\sigma). \tag{23.11}$$

To derive the likelihood function for the data recall that there are two types of endogenous outcomes: the micro choice data and the macro shares data. Thus the likelihood function is $L(y_1, \ldots, y_N, s_1, \ldots, s_J|\theta, Q, P, Z)$, where y_i records the choice made by person i in the micro sample and Q, P, and Z are data matrices holding the explanatory variables q, p and z, respectively. With this likelihood function the posterior distribution has the general form

$$\pi(\theta|y_1,\ldots,y_N,s_1,\ldots,s_J,Q,P,Z) \propto L(y_1,\ldots,y_N,s_1,\ldots,s_J|\theta,Q,P,Z)\times\pi(\theta). \tag{23.12}$$

To gain some insight on how we can characterize the posterior it is useful to consider the likelihood function in more detail. Dropping for convenience the explicit conditioning on Q, P, and Z note that we can rewrite $L(\cdot)$ as

$$L(y_1,\ldots,y_N,s_1,\ldots,s_J|\theta) = \Pr(y_1,\ldots,y_N|s_1,\ldots,s_J,\theta)\times\Pr(s_1,\ldots,s_J|\theta), \tag{23.13}$$

which allows us to consider the contributions of the micro and macro data individually. In particular, the second term on the right hand side of (23.13) is derived from the model restriction

---

[2] In some applications it may make sense to use a qualitative prior for α that assigns zero probability to negative values (i.e., values that result in positive price effects) and uniform non-zero probability for all positive values. This approach is fairly easy to accommodate in our framework.

$$s_j = \frac{1}{N}\sum_{i=1}^{N}\frac{\exp(\delta_j - \alpha p_{ij} + \alpha\omega_z z_i q_j)}{\sum_{k-1}^{J}\exp(\delta_k - \alpha p_{ik} + \alpha\omega_z z_i q_k)} = h_j(\delta_1, \ldots, \delta_J, \alpha, \omega_z), \quad j = 1, \ldots, J. \tag{23.14}$$

From (23.14) we can see that conditional on (α, ω_z) and the explanatory variables, s_j inherits its randomness from the distribution of the δ_j's, which in turn depends on the distribution of the ξ_j's. Also, as shown by Berry (1994), Eq. (23.14) has a unique mapping between (δ_1, …, δ_J) and (s_1, …, s_J). This mapping means it is possible to solve for δ_j as a function of the shares so that

$$\delta_j = h_j^{-1}(s_1, \ldots, s_J, \alpha, \omega_z). \tag{23.15}$$

From (23.10) and the assumptions on ξ_j we can see that $\delta_j \sim N(\mu + \alpha\omega q_j, \sigma)$ for $j = 1, \ldots, J - 1$. By the change of variables theorem it follows that

$$\Pr(s_1, \ldots, s_J|\theta) = \phi\left(h_1^{-1}(s_1, \ldots, s_J|\alpha, \omega_z), \ldots, h_{J-1}^{-1}(s_1, \ldots, s_J|\alpha, \omega_z)\right) \times |J_{s\to\delta}|^{-1}, \tag{23.16}$$

where $\phi(\cdot)$ is the normal distribution for δ_1, …, δ_{J-1} and $J_{s\to\delta}$ is the Jacobian transformation from s_1, …, s_J to δ_1, …, δ_{J-1}. These derivations are analyzed in more detail by Jiang et al. (2009), who describe a version of this model in an industrial organization application that limits attention to the aggregate shares.

With the likelihood for the shares in hand it is straightforward to write the likelihood of the micro data conditional on the macro shares as

$$\Pr(y_1, \ldots, y_N|s_1, \ldots, s_J, \theta) = \prod_{i=1}^{N}\prod_{j=1}^{J}\left(\frac{\exp(\delta_j - \alpha p_{ij} + \alpha\omega_z z_i q_j)}{\sum_{k=1}^{J}\exp(\delta_k - \alpha p_{ik} + \alpha\omega_z z_i q_k)}\right)^{y_{ij}}, \tag{23.17}$$

where $y_{ij} = 1$ if alternative j was selected and 0 otherwise. Note that the dependence on the aggregate shares in the micro data likelihood arises via the δ_j's, which from (23.15) are functions of the shares, utility function parameters, and data. With (23.16) and (23.17) we can write the posterior distribution as

$$\begin{aligned} &\pi(\theta|y_1, \ldots, y_N, s_1, \ldots, s_J) \\ &\quad \propto \prod_{i=1}^{N}\prod_{j=1}^{J}\left(\frac{\exp(\delta_j - \alpha p_{ij} + \alpha\omega_z z_i q_j)}{\sum_{k=1}^{J}\exp(\delta_k - \alpha p_{ik} + \alpha\omega_z z_i q_k)}\right)^{y_{ij}} \\ &\quad \times \phi\left(h_1^{-1}(s_1, \ldots, s_J|\alpha, \omega_z), \ldots, h_{J-1}^{-1}(s_1, \ldots, s_J|\alpha, \omega_z)\right) \\ &\quad \times |J_{s\to\delta}|^{-1} \times \pi(\theta). \end{aligned} \tag{23.18}$$

In the following section we examine in detail how to characterize this posterior for the transfer task. By way of summary here, note the specific ways that the data and assumptions combine to inform us about the utility function parameters. Informally, the micro choice data are the main sources of information about α; the micro choices also contribute variability that is used to estimate ω_z. This latter parameter is also driven by its prior distribution, which for small N is likely to be the main source of information. The close link between the aggregate shares and the alternative specific constants illustrates the central role played by the former in estimating the latter. Thus high quality information on the proportion of total annual visits that each destination receives can significantly improve the performance of the model. Finally, Eq. (23.10) shows that the estimate of ω depends on two things: how much of the variability in the δ_j's is explained by the q_j's, and the precision of the prior information on ω gleaned from previous studies. For small J the latter is likely to be the main information source. Thus, multiple sources of information—previous literature, locally available data, and aggregate summary statistics—are systematically combined as part of the Bayesian transfer exercise.

23.2.3 Characterizing the Posterior

Recall that the estimation objective for our Bayesian benefit transfer problem is to summarize the posterior distribution for θ. Regardless of how priors for θ are specified this distribution has a nonstandard form, meaning that analytical expressions for the posterior means, variances, and other moments are not available. Thus we need to simulate draws from the posterior and use the resulting empirical parameter distributions to calculate the posterior moments. For this procedure it is useful to examine the full set of conditional posterior distributions that, when multiplied together, result in the full (unconditional) posterior. In particular we are interested in examining the properties of the following conditional distributions:

$$\begin{aligned} &\Pr(\alpha, \omega_z | s_1, \ldots, s_J, y_1, \ldots, y_N, \omega, \mu, \sigma) \\ &\Pr(\mu, \omega | s_1, \ldots, s_J, y_1, \ldots, y_N, \alpha, \omega_z, \sigma) \\ &\Pr(\sigma | s_1, \ldots, s_J, y_1, \ldots, y_N, \alpha, \omega, \omega_z, \mu). \end{aligned} \tag{23.19}$$

The computational objective is to design techniques for drawing realizations from each of these conditional distributions. We will then use a Markov Chain Monte Carlo (MCMC) method to construct the empirical distribution. In the following sections we discuss the needed steps.

23.2.3.1 Distribution for α and ω_z

To derive conditional distribution for α and ω_z recall from (23.15) that the dependence on $s_1, \ldots, s_J$ and (μ, ω, σ) operates through $\delta_1, \ldots, \delta_J$. Given values for the δ_j's we can then use (23.18) to write

$$\pi(\alpha, \omega_z | y_1, \ldots, y_N, \delta_1, \ldots, \delta_J) \propto \prod_{i=1}^{N} \prod_{j=1}^{J} \left(\frac{\exp(\delta_j - \alpha p_{ij} + \alpha \omega_z z_i q_j)}{\sum_{k=1}^{J} \exp(\delta_k - \alpha p_{ik} + \alpha \omega_z z_i q_k)} \right)^{y_{ij}} \times \pi(\omega_z), \tag{23.20}$$

where the remaining terms from (23.18) are constant via the conditioning on $\delta_1, \ldots, \delta_J$, and therefore can be dropped in the proportional statement. Also, since we have assumed a flat (constant) prior for α, only the marginal prior for ω_z carries through to (23.20). The distribution has a nonstandard form, but it is easy to sample from using a Metropolis-Hastings algorithm, given that we can readily compute the value of the distribution for specific values of α, ω_z, and the conditioning variables and parameters. Train (2009, p. 302) details the steps that are needed for the algorithm.

23.2.3.2 Distributions for μ, ω and σ

Given values for α and ω_z we can use Eq. (23.15) to obtain conditional values for $\delta_1, \ldots, \delta_{J-1}$. Since μ and ω only appear in the δ_j's the conditional distribution will be based on the linear equation in (23.10). If the prior distributions for $\tau = (\mu, \omega)$ and σ are conjugate this becomes a standard Bayesian linear regression model. In particular, Koop (2003, pp. 34–38) describes the characteristics of this model when τ has a normal distribution prior and σ has an inverse gamma distribution prior, and Jiang et al. (2009) implement it in a similar model structure. Following this approach, we assume the prior distributions for τ and σ are

$$\begin{aligned} \tau | \sigma &\sim N(\underline{\tau}, \sigma \underline{B}) \\ \sigma &\sim IG(\underline{s}, \underline{v}), \end{aligned} \tag{23.21}$$

where *IG* denotes the inverse gamma distribution and the terms with underbars are known hyper parameters chosen to reflect the analyst's prior beliefs about the utility function parameters. In our case these are set to produce an informative prior for ω and uninformative priors for μ and σ. Under these assumptions the posterior distribution for $\tau = (\mu, \omega)$ conditional on the other model unknowns is

$$\tau | \sigma \sim t(\bar{\tau}, \bar{s}\bar{B}, \bar{v}), \tag{23.22}$$

where $t(\cdot)$ is the student's t distribution,

$$\begin{aligned}\bar{B} &= (-\underline{B}^{-1} + W'W)^{-1} \\ \bar{\tau} &= \bar{B}(\underline{B}^{-1}\underline{\tau} + W'W\hat{\tau}) \\ \hat{\tau} &= (W'W)^{-1}W^{-1}\delta \\ \bar{\nu} &= \underline{\nu} + J - 1,\end{aligned} \tag{23.23}$$

W is a matrix with each of $J - 1$ rows holding a W_j, $\delta = (\delta_1, \ldots, \delta_{J-1})$,

$$\overline{\nu s} = \underline{\nu s} + (J-3)s + (\hat{\tau} - \underline{\tau})'\left[\underline{B} + (W'W)^{-1}\right]^{-1}(\hat{\tau} - \underline{\tau}), \tag{23.24}$$

and

$$s = \frac{(\delta - W\hat{\tau})'(\delta - W\hat{\tau})}{J - 1 - 2}. \tag{23.25}$$

Though the expressions appear complicated, the conditional posterior distribution for τ has several intuitive properties. In particular, the second expression in (23.23) shows that the posterior mean is a weighted average of the prior distribution mean and the OLS estimate of τ (denoted by $\hat{\tau}$ in (23.23) above) obtained using W and δ. The weights depend on the prior variance—the confidence the analyst has in her beliefs—and the variability in the characteristics of the policy sites. As such the usual intuitive interpretation of the Bayesian linear regression model (e.g., Kennedy 2008) as producing estimates that depend on prior beliefs and data—weighted according to the information content of each—is valid in our context.

The conditional posterior distribution for σ is inverse gamma, with the scale and degrees of freedom parameters being a combination of the hyper parameters and functions of the data and conditioning parameters. In our transfer exercise we do not need to characterize the (marginal) posterior distribution for σ since it does not explicitly enter the utility function, and so we do not discuss its particular form further. We note nonetheless that it is easy to sample from and, if estimates of σ were of interest, incorporating this parameter into an MCMC routine is straightforward.

23.2.3.3 An MCMC Sampler

The following MCMC algorithm can be use used to construct an empirical distribution of draws from the posterior distribution of the unknown parameters:

1. Set initial values for α and ω_z, denoted by α^0 and ω_z^0 respectively. Compute the initial values for the alternative specific constants $\delta_1^0, \ldots, \delta_J^0$ using Eq. (23.14) and α^0 and ω_z^0.

2. For $t = 1, \ldots, R$ complete the following steps:
   - Use a Metropolis-Hastings method to draw updated values α^t and ω_z^t from the conditional distribution for α and ω_z as shown in (23.20).[3]
   - Compute $\delta_1^t, \ldots, \delta_J^t$ using (23.14) and α^t and ω_z^t.
   - Draw realizations ω^t and μ^t from the conditional distribution shown in (23.22), given $\delta_1^t, \ldots, \delta_J^t$ and α^t.
3. For draws of the unknown parameters $t > b$ (the "burn in" point) α^t, ω_z^t, ω^t and μ^t constitute draws from the full posterior distribution. The following estimates of the posterior means serve as point estimates of the utility function parameters:

$$\begin{aligned} \bar{\alpha} &= \frac{1}{R-b}\sum_{t=b+1}^{R} \alpha^t \quad & \bar{\omega} &= \frac{1}{R-b}\sum_{t=b+1}^{R} \omega^t \\ \bar{\omega}_z &= \frac{1}{R-b}\sum_{t=b+1}^{R} \omega_z^t \quad & \bar{\mu} &= \frac{1}{R-b}\sum_{t=b+1}^{R} \mu^t. \end{aligned} \tag{23.26}$$

Each of these steps is relatively standard in the applied Bayesian econometrics literature, and the computations are generally mechanical enough that canned routines can be made available for the non-expert analyst who wishes to implement the routines.

23.3 Links to Prior Literature

The ideas we have presented thus far have borrowed liberally from several different strands of literature. The most obvious connection is to the handful of papers that have explicitly considered Bayesian methods in benefit transfer. For example, Parsons and Kealy (1994) anticipated many of the ideas we have developed here in an experiment illustrating the use of RUM models for benefit transfer. In their application to water quality and recreation the authors estimate a "policy" RUM model using the limited data available for their policy site. The parameter estimates from this model are combined with estimates from a larger scale "study" RUM model to produce weighted average estimates of the utility function parameters; the weights are based on the sampling error from the two models. More recently, Leon et al. (2002) discuss the formal use of Bayesian methods in benefit transfer. In their context the analyst constructs a prior distribution on the WTP for a policy outcome,

[3]This involves drawing candidate values $\tilde{\alpha}^t$ and $\tilde{\omega}_z^t$ and then evaluating the likelihood of the candidates relative to α^{t-1} and ω_z^{t-1}. For this comparison is also necessary to compute $\tilde{\delta}_1^t, \ldots, \tilde{\delta}_J^t$ at the candidate values $\tilde{\alpha}^t$ and $\tilde{\omega}_z^t$ and use them in the comparison step.

and then combines this with policy-site specific dichotomous choice contingent valuation data to construct a posterior distribution for WTP. Following up on this study, in Leon et al. (2003) the same research group considers alternative ways that experts' knowledge of the relevant literature can be organized into prior distributions for Bayesian benefit transfer. The two Leon et al. papers are similar in spirit to our approach in that they use formal Bayesian methods to combine different information sources as part of the transfer exercise. The frameworks are complementary in that we focus on a multinomial choice context in which the objective is to estimate a preference function, while Leon et al. focus on characterizing the distribution of WTP based on a referendum context.

The literature on using discrete choice methods for benefit transfer is also relevant for our ideas. For example, Morrison et al. (2002) provide a case study examining the performance of choice experiments for benefit transfer. The authors show for their application that the transfer of implicit prices from study to policy contexts is usually convergent valid, though transfers of surplus estimates are usually invalid. Morrison and Bergland (2006) discuss benefit transfer and choice models more generally. They conclude that direct transfers of estimates are unlikely to be of sufficient accuracy for policy use. Instead, methods are needed that can tailor the existing study site estimate to the specific policy context. Both of these papers lend support to our strategy of using the existing literature to summarize marginal WTP for policy site attributes, while relying on locally available information to calibrate the preference function to local characteristics of the resource.

Our analysis has also exploited two literature strands more generally associated with discrete choice econometrics. First, the large literature on combining revealed and stated preference data in discrete choice models provides a classical analog to our Bayesian model. For example, von Haefen and Phaneuf (2008) cast the combined RP/SP approach the use of stated preference data to identify the marginal WTP for non-price attributes, the revealed preference data to identify the coefficient on price, and the choice frequencies to identify alternative specific constants. Similarly Berry et al. (2004) examine new car market demand by using micro-level product choice data (both revealed and stated) to estimate marginal utilities for automobile attributes and aggregate market share data to calibrate the alternative specific constants.

The Berry et al. (2004) paper is also an example of how we have borrowed ideas from the empirical industrial organization literature. Beginning with Berry et al. (1995), estimation of the demand for differentiated products in industrial organization has used a two-stage model. In the first stage a full set of alternative specific constants for all of the elements in the choice set is estimated in a typical discrete choice setting. In the second stage a linear regression is used to decompose the alternative specific constants into observable (i.e., determined by measured variables) and unobservable components, usually while accounting for endogenous regressors. Berry (1994) discusses how this strategy transforms a difficult non-linear problem with endogenous variables into a familiar linear one. In our context the same logic allows us to use a simple closed form Bayesian linear regression as a means of combining prior beliefs on the (population average) marginal WTP with the available policy site specific data. Bayesian versions of the Berry et al. (1995) framework from Jiang et al. (2009) and Yang et al. (2003) are particularly relevant for this approach.

The final strand of literature that is relevant for our study is the small collection of papers on "preference calibration" (Smith et al. 2002; Van Houtven and Poulos 2009; Van Houtven et al. 2011). These papers employ the same basic objective of calibrating the parameters of an assumed preference function; however, the operational strategy is distinct in that it relies on matching analytical expressions for economic values implied by the function to literature-gleaned estimates of these values. As envisioned the preference calibration approach does not have a role for primary data, though the method is flexible enough to accommodate primary data if available.

23.4 Case Study

In this section we demonstrate the proposed approach using an application to beach recreation in North Carolina. The quality measure of interest is beach width, which is of policy relevance because many locations rely on beach nourishment to maintain the width necessary to support recreational use and to maintain coastal property values. Landry (2011) summarizes the economics of beach erosion management. For this demonstration we assume that the WTP to maintain beach width at a policy site is needed to support cost-benefit analysis of proposed nourishment activities.

Our policy site is Wrightsville Beach, a popular destination in southern North Carolina near Wilmington. For illustration we define a choice set of 17 beaches that stretches from near the border with South Carolina north to the southern tip of the Outer Banks. Wrightsville Beach lies near the geographic center of the choice set. The beaches in our choice set are listed alphabetically in Table 23.1. We selected this area because a recent recreation study by Whitehead et al. (2008, 2010) focused on these 17 beaches, and provided useful background and data for our analysis. Table 23.1 shows the average width of the choice set beaches at the time of the study, as well as the frequency with which each destination was selected by recreationists in the full Whitehead et al. revealed preference data. The last column in Table 23.1 shows the macro share assumption that we will use in our transfer exercise. To mimic the type of information that would most likely be available to an analyst familiar with the policy choice set, we made assertions about the share frequencies for the three beaches (Atlantic, Carolina, and Wrightsville) that are known to be the most popular, and then divided the remaining part of the unit interval evenly among the remaining 14 sites. An alternative approach would be to use the observed frequencies from the actual revealed preference data, though this level of detail is unlikely to be available in most transfer contexts.

The first step in our exercise is to survey the literature to identify studies looking at beach recreation generally, and reporting information on beach width valuation in

Table 23.1 Policy sites for case study

Beach	Average travel cost (micro samples)	Baseline width (m)	Actual shares	Assumed shares
Atlantic Beach	$112.44	41	0.13	0.10
Carolina Beach	$132.80	56	0.14	0.15
Caswell Beach	$158.58	24	0.02	0.04
Emerald Isle	$113.64	40	0.11	0.04
Fort Fisher	$137.91	122	0.03	0.04
Fort Macon	$112.17	27	0.03	0.04
Holden Beach	$160.84	27	0.03	0.04
Kure Beach	$135.97	40	0.02	0.04
North Topsail	$111.57	25	0.08	0.04
Oak Island	$162.90	37	0.02	0.04
Ocean Isle	$166.78	26	0.06	0.04
Pine Knoll	$112.31	34	0.03	0.04
Salter Path	$112.18	27	0.02	0.04
Sunset	$172.14	35	0.02	0.04
Surf City	$118.29	27	0.04	0.04
Topsail Beach	$125.53	34	0.05	0.04
Wrightsville Beach	$120.00	49	0.20	0.20

particular. Table 23.2 lists eight studies that were published in 2000 or later and contained sufficient information for us to calculate a measure of marginal WTP per meter of beach width in a recreation context. The list includes studies using a wide range of approaches, and each was unique in how it defined the beach attribute and reported values. Columns three through five in the table describe how we used these primary study results to produce a measure of marginal WTP to inform our prior distribution. Judgments such as these are often a critical part of benefit transfer; our goal is to make these transparent. Our interpretations (Table 23.2) led to per meter, per trip values ranging from $0.44 to $3.82 (both from contingent valuation studies). The latter is an outlier that we decided to exclude in forming our prior beliefs, given that it is more than twice as large as any of the other point estimates. This subjective decision is consistent with the notion of an expert using available information and judgment to summarize her subjective prior beliefs, and therefore does not constitute "selection bias" in the narrow, classical sense of the term. Nonetheless other experts' may reasonably formulate a prior that places non-zero prior probability weight on this outlier. As we discuss in the conclusion, different priors can be examined as a way to gauge the robustness of policy predictions to these types of decisions. In our case the remaining seven studies have a mean of $0.91 per meter per trip and a standard deviation of 0.42. We proceed using a normal prior for marginal WTP such that, using the notation from above, $\omega \sim N(0.91, 0.42^2)$. Figure 23.1 displays a histogram of the seven studies that form the basis for our priors, with the actual prior distribution superimposed on top.

Table 23.2 Studies used for prior elicitation

Paper	Type	Original amenity definition	Original valuation	Manipulation	MWTP per trip per meter beach width
Parsons et al. (1999)	RP RUM model; 1997 behavior	Narrow beach: <75 feet Wide beach: >200 feet	Implicit price narrow beach: −10.09 Implicit price wide beach: −21.54	Assume 10.09 is per trip value for a 50 foot improvement to narrow beaches	$0.66
Landry et al. (2003)	CV per year; 1996 dollars	Choice scenario: status quo versus designed alternative to preserve barrier island beach width	Average WTP per trip of $7.43 for 1 m width increase across length of barrier island beach. Average WTP per trip of $9.56 for 2.5 m width increase across length of barrier island beach	Compute per meter value using the 2.5 m scenario	$3.82
Shivlani et al. (2003)	CV per beach visit; 1999 dollars	Maintenance of beach width with per trip parking fee payment vehicle	Per trip WTP for width maintenance: $1.69	Interpret no change value as WTP to prevent any change in beach erosion	$1.69
Kriesel et al. (2004)	CV per year; 1998 dollars	Choice scenario: status quo mixture of wide and narrow beaches versus designed alternative improving narrow beaches to wide	Average WTP per trip $6.06 to improve 2 miles of beach width from 0 to 20 years and 1 mile of beach width from 10 to 20 yards	Compute a weighted average willingness to pay for 1 yard extra of $0.40	$0.44
Huang et al. (2007)	Choice experiment per year; 2000 dollars	Preservation of 1, 2, 3, 4 miles of beach with annual fee payment vehicle	Basic logit: 2.95 per mile of beach preservation. Mean trips 7.5. Median not reported	Interpret no change value as WTP to prevent 1 m of beach erosion. Guess median trips per year is 3	$0.98

(continued)

Table 23.2 (continued)

Paper	Type	Original amenity definition	Original valuation	Manipulation	MWTP per trip per meter beach width
Whitehead et al. (2010)	RP/SP contingent behavior; 2003 dollars	Behavioral response to 100 foot increase in width at each of the 17 beaches in the choice set	Annual WTP for beach width increase of $106 Mean trips 7.61. Median not reported	Estimate median trips per year is 3. Median per trip willingness to pay for 1 extra foot $0.35	$1.15
Whitehead et al. (2010)	Multiple site RP model; 2003 dollars	Continuous beach width in feet included as explanatory variable in demand system model	Implicit price 1 foot beach width (baseline width 125 feet) computed using ratios of demand system parameters from semi-log model	Use directly	$0.60
Pendleton et al. (2012)	RP RUM model; 2000 behavior	Continuously measured beach width entered as non-linear variables	Ignoring higher order polynomials marginal WTP per 1 m width for water activities $0.88	Use directly	$0.88

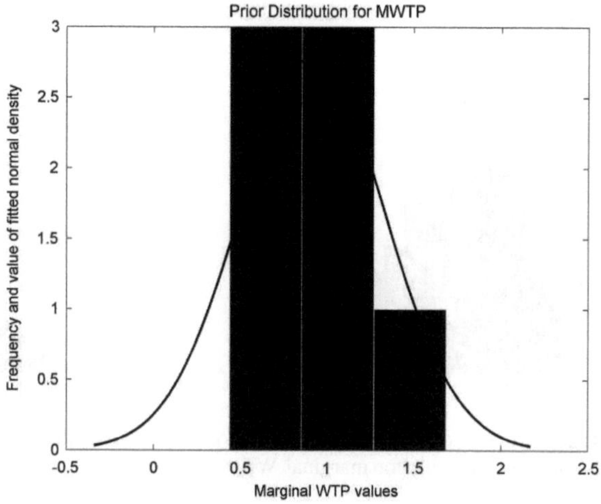

Fig. 23.1 Histogram of marginal WTP points and fitted normal distribution

The next step in our exercise is to consider potential sources of micro choice data. For this we rely on the 2000 National Survey of Recreation and the Environment (NSRE). The NSRE is a periodic survey done by the US Forest Service to measure Americans' participation in outdoor recreation and their use of environmental resources. The 2000 survey sampled over 25,000 households across the country, with stratification sufficient to provide some representation in each state. Among the full set of survey respondents, 4129 completed the *saltwater recreation module*, which solicits information on visits to coastal areas for outdoor recreation. From these respondents we selected the residents of North Carolina, South Carolina, and Virginia as being the likely population of users of the sites in our choice set. This selection resulted in 321 respondents from the three states who completed the module. Among these, 255 people reported making at least one visit during the previous year to a saltwater destination. By searching the self-reported names of the destinations we identified N = 17 people who visited at least one of the J = 17 beaches in our choice set. This small collection of observed users constitutes the micro choice data that we integrate into our analysis. The travel distances and time from each person's zip code to each of the beaches was computed using GIS software; based on these travel costs were imputed using an estimate of \$0.40 per mile out of pocket costs[4] and one third the wage rate as a proxy for the opportunity cost of time. Table 23.1 displays the average travel costs across the micro sample for each of the beaches in the choice set.

Implementing the MCMC algorithm requires several practical decisions. We obtained an initial value for α by running a conditional logit model using the micro sample and a specification that only included the travel cost variable. With this we obtained $\alpha^0 = 0.059$,

[4]Assuming \$3 per gallon for gas, 20 miles per gallon average gas mileage, and \$0.25 per mile for depreciation cost.

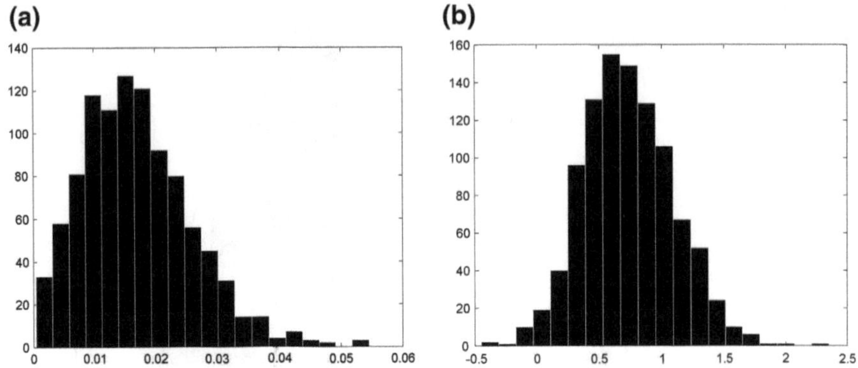

Fig. 23.2 Histograms for marginal posterior distributions. **a** Posterior for marginal utility of income. **b** Posterior for marginal WTP

and used it to obtain initial values for each δ_j.[5] Implementing the Metropolis-Hastings step requires a jumping distribution that generates the candidate value $\tilde{\alpha}^t$. We used $\tilde{\alpha}^t = \alpha^{t-1} + \eta^t/10$, where η^t is a draw from a standard normal distribution. We nested an accept/reject step in this process, in which negative values of α (implying a positive price effect) were rejected. We ran the chain out to t = 200,000 before assuming it had reached a stationary point (i.e., the simulated draws are from the posterior distribution); graphical diagnostics suggested convergence had occurred by this time. Following the burn in we kept 5000 draws from the posterior, and used every fifth realization (to reduce serial correlation) for our summaries.

Figure 23.2 presents histograms for the two utility function parameters of primary interest (we do not display results for μ). Point estimates for the posterior mean and standard deviation of ω (the marginal WTP for beach width) are \$0.74 and \$0.36, respectively. Note that updating our prior beliefs with the available micro data and aggregate shares decreased our point estimate of marginal WTP for the policy sites. Point estimates for the posterior mean and standard deviation of α (the marginal utility of income) are 0.017 and 0.009, respectively. Each value of α drawn from the posterior implies a unique value for each δ_j; combined with the corresponding draws of ω and μ (posterior mean −0.685; posterior standard deviation 0.546) we are able to back out the implied value for each ξ_j. Thus we obtain a full characterization of the utility function

$$\begin{aligned} V_{ij} &= \delta_j - \alpha p_{ij} + \varepsilon_{ij}, \\ \delta_j &= \mu + \alpha\omega q_j + \xi_j, \quad j = 1,\ldots,J-1, \quad \xi_J = -\alpha\omega q_J, \end{aligned} \tag{23.27}$$

and the necessary platform for conducting welfare analysis at the policy site.

[5]This value cannot be interpreted as a maximum likelihood estimate with known properties, given the small sample. It does, however, provide a useful initial value that is based on the probability structure of the model.

We use the characterization of the utility function to examine two welfare scenarios that might be of interest in a beach nourishment cost-benefit analysis:

- *Scenario 1*: Erosion of the beach width at Wrightsville Beach to 75 % of its current width.
- *Scenario 2*: Loss of access to Wrightsville Beach as a recreation destination due to complete erosion.

Because we do not have a meaningful micro sample to compute welfare measures for specific individuals, we use the average travel costs in Table 23.1 as our baseline prices for trips to each of the 17 beaches. We use the formula for per trip welfare effects in Eq. (23.5), and compute the formula for each draw of the parameters from the posterior distribution. In this way we are able to present a distribution for the welfare measures, rather than a point estimate. Using this approach we find a mean welfare loss per trip of $1.74 (0.82) for scenario 1, with a median of $1.71. The inter-quartile range for the welfare loss is $1.19, $2.27. For scenario 2, we find a mean welfare loss per trip of $22.22 and a median of $14.34. A few large outliers skew the distribution right and inflate the standard deviation to 34.36. Nonetheless, the bulk of the distribution is concentrated near the median, as shown by an inter-quartile range of $10.55, $21.55.

23.5 Conclusion

Our objective in this chapter has been to present a modeling framework and operational strategy for implementing a Bayesian approach to benefit transfer within a structural context. We have argued that the Bayesian paradigm is a natural way to approach the problem of benefit transfer, and have demonstrated its potential with a recreation site choice application. A key insight from our approach is that specifying the underlying utility function in WTP space provides a connection between the analytical structure of the model and values reported in the literature. Thus we were able to use existing studies to construct a prior distribution for the key structural parameter(s) in our policy utility function. This information was combined systematically with available information on aggregate choice frequencies and micro data to characterize the policy utility function. We demonstrated the potential of our approach using a case study of North Carolina beaches, which used existing data sources and a systematic review of existing beach-width valuation studies to examine the welfare consequences of beach erosion at a policy site. This demonstration does not constitute a full-blown policy analysis, however, in that we have not illustrated how robust our predictions are to alternative assumptions. A more complete analysis could examine different prior distributions (perhaps using the outlier point estimate that we excluded), different assertions on the aggregate shares, and perhaps compare the results of our structural approach with predictions arising from a meta-analysis framework. We stress that the predictions from a Bayesian model may be more or less robust to analysts' assumptions than those from a competing approach, and so greater robustness is not necessarily a characteristic of the approach. Instead, in a well-executed Bayesian approach the role of these assumptions is made explicit, which allows the analyst's judgments to serve as a focal point for gauging the usefulness of her predictions.

What are the advantages of the ideas we have pursued here? Benefit transfer exercises almost always rely implicitly on the analyst's subjective interpretations of existing studies. These interpretations include judgments on study quality and suitability, as well as on the manipulations and assumptions that are needed to derive estimates of value at the policy site from the published results. Thus the explicit role for subjective beliefs in the Bayesian paradigm formalizes what is already a defining feature of the task. The requirement that the analyst's interpretations of existing studies be distilled into a formal prior systematizes the way her judgments and beliefs are used, and as argued above makes the inevitably subjective nature of benefit transfer more transparent. An approach that systematically combines disparate information sources—the analyst's beliefs and locally available data—also seems to be a desirable feature of a benefit transfer protocol. Of course the opposite might also be true: if any source of information is fair game and anyone's prior beliefs are acceptable, then any desired outcome can be supported by some type of analysis. Thus as with any secondary data policy analysis, the prediction—regardless of the paradigm—is only as good as what goes into constructing it. If the Bayesian paradigm provides more access to information of acceptable quality, its flexibility is an advantage. If it enables the expanded use of poor quality information this flexibility may be a disadvantage.

There are of course challenges to operationalizing the ideas we have explored here. The first of these challenges is technical. The estimation approach used here is more involved than many currently used methods. Indeed some concern has been expressed that policy analysts have not embraced existing methods that go beyond simple value transfers (Johnston and Rosenberger 2010). While this is a valid concern, it is likely that the mechanical steps needed for implanting the MCMC algorithm used here could be automated and packaged for use by non-experts. The second and perhaps larger concern is that most policy analysts have little experience with the process of prior elicitation, and our sense is that research in environmental economics has not provided much insight on this task (Leon et al. 2003 is the one exception that we are aware of). Thus one obvious research need involves investigating the best ways of organizing insights gained from existing studies into tractable probability distributions, and evaluating the robustness of the attendant results.

A final point concerns the types of data infrastructure that are most useful for policy analysis. Applied non-market valuation has been case-study focused in the sense that most studies are spatially and contextually specific, which leads to high quality results for the study location but challenges for transferring the findings to other needs. An alternative approach would be to invest in two more general types of databases and studies. The first type is large nationally representative surveys of behaviors that can inform environmental benefits estimation (e.g., recreation, residential location choice). Although these data often seem too limited for use in stand-alone studies of specific contexts, our North Carolina example has demonstrated the potential usefulness of this type of data when combined with other information. Behavioral databases that cover a wide geographical range with small to moderate numbers of observations at specific points in space may, when viewed from a different perspective, have substantial policy relevance. Second, complementary to these data collection efforts would be studies that focus on establishing transferable conditional distributions of marginal WTP for common environmental amenities. For example, in a recently completed study Phaneuf et al. (2013) designed a choice experiment that was designed to measure the marginal value of freshwater quality in recreation across a broad geographical area. This study, when combined with local information on freshwater recreation choices by a small sample of people, provides the information needed to measure the WTP for water quality in a wide range of spatial and policy contexts.

Acknowledgments This work was funded by a grant from the U.S. Environmental Protection Agency (No. RD-83346101).

References

Abidoye, B., Herriges, J. A., & Tobias, J. (2010). *Controlling for observed and unobserved site characteristics in RUM models of recreation demand.* Ames, Iowa: Iowa State University Department of Economics (Working Paper No. 10011).

Berry, S. T. (1994). Estimating discrete choice models of product differentiation. *RAND Journal of Economics, 25*, 242–262.

Berry, S., Levinsohn, J., & Pakes, A. (1995). Automobile prices in market equilibrium. *Econometrica, 63*, 841–890.

Berry, S., Levinsohn, J., & Pakes, A. (2004). Differentiated products demand systems from a combination of micro and macro data: The new car markets. *Journal of Political Economy, 112*, 68–105.

Huang, J. C., Poor, P. J., & Zhao, M. Q. (2007). Economic valuation of beach erosion control. *Marine Resource Economics, 22*, 221–238.

Jiang, R., Manchanda, P., & Rossi, P. E. (2009). Bayesian analysis of random coefficient logit models using aggregate data. *Journal of Econometrics, 149*, 136–148.

Johnston, R. J., & Moeltner, K. (2014). Meta-modeling and benefit transfer: The empirical relevance of source-consistency in welfare measures. *Environmental and Resource Economics, 59*, 337–361. doi:10.1007/s10640-013-9730-3.

Johnston, R. J., & Rosenberger, R. S. (2010). Methods, trends and controversies in contemporary benefit transfer. *Journal of Economic Surveys, 24*, 479–510.

Kennedy, P. (2008). *A guide to econometrics* (6th ed.). Oxford: Blackwell Publishing.

Klaiber, H. A., & Phaneuf, D. J. (2010). Valuing open space in a residential sorting model of the Twin Cities. *Journal of Environmental Economics and Management, 60*, 57–77.

Koop, G. (2003). *Bayesian econometrics*. West Sussex, UK: Wiley.

Kriesel, W., Keeler, A., & Landry, C. (2004). Financing beach improvements: Comparing two approaches on the Georgia coast. *Coastal Management, 32*, 433–447.

Landry, C. E. (2011). Coastal erosion as a natural resource management problem: An economic perspective. *Coastal Management, 39*, 259–281.

Landry, C. E., Keeler, A. G., & Kriesel, W. (2003). An economic evaluation of beach erosion management alternatives. *Marine Resource Economics, 18*, 105–127.

Leon, C. J., Vazquez-Polo, F. J., & Gonzalez, R. G. (2003). Elicitation of expert opinion in benefit transfer of environmental goods. *Environmental and Resource Economics, 26*, 199–210.

Leon, C. J., Vazquez-Polo, F. J., Guerra, N., & Riera, P. (2002). A Bayesian model for benefit transfer: Application to national parks in Spain. *Applied Economics, 34*, 749–757.

León-Gonzalez, R., & Scarpa, R. (2008). Improving multi-site benefit functions via Bayesian model averaging: A new approach to benefit transfer. *Journal of Environmental Economics and Management, 56*, 50–68.

Moeltner, K., Boyle, K. J., & Paterson, R. W. (2007). Meta-analysis and benefit transfer for resource valuation—addressing classical challenges with Bayesian modeling. *Journal of Environmental Economics and Management, 53*, 250–269.

Moeltner K., & Rosenberger R. S. (2008). Predicting resource policy outcomes via meta-regression: Data space, model space, and the quest for 'optimal scope'. *BE Journal of Economics Analysis and Policy*, Contributions 8, Article 31.

Morrison, M., Bennett, J., Blamey, R., & Louviere, J. (2002). Choice modeling and tests of benefit transfer. *American Journal of Agricultural Economics, 84*, 161–170.

Morrison, M., & Bergland, O. (2006). Prospects for the use of choice modeling for benefit transfer. *Ecological Economics, 60*, 420–428.

Murdock, J. (2006). Handling unobserved site characteristics in random utility models of recreation demand. *Journal of Environmental Economics and Management, 51*, 1–25.

Parsons, G. R., & Kealy, M. J. (1994). Benefit transfer in a random utility model of recreation. *Water Resources Research, 30*, 2477–2484.

Parsons, G. R., Massey, D. M., & Tomasi, T. (1999). Familiar and favorite sites in a random utility model of beach recreation. *Marine Resource Economics, 14*, 299–315.

Pendleton, L., Mohn, C., Vaughn, R. K., King, P., & Zoulas, J. G. (2012). Size matters: The economic value of beach erosion and nourishment in Southern California. *Contemporary Economic Policy, 30*, 223–237.

Phaneuf, D. J., von Haefen, R. H., Mansfield, C. A., & Van Houtven, G. (2013). *Measuring nutrient reduction benefits for policy analysis using linked non-market valuation and environmental assessment models: Final report on stated preference surveys*. Report to U.S. EPA, Office of Water, Grant #X7-83381001-0.

Shivlani, M. P., Letson, D., & Theis, M. (2003). Visitor preferences for public beach amenities in South Florida. *Coastal Management, 31*, 367–385.

Smith, V., Pattanayak, S. K., & Van Houtven, G. L. (2006). Structural benefit transfer: An example using VSL estimates. *Ecological Economics, 60*, 361–371.

Smith, V. K., Van Houtven, G. L., & Pattanayak, S. K. (2002). Benefit transfer via preference calibration prudential algebra for policy. *Land Economics, 78*, 132–152.

Swait, J., & Louviere, J. (1993). The role of the scale parameter in the estimation and comparison of multinomial logit models. *Journal of Marketing Research, 30*, 305–314.

Train, K. E. (2009). *Discrete choice methods with simulation* (2nd ed.). New York: Cambridge University Press.

Train, K., & Weeks, M. (2005). Discrete choice models in preference space and willingness-to-pay-space. In A. Alberini & R. Scarpa (Eds.), *Applications of simulation methods in environmental and resource economics* (pp. 1–16). Dordrecht: Springer.

Van Houtven, G., Pattanayak, S. K., Patil, S., & Depro, B. (2011). Benefits transfer of a third kind: An examination of structural benefits transfer. In J. Whitehead, T. Haab, & J. Huang (Eds.), *Preference data for environmental evaluation: Combining revealed and stated approaches* (pp. 303–321). New York: Routledge.

Van Houtven, G., & Poulos, C. (2009). Valuing welfare impacts of beach erosion: An application of the structural benefit transfer method. *American Journal of Agricultural Economics, 91*, 1343–1350.

von Haefen, R., & Phaneuf, D. (2008). Identifying demand parameters in the presence of unobservables: A combined revealed and stated preference approach. *Journal of Environmental Economics and Management, 56*, 19–32.

Whitehead, J. C., Dumas, C. F., Herstine, J., Hill, J., & Buerger, B. (2008). Valuing beach access and width with revealed and stated preference data. *Marine Resource Economics, 23*, 119–135.

Whitehead, J. C., Phaneuf, D., Dumas, C. F., Herstine, J., Hill, J., & Buerger, B. (2010). Convergent validity of revealed and stated preference behavior with quality change: A comparison of multiple and single site demands. *Environmental and Resource Economics, 45*, 91–112.

Yang, S., Chen, Y., & Allenby, G. M. (2003). Bayesian analysis of simultaneous demand and supply. *Quantitative Marketing and Economics, 1*, 251–275.

Part VI
Status and Prospects

Chapter 24
Benefit Transfer: The Present State and Future Prospects

Robert J. Johnston, Randall S. Rosenberger, John Rolfe and Roy Brouwer

Abstract The goal of this final chapter is to provide a brief perspective on the state of the art and future prospects in benefit transfer. It begins with a brief summary of what we know (primary areas of consensus), drawn from prior chapters and two decades of published work. This summary is followed by a discussion of what we do not yet know, including areas where consensus has been elusive, research results have been equivocal, or current work is otherwise insufficient to answer central questions. This is followed by a set of proposed research questions designed to address these knowledge and consensus gaps. The chapter concludes with a discussion of future prospects for benefit transfer research and practice.

Keywords Unit value transfer · Benefit function transfer · Meta-analysis · Structural benefit transfer · Accuracy · Validity · Non-market valuation · Methodology · Research needs · Benefits transfer

R.J. Johnston (✉)
Department of Economics and George Perkins Marsh Institute, Clark University, Worcester, MA, USA
e-mail: rjohnston@clarku.edu

R.S. Rosenberger
Department of Forest Ecosystems and Society, Oregon State University, Corvallis, OR, USA
e-mail: R.Rosenberger@oregonstate.edu

J. Rolfe
School of Business and Law, CQUniversity, Rockhampton, QLD, Australia
e-mail: j.rolfe@cqu.edu.au

R. Brouwer
Department of Environmental Economics, Institute for Environmental Studies, Vrije Universiteit, Amsterdam, Netherlands
e-mail: roy.brouwer@vu.nl

R.J. Johnston et al. (eds.), *Benefit Transfer of Environmental and Resource Values*, The Economics of Non-Market Goods and Resources 14,
DOI 10.1007/978-94-017-9930-0_24

24.1 Introduction

Benefit transfer methods have advanced significantly since the pioneering 1992 Association of Environmental and Resource Economics (AERE) and U.S. EPA workshop and subsequent special section of *Water Resources Research*, 28(3), widely credited with launching contemporary research in this area. Available methods now include approaches such as meta-analysis, structural benefit transfers and a suite of Bayesian approaches that were either undeveloped or not widely recognized at the time of the pioneering workshop.[1] Two decades later, benefit transfer is the most frequently applied non-market valuation method, particularly for practical policy applications, and is a central component of nearly all major cost-benefit analyses (CBAs) (Johnston and Rosenberger 2010; Smith et al. 2002). It is also the subject of a large and expanding literature, much of which has been discussed in earlier chapters.[2] The prospects for future use look strong, as governments increasingly evaluate options for managing resources (with limited time and budgets for these evaluations) and the number of available source valuation studies steadily grows.

At the same time, benefit transfer is often the least accurate way to estimate theoretically defensible welfare estimates.[3] It will never be able to outperform high quality primary valuation studies. Primary studies are therefore always preferred where feasible (Allen and Loomis 2008). The ubiquity of benefit transfer in policy analysis belies official guidance that such methods "should only be used as a last resort" and only when "a clear justification for using this approach over conducting original valuation studies" is provided (U.S. EPA 2010, pp. 7–51). Benefit transfer errors, while averaging approximately 36 and 45 % for function and unit value transfers (see Chap. 14), respectively, can exceed hundreds of percent in some cases (Boyle et al. 2010; Rosenberger and Stanley 2006). Methods associated with the largest transfer errors are often the most commonly used for applied policy analysis (Johnston and Rosenberger 2010). Moreover, only a small proportion of studies in the published valuation literature incorporate both the empirical quality and data reporting ideally suited to benefit transfer applications (Loomis and Rosenberger 2006). This situation had led to a perfect storm of reasons why continued work in

---

[1]An example is valuation meta-analysis, which was presented by Walsh et al. (1992) as a potential tool for benefit transfer but was not widely recognized and applied until the 2000s. Similarly, Atkinson and Crocker (1992) discussed Bayesian exchangeability within the context of hedonic models, but the majority of work on Bayesian transfer methods has occurred more recently (e.g., Johnston and Moeltner 2014; León et al. 2002, 2007; Leon-Gonzalez and Scarpa 2008; Moeltner and Rosenberger 2008; Moeltner and Woodward 2009; Moeltner et al. 2007).

[2]Boyle et al. (2010) and Johnston and Rosenberger (2010) provide recent summaries of this literature.

[3]Our qualification for this statement is that poorly constructed primary valuation studies may provide invalid or highly inaccurate estimates and lead to erroneous inferences. However, well-constructed primary valuation studies should always provide more defensible estimates than benefit transfer, even when they result in the same informational gains.

benefit transfer is required.[4] These include: (1) the frequency with which benefit transfer is applied, (2) the large errors that can result, (3) the common use of inaccurate methods by practitioners, (4) scarcity of suitable empirical studies with sufficient data reporting, and (5) a continuing lack of consensus surrounding important areas of methodology.

The goal of this final chapter is to provide a broad perspective on the state of the art and future prospects in benefit transfer, focusing on areas of consensus and research needs. It begins with a summary of what we know (or areas of consensus) drawn from prior chapters and two decades of published work. This summary is followed by a discussion of what we do not yet know, including areas where consensus has been elusive, research results have been equivocal, or current work is insufficient to answer central questions. We then propose a set of research questions designed to address these knowledge and consensus gaps. The chapter concludes with a discussion of future prospects for research and practice.

24.2 What We Know: General Areas of Consensus

Boyle et al. (2010, p. 162) note that two primary agreements in the benefit transfer literature are (1) "study sites and policy sites should be similar" and (2) "equation transfers are more accurate than value transfers." Site similarity includes consistency across a broad range of physical site and population characteristics, along with geospatial factors such as the distribution of beneficiaries and the proximity of substitutes.[5] There is also broad consensus that temporal factors are relevant to benefit transfer. As summarized by Johnston and Rosenberger (2010), temporal reliability assessments suggest that non-market values are often stable over short periods of time but change significantly after time periods as short as a few years (e.g., Brouwer 2006; Liebe et al. 2012; McConnell et al. 1998; Zandersen et al. 2007). It is also recognized that site similarity, while arguably a necessary condition for reliable benefit transfer, is not a sufficient condition. Although transfer errors are often smaller when study and policy sites are more similar, transfer errors can still be large over seemingly similar study and policy sites (Rosenberger and Stanley 2006).

These are among the most agreed-upon concepts in benefit transfer. Yet underlying these general agreements there are other areas of consensus built upon the last two decades of work. Among these are the general strengths and weaknesses of alternative transfer methods. For ease of discussion, we group methods into four categories: unit value transfer, benefit function transfer (not including meta-analysis or structural transfers), meta-analytic benefit function transfer, and structural benefit

[4]Also see the related discussion in Brouwer (2000).

[5]Despite strong evidence of the importance of site similarity for most types of benefit transfer, there are certain areas such as the benefits of health impacts in which more universal, "context free" benefit transfers have been shown superior to those adjusting for differences across sites (e.g., Ready et al. 2004).

transfer. Despite the substantial quality differentials that can occur across different applications, the literature appears to have reached consensus on some of the primary advantages and disadvantages of each methodological category.

For example, advantages of unit value transfers include ease of use and minimal data requirements (Chap. 8). Because unit value transfers involve little or no modeling, they are also less sensitive to modeling assumptions.[6] However, it is also generally accepted that unit value transfers afford the least ability to adjust welfare measures according to characteristics of the policy site. Although unit value transfers can perform acceptably when policy and study sites (including population characteristics) are very similar, they are the least accurate transfer method on average (Chap. 14; Bateman et al. 2011; Boyle et al. 2010; Kaul et al. 2013). Benefit function transfer (not including meta-analytic or structural approaches), in contrast, trades increased complexity and data requirements for increased flexibility and capacity to adjust welfare measures (see Chaps. 7–13). However, although function transfer may relax the need for site and population similarity at least somewhat (Boyle et al. 2010; Loomis 1992), closely matching study and policy sites are still required (Johnston and Rosenberger 2010).[7]

Compared to unit value and benefit function transfer, meta-analytic benefit function transfer requires greater expertise and more extensive data (see Chaps. 15–17, 19, 20, 22). Results can be sensitive to statistical modeling methods (Chap. 17; Boyle et al. 2010; Nelson and Kennedy 2009). However, these methods also provide greater capacity to adjust benefit measures, and reduce the need to find a single, closely matching study site (Rosenberger and Johnston 2009).[8] As a result, limited evidence suggests that—at least when appropriately applied—these methods can reduce transfer errors relative to alternative transfer methods (Chaps. 14 and 22; Engel 2002; Rosenberger and Phipps 2007). For example, while not focusing solely on meta-analysis, Kaul et al. (2013) find that methods combining data from multiple studies generate more accurate transfers than those using data from a single study.

Structural benefit transfer methods (Chap. 23), in contrast, incorporate a strong theoretical structure absent from other transfer methods that combine data from multiple studies, while also incorporating some of the flexibility of reduced form benefit function transfers (Boyle et al. 2010; Smith et al. 2002). However, these

[6]However, unit value transfers can be highly sensitive to the selection of studies from which one derives unit values, along with the valuation methods used by those studies. Issues related to study selection are discussed below.

[7]See, for example, discussions in Bateman et al. (2011), Boyle et al. (2010), Johnston and Rosenberger (2010), Colombo and Hanley (2008), Johnston (2007), Loomis and Rosenberger (2006) and Rosenberger and Phipps (2007). Some others have argued that sufficiently general benefit function transfers can reduce needs for site similarity under certain conditions (e.g., Boyle et al. 2010; Loomis 1992).

[8]As noted by Rosenberger and Johnston (2009, p. 410), "the probability of finding a good fit between a … study site and a policy site … is usually low." Boyle and Bergstrom (1992), Spash and Vatn (2006) and others have raised similar concerns.

methods are among the most complex available to practitioners, and results can be sensitive to modeling assumptions (e.g., the functional form of the assumed utility function). Moreover, the effects of the strong utility structure on empirical transfer accuracy have yet to be demonstrated (Johnston and Rosenberger 2010).

A second area of consensus regards the importance of scope and spatial scale for transfer validity and reliability (Chaps. 12, 13, 18 and 20; Bateman et al. 2006; Rolfe and Wang 2011). Following prior chapters, we define scope as the quantity or quality of a commodity considered by a benefit transfer; scale is defined as the geographic area over which an analysis is conducted. As discussed in Chaps. 2, 12 and 18, scaling of benefit (or other) estimates to larger or smaller scopes and scales than those addressed within the primary source studies risks substantial transfer errors (Bockstael et al. 2000; Johnston and Duke 2009; Rolfe et al. 2011). Transfer accuracy can also depend on an ability to account for the effect of distance decay and other spatial effects on estimates to be transferred (Chap. 18; Bateman et al. 2006; Loomis and Rosenberger 2006; Rolfe and Windle 2012; Schaafsma and Brouwer 2013; Schaafsma et al. 2012 and 2013). These can include the effects of borders or thresholds on welfare estimates, for example as they relate to heterogeneous preferences for environmental outcomes that occur in different regions or nations (Chaps. 18–20; Brouwer et al. 2010; Johnston and Duke 2009; Johnston and Thomassin 2010); Lindhjem and Navrud 2008; Martin-Ortega et al. 2012; Ready and Navrud 2006).

Finally, there is agreement that the accuracy of benefit transfers in general requires a diverse, high-quality, unbiased and well-reported body of primary valuation research (Chaps. 2, 14, 15 and 17; Boyle et al. 2010; Hoehn 2006; Johnston and Rosenberger 2010; Loomis and Rosenberger 2006; Rosenberger and Johnston 2009). The quality of benefit transfers cannot exceed the quality of the underlying primary study data.[9] Among the factors affecting the quality and unbiasedness of available primary studies are selection effects; these reflect the extent to which the literature provides an unbiased sample of the population of empirical estimates and to which these estimates provide an unbiased representation of true values (Hoehn 2006; Rosenberger and Johnston 2009). Adequate reporting of data and methods is necessary to evaluate study quality and to characterize site, commodity and population consistency; it is also required in order to adjust estimates for differences between study and policy sites (Loomis and Rosenberger 2006). While valuation databases can make the literature more broadly visible, they cannot offset limitations or biases in the underlying research (Johnston and Thomassin 2009; McComb et al. 2006; Morrison 2001). Recognizing challenges to benefit transfer caused by a lack of adequate data, many authors have called for additional emphasis into the provision of high-quality, well-documented empirical estimates of non-market values (Chap. 14; Johnston and Rosenberger 2010; Loomis and Rosenberger 2006; Rosenberger and Johnston 2009).

[9]The ideal goal for meta-analytic benefit function transfer is to glean fragments of truth from a highly variable literature and construct a multidimensional valuation surface (Johnston and Rosenberger 2010; Rosenberger and Phipps 2007).

24.3 What We Don't Know: Questions and Debates

Despite two decades of research in benefit transfer, there remain many unanswered questions and areas in which consensus has been elusive. Many of these are similar to "technical criteria for valid value transfer" and "protocol(s) for good practice" identified by Brouwer (2000) as unmet research needs, but for which conclusive guidance remains unavailable. As noted by Boyle et al. (2010, p. 162), "although benefit transfers are now common practice, even a cursory review of the benefit-transfer literature displays a wide variety of implementation procedures, with no consensus on which procedure actually results in the lowest transfer error [beyond the observations that] study sites and policy sites should be similar, and equation transfers are more accurate than value transfers."

We group primary unanswered benefit transfer research questions into three categories. The first includes *methodological application and accuracy* questions related to the suitability, reliability and validity of benefit transfer relative to primary research, and how these characteristics depend on the transfer methods that are applied. For example, under what conditions is benefit transfer appropriate? What errors are expected, how are these influenced by the transfer method(s) that are applied, and how can these errors be minimized? What specific transfer methods are preferred under different circumstances and why? For what types of contexts (e.g., sites and populations) are benefit transfers expected to be more (or less) accurate? How are site and commodity consistency best defined? Questions such as these are often a primary concern of benefit transfer practitioners.

The second category includes *data quality and availability* questions related to relationships between benefit transfer and available information, including characteristics of the broader literature of primary studies. Related questions address such issues as the type of information required to support accurate benefit transfers, changes in research design and reporting necessary to enhance transfer accuracy, the amelioration of selection biases in the economic literature, the time lag between implementation of a primary study and its use for benefit transfer,[10] and evaluation of study quality for benefit transfer purposes (Brouwer 2012; Hoehn 2006; Johnston and Thomassin 2009; Loomis and Rosenberger 2006; McComb et al. 2006; Morrison 2001; Rosenberger and Johnston 2009; Stanley et al. 2013).

The third category includes *research versus practice* questions related to the links between scholarly research and common practice within policy analysis. As noted by Johnston and Rosenberger (2010), there is a divergence between transfer practices recommended by the scholarly literature and those applied within policy analysis. While some governmental analyses have been informed by sophisticated benefit transfer methods such as meta-analysis, a much larger number have

[10]In practice, values and value functions are often transferred over extended periods of time. This requires the strong and often unrealistic assumption that preferences, values and estimated coefficients in value functions remain constant over these long time periods (Brouwer 2012).

incorporated simpler methods such as the transfer of unadjusted unit values, administratively approved values, or values adjusted using expert opinion (e.g., Chaps. 3–6). The use of simplified methods by analysts continues, despite evidence suggesting that such methods are inaccurate for all but the most similar study and policy sites. Such practices have been further encouraged in recent years by the inclusion of relatively rudimentary transfer tools within software decision support tools or canned benefit transfer spreadsheets, often used for ecosystem service valuation (cf. Bagstad et al. 2013). Challenges to the expanded use of more sophisticated methods include the complexity and equivocal nature of the current research literature and the lack of universally accepted protocols (Boyle et al. 2010; Johnston and Rosenberger 2010).

24.3.1 Questions of Methodological Application and Accuracy

Most of the benefit transfer literature is dedicated to questions of methodological application and accuracy. These include hundreds of published studies illustrating applied or hypothetical transfers, proposing and testing alternative methods, and evaluating transfer errors (typically using a convergent validity framework). Despite this work, which is largely grounded in case studies, there is a paucity of general conclusions to guide applied transfers. Some recent activity has begun to address this need. For example, Kaul et al. (2013) meta-analyze prior studies of benefit transfer convergent validity in an attempt to identify systematic patterns linking methods to transfer errors. Among expected findings, such as "function transfers outperform value transfers," this analysis also identifies potentially useful results such as "transfers describing environmental quantity generate lower transfer errors than transfers describing quality," and "combining data from multiple studies tends to reduce transfer errors" (Kaul et al. 2013, p. 90). In related work, Bateman et al. (2011) design a multisite experiment to test guidance principles for benefit transfers. Their findings provide insight into cases in which different types of transfer approaches are preferred, along with guidance on the specification of benefit functions. Drawing from the prior literature in this area, Boyle et al. (2009, 2010) propose general theoretical principles designed to inform transfer practice. Despite the findings of such work, there are many remaining questions regarding transfer methodology and performance.

24.3.1.1 Benefit Transfer Methodology, Accuracy and Valuation Context

Despite significant work in this area, there is still uncertainty regarding the type of transfer methods likely to be more (or most) accurate in different types of benefit transfer applications. This uncertainty extends to many areas, such as the relative performance of different types of benefit function transfer (e.g., single-site, meta-analytic, structural) in different situations, the impact of different types of adjustments on transfer accuracy, and the formal definition of consistency across sites, populations and commodities.

A key area of ambiguity relates to the relative performance of different types of benefit function transfers. Despite recent evidence that meta-analytic transfers can outperform single-site function transfers in some applications (Kaul et al. 2013; Rosenberger and Phipps 2007), evaluation of relative accuracy has been stymied by ambiguities in testing procedures. For example, when comparing meta-analytic to single-site function transfer for a single policy site, it is often unclear which studies or sites should be selected to test the accuracy of single-site transfers (Engel 2002). Testing accuracy over all possible sites (e.g., all those included in the metadata used for comparison) can be misleading, because these may include very dissimilar sites that would rarely be chosen for single-site function transfer. However, choosing a small number of test sites that are "most similar" to the policy site based on an ad hoc set of criteria may also be misleading, because results can be sensitive to the specific study sites chosen for analysis.[11] As a result, accuracy tests comparing meta-analysis to other types of benefit function transfer often lead to tentative results (cf. Engel 2002). Similarly, although the potential theoretical advantages of structural benefit transfers have been established (Bergstrom and Taylor 2006; Smith and Pattanayak 2002; Smith et al. 2002, 2006), the relative empirical accuracy of these methods has yet to be established and compared systematically to alternative methods (Johnston and Rosenberger 2010). The result has been a lack of definitive evidence supporting the use or relative accuracy of different types of benefit function transfer methods. Related ambiguities over testing methods are discussed by Boyle et al. (2010).

Another area in which evidence is beginning to emerge—but in which consensus is not yet established—is the relative performance of alternative transfer methods for different types of study and policy sites. There have been arguments that meta-analysis may be superior in cases for which a closely matching study and policy site cannot be found, given the capacity of meta-analysis to generate more generally applicable benefit functions (e.g., Downing and Ozuna 1996; Engel 2002; Moeltner et al. 2007; Rosenberger and Johnston 2009; Rosenberger and Phipps 2007; Stapler and Johnston 2009; Vandenberg et al. 2001). However, as noted above, this has yet to be demonstrated in a systematic manner. Recent evidence also suggests that "when transferring across relatively similar sites, simple mean value transfers are to be preferred but … when sites are relatively dissimilar … value function transfers will yield lower errors" (Bateman et al. 2011, p. 365). Although such results support expectations, they have yet to be established conclusively by a wide range of supporting evidence.

---

[11]Choosing the most accurate single-site transfer to illustrate comparative performance is also problematic, because of the endogenous nature of the test. That is, the test site is chosen based on its superior performance in the test.

Some of the work in this book speaks directly or indirectly to this issue. For example, Chap. 22 provides methods that may be used to evaluate optimal scope (or combinations of data) for different types of transfers. A number of other chapters evaluate the relative performance of different types of transfers in different contexts. These include Chap. 14, which draws some general conclusions on benefit transfer reliability across different types of applications in the literature. Despite this and other evidence, the literature still lacks clear, operational protocols to guide practitioners seeking the most appropriate types of transfer methods for different types of study and policy sites.

A final area in which consensus is lacking relates to ambiguity in operational definitions of site, population and commodity consistency. Despite detailed discussions of site and commodity consistency in the literature (e.g., Loomis and Rosenberger 2006), the literature has yet to agree on a concise set of criteria for consistency (Johnston and Rosenberger 2010). For example, there are different ways to quantify and reconcile quantity and quality measures across sites (e.g., Johnston et al. 2005; Van Houtven et al. 2007), and different ways to define site similarity (Bateman et al. 2011; Colombo and Hanley 2008; Johnston 2007). This issue is particularly obvious for meta-analyses that pool data from multiple prior sites and studies (Engel 2002; Johnston et al. 2005; Smith and Pattanayak 2002), but affects the accuracy and potential validity of all benefit transfers. Nonetheless, and despite almost universal guidance that benefit transfers require consistency among sites, populations and commodities, there are still no systematic criteria for these types of consistency along with a body of research definitively linking these criteria to changes in transfer accuracy. Some of this ambiguity is unavoidable given the myriad ways that sites can correspond or differ, and that commodities can be valued across various sites. Yet additional knowledge and protocols in this area are needed to help ensure that transfers meet minimum standards. Future work in meta-analysis and evaluations of data pooling offers promise as a means to address this challenge.

24.3.1.2 Theory Versus Empirics

Although often implicit and sometimes poorly articulated, there is an unresolved tension in the benefit transfer literature regarding the relative importance of strong theoretical structure and consistency (often based on a priori or *ex ante* expectations) versus empirical performance and accuracy (often tested *ex post* using convergent validity or other mechanisms). Simply put, what if more accurate empirical transfers can be accomplished using methods that relax restrictions or structures suggested by economic theory? This tension is perhaps best exemplified by debates concerning the relative merits of structural benefit transfers compared to alternative methods, but it appears in many other important areas. To the extent that these debates can be resolved, the solutions could have profound implications for benefit transfer practice, including current practices.

Tradeoffs between the empirical accuracy improvements possible using reduced form meta-analytic benefit transfer and the concomitant loss of micro-level theoretical foundation are well-established (Engel 2002; Smith and Pattanayak 2002). Similar debates have occurred concerning the appropriateness of pooling otherwise commensurable Marshallian and Hicksian welfare measures within a single meta-regression model (Bergstrom and Taylor 2006; Johnston and Moeltner 2014; Londoño and Johnston 2012; Nelson and Kennedy 2009; Smith and Pattanayak 2002). This issue pits theoretical concerns that Marshallian and Hicksian welfare measures are theoretically distinct against empirical evidence that suggests pooling these often similar measures can sometimes enhance benefit transfer accuracy.

The latter debate is an example of recent empirical considerations of optimal scope in meta-analysis and benefit transfer (Engel 2002; Johnston and Moeltner 2014; León-Gonzalez and Scarpa 2008; Moeltner and Rosenberger 2008, 2014). These analyses evaluate empirical value distributions within alternative data pools to identify those groupings of sites, studies or valuation contexts that provide the greatest gains in transfer efficiency and reliability. The general approach is based on the observation that "value distributions can converge across contexts despite differences in site characteristics, population features, or preferences" (Moeltner and Rosenberger 2014, p. 19). This perspective, while demonstrating clear empirical advantages, contravenes common approaches that prioritize similarity across these characteristics, based on theoretical considerations. These analyses raise the question of whether the comparability and scope of data for benefit transfers should be based primarily on empirical or theoretical grounds. It is yet to be seen whether and how this work will influence benefit transfer practice.

At its core, the theory versus empirics debate may be linked to different approaches to reasoning—deductive versus inductive. Deductive reasoning postulates theory and then derives conclusions based on empirical evidence in support or opposition to said theory (e.g., economic values are defined by the utility-theoretic foundation from which they are derived). Inductive reasoning, in contrast, begins with empirical evidence and then derives theory or conclusions based on patterns that emerge from said evidence.[12] While not made explicit, arguments over the relative importance of theoretical structure and strict theoretical consistency in benefit transfer are often grounded in different perspectives on the relevance of deductive versus inductive reasoning in welfare economics.

Considering these and other ongoing debates in the benefit transfer literature, it is clear that researchers have yet to reach consensus on the relative importance of strong structural (utility theoretic) foundations and strict adherence to theoretical expectations versus that of empirical relationships that emerge from valuation data. Bergstrom and Taylor (2006) characterize the divergence as between weak structural and strong structural approaches. The inability to resolve this debate has led to sometimes disparate and conflicting guidance across the benefit transfer literature.

[12]Meta-analysis works in both directions; it can be used to test theory based on evidence or to postulate theory based on identified patterns in evidence.

24.3.1.3 Scope and Scale

Although there is widespread agreement on the relevance of scope and scale for accurate benefit transfer, methods to incorporate scope and scale vary. For example, repeated empirical observations demonstrate spatial heterogeneity in welfare estimates and the relevance of this heterogeneity for benefit aggregation and transfer over different types of economic and political jurisdictions (Chaps. 18 and 20; Bateman et al. 2006; Campbell et al. 2009; Johnston and Duke 2009; Johnston and Ramachandran 2014; Loomis 2000; Rolfe and Windle 2012; Schaafsma et al. 2012). Despite these findings, perhaps the most common approach to benefit transfers simply multiplies a measure of central tendency (e.g., mean WTP) by the number of individuals or households in a predetermined region. As noted by Bateman et al. (2006), the potential errors that can be introduced by such approaches can overwhelm other sources of econometric or other error in unit value estimation given much greater attention in the literature.

Similar concerns relate to adjustments for scope. It is widely recognized that adjusting for commodity scope is critical for accurate benefit transfers. However, treatment of scope in both primary studies and benefit transfers can be more complex than is frequently assumed (Heberlein et al. 2005; Rolfe and Wang 2011). Moreover, methods often proposed for benefit function transfers, including choice experiments (Morrison and Bergland 2006; Rolfe and Bennett 2006), typically provide linear benefit functions that—while appropriate for the small scale of changes often modeled in these studies—do not allow for the diminishing marginal utility expected for larger improvements. Although choice experiments can and do incorporate nonlinear benefit functions of the type that could incorporate downward sloping demand (e.g., Johnston et al. 2002), linear preference functions remain the norm in the choice experiment literature. As a result, benefit transfers relying on these studies often incorporate simplistic treatments of scope (e.g., assuming that benefits increase as a linear function of scope) that can lead to errors when the scope of change at the policy site differs from that at the study site. This is an example of limitations in benefit transfer caused by limitations in the underlying primary valuation study; other examples of this issue are discussed by Loomis and Rosenberger (2006).

It is clear that adjustments for scope and scale are critical elements of accurate benefit transfer, yet findings and specific guidance regarding these adjustments and associated protocols remain inconsistent, disconnected and scattered across the literature. The lack of cohesive guidance has contributed to significant variability in treatments of scope and scale across applied benefit transfers. This is another area in which knowledge is lacking and research is required—particularly research that coordinates and reconciles disconnected results across the literature.

24.3.1.4 The Welfare Architecture of Benefit Transfer

To a large extent the above questions reflect broader unknowns regarding the welfare architecture of benefit transfer. This relates to more general concepts concerning the motivations of non-market values (Johnston and Thomassin 2009), as well as the distribution of property rights attached to non-market goods and baseline conditions (Bateman et al. 2006). That is, accurate transfers require an understanding of the specific welfare-relevant quantities or qualities (definitions of non-market goods and the change in their provision levels, informing relevant welfare measures such as compensating or equivalent surplus) at affected sites, both in primary studies from which values are estimated and in transfer sites for which value estimates are needed. Even studies of seemingly similar biophysical changes (e.g., a specific change in the biophysical properties of water, or water quality) may estimate values associated with differing underlying quantities or qualities (e.g., the quality of water for drinking, fishing for different species, aesthetics, or as an indicator of ecosystem health). The direction of change is also relevant (e.g., improvement and willingness to pay to secure the improvement, or a deterioration and willingness to pay to prevent this deterioration), as are implications for assumed property rights (e.g., related to the choice of WTP versus WTA as an appropriate measure of value). To the extent that these are poorly defined or unknown in primary studies, they cannot be incorporated within subsequent benefit transfers.

Although it is sometimes possible to reconcile commodity definitions across studies, reconciliation that promotes sufficient uniformity is not always feasible (Smith et al. 2002; Van Houtven et al. 2007), and analysts are often "delinquent" in such areas (Nelson and Kennedy 2009, p. 346). The task is made more difficult "as [the] complexity of changes in environmental quality and natural resources increase [s]" (Navrud and Ready 2007b, p. 3). In many cases, the assumptions required to reconcile commodity definitions across sites are not well-specified. Chapters 12 and 16 discuss this issue with regard to benefit transfers of ecosystem service values and the construction of metadata, respectively, but similar concerns apply to virtually all transfers of environmental values.

Value estimates may also diverge when different valuation methods are used (Johnston et al. 2006; Moeltner et al. 2007; Stapler and Johnston 2009). Although this pattern is frequently consistent with theoretical expectations (Smith and Pattanayak 2002; Smith et al. 2002), it can complicate benefit transfers. Moreover, as discussed by Brouwer (2000) and Johnston and Duke (2008), among others, households' WTP for seemingly similar biophysical changes can be sensitive to such factors as the policy process through which change is realized, attributes of local context, motivations of individuals, and other sometimes latent factors that add to the complexity of producing generally applicable benefit functions. Many of these effects are context-specific and hence difficult to quantify or compare across sites. Variations such as these may help explain common observations of reduced transfer accuracy and increased conceptual difficulties associated with the transfer of nonuse values (Brouwer 2000; Kaul et al. 2013; Navrud and Ready 2007a), given that these values may be influenced by a wide range of latent or unobservable motivations.

Concerns related to commodity consistency are particularly apparent for transfers involving meta-analysis. All meta-analyses in the valuation literature make explicit or implicit assumptions regarding the commensurability or reconcilability of welfare measures from a number of valuation contexts (Engel 2002; Johnston et al. 2005;

Loomis and Rosenberger 2006; Nelson and Kennedy 2009; Smith and Pattanayak 2002). Chapter 16 addresses these issues explicitly for a case study of river condition. Although appropriately specified meta-analyses may be able to account for systematic patterns relating theoretically distinct welfare measures for otherwise similar commodities, meta-analysis cannot by itself solve challenges related to commensurability (Bergstrom and Taylor 2006; Johnston and Rosenberger 2010). For example, in a meta-analysis of a specific subset of wetland ecosystem services using cost estimates only, which are generally considered more reliable than benefit estimates even if theoretically inferior, Brander et al. (2013) demonstrate that prediction errors can still be high even at a relatively high explanatory power of the estimated cost model. Similar findings of a mismatch between construct validity and prediction error were reported already by Brouwer and Spaninks (1999).

The conceptual issue was described by Brouwer (2000, p. 143) over ten years ago, but remains a challenge today: "Investigating the process of value formation, articulation and elicitation in order to better understand the values themselves means that one of the core assumptions underlying economics has to be revisited; namely, that preferences are given and that it does not matter why people value things. This core assumption deserves much more attention from environmental economists. Differences in underlying reasons and motives may enable us to better explain differences in valuation outcomes and hence come up with a model which has a sufficiently high explanatory power to validly and reliably predict values across sites and groups of people."

24.3.2 Questions of Data Quality and Availability

Benefit transfer relies upon the quality and applicability of underlying primary study data. Yet current academic and other incentives do not commonly reward the development of empirical valuations suitable for benefit transfer (Johnston and Rosenberger 2010; Rosenberger and Johnston 2009; Smith and Pattanayak 2002). As stated by McComb et al. (2006, p. 471), "[p]ressure from publications to create novel methods or formulations has resulted in an abundance of studies that are distant from the day-to-day needs of policy makers…". Despite efforts among economists to improve valuation methodologies, and a correspondingly large methodological literature, benefit transfer analysts nonetheless face a "lack of adequate [empirical] studies for benefit transfer" (Loomis and Rosenberger 2006, p. 344). Similarly, Van Houtven et al. (2007, p. 225) note "a continued need for … valuation research that can be used to address the requirements of national and regional-scale benefit assessments."

Compounding this problem, benefit transfer itself can threaten incentives for underlying valuation research. If decision-makers regularly choose benefit transfer as an easy alternative to primary research (despite the potential errors), the demand for—and resources to support—high-quality primary research can decline. Despite widespread recognition of concerns related to the size, diversity and quality of the empirical valuation literature, the problem remains unsolved. Academics and others who support valuation research have thus far been unable to promote development of a large set of empirical valuation studies specifically designed and documented for benefit transfer applications.

Related to this broader data-availability concern are a number of specific questions regarding the types of studies and data required to support accurate benefit transfers. These also concern potential tradeoffs between study quality and selection effects. This is an important area for future attention and research.

24.3.2.1 Data Required to Support Benefit Transfers

With some notable counter-examples (e.g., work funded through the EU Water Framework Directive or in support of U.S. government CBAs or natural resource damage assessments), much of the published research available for transfer has been a side effect of research funded for methodological purposes (Johnston and Rosenberger 2010; McComb et al. 2006). These studies often lack one or more of the elements necessary for ideal transfer applications, including sufficient reporting of empirical methods, data and site/population characteristics (Loomis and Rosenberger 2006). Although many benefit transfer researchers have commented on the need for additional, high-quality and well-documented empirical studies of resource values, the specific resources or sites that should be targeted by these proposed efforts remain largely unspecified. Practical methods to increase incentives or support for such research have also been elusive. In short, there is a gap between empirical research needs identified by the benefit transfer literature and practical solutions to address these needs. Lack of solutions to this problem threatens the long-term availability of source material for benefit transfers.

There are also questions regarding specific improvements in the research design and reporting necessary to enhance transfer accuracy. Loomis and Rosenberger (2006) discuss a broad set of requirements for primary study design and reporting. At the same time, they note (p. 344), "…we recognize that sermonizing about these external benefits is likely to fall on deaf ears. Incentives to undertake the extra cost to more completely document results and data are needed." Moreover, benefit transfer research is equivocal regarding the relevance of different types of data for accuracy-enhancing adjustments to welfare measures. For example, it is widely expected that "differences in income and other socio-demographic characteristics of consumers would affect a resource's economic value" (Loomis and Rosenberger 2006, p. 345). Yet evidence that socioeconomic adjustments reduce transfer errors has been mixed; in some cases these adjustments increase errors (Brouwer 2000; Johnston and Duke 2010; Spash and Vatn 2006). Recent work also emphasizes the potential advantages of minimally parameterized benefit functions with only core variables suggested by economic theory, thereby reducing some data and reporting needs (Bateman et al. 2011).

The lack of unequivocal evidence regarding "which procedure actually results in the lowest transfer error" (Boyle et al. 2010, p. 162) translates to a similar lack of certainty regarding what types of data requirements should be expected of empirical work, in order to improve subsequent benefit transfers. Although the need for improved primary study research design and reporting is clear, the ways that this should be accomplished are less so.

24.3.2.2 Study Quality and Selection Effects

When one selects primary studies for benefit transfer, there are implicit assumptions that the underlying literature provides an unbiased sample of the population of empirical estimates and that these estimates provide an unbiased representation of true resource values. If these assumptions do not hold, the result will be systematic errors in benefit transfer, often called selection biases. Types of selection biases include research priority selection, methodology selection, publication selection, and sample selection (Chaps. 2 and 14; Hoehn 2006; Rosenberger and Johnston 2009). Methods to identify and correct for selection biases include a variety of statistical processes, along with careful procedures to identify published and unpublished literature from which to draw information (Chaps. 14–17; Florax et al. 2002; Hoehn 2006; Rosenberger and Johnston 2009; Stanley 2005, 2008). Some common statistical procedures used to evaluate selection biases, however, require data such as the standard error of estimates (e.g., the standard error of welfare estimates) not routinely provided by the valuation literature. Other approaches, such as the two-stage Heckman selection models outlined by Hoehn (2006) and Rosenberger and Johnston (2009) (cf. Florax et al. 2002; Smith and Huang 1993) require either additional data not frequently collected as part of benefit transfer applications, or assumptions regarding the form of the selection equation. As a result, although researchers agree that selection biases are a concern and are often overlooked—particularly publication and research priority selection—the literature has not agreed on a set of feasible consensus protocols or methods to address these biases in common benefit transfer applications.

Amelioration of selection biases can also be confounded by the methods used to screen primary studies for quality. As noted in Chap. 2, avoidance of measurement error in benefit transfers requires that primary studies are of a certain minimum quality. However, the same type of screening criteria used to evaluate quality in potential source studies for benefit transfer (e.g., peer review, publication, minimum methodological standards, statistical criteria) can also be the source of unanticipated selection biases. For example, there may be biases inherent in the types of papers accepted for publication or methods/results favored by peer reviewers, leading to systematic biases in the published literature (Stanley 2005, 2008). To date, procedures for primary study selection within applied benefit transfers often remain either unspecified or ad hoc; only the applied meta-analysis literature has begun to address these issues systematically (Rosenberger and Johnston 2009; Stanley 2005, 2008; Stanley et al. 2013). Moreover, the literature has yet to provide protocols that address the tradeoffs between screening for primary study quality and minimizing the possibility of attendant selection biases.

24.3.3 *Questions of Research Versus Practice*

Chapter 1 points out that benefit transfer practitioners often make informal and sometimes uninformed judgments about the applicability of transfer practices recommended in the research literature. This can lead to a gap between transfer practices recommended in the academic literature and those commonly applied within policy analysis (2010). For example, guidance from the U.S. EPA (2010) advised that "benefit function transfers are preferable to unit value transfers as they incorporate information relevant to the policy scenario" (also see U.S. OMB 2003). These guidelines also describe the basics of meta-analysis and structural benefit transfer, along with some of the relevant academic literature. However, despite such guidance documents, many policy analysts remain unaware of even the most basic findings from benefit transfer research. These findings are also unfamiliar to many scientists from other disciplines who may wish to use benefit transfer (e.g., for ecosystem service valuation; Chaps. 12 and 13). Key questions in this area relate to the pervasive gap between research and practice.

24.3.3.1 Narrowing the Gap Between Research and Practice

As discussed in Chap. 1, the technical knowledge needed to perform contemporary benefit transfer has increased, raising questions about whether it is possible for benefit transfer to be conducted by nonspecialists. For example, recent guidelines for valuation meta-analysis include rigorous econometric, evaluation and reporting procedures (Chaps. 15, 16 and 17; Nelson and Kennedy 2009; Stanley et al. 2013), some of which involve techniques requiring expertise common only among PhD economists or statisticians. Similarly specialized expertise is required for Bayesian and structural benefit transfer approaches (Chaps. 21–23). These recent, rapid (and some would argue needed) advances increase the complexity and cost of benefit transfer applications. The literature has yet to reconcile the continued development of ever-more-demanding and costly benefit transfer methods with the common need for benefit transfer as a "quick and easy" means to estimate values when primary studies are infeasible. In the absence of a solution to this problem, the gap between research and practice may widen further.

24.3.3.2 The Role of Databases and Decision Support Tools

As practitioners seek data and support for applied benefit transfers—particularly amidst the imposing complexity of some benefit transfer methods—they often turn to valuation databases and decision support tools. As described in Chap. 2, valuation databases such as the Environmental Valuation Reference Inventory (EVRI, http://www.evri.ca) can often help practitioners identify research studies suitable for transfer (Johnston and Thomassin 2009; McComb et al. 2006; Morrison 2001).

Yet these databases cannot substitute for the expertise and detailed analysis of original primary studies: "analysts should not expect to be able to simply download value estimates for a cost-benefit analysis from these [valuation] databases, unless the cost-benefit analysis is particularly rudimentary and of little policy significance" (Morrison 2001, p. 54). A survey of benefit transfer experts by Johnston and Thomassin (2009) notes the perceived risk of such database misuse by policy analysts. Yet a corresponding survey of EVRI users showed strong support for the inclusion of automated benefit transfer spreadsheet tools in the database, implying that many policy analysts support the use of valuation databases in ways that contravene the guidance of benefit transfer experts (Johnston and Thomassin 2009).

These concerns are even more pointed for off-the-shelf decision support tools that forecast and often map economic values generated using internal (and sometimes proprietary) benefit transfer algorithms (Chap. 12; Bagstad et al. 2013). Such tools are often used to quantify economic values for ecosystem services (Chaps. 4 and 12). They generally rely on rudimentary unit value or single-site benefit function transfers, or spreadsheet-type calculations that are likely to be inaccurate for most applications. Others generate estimates of economic value with little or no basis in economic theory, or that aggregate values in a naïve manner, without recognition of central issues such as distance decay, substitutability and downward sloping demand. Yet despite the increasing application of these tools by policy analysts and others, the benefit transfer literature provides no clear protocols or guidance to distinguish appropriate and inappropriate (or accurate versus inaccurate) uses of these tools. The literature also lacks a systematic evaluation of these tools with regard to the validity and expected accuracy of the embedded benefit transfer methods.

Given the increasing interest of policy analysts and government agencies in databases and decision support tools, benefit transfer researchers no longer have the luxury of (largely) ignoring these instruments. In the absence of dedicated work to evaluate and improve the most promising of these databases/tools and provide consensus protocols regarding the role and use of these instruments, an increasing proportion of applied benefit transfers will likely rely on off-the-shelf mechanisms viewed by economists as either inaccurate or invalid. The lack of engagement of benefit transfer researchers in this area has led to a knowledge gap that threatens to undermine the quality of benefit transfer practices (and the quality of economic information used to inform decisions) worldwide.

24.4 Future Prospects

It has been just over two decades since the original 1992 workshop on benefit transfer. Methods have advanced significantly since that time, but many central questions remain. While many participants in the original workshop are still driving advances in benefit transfer today, new advances are being motivated by subsequent generations of researchers. Amidst the flurry of recent activity in benefit transfer research, future prospects depend largely on the extent to which the uncertainties and inconsistencies discussed above—along with related issues discussed by prior publications (e.g., Boyle et al. 2010; Brouwer 2000; Johnston and Rosenberger 2010)—can be addressed by future work. It is almost certain that benefit transfer will continue to be a central component of policy analysis worldwide. What is less clear is the ways in which these transfers will be conducted, and whether the resulting economic information will be of sufficient relevance and accuracy to improve policy decisions.

There are reasons for hope as well as concern. Among the more worrying signs is the rapid proliferation of computerized decision support tools that promise quick and easy access to information on economic values, yet estimate these values using methods that are unknown, inaccurate and/or invalid. Despite frequent marketing as methods for economic valuation, these tools have been developed largely independent of the valuation and benefit transfer literature, and generally prioritize biophysical and GIS elements over the accurate estimation of economic values. Nonetheless, these tools are rapidly being considered and adopted for use by government agencies, often with considerable marketing from tool developers (including private firms and nongovernmental organizations). To the extent that these tools are inadequate and supplant more sophisticated and accurate benefit transfers, decisions may be made based on gross misunderstandings of economic benefits and costs.

At the same time, there are promising signs of improvement in benefit transfer practices. There is an increasing call among researchers for more systematic and consistent benefit transfer protocols. Recent work has begun to clarify what some of those protocols might be. Dedicated outreach is also being conducted to help ensure that usable knowledge from the literature reaches policymakers outside of academia, leading to increased use of more sophisticated benefit transfer methods. This combination of research and engagement seeks to reduce the burden and uncertainty associated with the conduct of accurate and defensible benefit transfers within the funding and time limitations of the policy process. To the extent that future research can inform improved, consistent and practical benefit transfer methods, and these methods are made accessible for use within the policy process, there is the potential for substantial improvements in the economic analyses now used to guide policy. In this sense, benefit transfer may be among the most impactful and socially beneficial areas within which environmental economists can work. Among the goals of this book is to promote this future work.

References

Allen, B. P., & Loomis, J. B. (2008). The decision to use benefit transfer or conduct original valuation research for benefit-cost and policy analysis. *Contemporary Economic Policy, 26*, 1–12.

Atkinson, S. E., & Crocker, T. D. (1992). The exchangeability of hedonic property price studies. *Journal of Regional Science, 32*, 69–83.

Bagstad, K. J., Semmens, D. J., Waage, S., & Winthrop, R. (2013). A comparative assessment of decision-support tools for ecosystem services quantification and valuation. *Ecosystem Services, 5*, e27–e39.

Bateman, I. J., Day, B. H., Georgiou, S., & Lake, I. (2006). The aggregation of environmental benefit values: Welfare measures, distance decay and total WTP. *Ecological Economics, 60*, 450–460.

Bateman, I. J., Brouwer, R., Ferrini, S., Schaafsma, M., Barton, D. N., Dubgaard, A., et al. (2011). Making benefit transfers work: Deriving and testing principles for value transfers for similar and dissimilar sites using a case study of the non-market benefits of water quality improvements across Europe. *Environmental & Resource Economics, 50*, 365–387.

Bergstrom, J. C., & Taylor, L. O. (2006). Using meta-analysis for benefits transfer: Theory and practice. *Ecological Economics, 60*, 351–360.

Bockstael, N. E., Freeman, A. M., Kopp, R. J., Portney, P. R., & Smith, V. K. (2000). On measuring economic values for nature. *Environmental Science and Technology, 34*, 1384–1389.

Boyle, K. J., & Bergstrom, J. C. (1992). Benefit transfer studies: Myths, pragmatism and idealism. *Water Resources Research, 28*, 657–663.

Boyle, K. J., Kuminoff, N. V., Parmeter, C. F., & Pope, J. C. (2009). Necessary conditions for valid benefit transfers. *American Journal of Agricultural Economics, 91*, 1328–1334.

Boyle, K. J., Kuminoff, N. V., Parmeter, C. F., & Pope, J. C. (2010). The benefit-transfer challenges. *Annual Review of Resource Economics, 2*, 161–182.

Brouwer, R. (2000). Environmental value transfer: State of the art and future prospects. *Ecological Economics, 32*, 137–152.

Brouwer, R. (2006). Do stated preference methods stand the test of time? A test of the stability of contingent values and models for health risks when facing an extreme event. *Ecological Economics, 60*, 399–406.

Brouwer, R. (2012). Constructed preference stability: A test-retest. *Journal of Environmental Economics and Policy, 1*, 70–84.

Brouwer, R., & Spaninks, F. A. (1999). The validity of environmental benefits transfer: Further empirical testing. *Environmental & Resource Economics, 14*, 95–117.

Brouwer, R., Martín-Ortega, J., & Berbel, J. (2010). Spatial preference heterogeneity: A choice experiment. *Land Economics, 86*, 552–568.

Brander, L., Brouwer, R., & Wagtendonk, A. (2013). Economic valuation of regulating services provided by wetlands in agricultural landscapes: A meta-analysis. *Ecological Engineering, 56*, 89–96.

Campbell, D., Hutchinson, W. G., & Scarpa, R. (2009). Using choice experiments to explore the spatial distribution of willingness to pay for rural landscape improvements. *Environment and Planning A, 41*, 97–111.

Colombo, S., & Hanley, N. (2008). How can we reduce the errors from benefits transfer? An investigation using the choice experiment method. *Land Economics, 84*, 128–147.

Downing, M., & Ozuna, T. (1996). Testing the reliability of the benefit function transfer approach. *Journal of Environmental Economics and Management, 30*, 316–322.

Engel, S. (2002). Benefit function transfer versus meta-analysis as policy tool: A comparison. In R. J. G. M. Florax, P. Nijkamp, & K. G. Willis (Eds.), *Comparative environmental economics assessment* (pp. 133–153). Chelthenham, UK: Edward Elgar.

Florax, R. J. G. M., Nijkamp, P., & Willis, K. G. (Eds.). (2002). *Comparative environmental economics assessment*. Chelthenham, UK: Edward Elgar.

Heberlein, T. A., Wilson, M. A., Bishop, R. C., & Schaeffer, N. C. (2005). Rethinking the scope test as a criterion in contingent valuation. *Journal of Environmental Economics and Management, 50*, 1–22.

Hoehn, J. P. (2006). Methods to address selection effects in the meta regression and transfer of ecosystem values. *Ecological Economics, 60*, 389–398.

Johnston, R. J. (2007). Choice experiments, site similarity and benefits transfer. *Environmental & Resource Economics, 38*, 331–351.

Johnston, R. J., & Duke, J. M. (2008). Benefit transfer equivalence tests with non-normal distributions. *Environmental & Resource Economics, 41*, 1–23.
Johnston, R. J., & Duke, J. M. (2009). Willingness to pay for land preservation across states and jurisdictional scale: Implications for benefit transfer. *Land Economics, 85*, 217–237.
Johnston, R. J., & Duke, J. M. (2010). Socioeconomic adjustments and choice experiment benefit function transfer: Evaluating the common wisdom. *Resource and Energy Economics, 32*, 421–438.
Johnston, R. J., & Moeltner, K. (2014). Meta-modeling and benefit transfer: The empirical relevance of source-consistency in welfare measures. *Environmental & Resource Economics, 59*, 337–361.
Johnston, R. J., & Ramachandran, M. (2014). Modeling spatial patchiness and hot spots in stated preference willingness to pay. *Environmental & Resource Economics, 59*, 363–387.
Johnston, R. J., & Rosenberger, R. S. (2010). Methods, trends and controversies in contemporary benefit transfer. *Journal of Economic Surveys, 24*, 479–510.
Johnston, R. J., & Thomassin, P. J. (2009). Evaluating the Environmental Valuation Reference Inventory (EVRI): Results from a survey of actual and potential users. *AERE (Association of Environmental and Resource Economists) Newsletter, 29*, 33–38.
Johnston, R. J., & Thomassin, P. J. (2010). Willingness to pay for water quality improvements in the United States and Canada: Considering possibilities for international meta-analysis and benefit transfer. *Agricultural and Resource Economics Review, 39*, 114–131.
Johnston, R. J., Swallow, S. K., Allen, C. W., & Smith, L. A. (2002). Designing multidimensional environmental programs: Assessing tradeoffs and substitution in watershed management plans. *Water Resources Research, 38*, 1099–1105.
Johnston, R. J., Besedin, E. Y., Iovanna, R., Miller, C., Wardwell, R., & Ranson, M. (2005). Systematic variation in willingness to pay for aquatic resource improvements and implications for benefit transfer: A meta-analysis. *Canadian Journal of Agricultural Economics, 53*, 221–248.
Johnston, R. J., Besedin, E. Y., & Ranson, M. H. (2006). Characterizing the effects of valuation methodology in function-based benefits transfer. *Ecological Economics, 60*, 407–419.
Kaul, S., Boyle, K. J., Kuminoff, N. V., Parmeter, C. F., & Pope, J. C. (2013). What can we learn from benefit transfer errors? Evidence from 20 years of research on convergent validity. *Journal of Environmental Economics and Management, 66*, 90–104.
León, C., Leon, R., & Vazquez-Polo, F. (2007). The application of Bayesian methods to benefit transfer. In S. Navrud & R. Ready (Eds.), *Environmental value transfer: Issues and methods* (pp. 227–240). Dordrecht: Springer.
León, C. J., Vazquez-Polo, F. J., Guerra, N., & Riera, P. (2002). A Bayesian model for benefit transfer: Application to national parks in Spain. *Applied Economics, 34*, 749–757.
Leon-Gonzalez, R., & Scarpa, R. (2008). Improving multi-site benefit functions via Bayesian model averaging: A new approach to benefit transfer. *Journal of Environmental Economics and Management, 56*, 50–68.
Liebe, U., Meyerhoff, J., & Hartje, V. (2012). Test-retest reliability of choice experiments in environmental valuation. *Environmental & Resource Economics, 53*, 389–407.
Lindhjem, H., & Navrud, S. (2008). How reliable are meta-analyses for international benefit transfers? *Ecological Economics, 66*, 425–435.
Londoño, L. M., & Johnston, R. J. (2012). Enhancing the reliability of benefit transfer over heterogeneous sites: A meta-analysis of international coral reef values. *Ecological Economics, 78*, 80–89.
Loomis, J. B. (1992). The evolution of a more rigorous approach to benefit transfer: Benefit function transfer. *Water Resources Research, 28*, 701–705.
Loomis, J. B. (2000). Vertically summing public good demand curves: An empirical comparison of economic versus political jurisdictions. *Land Economics, 76*, 312–321.
Loomis, J. B., & Rosenberger, R. S. (2006). Reducing barriers in future benefit transfers: Needed improvements in primary study design and reporting. *Ecological Economics, 60*, 343–350.

Martin-Ortega, J., Brouwer, R., Ojea, E., & Berbel, J. (2012). Benefit transfer and spatial heterogeneity of preferences for water quality improvements. *Journal of Environmental Management, 106*, 22–29.

McComb, G., Lantz, V., Nash, K., & Rittmaster, R. (2006). International valuation databases: Overview, methods and operational issues. *Ecological Economics, 60*, 461–472.

McConnell, K. E., Strand, I. E., & Valdes, S. (1998). Testing temporal reliability and carryover effect: The role of correlated responses in test–retest reliability studies. *Environmental & Resource Economics, 12*, 357–374.

Moeltner, K., & Rosenberger, R. S. (2008). Predicting resource policy outcomes via meta-regression: Data space, model space, and the quest for 'optimal scope'. *The B. E. Journal of Economic Analysis & Policy*. doi:10.2202/1935-1682.2028.

Moeltner, K., & Rosenberger, R. S. (2014). Cross-context benefit transfer: A Bayesian search for information pools. *American Journal of Agricultural Economics*. Advance Access published January 22, 2014.

Moeltner, K., & Woodward, R. (2009). Meta-functional benefit transfer for wetland valuation: Making the most of small samples. *Environmental & Resource Economics, 42*, 89–108.

Moeltner, K., Boyle, K. J., & Paterson, R. W. (2007). Meta-analysis and benefit transfer for resource valuation—Addressing classical challenges with Bayesian modeling. *Journal of Environmental Economics and Management, 53*, 250–269.

Morrison, M. (2001). Non-market valuation databases: How useful are they? *Economic Analysis & Policy, 31*, 33–55.

Morrison, M., & Bergland, O. (2006). Prospects for the use of choice modelling for benefit transfer. *Ecological Economics, 60*, 420–428.

Navrud, S., & Ready, R. (2007a). Lessons learned for environmental value transfer. In S. Navrud & R. Ready (Eds.), *Environmental value transfer: Issues and methods* (pp. 283–290). Dordrecht: Springer.

Navrud, S., & Ready, R. (2007b). Review of methods for value transfer. In S. Navrud & R. Ready (Eds.), *Environmental value transfer: Issues and methods* (pp. 1–10). Dordrecht: Springer.

Nelson, J. P., & Kennedy, P. E. (2009). The use (and abuse) of meta-analysis in environmental and resource economics: An assessment. *Environmental & Resource Economics, 42*, 345–377.

Ready, R., & Navrud, S. (2006). International benefit transfer: Methods and validity test. *Ecological Economics, 60*, 429–434.

Ready, R., Navrud, S., Day, B., Dubourg, R., Machado, F., Mourato, S., et al. (2004). Benefit transfer in Europe: How reliable are transfers between countries? *Environmental & Resource Economics, 29*, 67–82.

Rolfe, J., & Bennett, J. (Eds.). (2006). *Choice modelling and the transfer of environmental values*. Cheltenham, UK: Edward Elgar.

Rolfe, J., & Wang, X. (2011). Dealing with scale and scope issues in stated preference experiments. In J. Bennett (Ed.), *The international handbook on non-market environmental valuation* (pp. 254–272). Cheltenham, UK: Edward Elgar.

Rolfe, J., & Windle, J. (2012). Distance decay functions for iconic assets: Assessing national values to protect the health of the Great Barrier Reef in Australia. *Environmental & Resource Economics, 53*, 347–365.

Rolfe, J., Bennett, J., Johnston, R. J., & Kerr, G. (2011). 'Yes we can…': Using benefit transfer to inform environmental policymaking. EERH Policy Brief. Canberra: Environmental Economics Research Hub, Australian National University.

Rosenberger, R. S., & Johnston, R. J. (2009). Selection effects in meta-analysis and benefit transfer: Avoiding unintended consequences. *Land Economics, 85*, 410–428.

Rosenberger, R. S., & Phipps, T. T. (2007). Correspondence and convergence in benefit transfer accuracy: Meta-analytic review of the literature. In S. Navrud & R. Ready (Eds.), *Environmental value transfer: Issues and methods* (pp. 23–43). Dordrecht: Springer.

Rosenberger, R. S., & Stanley, T. D. (2006). Measurement, generalization and publication: Sources of error in benefit transfers and their management. *Ecological Economics, 60*, 372–378.

Schaafsma, M., & Brouwer, R. (2013). Testing geographical framing and substitution effects in spatial choice experiments. *Journal of Choice Modelling, 8*, 32–48.

Schaafsma, M., Brouwer, R., Gilbert, A., van den Bergh, J. C. J. M., & Wagtendonk, A. (2013). Estimation of distance-decay functions to account for substitution and spatial heterogeneity in stated preference research. *Land Economics, 89*, 514–537.

Schaafsma, M., Brouwer, R. & Rose, J. (2012). Directional heterogeneity in WTP models for environmental valuation. *Ecological Economics*, *79*, 21–31

Smith, V. K., & Huang, J. (1993). Hedonic models and air pollution: Twenty-five years and counting. *Environmental & Resource Economics, 3*, 381–394.

Smith, V. K., & Pattanayak, S. K. (2002). Is meta-analysis a Noah's Ark for non-market valuation? *Environmental & Resource Economics, 22*, 271–296.

Smith, V. K., van Houtven, G., & Pattanayak, S. K. (2002). Benefit transfer via preference calibration: "Prudential algebra" for policy. *Land Economics, 78*, 132–152.

Smith, V. K., Pattanayak, S. K., & Van Houtven, G. (2006). Structural benefit transfer: An example using VSL estimates. *Ecological Economics, 60*, 361–371.

Spash, C. L., & Vatn, A. (2006). Transferring environmental value estimates: Issues and alternatives. *Ecological Economics, 60*, 379–388.

Stanley, T. D. (2005). Beyond publication bias. *Journal of Economic Surveys, 19*, 309–345.

Stanley, T. D. (2008). Meta-regression methods for detecting and estimating empirical effect in the presence of publication selection. *Oxford Bulletin of Economics and Statistics, 70*, 103–127.

Stanley, T. D., Doucouliagos, H., Giles, M., Heckemeyer, J. H., Johnston, R. J., Laroche, P., et al. (2013). Meta-analysis of economics reporting guidelines. *Journal of Economic Surveys, 27*, 390–394.

Stapler, R. W., & Johnston, R. J. (2009). Meta-analysis, benefit transfer, and methodological covariates: Implications for transfer error. *Environmental & Resource Economics, 42*, 227–246.

U.S. OMB (Office of Management and Budget) (2003). *Circular A-4: Regulatory analysis.* http://www.whitehouse.gov/OMB/circulars/a004/a-4.pdf.

U.S. EPA (Environmental Protection Agency) (2010). *Guidelines for preparing economic analyses.* EPA 240-R-10-001. U.S. EPA, Office of the Administrator, December. Washington, DC.

Vandenberg, T. P., Poe, G. L., & Powell, J. R. (2001). Accessing the accuracy of benefits transfer: Evidence from a multi-site contingent valuation study of groundwater quality. In J. C. Bergstrom, K. J. Boyle, & G. L. Poe (Eds.), *The economic value of water quality* (pp. 100–120). Cheltenham, UK: Edward Elgar.

Van Houtven, G., Powers, J., & Pattanayak, S. K. (2007). Valuing water quality improvements in the United States using meta-analysis: Is the glass half-full or half-empty for national policy analysis? *Resource and Energy Economics, 29*, 206–228.

Walsh, R. G., Johnson, D. M., & McKean, J. R. (1992). Benefit transfer of outdoor recreation demand studies, 1968–1988. *Water Resources Research, 28*, 707–713.

Zandersen, M., Termansen, M., & Jensen, F. S. (2007). Testing benefits transfer of forest recreation values over a 20-year time horizon. *Land Economics, 83*, 412–440.

Index

R.J. Johnston et al. (eds.), *Benefit Transfer of Environmental and Resource Values*, The Economics of Non-Market Goods and Resources 14,
DOI 10.1007/978-94-017-9930-0

MIX
Papier aus verantwortungsvollen Quellen
Paper from responsible sources
FSC® C105338

If you have any concerns about our products,
you can contact us on
ProductSafety@springernature.com

In case Publisher is established outside the EU,
the EU authorized representative is:
Springer Nature Customer Service Center GmbH
Europaplatz 3, 69115 Heidelberg, Germany

Printed by Libri Plureos GmbH
in Hamburg, Germany